WORKING GUIDE TO DRILLING EQUIPMENT AND OPERATIONS

WILLIAM C. LYONS

ELSEVIER SCIENCE

AMSTERDAM • BOSTON • HEIDELBERG • LONDON
NEW YORK • OXFORD • PARIS • SAN DIEGO
SAN FRANCISCO • SINGAPORE • SYDNEY • TOKYO
Gulf Professional Publishing is an imprint of Elsevier

GPP

Gulf Professional Publishing is an imprint of Elsevier
30 Corporate Drive, Suite 400, Burlington, MA 01803, USA,
The Boulevard, Langford Lane, Oxford OX5 1GB

First edition 2010

Library of Congress Cataloging in Publication Data
A catalog record for this book is available from the Library of Congress

British Library Cataloguing in Publication Data
A catalogue record for this book is available from the British Library

ISBN: 978-1-85617-843-3

For information on all Elsevier publications
visit our website at *elsevierdirect.com*

Typeset by: diacriTech, India

Printed and bound in United States of America
10 11 12 13 11 10 9 8 7 6 5 4 3 2 1

WORKING GUIDE TO DRILLING EQUIPMENT AND OPERATIONS

Contents

Full Contents

Chapter 1

DRILLING MUDS AND COMPLETION SYSTEMS

Chapter 2

DRILL STRING: COMPOSITION AND DESIGN

Chapter 3

AIR AND GAS DRILLING

Chapter 4

DIRECTIONAL DRILLING

Chapter 5

SELECTION OF DRILLING PRACTICES

Chapter 6

WELL PRESSURE CONTROL

Chapter 7

FISHING OPERATIONS AND EQUIPMENT

Chapter 8

CASING AND CASING STRING DESIGN

Chapter 9

WELL CEMENTING

Chapter 10

TUBING AND TUBING STRING DESIGN

Chapter 11

ENVIRONMENTAL CONSIDERATIONS FOR DRILLING OPERATIONS

CHAPTER

1

Drilling Muds and Completion Systems

OUTLINE

1.1 FUNCTIONS OF DRILLING MUDS

1.1.1 Drilling Fluid Definitions and General Functions

Results of research has shown that penetration rate and its response to weight on bit and rotary speed is highly dependent on the hydraulic horsepower reaching the formation at the bit. Because the drilling fluid flow rate sets the system pressure losses and these pressure losses set the hydraulic horsepower across the bit, it can be concluded that the drilling fluid is as important in determining drilling costs as all other "controllable" variables combined. Considering these factors, an optimum drilling fluid is properly formulated so that the flow rate necessary to clean the hole results in the proper hydraulic horsepower to clean the bit for the weight and rotary speed imposed to give the lowest cost, provided that this combination of variables results in a stable borehole which penetrates the desired target. This definition incorporates and places in perspective the five major functions of a drilling fluid.

1.1.2 Cool and Lubricate the Bit and Drill String

Considerable heat and friction is generated at the bit and between the drill string and wellbore during drilling operations. Contact between the drill string and wellbore can also create considerable torque during rotation and drag during trips. Circulating drilling fluid transports heat away from these frictional sites, reducing the chance of premature bit failure and pipe damage. The drilling fluid also lubricates the bit tooth penetration through the bottom hole debris into the rock and serves as a lubricant between the wellbore and drill string, reducing torque and drag.

1.1.3 Clean the Bit and the Bottom of the Hole

If the cuttings generated at the bit face are not immediately removed and started toward the surface, they will be ground very fine, stick to the bit, and in general retard effective penetration into uncut rock.

1.1.4 Suspend Solids and Transport Cuttings and Sloughings to the Surface

Drilling fluids must have the capacity to suspend weight materials and drilled solids during connections, bit trips, and logging runs, or they will settle to the low side or bottom of the hole. Failure to suspend weight materials can result in a reduction in the drilling fluids density, which can lead to kicks and potential of a blowout.

The drilling fluid must be capable of transporting cuttings out of the hole at a reasonable velocity that minimizes their disintegration and incorporation as drilled solids into the drilling fluid system and able to release the cuttings at the surface for efficient removal. Failure to adequately clean the hole or to suspend drilled solids can contribute to hole problems such as fill on bottom after a trip, hole pack-off, lost returns, differentially stuck pipe, and inability to reach bottom with logging tools.

Factors influencing removal of cuttings and formation sloughings and solids suspension include

- Density of the solids
- Density of the drilling fluid
- Rheological properties of the drilling fluid
- Annular velocity

- Hole angle
- Slip velocity of the cuttings or sloughings

1.1.5 Stabilize the Wellbore and Control Subsurface Pressures

Borehole instability is a natural function of the unequal mechanical stresses and physical-chemical interactions and pressures created when supporting material and surfaces are exposed in the process of drilling a well. The drilling fluid must overcome the tendency for the hole to collapse from mechanical failure or from chemical interaction of the formation with the drilling fluid. The Earth's pressure gradient at sea level is 0.465 psi/ft, which is equivalent to the height of a column of salt water with a density (1.07 SG) of 8.94 ppg.

In most drilling areas, the fresh water plus the solids incorporated into the water from drilling subsurface formations is sufficient to balance the formation pressures. However, it is common to experience abnormally pressured formations that require high-density drilling fluids to control the formation pressures. Failure to control downhole pressures can result in an influx of formation fluids, resulting in a kick or blowout. Borehole stability is also maintained or enhanced by controlling the loss of filtrate to permeable formations and by careful control of the chemical composition of the drilling fluid.

Most permeable formations have pore space openings too small to allow the passage of whole mud into the formation, but filtrate from the drilling fluid can enter the pore spaces. The rate at which the filtrate enters the formation depends on the pressure differential between the formation and the column of drilling fluid and the quality of the filter cake deposited on the formation face.

Large volumes of drilling fluid filtrate and filtrates that are incompatible with the formation or formation fluids may destabilize the formation through hydration of shale and/or chemical interactions between components of the drilling fluid and the wellbore.

Drilling fluids that produce low-quality or thick filter cakes may also cause tight hole conditions, including stuck pipe, difficulty in running casing, and poor cement jobs.

1.1.6 Assist in the Gathering of Subsurface Geological Data and Formation Evaluation

Interpretation of surface geological data gathered through drilled cuttings, cores, and electrical logs is used to determine the commercial value of the zones penetrated. Invasion of these zones by the drilling fluid, its filtrate (oil or water) may mask or interfere with interpretation of data retrieved or prevent full commercial recovery of hydrocarbon.

1.1.7 Other Functions

In addition to the functions previously listed, the drilling fluid should be environmentally acceptable to the area in which it is used. It should be noncorrosive to tubulars being used in the drilling and completion operations. Most importantly, the drilling fluid should not damage the productive formations that are penetrated.

The functions described here are a few of the most obvious functions of a drilling fluid. Proper application of drilling fluids is the key to successfully drilling in various environments.

1.2 CLASSIFICATIONS

A generalized classification of drilling fluids can be based on their fluid phase, alkalinity, dispersion, and type of chemicals used in the formulation and degrees of inhibition. In a broad sense, drilling fluids can be broken into five major categories.

1.2.1 Freshwater Muds—Dispersed Systems

The pH value of low-pH muds may range from 7.0 to 9.5. Low-pH muds include spud muds, bentonite-treated muds, natural muds, phosphate-treated muds, organic thinned muds (e.g., red muds, lignite muds, lignosulfonate muds), and organic colloid–treated muds. In this case, the lack of salinity of the water phase and the addition of chemical dispersants dictate the inclusion of these fluids in this broad category.

1.2.2 Inhibited Muds—Dispersed Systems

These are water-base drilling muds that repress the hydration and dispersion of clays through the inclusion of inhibiting ions such as calcium and salt. There are essentially four types of inhibited muds: lime muds (high pH), gypsum muds (low pH), seawater muds (unsaturated saltwater muds, low pH), and saturated saltwater muds (low pH). Newer-generation inhibited-dispersed fluids offer enhanced inhibitive performance and formation stabilization; these fluids include sodium silicate muds, formate brine-based fluids, and cationic polymer fluids.

1.2.3 Low Solids Muds—Nondispersed Systems

These muds contain less than 3–6% solids by volume, weight less than 9.5 lb/gal, and may be fresh or saltwater based. The typical low-solid systems are selective flocculent, minimum-solids muds, beneficiated clay muds, and low-solids polymer muds. Most low-solids drilling fluids are composed of water with varying quantities of bentonite and a polymer. The difference among low-solid systems lies in the various actions of different polymers.

1.2.4 Nonaqueous Fluids

Invert Emulsions Invert emulsions are formed when one liquid is dispersed as small droplets in another liquid with which the dispersed liquid is immiscible. Mutually immiscible fluids, such as water and oil, can be emulsified by shear and the addition of surfactants. The suspending liquid is called the *continuous phase,* and the droplets are called the *dispersed* or *discontinuous phase.* There are two types of emulsions used in drilling fluids: oil-in-water emulsions that have water as the continuous phase and oil as the dispersed phase and water-in-oil emulsions that have oil as the continuous phase and water as the dispersed phase (i.e., invert emulsions).

Oil-Base Muds (nonaqueous fluid [NAF]) Oil-base muds contain oil (refined from crude such as diesel or synthetic-base oil) as the continuous phase and trace amounts of water as the dispersed phase. Oil-base muds generally contain less than 5% (by volume) water (which acts as a polar activator for organophilic clay), whereas invert emulsion fluids generally have more than 5% water in mud. Oil-base muds are usually a mixture of base oil, organophilic clay, and lignite or asphalt, and the filtrate is all oil.

1.3 TESTING OF DRILLING SYSTEMS

To properly control the hole cleaning, suspension, and filtration properties of a drilling fluid, testing of the fluid properties is done on a daily basis. Most tests are conducted at the rig site, and procedures are set forth in the API RPB13B. Testing of water-based fluids and nonaqueous fluids can be similar, but variations of procedures occur due to the nature of the fluid being tested.

1.3.1 Water-Base Muds Testing

To accurately determine the physical properties of water-based drilling fluids, examination of the fluid is required in a field laboratory setting. In many cases, this consists of a few simple tests conducted by the derrickman or mud Engineer at the rigsite. The procedures for conducting all routine drilling fluid testing can be found in the American Petroleum Institute's API RPB13B.

Density Often referred to as the mud weight, density may be expressed as pounds per gallon (lb/gal), pounds per cubic foot (lb/ft^3), specific gravity (SG) or pressure gradient (psi/ft). Any instrument of sufficient accuracy within $\pm 0.1\,\mathrm{lb/gal}$ or $\pm 0.5\,\mathrm{lb/ft^3}$ may be used. The mud balance is the instrument most commonly used. The weight of a mud cup attached to one end of the beam is balanced on the other end by a fixed counterweight and a rider free to move along a graduated scale. The density of the fluid is a direct reading from the scales located on both sides of the mud balance (Figure 1.1).

Marsh Funnel Viscosity Mud viscosity is a measure of the mud's resistance to flow. The primary function of drilling fluid viscosity is a to transport cuttings to the surface and suspend weighing materials. Viscosity must be high enough that the weighting material will remain suspended but low enough to permit sand and cuttings to settle out and entrained gas to escape at the surface. Excessive viscosity can create high pump pressure, which magnifies the swab or surge effect during tripping operations. The control of equivalent circulating density (ECD) is always a prime concern when managing the viscosity of a drilling fluid. The Marsh funnel is a rig site instrument used to measure funnel viscosity. The funnel is dimensioned so that by following standard procedures, the outflow time of 1 qt (946 ml) of freshwater at a temperature of $70 \pm 5°\mathrm{F}$ is 26 ± 0.5 seconds (Figure 1.2). A graduated cup is used as a receiver.

Direct Indicating Viscometer This is a rotational type instrument powered by an electric motor or by a hand crank (Figure 1.3). Mud is contained in the annular space between two cylinders. The outer cylinder or rotor sleeve is driven at a constant rotational velocity; its rotation in the mud produces a torque on the inner cylinder or bob. A torsion spring restrains the movement of the bob. A dial attached to the bob indicates its displacement on a direct reading scale. Instrument constraints have been adjusted

FIGURE 1.1 API mud balance.

FIGURE 1.2 Marsh funnel.

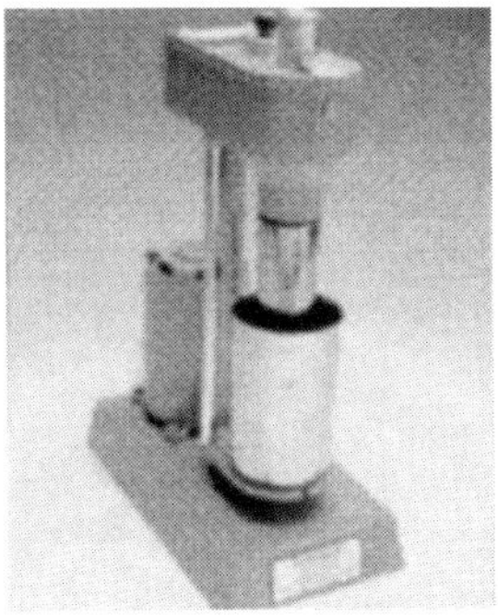

FIGURE 1.3 Variable speed viscometer.

so that plastic viscosity, apparent viscosity, and yield point are obtained by using readings from rotor sleeve speeds of 300 and 600 rpm.

Plastic viscosity (PV) in centipoise is equal to the 600 rpm dial reading minus the 300 rpm dial reading. Yield point (YP), in pounds per 100 ft^2, is equal to the 300-rpm dial reading minus the plastic viscosity. Apparent viscosity in centipoise is equal to the 600-rpm reading, divided by two.

Gel Strength Gel strength is a measure of the inter-particle forces and indicates the gelling that will occur when circulation is stopped. This property prevents the cuttings from setting in the hole. High pump pressure is generally required to "break" circulation in a high-gel mud. Gel strength is measured in units of lbf/100 ft^2. This reading is obtained by noting the maximum dial deflection when the rotational viscometer is turned at a low rotor speed (3 rpm) after the mud has remained static for some period of time (10 seconds, 10 minutes, or 30 minutes). If the mud is allowed to remain static in the viscometer for a period of 10 seconds, the maximum dial deflection obtained when the viscometer is turned on is reported as the *initial gel* on the API mud report form. If the mud is allowed to remain static for 10 minutes, the maximum dial deflection is reported as the *10-min gel*. The same device is used to determine gel strength that is used to determine the plastic viscosity and yield point, the Variable Speed Rheometer/Viscometer.

API Filtration A standard API filter press is used to determine the filter cake building characteristics and filtration of a drilling fluid (Figure 1.4). The API filter press consists of a cylindrical mud chamber made of materials resistant to strongly alkaline solutions. A filter paper is placed on the bottom of the chamber just above a suitable support. The total filtration area is 7.1 ($\pm$ 0.1) $in.^2$. Below the support is a drain tube for discharging the filtrate into a graduated cylinder. The entire assembly is supported by a stand so 100-psi pressure can be applied to the mud sample in the chamber. At the end of the 30-minute filtration time, the volume of filtrate is reported as API filtration

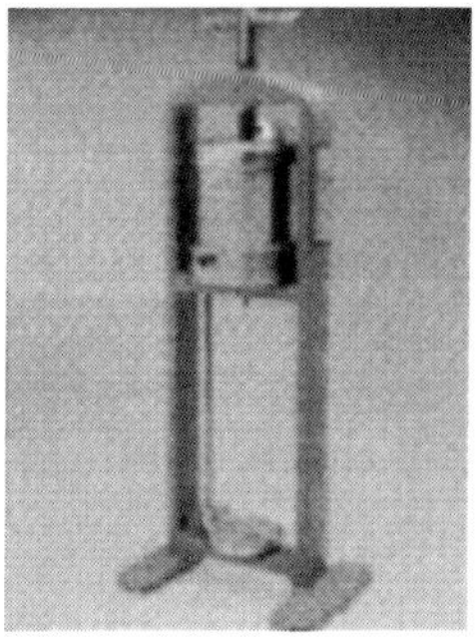

FIGURE 1.4 API style filter press.

FIGURE 1.5 Sand content kit.

in milliliters. To obtain correlative results, one thickness of the proper 9-cm filter paper—Whatman No. 50, S&S No. 5765, or the equivalent—must be used. Thickness of the filter cake is measured and reported in 32nd of an inch. The cake is visually examined, and its consistency is reported using such notations as "hard," "soft," tough," "rubbery," or "firm."

Sand Content The sand content in drilling fluids is determined using a 200-mesh sand sieve screen 2 inches in diameter, a funnel to fit the screen, and a glass-sand graduated measuring tube (Figure 1.5). The measuring tube is marked to indicate the volume of "mud to be added," water to be added and to directly read the volume of sand on the bottom of the tube.

Sand content of the mud is reported in percent by volume. Also reported is the point of sampling (e.g., flowline, shale shaker, suction pit). Solids other than sand may be retained on the screen (e.g., lost circulation material), and the presence of such solids should be noted.

Liquids and Solids Content A mud retort is used to determine the liquids and solids content of a drilling fluid. Mud is placed in a steel container and heated at high temperature until the liquid components have been distilled off and vaporized (Figure 1.6). The vapors are passed through a condenser and collected in a graduated cylinder. The volume of liquids (water and oil) is then measured. Solids, both suspended and dissolved, are determined by volume as a difference between the mud in container and the distillate in graduated cylinder. Drilling fluid retorts are generally designed to distill 10-, 20-, or 50-ml sample volumes.

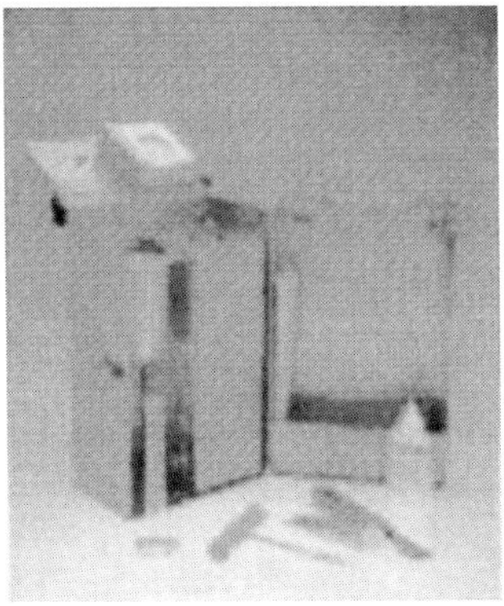

FIGURE 1.6 Retort kit (10 ml).

TABLE 1.1 High- and Low-Gravity Solids in Drilling Fluids

Specific Gravity of Solids	Barite, Percent by Weight	Clay, Percent by Weight
2.6	0	100
2.8	18	82
3.0	34	66
3.2	48	52
3.4	60	40
3.6	71	29
3.8	81	19
4.0	89	11
4.3	100	0

For freshwater muds, a rough measure of the relative amounts of barite and clay in the solids can be made (Table 1.1). Because both suspended and dissolved solids are retained in the retort for muds containing substantial quantities of salt, corrections must be made for the salt. Relative amounts of high- and low-gravity solids contained in drilling fluids can be found in Table 1.1.

pH Two methods for measuring the pH of drilling fluid are commonly used: (1) a modified colorimetric method using pH paper or strips and (2) the electrometric method using a glass electrode (Figure 1.7). The paper strip test may not be reliable if the salt concentration of the sample is high. The electrometric method is subject to error in solutions containing high concentrations of sodium ions unless a special glass electrode is used or unless suitable correction factors are applied if an ordinary electrode is used. In addition, a temperature correction is required for the electrometric method of measuring pH.

The paper strips used in the colorimetric method are impregnated with dyes so that the color of the test paper depends on the pH of the medium in

FIGURE 1.7 pH Meter.

which the paper is placed. A standard color chart is supplied for comparison with the test strip. Test papers are available in a wide range, which permits estimating pH to 0.5 units, and in narrow range papers, with which the pH can be estimated to 0.2 units.

The glass electrode pH meter consists of a glass electrode, an electronic amplifier, and a meter calibrated in pH units. The electrode is composed of (1) the glass electrode, a thin-walled bulb made of special glass within which is sealed a suitable electrolyte and an electrode, and (2) the reference electrode, which is a saturated calomel cell. Electrical connection with the mud is established through a saturated solution of potassium chloride contained in a tube surrounding the calomel cell. The electrical potential generated in the glass electrode system by the hydrogen ions in the drilling mud is amplified and operates the calibrated pH meter.

Resistivity Control of the resistivity of the mud and mud filtrate while drilling may be desirable to permit enhanced evaluation of the formation characteristics from electric logs. The determination of resistivity is essentially the measurement of the resistance to electrical current flow through a known sample configuration. Measured resistance is converted to resistivity by use of a cell constant. The cell constant is fixed by the configuration of the sample in the cell and id determined by calibration with standard solutions of known resistivity. The resistivity is expressed in ohm-meters.

Filtrate Chemical Analysis Standard chemical analyses have been developed for determining the concentration of various ions present in the mud. Tests for the concentration of chloride, hydroxyl, and calcium ions are required to fill out the API drilling mud report. The tests are based on filtration (i.e., reaction of a known volume of mud filtrate sample with a standard solution of known volume and concentration). The end of chemical reaction is usually indicated by the change of color. The concentration of the ion being tested can be determined from a knowledge of the chemical reaction taking place.

Chloride The chloride concentration is determined by titration with silver nitrate solution. This causes the chloride to be removed from the solution as $AgCl^-$, a white precipitate. The endpoint of the titration is detected using a potassium chromate indicator. The excess Ag present after all Cl^- has been removed from solution reacts with the chromate to form Ag_9CrO_4, an orange-red precipitate. Contamination with chlorides generally results from drilling salt or from a saltwater flow. Salt can enter and contaminate the mud system when salt formations are drilled and when saline formation water enters the wellbore.

Alkalinity and Lime Content Alkalinity is the ability of a solution or mixture to react with an acid. The *phenolphthalein alkalinity* refers to the amount of acid required to reduce the pH of the filtrate to 8.3, the phenolphthalein end point. The phenolphthalein alkalinity of the mud and mud filtrate is called the P_m and P_f, respectively. The P_f test includes the effect of only dissolved bases and salts, whereas the P_m test includes the effect of both dissolved and suspended bases and salts. The $_m$ and $_f$ indicate if the test was conducted on the whole mud or mud filtrate. The M_f alkalinity refers to the amount of acid required to reduce the pH to 4.3, the methyl orange end point. The methyl orange alkalinity of the mud and mud filtrate is called the M_m and M_f, respectively. The API diagnostic tests include the determination of P_m, P_f, and M_f. All values are reported in cubic centimeters of 0.02 N (normality = 0.02) sulfuric acid per cubic centimeter of sample. The lime content of the mud is calculated by subtracting the P_f from the P_m and dividing the result by 4.

The P_f and M_f tests are designed to establish the concentration of hydroxyl, bicarbonate, and carbonate ions in the aqueous phase of the mud. At a pH of 8.3, the conversion of hydroxides to water and carbonates to bicarbonates is essentially complete. The bicarbonates originally present in solution do not enter the reactions. As the pH is further reduced to 4.3, the acid reacts with the bicarbonate ions to form carbon dioxide and water.

$$\text{ml N/50}\, H_2SO_4 \text{ to reach pH} = 8.3$$

$$CO_3^{2-} + H_2SO_4 \rightarrow HCO_3^- + HSO_4$$

$$\text{carbonate} + \text{acid} \rightarrow \text{bicarbonate} + \text{bisulfate}$$

$$OH^- + H_2SO_4 \rightarrow HOH + SO_4 =$$

$$\text{hydroxyl} + \text{acid} \rightarrow \text{water} + \text{sulfate salt}$$

The P_f and P_m test results indicate the reserve alkalinity of the suspended solids. As the $[OH^-]$ in solution is reduced, the lime and limestone suspended in the mud will go into solution and tend to stabilize the pH (Table 1.2). This reserve alkalinity generally is expressed as an excess lime concentration, in lb/bbl of mud. The accurate testing of P_f, M_f, and P_m are needed to determine the quality and quantity of alkaline material present in the drilling fluid. The chart below shows how to determine the hydroxyl, carbonate, and bicarbonate ion concentrations based on these titrations.

TABLE 1.2 Alkalinity

Criteria	OH^- (mg/L)	CO_3^{2-} (mg/L)	HCO_3^- (mg/L)
$P_f = 0$	0	0	1,220 M_f
$2P_f < M_f$	0	1,200 P_f	1,220 $(M_f - 2P_f)$
$2P_f = M_f$	0	1,200 P_f	0
$2P_f < M_f$	340 $(2P_f - M_f)$	1,200 $(M_f - P_f)$	0
$P_f = M_f$	340 M_f	0	0

Total Hardness The total combined concentration of calcium and magnesium in the mud-water phase is defined as total hardness. These contaminants are often present in the water available for use in the drilling fluid makeup. In addition, calcium can enter the mud when anhydrite ($CaSO_4$) or gypsum ($CaSO_4 \cdot 2H_2O$) formations are drilled. Cement also contains calcium and can contaminate the mud. The total hardness is determined by titration with a standard (0.02 N) versenate hardness titrating solution (EDTA). The standard versenate solution contains sodium versenate, an organic compound capable of forming a chelate when combined with Ca^2 and Mg^2.

The hardness test sometimes is performed on the whole mud as well as the mud filtrate. The mud hardness indicates the amount of calcium suspended in the mud and the amount of calcium in solution. This test usually is made on gypsum-treated muds to indicate the amount of excess $CaSO_4$ present in suspension. To perform the hardness test on mud, a small sample of mud is first diluted to 50 times its original volume with distilled water so that any undissolved calcium or magnesium compounds can go into solution. The mixture then is filtered through hardened filter paper to obtain a clear filtrate. The total hardness of this filtrate then is obtained using the same procedure used for the filtrate from the low-temperature, low-pressure API filter press apparatus.

Methylene Blue Capacity (CEC or MBT) It is desirable to know the cation exchange capacity (CEC) of the drilling fluid. To some extent, this value can be correlated to the bentonite content of the mud. The test is only qualitative because organic material and other clays present in the mud also absorb methylene blue dye. The mud sample is treated with hydrogen peroxide to oxidize most of the organic material. The cation exchange capacity is reported in milliequivalent weights (mEq) of methylene blue dye per 100 ml of mud. The methylene blue solution used for titration is usually 0.01 N, so that the cation exchange capacity is numerically equal to the cubic centimeters of methylene blue solution per cubic centimeter of sample required to reach an end point. If other adsorptive materials are not present in significant quantities, the montmorillonite content of the mud in pounds per barrel is calculated to be five times the cation exchange capacity. The methylene blue test can also be used to determine cation exchange capacity of clays and shales. In the test, a weighed amount of clay is dispersed into water by a high-speed stirrer or mixer. Titration is carried out as for drilling muds, except that hydrogen peroxide is not added. The cation exchange capacity of clays is expressed as milliequivalents of methylene blue per 100 g of clay.

1.3.2 Oil-Base and Synthetic-Base Muds (Nonaqueous Fluids Testing)

The field tests for rheology, mud density, and gel strength are accomplished in the same manner as outlined for water-based muds. The main difference is that rheology is tested at a specific temperature, usually 120°F or 150°F. Because oils tend to thin with temperature, heating fluid is required and should be reported on the API Mud Report.

Sand Content Sand content measurement is the same as for water-base muds except that the mud's base oil instead of water should be used for dilution. The sand content of oil-base mud is not generally tested.

HPHT Filtration The API filtration test result for oil-base muds is usually zero. In relaxed filtrate oil-based muds, the API filtrate should be all oil. The API test does not indicate downhole filtration rates. The alternative high-temperature–high pressure (HTHP) filtration test will generally give a better indication of the fluid loss characteristics of a fluid under downhole temperatures (Figure 1.8).

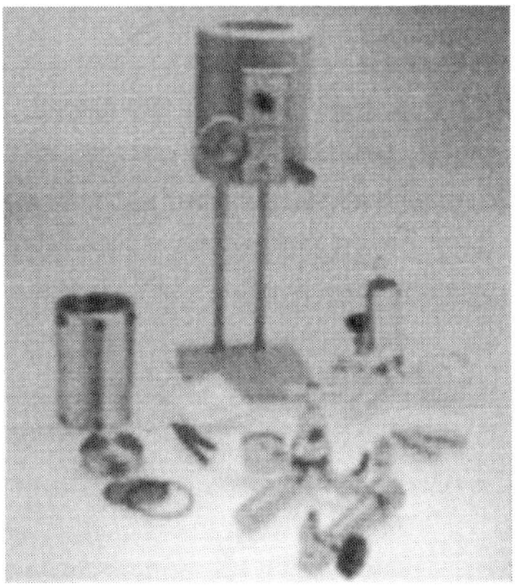

FIGURE 1.8 HPHT fluid loss testing device.

The instruments for the HTHP filtration test consists essentially of a controlled pressure source, a cell designed to withstand a working pressure of at least 1,000 psi, a system for heating the cell, and a suitable frame to hold the cell and the heating system. For filtration tests at temperatures above 200°F, a pressurized collection cell is attached to the delivery tube. The filter cell is equipped with a thermometer well, oil-resistant gaskets, and a support for the filter paper (Whatman no. 50 or the equivalent). A valve on the filtrate delivery tube controls flow from the cell. A nonhazardous gas such as nitrogen or carbon dioxide should be used as the pressure source. The test is usually performed at a temperature of 220 – 350°F and a pressure of 500 psi (differential) over a 30-minute period. When other temperatures, pressures, or times are used, their values should be reported together with test results. If the cake compressibility is desired, the test should be repeated with pressures of 200 psi on the filter cell and 100 psi back pressure on the collection cell. The volume of oil collected at the end of the test should be doubled to correct to a surface area of 7.1 inches.

Electrical Stability The electrical stability test indicates the stability of emulsions of water in oil mixtures. The emulsion tester consists of a reliable circuit using a source of variable AC current (or DC current in portable units) connected to strip electrodes (Figure 1.9). The voltage imposed across the electrodes can be increased until a predetermined amount of current flows through the mud emulsion-breakdown point. Relative stability is indicated as the voltage at the breakdown point and is reported as the electric stability of the fluid on the daily API test report.

Liquids and Solids Content Oil, water, and solids volume percent is determined by retort analysis as in a water-base mud. More time is required to get a complete distillation of an oil mud than for a water mud. The corrected water phase volume, the volume percent of low-gravity solids, and the oil-to-water ratio can then be calculated.

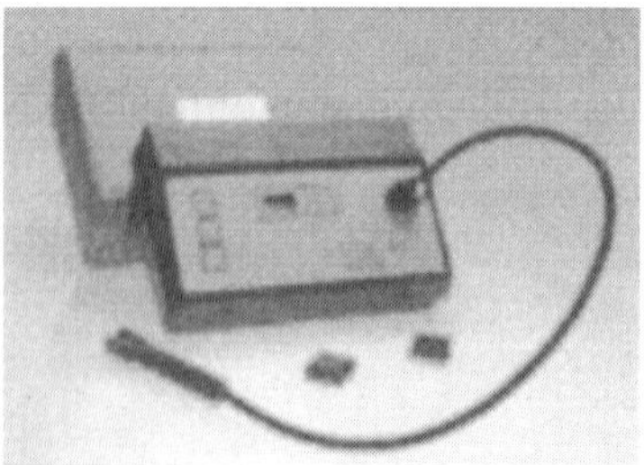

FIGURE 1.9 Electrical stability meter.

The volume oil-to-water ratio can be found from the procedure below:

$$\text{Oil fraction } 100 \times \frac{\%\text{ by volume oil or synthetic oil}}{\%\text{ by volume oil or synthetic oil} - \%\text{ by volume water}}$$

Chemical analysis procedures for nonaqueous fluids can be found in the API 13B bulletin available from the American Petroleum Institute.

Alkalinity and Lime Content (NAF) The whole mud alkalinity test procedure is a titration method that measures the volume of standard acid required to react with the alkaline (basic) materials in an oil mud sample. The alkalinity value is used to calculate the pounds per barrel of unreacted, "excess" lime in an oil mud. Excess alkaline materials, such as lime, help to stabilize the emulsion and neutralize carbon dioxide or hydrogen sulfide acidic gases.

Total Salinity (Water-Phase Salinity [WAF] for NAF) The salinity control of NAF fluids is very important for stabilizing water-sensitive shales and clays. Depending on the ionic concentration of the shale waters and of the mud water phase, an osmotic flow of pure water from the weaker salt concentration (in shale) to the stronger salt concentration (in mud) will occur. This may cause dehydration of the shale and, consequently, affect its stabilization (Figure 1.10).

1.3.3 Specialized Tests

Other, more advanced laboratory-based testing is commonly carried out on drilling fluids to determine treatments or to define contaminants. Some of the more advanced analytical tests routinely conducted on drilling fluids include:

Advanced Rheology and Suspension Analysis

FANN 50 — A laboratory test for rheology under temperature and moderate pressure (up to 1,000 psi and 500°F).

FANN 70 — Laboratory test for rheology under high temperature and high pressure (up to 20,000 psi and 500°F).

FANN 75 — A more advanced computer-controlled version of the FANN 70 (up to 20,000 psi and 500°F).

High-Angle Sag Test (HAST) A laboratory test device to determine the suspension properties of a fluid in high-angle wellbores. This test is designed to evaluate particle setting characteristics of a fluid in deviated wells.

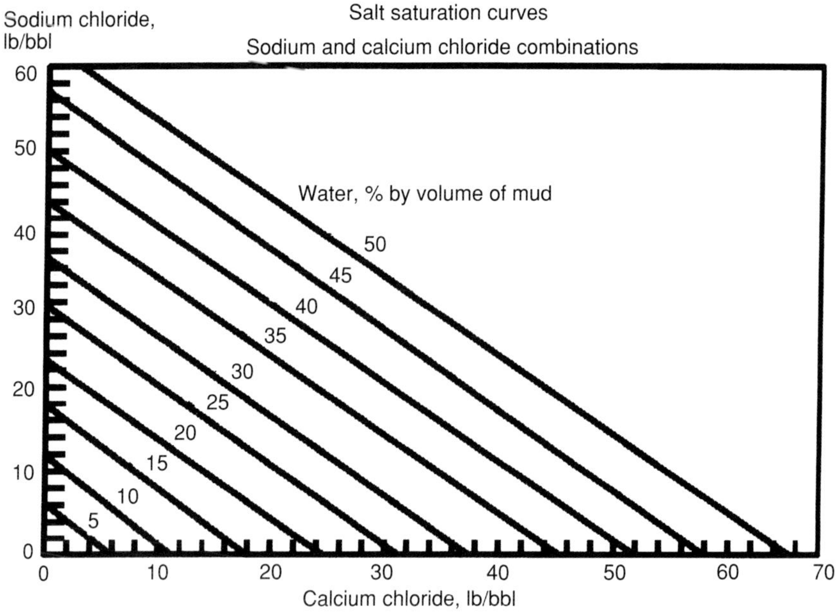

FIGURE 1.10 Salt saturation curves.

Dynamic HAST Laboratory test device to determine the suspension properties of a drilling fluid under high angle and dynamic conditions.

1.3.4 Specialized Filtration Testing

FANN 90 Dynamic filtration testing of a drilling fluid under pressure and temperature. This test determines if the fluid is properly conditioned to drill through highly permeable formations. The test results include two numbers: the dynamic filtration rate and the cake deposition index (CDI). The dynamic filtration rate is calculated from the slope of the curve of volume versus time. The CDI, which reflects the erodability of the wall cake, is calculated from the slope of the curve of volume/time versus time. CDI and dynamic filtration rates are calculated using data collected after twenty minutes. The filtration media for the FAN 90 is a synthetic core. The core size can be sized for each application to optimize the filtration rate.

Particle-Plugging Test (PPT) The PPT test is accomplished with a modified HPHT cell to examine sealing characteristics of a drilling fluid. The PPT, sometimes known as the PPA (particle-plugging apparatus), is key when drilling in high-differential-pressure environments.

Aniline Point Test Determine the aniline point of an oil-based fluid base oil. This test is critical to ensure elastomer compatibility when using non-aqueous fluids.

Particle-Size Distribution (PSD) Test The PSD examines the volume and particle size distribution of solid sin a fluid. This test is valuable in determining the type and size of solids control equipment that will be needed to properly clean a fluid of undesirable solids.

Luminescence Fingerprinting This test is used to determine if contamination of a synthetic-based mud has occurred with crude oil during drilling operations.

Lubricity Testing Various lubricity meters and devices are available to the industry to determine how lubricous a fluid is when exposed to steel or shale. In high-angle drilling applications, a highly lubricious fluid is desirable to allow proper transmission of weight to the bit and reduce side wall sticking tendencies.

1.3.5 Shale Characterization Testing

Capillary Suction Time (CST) Inhibition testing looks at the inhibitive nature of a drilling fluid filtrate when exposed to formation shale samples. The CST is one of many tests that are run routinely on shale samples to optimize the mud chemistry of a water-base fluid.

Linear-Swell Meter (LSM) Another diagnostic test to determine the inhibitive nature of a drilling fluid on field shale samples. The LSM looks at long-term exposure of a fluid filtrate to a formation shale sample. Test times for LSM can run up to 14 days.

Shale Erosion Shale inhibition testing looks at the inhibitive nature of a drilling fluid and examines the erodability of a shale when exposed to a drilling fluid. Various tests procedures for this analytical tool.

Return Permeability Formation damage characterization of a fluid through an actual or simulated core is accomplished with the return permeability test. This test is a must when designing specialized reservoir drilling fluids to minimize formation impairment.

Bacteria Testing Tests for the presence of bacteria in water-base muds; this is especially important in low-pH fluids because bacterial growth is high in these types of fluids.

Static Aging The aging test is used to determine how bottom-hole conditions affect mud properties. Aging cells were developed to aid in predicting the performance of drilling mud under static, high-temperature conditions. If the bottom-hole temperature is greater than 212°F, the aging cells can be pressurized with nitrogen, carbon dioxide, or air to a desired pressure to prevent boiling and vaporization of the mud.

After the aging period, three properties of the aged mud are determined before the mud is agitated or stirred: shear strength, free oil (top oil separation in NAF), and solids setting. Shear strength indicates the gelling tendencies of fluid in the borehole. Second, the sample should be observed to determine if free oil is present. Separation of free oil is a measure of emulsion instability in the borehole and is expressed in 32nd of an inch. Setting of mud solids indicates the formation of a hard or soft layer or sediment in the borehole. After the unagitated sample has been examined, the sample is sheared, and the usual tests for determining rheological and filtration properties are performed.

1.3.6 Drilling Fluid Additives

Each drilling fluid vendor provides a wide array of basic and specialty chemicals to meet the needs of the drilling industry. The general classification of drilling fluid additives below is based on the definitions of the International Association of Drilling Contractors (IADC):

A. Alkalinity or pH control additives are products designed to control the degree of acidity or alkalinity of a drilling fluid. These additives include lime, caustic soda, and bicarbonate of soda.
B. Bactericides reduce the bacteria count of a drilling fluid. Paraformaldehyde, caustic soda, lime, and starch are commonly used as preservatives.
C. Calcium removers are chemicals used to prevent and to overcome the contaminating effects of anhydride and gypsum, both forms of calcium sulfate, which can wreck the effectiveness of nearly any chemically treated mud. The most common calcium removers are caustic soda, soda ash, bicarbonate of soda, and certain polyphosphates.
D. Corrosion inhibitors such as hydrated lime and amine salts are often added to mud and to air-gas systems. Mud containing an adequate percentage of colloids, certain emulsion muds, and oil muds exhibit, in themselves, excellent corrosion-inhibiting properties.
E. Defoamers are products designed to reduce foaming action, particularly that occurring in brackish water and saturated saltwater muds.
F. Emulsifiers are used for creating a heterogeneous mixture of two liquids. These include modified lignosulfonates, certain surface-active agents, anionic and nonionic (negatively charged and noncharged) products.

G. Filtrate, or fluid loss, reducers such as bentonite clays, sodium carboxymethyl cellulose (CMC), and pregelatinized starch serve to cut filter loss, a measure of the tendency of the liquid phase of a drilling fluid to pass into the formation.
H. Flocculants are used sometimes to increase gel strength. Salt (or brine), hydrated lime, gypsum, and sodium tetraphosphates may be used to cause the colloidal particles of a suspension to group into bunches of "flocks," causing solids to settle out.
I. Foaming agents are most often chemicals that also act as surfactants (surface-active agents) to foam in the presence of water. These foamers permit air or gas drilling through water-production formations.
J. Lost circulation materials (LCM) include nearly every possible product used to stop or slow the loss of circulating fluids into the formation. This loss must be differentiated from the normal loss of filtration liquid and from the loss of drilling mud solids to the filter cake (which is a continuous process in an open hole).
K. Extreme-pressure lubricants are designed to reduce torque by reducing the coefficient of friction and thereby increase horsepower at the bit. Certain oils, graphite powder, and soaps are used for this purpose.
L. Shale control inhibitors such as gypsum, sodium silicate, chrome lignosulfonates, as well as lime and salt are used to control caving by swelling or hydrous disintegration of shales.
M. Surface-active agents (surfactants) reduce the interfacial tension between contacting surfaces (e.g., water—oil, water—solid, water—air); these may be emulsifiers, de-emulsifiers, flocculants, or defloccuLents, depending upon the surfaces involved.
N. Thinners and dispersants modify the relationship between the viscosity and the percentage of solids in a drilling mud and may further be used to vary the gel strength and improve "pumpability." Tannins (quebracho), various polyphosphates, and lignitic materials are chosen as thinners or as dispersants, because most of these chemicals also remove solids by precipitation or sequestering, and by deflocculation reactions.
O. Viscosifiers such as bentonite, CMC, Attapulgite clays, sub-bentonites, and asbestos fibers are employed in drilling fluids to ensure a high viscosity–solids ratio.
P. Weighting materials, including barite, lead compounds, iron oxides, and similar products possessing extraordinarily high specific gravities, are used to control formation pressures, check caving, facilitate pulling dry drill pipe on round trips, and aid in combating some types of circulation loss.

The most common commercially available drilling mud additives are published annually by *World Oil*. The listing includes names and descriptions of more than 2,000 mud additives.

1.3.7 Clay Chemistry

Water-base drilling fluids normally contain a number of different types of clays. Most of the clays are added to attain certain physical properties (e.g., fluid loss, viscosity, yield point) and eliminate hole problems.

The most common clays incorporated into the drilling fluid from the formation (in the form of drill solids) are calcium montmorillonite, illites, and kaolinites. The most used commercial clay is sodium montmorillonite.

Bentonite is added to water-base drilling fluids to increase the viscosity and gel strength of the fluid. This results in quality suspension properties for weight materials and increases the carrying capacity for removal of solids from the well. The most important function of bentonite is to improve the filtration and filter cake properties of the water-base drilling fluid.

Clay particles are usually referred to as clay platelets or sheets. The structure of the sodium montmorillonite platelet has sheets consisting of three layers. The platelet, if looked at under an electron microscope, reveals that the sections are honeycombed inside the three layers. The three-layered (sandwich-type) sheet is composed of two silica tetrahedral layers with an octahedral aluminum center core layer between them. The section layers are bonded together in a very intricate lattice-type structure.

Cations are absorbed on the basal surface of the clay crystals to form a natural forming structure. This occurred in the earth over a period of 100 million years. The positive sodium or calcium cations compensate for the atomic substitution in the crystal structure (the isomorphic substitution that took place in forming of the clay). This is the primary way that sodium clays are differentiated from calcium clays.

Sodium montmorillonite absorbs water through expansion of the lattice structure. There are two mechanisms by which hydration can occur:

1. Between the layers (osmotic). The exposure of the clay to water vapor causes the water to condense between the layers, expanding them. The lower the concentration of sodium and chloride in the water, the greater the amount of water that can be absorbed into the clay lattice structure.
2. Around layers (crystalline). There is a layer of water that surrounds the clay particles (a cloud of Na^+ with water molecules held to the platelet by hydrogen bonding to the lattice network by the oxygen on the face of the platelet). The structure of water and clay is commonly called an envelope. (It must be remembered that the water envelope has viscosity.)

Aggregation Clays are said to be in the aggregated state when the platelets are stacked loosely in bundles. When clay is collapsed and its layers are parallel, the formation is like a deck of cards stacked in a box. This is the state of sodium bentonite in the sack having a moisture content of 10%. When added to freshwater (does not contain a high concentration of chlorides), diffuses of water into the layers occurs, and swelling or dispersion results.

In solutions with high chloride concentrations, the double layer is compressed still further, and aggregation occurs. Consequently, the size of the particle is reduced, and the total particle area per unit volume decreases. This occurs because the chloride ion has a strong bond with the H_2O, and free water is not available to enter the clay and hydrate effectively. In muds in which the clay is aggregated, the viscosity is low.

The relationship between the type and concentration of the salt in the water determines the point at which aggregation (inhibition) will occur:

- Sodium chloride (NaCl) 400 mEq/L
- Calcium chloride ($CaCl_2$) 20 mEq/L
- Aluminum chloride ($AlCl_2$) 20 mEq/L

It may be inferred that the higher the chloride content and the higher the valence of the cation salts in solution, the more the clay will be inhibited from swelling. It is also true that the tendency of the dispersed clays to revert to an aggregated (inhibited) state is measurable.

Dispersion The subdivision of particles from the aggregated state in a fluid (water) to a hydrated colloid particle is the dispersing of that particle. In freshwater dispersion, the clay platelets drift about in an independent manner or in very small clusters. There are times when the platelets configure in random patterns. This usually occurs in a static condition and is termed *gel strength of dispersed day*. The random movement and drifting of a positively charged edge toward a negatively changed face happens slowly in a dispersed state. When bentonite is in a dispersed state, the positive ion cloud presents an effective "shield" around the clay and sometimes slows this effect. The ionized Na^+ surrounds the clay to form a weak crystalline barrier.

Dispersed clay state is characterized by

- High viscosity
- High gel strength
- Low filtrate

Flocculation (NaCl) The most common cause of flocculation of clays in the field is the incorporation of NaCl in to a fresh water mud. When the Na^+ content is raised toward 1%, the water becomes more positively charged. The ionized envelope cloud that "protected" the platelet is of a lower charge than the bulk water. The positive Al^{3+} edge joins with the oxygen face, and the drift of edge to face is accelerated.

The viscosity rises, and water loss is uncontrollable when the clay flocculates edge to face in a "House of Cards structure," and the increase in viscosity and water loss is dramatic. As the NaCl content increases to 5%, the free water is tied up by the chloride ion, and the ion and the clays collapse and revert to the aggregated state. The water is removed from the clay platelet body.

When the NaCl content increases to 15% to 30% by weight, the agglomerates flocculate into large edge to face groups. This leads to extreme viscosities and very poor fluid loss control. This also depends on the solids content. In diluted suspensions, the viscosity usually is reduced by increasing salt concentrations, and clay platelets are in the aggregated state. Viscosity will go through a "hump."

Deflocculation (Chemical Dispersion) One way to deflocculate, or chemically disperse, a clay platelet is with a large molecule having many carboxyl and sulfonate anions at scattered intervals on the cellulose chain. In deflocculated or chemically dispersed muds, the viscosity will be lower than it was in the flocculated state.

Lignosulfonate works to deflocculate by the anionic charges that latch onto the positive edges of the clay platelet. The remainder of this huge (flat) cellulosic molecule is repelled from the negative clay face and rolls out from the edges.

The edge-to-face flocculation that occurred becomes virtually impossible. The polyanionic encapsulator can be rendered neutral if the pH drops below 9.5. The NaCl flocculant is still present in the solution, but its flocculating effects are rendered ineffective if the pH is maintained above 9.5.

Flocculation (Calcium) When calcium is induced into a drilling fluid, its solubility depends on the pH of the water in the fluid. The double-positive charge on the calcium ion will attract itself to the face of the bentonite platelet at an accelerated rate, because this attraction is far superior to the sodium's ability to retain its place on the clay face. The divalent calcium ions will still partially hydrated, but the amount of water is less around the clay platelet. This will allow flocculation to occur much faster, because there is little water structure around the clay in this situation.

Calcium can cause flocculation in the same manner as salt (NaCl) in that edge-to-face groupings are formed. Calcium is a divalent cation, so it holds onto two platelet faces, which causes large groups to form, and then the edge-to-face grouping to take hold. Because calcium (Ca^{2+}) has a valence of 2, it can hold two clay platelets tightly together, and the flocculation reaction starts to happen at very low concentrations. To achieve flocculation with salt (NaCl), it takes 10 times the concentration for the edge-to-face groupings to form.

In the flocculated state, a dispersant (thinner) will work to separate the flocculating ions and encapsulate the platelets by mechanical shear. This is a short-term answer to the problem, however, because the contaminating ion is still active in the system, and it must be reduced to a normal active level for drilling to continue.

Deflocculation (Calcium Precipitation) The most effective way to remove the flocculating calcium ion from the system is to chemically precipitate it. Two common chemicals can be used to accomplish removal of the calcium ion. They are Na_2Co_3 (soda ash) and $NaHCO_3$ (bicarbonate of soda). Because calcium is lodged between two platelets and holding them together, the two chemicals will, with mechanical help, bond together with the flocculant calcium as shown in the formula below:

$$Ca^{2+}(OH)_2 + Na_2HCO_3 \rightarrow CaCO_3 + NaOH + H_2O$$

Lime + Sodium Bicarbonate
→ Calcium Carbonate + Caustic + Water

$$CaSO_4 + Na_2CO_3 \rightarrow CaCO_3 + Na_2SO_4$$

Calcium sulfate + Sodium carbonate
→ Calcium carbonate + Sodium sulfate

In the previous chemical equation, calcium is precipitated and rendered inert. There is no longer a possible flocculating calcium ion to deal with.

Inhibition (NaCl) When a water solution contains more than 12,000 mg/l of NaCl, it can inhibit clays from swelling or hydrating. This happens because the sodium ion content is high in the water, and the sodium ions on the clay face cannot leave to allow space for the water to enter the clay platelet. The chloride ion has an ability to tightly hold onto water molecules, which leaves few free ions to envelope or surround the clay. When the clay (aggregation) platelet does not hydrate, the state is the same as it is in the sack. In this instance, the ion is controlling the swelling of clays and is referred to as *inhibition*.

Controlling these various clay states in water-base drilling fluids is important for the success of any well using this chemistry. It can be said that flocculation causes and increases viscosity and that aggregation and deflocculation decrease viscosity.

1.3.8 Water-Base Muds

A water-base drilling fluid is one that has water as its continuous or liquid phase. The types of drilling fluids are briefly described in the following sections.

Freshwater muds are generally lightly treated or untreated muds having a liquid phase of water, containing small concentrations of salt, and having a pH ranging from 8.0 to 10.5. Freshwater muds include the following types.

Spud Muds These muds are prepared with available water and appropriate concentrations of bentonite and/or premium commercial clays. They are generally untreated chemically, although lime, cement, or caustic soda is occasionally added to increase viscosity and give the mud a fluff to seal possible lost return zones in unconsolidated upper hole surface formations. Spud muds are used for drilling the surface hole. Their tolerance for drilled solids and contaminants is very limited.

Natural Mud Natural or native muds use native drilled solids incorporated into the mud for viscosity, weight, and fluid loss control. They are often supplemented with bentonite for added stability and water loss control. Surfactants can be used to aid in controlling mud weight and solids buildup. Natural muds are generally used in top hole drilling to mud-up or to conversion depth. They have a low tolerance for solids and contamination.

Saltwater Muds Muds ordinarily are classified as saltwater muds when they contain more than 10,000 mg/L of chloride. They may be further classified according to the amount of salt present and/or the source of makeup water (see Table 1.3):

Amount of chloride in mg/L

1. Saturated salt muds (315,000 ppm as sodium chloride)
2. Salt muds (over 10,000 mg/L chloride but not saturated)

Source of make-up water

A. Brackish water
B. Sea Water

Saltwater muds may be purposely prepared, or they may result from the use of salty makeup water, from drilling into salt domes or stringers, or when saltwater flows are encountered. Saltwater muds include the following types.

TABLE 1.3 Seawater Composition

Constituent	Parts per Million	Equivalent Parts per Million
Sodium	10440	454.0
Potassium	375	9.6
Magnesium	1270	104.6
Calcium	410	20.4
Chloride	18970	535.0
Sulfate	2720	57.8
Carbon dioxide	90	4.1
Other constituents	80	n/a

Seawater or Brackish Water Muds These muds are prepared with available makeup water, both commercial and formation clay solids, caustic soda, and lignite and/or a lignosulfonate. CMC is usually used for fluid loss control, although concentration of lignites and lignosulfonates are also often used for this purpose. Viscosity and gel strength are controlled with caustic soda, lignosulfonate, and/or lignites. Soda ash is frequently used to lower the calcium concentration. CMC or lignosulfonates are used for water loss control, and pH is controlled between 8.5 to 11.0 with caustic soda. Seawater muds and brackish or hard water muds are used primarily because of the convenience of makeup water, usually open sea or bays. The degree of inhibitive properties varies with the salt and calcium concentration in the formulated fluid.

Saturated Salt Muds Saturated salt water (natural or prepared) is used as makeup water in these fluids. Prehydrated bentonite (hydrated in freshwater) is added to give viscosity, and starch is commonly used to control fluid loss. Caustic soda is added to adjust the pH, and lignosulfonates are used for gel strength control. Occasionally, soda ash may be used to lower filtrate calcium and adjust the pH. Saturated salt muds are used to drill massive salt sections (composed mainly of NaCl) to prevent washouts and as a work-over or completion fluid. Freshwater bentonite suspensions are converted by adding NaCl to reach saturation. Conversion is carried out by diluting the freshwater mud to reduce the viscosity "hump" seen in breakovers. Saturated salt muds usually are used at mud weights below 14.0 lb/gal.

Composition of NaCl mud

- Brine NaCl
- Density — salt, barite, calcium carbonate or hematite
- Viscosity — CMC HV, Prehydrated bentonite, XC-polymer (xanthan gum)
- Rheology — lignosulfonate
- Fluid Loss — CMC LV or PAC (polyanionic cellulose)
- pH – P_f (alkalinity) — caustic potash or caustic soda

Chemically Treated Mud (No Calcium Compounds) This type of mud is made up of a natural mud that has been conditioned with bentonite and treated with caustic soda and lignite or lignosulfonate (organic thinner). No inhibiting ions are found in this type of fluid.

Lignite/Lignosulfonate Mud This fluid is prepared from freshwater and conditioned with bentonite. Lignosulfonate is added as a thinner and lignite for filtration control and increased temperature stability. CMC or PAC may be used for additional filtration control when the bottom-hole temperature does not exceed 121°C (250°F). This type of mud is applied at all mud weights and provides a relatively low pH system (pH values for calcium lignosulfonates will be 10.0–11.0). This type of fluid is stable at reasonably high temperatures (325°F) and has good resistance to contamination.

Calcium Treated Muds Calcium-treated fluids are prepared from any low or high pH mud by the addition of appropriate amounts of lime or gypsum, caustic soda, and thinner (lignite or lignosulfonate). Calcium-treated muds include lime and gypsum muds.

Lime Muds Lime muds include low- and high-lime muds. They are prepared from available muds by adding calcium lignosulfonate, lignite, caustic soda or KOH, lime, and a filtration-control material, PAC or starch. Caustic soda is used to maintain the filtrate alkalinity (P_f values) and lime to control the mud alkalinity (P_m values) and excess lime. Lime muds offer resistance to salt, cement, or anhydrite contamination even at high mud weights.

Gypsum Mud Commonly called "gyp muds," they are prepared from freshwater and conditioned with bentonite or from available gel and water mud. Caustic soda is added for pH control. Gypsum, lignosulfonate, and additional caustic soda are added simultaneously to the mud. CMC may be added for filtration control. This fluid is used for drilling in mildly reactive shale or where gypsum or anhydrite must be drilled. It resists contamination from cement or salt. Use is limited by the temperature stability of the filtration control materials, CMC (250°F ±).

1.3.9 Special Muds

In addition to the most common mud systems discussed previously, there are other muds that do not fall neatly into one category or another in the classification scheme.

Low-Density Fluids and Gaseous Drilling Mud (Air-Gas Drilling Fluids) The basic gaseous drilling fluids and their characteristics are presented in Table 1.4.

This system involves injecting air or gas downhole at the rates sufficient to attain annular velocity of 2,000 to 3,000 ft/min. Hard formations that are relatively free from water are most desirable for drilling with air-gas drilling fluids. Small quantities of water usually can be dried up or sealed off by various techniques.

Air-gas drilling usually increases drilling rate by three or four times over that when drilling with mud, as well as one-half to one-fourth the number of bits are required. In some areas, drilling with air is the only solution;

TABLE 1.4 Gaseous Drilling Mud Systems

Type of Mud	Density, ppg	PH	Temp. Limit °F	Application Characteristics
Air/gas	0	—	500	High-energy system. Fastest drilling rate in dry, hard formations. Limited by water influx and hole size.
Mist	0.13–0.8	7–11	300	High-energy system. Fast penetration rates. Can handle water intrusions. Stabilizes unstable holes (mud misting).
Foam	0.4–0.8	4–10	400	Very-low-energy system. Good penetration rates. Excellent cleaning ability regardless of hole size. Tolerates large water influx.

these are (1) severe lost circulation, (2) sensitive producing formation that can be blocked by drilling fluid (skin effect), and (3) hard formations near the surface that require the use of an air hammer to drill.

There are two major limitations with using air as a drilling fluid: large volumes of free water and size of the hole. Large water flows generally necessitate converting to another type of drilling fluid (mist or foam). Size of the hole determines a volume of air required for good cleaning. Lift ability of air depends annular velocity entirely (no viscosity or gel strength). Therefore, large holes require an enormous volume of air, which is not economical.

Mist Drilling Fluids Misting involves the injection of air and mud or water and foam-making material. In the case of "water mist," only enough water and foam is injected into the air stream to clear the hole of produced fluids and cuttings. This unthickened water can cause problems due to the wetting of the exposed formation, which can result in sloughing and caving of water-sensitive shale into the wellbore. Mud misting, on the other hand, coats the walls of the hole with a thin film and has a stabilizing effect on water-sensitive formations. A mud slurry that has proved adequate for most purposes consists of 10 ppb of bentonite, 1 ppb of soda ash, and less than 0.5 ppb of foam-stabilizing polymer such as high-viscosity CMC. If additional foam stability is needed, additional foamer is used.

Nondispersed (Low-Solids) Muds The term *low-solids mud* covers a wide variety of mud types, including clear water (fresh, salt, or brine), oil-in-water emulsions, and polymer or biopolymer fluids (muds with polymer and no other additives).

Extended Bentonite Muds Low-solids nondispersed mud is generally prepared from freshwater with little or no drilled solids and bentonite, along with a dual-action polymer for extending the bentonite and flocculating drilled solids. This type of mud is designed for low-solids content and to have low viscosity at the bit for high drilling rates. The polymers used greatly increases the viscosity contributed by the bentonite and serve as flocculants for native clay solids, making them easier to remove by solids-control equipment. These polymers or bentonite extenders permit the desired viscosity to be maintained with about half of the amount of bentonite normally required. No deflocculant is used, so a flocculated system is maintained. The flocculation and lower solids content permit the mud to have a relatively low viscosity at the bit and at the bottom of the hole, where shear rates are high, and a relatively high viscosity at the lower shear rates in the annulus for good hole cleaning. One problem with this type of fluid is that filtration rates are fairly high, because the solids are flocculated and their quantity is low. This means that they do not pack tightly in the filter cake. Sodium polyacrylates or small amounts of CMC may be added for filtration control.

The temperature limitation of extended bentonite fluids is 200–275F°. Other benefits include improved hydraulics and less wear on bits and pump parts.

Inhibitive Salt/Polymer Muds An *inhibitive mud* is one that does not appreciably alter a formation once it has been cut by the bit. The term covers a large number of mud systems, among them saltwater muds with more than 10,000 mg/L of sodium chloride, calcium-treated muds (lime and gyp), and surfactant-treated muds. Under the category of inhibiting salt/polymer muds, however, we are speaking specifically about muds containing inhibitive salts such as KCl, NaCl, or diammonium phosphate along with complex, high-molecular-weight polymers. In these muds, prehydrated bentonite and polymer are added for viscosity and gel strength, polyanionic cellulose (PAC) or CMC are added for fluid loss control, and corrosion inhibitors and oxygen scavengers often are used to protect tubular goods. These muds are used for drilling and protecting water-sensitive formations and are good for minimizing formation damage due to filtrate invasion when the formation contains hydratable clay solids. Good hole cleaning and shear thinning are characteristics of these fluids. High-solids concentrations cannot be tolerated, however, making good solids control very important. Temperature limitations of 200–250°F are also characteristic. Among the muds of this type is KCl/lime mud. This mud system uses pre-hydrated bentonite or KCl for inhibition, lignosulfonate and/or lignite as a thinner, KOH (caustic potash) or caustic soda for alkalinity, lime for alkalinity and inhibition, and polymers such as CMC or PAC for filtration control.

Surfactant Muds Surfactant muds were developed primarily to replace calcium-treated muds when high temperature becomes a problem. The term *surfactant* means surface-acting agent, or a material that is capable of acting on the surface of a material. In drilling muds, surfactants are additives that function by altering the surface properties of the liquid and solid phases of the mud or by imparting certain wetting characteristics to the mud. The composition of the surfactant mud system tends to retard hydration or dispersion of formation clays and shales. The pH of these muds is kept from 8.5 to 10.0 to give a more stable mud at higher temperatures.

The surfactant mud usually encountered is a lignite surfactant mud system. This mud is made up from freshwater using bentonite, lignite, and the surfactant. Small amounts of defoamer may be required with the addition of the lignite. The pH of this mud is maintained within closely fixed limited (8.5 – 10.0) for maximum solubility of the thinner (lignite). Tolerance to salt, gyp, and cement contamination is limited. To retain satisfactory flow properties at high temperatures, the clay content of the mud must be kept low (1–1.6 CEC capacity) through the use of dilution and solids-control equipment. The combination of lignite with surfactant in this mud enables its use at extremely high bottom-hole temperatures. This is due to the temperature stability of lignite and the effect of the surfactant in providing viscosity control and minimizing gel strength development at higher temperatures.

High-Temperature Polymer Muds Development of a high-temperature polymer system evolved from a need for a mud system with low solids and nondispersive performance at higher temperatures.

System capabilities:

- Good high-temperature stability
- Good contaminant tolerance
- Can formulate temperature stable nondispersed polymer mud system
- Can be used in wide variety of systems for good shale stability
- Minimum dispersion of cuttings and clays
- Flexibility of general application

Application of the high-temperature polymer system primarily consists of five products: (1) polymeric deflocculant, (2) acrylamide copolymer, (3) bentonite, (4) caustic soda or potassium hydroxide, and (5) oxygen scavenger. Barite, calcium carbonate, or hematite is then used as a weighting agent.

The polymeric deflocculant is a low-molecular-weight, modified polyacrylate deflocculant used to reduce rheological properties of the system. If differs from lignosulfonates in that it does not require caustic soda or an alkaline environment to perform. Limited amounts of the polymer may be used in low-mud-weight systems, but larger additions will be needed at higher mud weights and when adding barite to increase the fluid density.

The backbone of the system is an acrylamide copolymer used to control fluid loss. In freshwater systems, 1 to 2 lb/bb will be the range required to control the API fluid loss. In seawater systems, the concentration will range from 4 to 5 lb/bbl. HPHT fluid loss can also be controlled with the polymer. It is not affected by salinity or moderate levels of calcium. At higher concentrations of contaminants, some increase in viscosity will result.

Caustic soda and/or potassium hydroxide are alkaline agents used to control the pH of the system. Either is used to maintain the system pH between 8.3 and 9.0.

Oxygen scavengers serve two purposes in this system. First, because of the low pH characteristic of the system, it should be added to protect the drill pipe. (Run corrosion rings in the drill pipe to determine treatment rates for the corrosion that may be occurring.) Second, as the temperature of the mud exceeds 300°F, any oxygen present will react with the polymers and reduce their efficiency. Additional treatment will be required to replace affected or degraded polymers.

New-Generation Water-Based Chemistry Several companies have developed water-base fluids that provide the inhibition formerly seen only when using oil-base fluids. Novel chemistry such as sodium silicates, membrane-efficient water-base muds, and highly inhibiting encapsulating polymers make these new systems unique and high in performance. Product development in the area of highly inhibitive polymers will no doubt result in the total replacement of invert emulsions. The need to provide more environmentally acceptable products drive the research and development of many drilling fluids by vendors around the world.

Oil-Base Mud Systems and Nonaqueous Fluids (NAF) Oil-base muds are composed of oil as the continuous phase, water as the dispersed phase, emulsifiers, wetting agents, and gellants. Other chemicals are used for oil-base mud treatment, such as degellants, filtrate reducers, and weighting agents.

The oil for an oil-base mud can be diesel oil, kerosene, fuel oil, selected crude oil, mineral oil, vegetable esters, linear paraffins, olefins, or blends of various oils. There are several desired performance requirements for any oil:

- API gravity = 36° – 37°
- Flash point = 180°F or above
- Fire point = 200°F or above
- Aniline point = 140°F or above

Emulsifiers are very important in oil-base mud because water contamination on the drilling rig is very likely and can be detrimental to oil mud. Thinners, on the other hand, are far more important in waterbase mud than in oil-base mud; oil is dielectric, so there are no interparticle electric forces to be nullified.

The water phase of oil-base mud can be freshwater or various solutions of calcium chloride ($CaCl_2$), sodium chloride (NaCl), or formates. The concentration and composition of the water phase in oil-base mud determines its ability to solve the hydratable shale problem.

The external phase of oil-base mud is oil and does not allow the water to contact the formation; the shales are thereby prevented from becoming wet with water and dispersing into the mud or caving into the hole.

The stability of an emulsion mud is an important factor that has to be closely monitored while drilling. Poor stability results in coalescence of the dispersed phase, and the emulsion will separate into two distinct layers. Presence of any water in the HPHT filtrate is an indication of emulsion instability.

The advantages of drilling with emulsion muds rather than with water-base muds are

- High penetration rates
- Reduction in drill pipe torque and drag
- Less bit balling
- Reduction in differential sticking

Oil-base muds are generally expensive and should be used when conditions justify their application. As in any situation, a cost-benefit analysis should be done to ensure that the proper mud system is selected. Oil-based fluids are well suited for the following applications:

- Drilling troublesome shales that swell (hydrate) and disperse (slough)
- Drilling deep, high-temperature holes in which water-base muds solidify
- Drilling water-soluble formations such as salt, anhydride, camallite, and potash zones
- Drilling the producing zones

For additional applications, oil muds can be used

- As a completion and workover fluid
- As a spotting fluid to relieve stuck pipe
- As a packer fluid or a casing pack fluid

Drilling in younger formations such as "gumbo," a controlled salinity invert fluid is ideally suited. Gumbo, or plastic, flowing shale encountered in offshore Gulf of Mexico, the Oregon coast, Wyoming, West Africa, Venezuela, the Middle East, Western Asia, and the Sahara desert, benefits from a properly designed salinity program. Drilling gumbo with water-base mud shale disperses into the mud rapidly, which reduces the drilling rate and causes massive dilution of the mud system to be required. In some cases, the

ROP must be controlled to prevent plugging of the flowline with hydrated "gumbo balls." Solids problems also are encountered with water-based fluid drilling gumbo such as bit balling, collar balling, stuck pipe, and shaker screens plugging.

Properly designed water-phase salinity invert fluids will pull water from the shale (through osmosis), which hardens the shale and stabilizes it for long-term integrity.

Generally, oil-base mud is to delivered to the rig mixed to the desired specifications. In some cases, the oil-base mud can be mixed on location, but this process can cost expensive rig time. In the latter case, the most important principles are (1) to ensure that ample energy in the form of shear is applied to the fluid and (2) to strictly follow a prescribed order of mixing. The following mixing procedure is usually recommended:

1. Pump the required amount of oil into the tank.
2. Add the calculated amounts of emulsifiers and wetting agent. Stir, agitate, and shear these components until adequate dispersion is obtained.
3. Mix in all of the water or the $CaCl_2$-water solution that has been premixed in the other mud tank. This requires shear energy. Add water slowly through the submerged guns; operation of a gun nozzle at 500 psi is considered satisfactory. After emulsifying all the water into the mud, the system should have a smooth, glossy, and shiny appearance. On close examination, there should be no visible droplets of water.
4. Add all the other oil-base mud products specified.
5. Add the weighting material last; make sure that there are no water additions while mixing in the weighting material (the barite could become water wet and be removed by the shale shakers).

When using an oil-base mud, certain rig equipment should be provided to control drilled solids in the mud and to reduce the loss of mud at the surfaces:

- Kelly valve—a valve installed between the Kelly and the drill pipe will save about one barrel per connection.
- Mud box—to prevent loss of mud while pulling a wet string on trips and connections; it should have a drain to the bell nipple and flow line.
- Wiper rubber—to keep the surface of the pipe dry and save mud.

Oil-base mud maintenance involves close monitoring of the mud properties, the mud temperature, and the chemical treatment (in which the order of additions must be strictly followed). The following general guidelines should be considered:

A. The mud weight of an oil mud can be controlled from 7 lb/gal (aerated) to 22 lb/gal. A mud weight up to 10.5 lb/gal can be achieved with sodium chloride or with calcium chloride. For densities above

TABLE 1.5 Estimated Requirements for Oil Mud Properties

Mud Weight, ppg	Plastic Viscosity, cP	Yield Point, lbs/sq ft^2	Oil-Water Ratio	Electrical Stability
8–10	15–30	5–10	65/35–75/25	200–300
10–12	20–40	6–14	75/25–80/20	300–400
12–14	25–50	7–16	80/20–85/15	400–500
14–16	30–60	10–19	85/15–88/12	500–600
16–18	40–80	12–22	88/15–92/8	Above 600

10.5 lb/gal, barite, hematite, or ground limestone can be used. Calcium carbonate can be used to weight the mud up to 14 lb/gal; it is used when an acid-soluble solids fraction is desired, such as in drill-in fluids or in completion/workover fluids. Iron carbonate may be used to obtain weights up to 19.0 lb/gal when acid solubility is necessary (Table 1.5).

B. Mud rheology of oil-base mud is strongly affected by temperature. API procedure recommends that the mud temperature be reported along with the funnel viscosity. The general rule for maintenance of the rheological properties of oil-base muds is that the API funnel viscosity, the plastic viscosity, and the yield point should be maintained in a range similar to that of comparable-weight water muds. Excessive mud viscosity can be reduced by dilution with a base oil or with specialized thinners. Insufficient viscosity can be corrected by adding water (pilot testing required) or by treatment with a gallant, usually an organophilic clay or surfactant.
C. Low-gravity solids contents of oil-base muds should be kept at less than 6% v/v. Although oil muds are more tolerant for solids contamination, care must be taken to ensure that solids loading does not exceed the recommended guidelines. A daily log of solids content enables the engineer to quickly determine a solids level at which the mud system performs properly.
D. Water-wet solids is a very serious problem; in severe cases, uncontrollable barite setting may result. If there are any positive signs of water-wet solids, a wetting agent should be added immediately. Tests for water-wet solids should be run daily.
E. Temperature stability and emulsion stability depend on the proper alkalinity maintenance and emulsifier concentration. If the concentration of lime is too low, the solubility of the emulsifier changes, and the emulsion loses its stability. Lime maintenance has to be established and controlled by alkalinity testing. The recommended range of lime content for oil-base muds is 0.1 to 4 lb/bbl, depending on base oil being used. Some of the newer ester-base muds have a low tolerance for hydroxyl ions; in this case, lime additions should be closely controlled.

F. $CaCl_2$ content should be checked daily to ensure the desired levels of inhibition are maintained.
G. The oil-to-water ratio influences funnel viscosity, plastic viscosity, and HTHP filtration of the oil-base mud. Retort analysis is used to detect any change in the oil-water ratio, because changes to the oil-water ration can indicate an intrusion of water.
H. Electrical Stability is a measure of how well the water is emulsified in the continuous oil phase. Because many factors affect the electrical stability of oil-base muds, the test does not necessarily indicate that a particular oil-base muds, the test does not necessarily indicate that a particular oil-base mud is in good or in poor condition. For this reason, values are relative to the system for which they are being recorded. Stability measurement should be made routinely and the values recorded and plotted so that trends may be noted. Any change in electrical stability indicates a change in the system.
I. HTHP filtration should exhibit a low filtrate volume (< 6 ml). The filtrate should be water free; water in the filtrate indicates a poor emulsion, probably caused by water wetting of solids.

1.3.10 Environmental Aspects of Drilling Fluids

Much attention has been given to the environmental aspects of the drilling operation and the drilling fluid components. Well-deserved concern about the possibility of polluting underground water supplies and of damaging marine organisms, as well as effects on soil productivity and surface water quality, has stimulated widespread studies on this subject.

Drilling Fluid Toxicity There are three contributing mechanisms of toxicity in drilling fluids: chemistry of mud mixing and treatment, storage and disposal practices, and drilled rock. The first group conventionally has been known the best because it includes products deliberately added to the system to build and maintain the rheology and stability of drilling fluids.

Petroleum, whether crude or refined products, needs no longer to be added to water-base muds. Adequate substitutes exist and are economically viable for most situations. Levels of 1% or more of crude oil may be present in drilled rock cuttings, some of which will be in the mud.

Common salt, or sodium chloride, is also present in dissolved form in drilling fluids. Levels up to 3,000 mg/L of chloride and sometimes higher are naturally present in freshwater muds as a consequence of the salinity of subterranean brines in drilled formations. Seawater is the natural source of water for offshore drilling muds. Saturated-brine drilling fluids become a necessity when drilling with water-base muds through salt zones to get to oil and gas reservoirs below the salt. In onshore drilling, there is no need for chlorides above these background levels. Potassium chloride has been added to some drilling fluids as an aid to controlling problem shale formations. Potassium acetate or potassium carbonate are acceptable substitutes in most of these situations.

Heavy metals are present in drilled formation solids and in naturally occurring materials used as mud additives. The latter include barite, bentonite, lignite, and mica (sometimes used to control mud losses downhole). There are background levels of heavy metals in trees that carry through into lignosulfonate made from them.

Attention has focused on heavy metal impurities found in sources of barite. Proposed U.S. regulations would exclude many sources of barite ore based on levels of contamination. European and other countries are contemplating regulations of their own.

Chromium lignosulfonates are the biggest contributions to heavy metals in drilling fluids. Although studies have shown minimal environmental impact, substitutes exist that can result in lower chromium levels in muds. The less-used chromium lignites (trivalent chromium complexes) are similar in character and performance, with less chromium. Nonchromium substitutes are effective in many situations. Typical total chromium levels in muds are 100–1000 mg/L.

Zinc compounds such as zinc oxide and basic zinc carbonate are used in some drilling fluids. Their function is to react out swiftly sulfide and bisulfide ions originating with hydrogen sulfide in drilled formations. Because human safety is at stake, there can be no compromising effectiveness, and substitutes for zinc have not seemed to be effective. Fortunately, most drilling situations do not require the addition of sulfide scavengers.

Indiscriminate storage and disposal practices using drilling mud reserve pits can contribute toxicity to the spent drilling fluid. The data in Table 1.6 is from the EPA survey of the most important toxicants in spent drilling fluids. The survey included sampling active drilling mud (in circulating system) and spent drilling mud (in the reserve pit). The data show that the storage disposal practices became a source of the benzene, lead, arsenic, and fluoride toxicities in the reserve pits because these components had not been detected in the active mud systems.

TABLE 1.6 Toxicity Difference between Active and Waste Drilling Fluids

Toxicant	Active Mud	Detection Rate	Reserve Pit	Detection Rate, %
Benzene	No	—	Yes	39
Lead	No	—	Yes	100
Barium	Yes	100	Yes	100
Arsenic	No	—	Yes	62
Fluoride	No	—	Yes	100

The third source of toxicity in drilling discharges are the cuttings from drilled rocks. A study of 36 cores collected from three areas (Gulf of Mexico, California, and Oklahoma) at various drilling depths (300 to 18,000 ft) revealed that the total concentration of cadmium in drilled rocks was more than five times greater than the cadmium concentration in commercial barites. It was also estimated, using a 10,000-ft model well discharge volumes, that 74.9% of all cadmium in drilling waste may be contributed by cuttings, but only 25.1% originate from the barite and the pipe dope.

Mud Toxicity Test for Water-Base Fluids The only toxicity test for water-base drilling fluids having an EPA approval is the Mysid shrimp bioassay. The test was developed in the mid-1970s as a joint effort of the EPA and the oil industry. The bioassay is a test designed to measure the effect of a chemical on a test population of marine organisms. The test is designed to determine the water-leachable toxicity of a drilling fluid or mud-coated cutting. The effect may be a physiological or biochemical parameter, such as growth rate, respiration, or enzyme activity. In the case of drilling fluids, lethality is the measured effect. For the Mysid test, all fluids must exceed a 30,000 concentration of whole mud mixed in a 9:1 ratio of synthetic seawater.

Nonaqueous Fluid (NAF) and Drilling Fluid Toxicity Until the advent of synthetic-based invert emulsion fluids in the early 1990s, the discharge of NAF was prohibited due the poor biodegradability of the base oils. In 1985, a major mud supplier embarked on a research program aimed at developing the first fully biodegradable base fluid. The base fluid would need to fulfill a number of criteria, regarded as critical to sustain drilling fluids performance while eliminating long-term impact on the environment:

- Technical performance — the fluid must behave like traditional oil-base muds and offer all of their technical advantages
- The fluid must be nontoxic, must not cause tainting of marine life, not have potential to bioaccumulate, and be readily biodegradable.

Research into alternative biodegradable base fluids began with common vegetable oils, including peanut, rapeseed, and soy bean oils. Fish oils such as herring oil were also examined. However, the technical performance of such oils was poor as a result of high viscosity, hydrolysis, and low temperature stability. Such performance could only be gained from a derivative of such sources, so these were then examined.

Esters were found to be the most suitable naturally derived base fluids in terms of potential for use in drilling fluids. Esters are exceptional lubricants, show low toxicity, and have a high degree of aerobic and anaerobic biodegradability. However, there are a vast number of fatty acids and alcohols from which to synthesize esters, each of which would have unique physical and chemical properties.

After 5 years of intensive research, an ester-based mud that fulfilled all of the design criteria was ready for field testing. This fluid provided the same shale stabilization and superior lubricity as mineral oil-based mud but also satisfied environmental parameters. The first trial, in February 1990, took place in Norwegian waters and was a technical and economic success. Since then, over 400 wells have been drilled world wide using this ester-base system, with full approval based on its environmental performance. This history of field use is unrivalled for any synthetic drilling fluid on a global basis, and no other drilling fluid has been researched in such depth. The research program included

- Technical performance testing using oil-base mud as a baseline
- Toxicity to six marine species, including water column and sediment reworker species
- Seabed surveys
- Fish taint testing
- Aerobic and anaerobic biodegradability testing
- Human health and safety factors

The release of ester-base fluids onto the market marked the beginning of the era of synthetic-base invert drilling fluids. Following the success of esters, other drilling fluids were formulated that were classed as synthetics, these fluids included base oils derived from ethylene gas and included linear alpha olefins, internal olefins, and poly-alpha olefins.

Summary of Flashpoint and Aromatic Data With the introduction of synthetic-base muds into the market, the EPA moved to provide guidelines on the quality and quantity of the synthetic oils being discharged into the Gulf of Mexico. In addition to the water column aquatic testing done for water-base fluids, the EPA set forth guidelines for examining toxicity to organisms living in the sediments of the seafloor. A Leptocheirus sedimentary reworker test was instituted in February of 2002 for all wells being drilled with synthetic-base muds to examine how oil-coated cuttings being discharged into the Gulf of Mexico would impact the organisms living on the seafloor.

Two standards were set forth to govern the discharge of synthetic-base muds. A *stock standard* test is required for the base oil looking at the biodegradability of a synthetic base oil and as well as a new test for the Leptocheirus sedimentary reworker. This stock standard is done once per year to certify that the base oils being used are in compliance with the regulation.

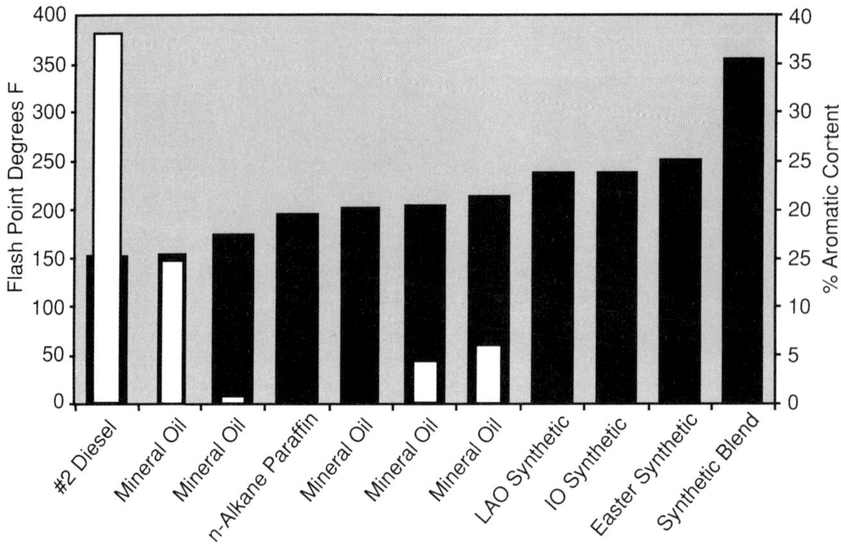

FIGURE 1.11 Base fluid type.

When a well is drilled with a synthetic-base mud, monthly and end-of-well tests are required for the *Mysid* and *Leptocheirus*, organisms to ensure that the synthetic-base oil being used meets a certain standard of environmental performance. There are two standards that can be used for these annual and well-to-well tests: an ester standard and a C 1618 Internal Olefin standard.

The test used as a standard is based on the type of base oil being tested against a similarly approved standard. Base oils that are less toxic and highly biodegradable would be compared with esters, while all others would have to meet or exceed the C1618 IO standard (Figure 1.11 and Table 1.7).

With synthetic-based muds being widely used in the Gulf of Mexico, especially in deep water, controlling the quality of these materials is extremely important to the environment.

From US-EPA 2001 NPDES General Permit for New and Existing Sources in the Offshore Subcategory of the Oil and Gas Extraction Category for the Western Portion of the Outer Continental Shelf of the Gulf of Mexico (GMC290000) 66 Fed. Reg. No. 243, p. 65209, December 18, 2001.

1.4 COMPLETION AND WORKOVER FLUIDS

Completion and workover fluids are any fluids used in the completion of a well or in a workover operation. These fluids range from low-density

TABLE 1.7 Aromatic Content Standards

	Fluids Meeting Internal Olefin Standard	Performance of New Blended-Base Fluid
Base fluid biodegradation	Equal to or better than a 65:35 blend of C16 C18 internal olefin in a 275-day test. Tested once per year. Ratio of IO result compared with base fluid must be calculated at 1.0 or less.	Ratio = 0.8
Leptocheirus base fluid toxicity	Ten-day Leptocheirus LC50 must be equal to or less toxic than a 65:35 blend of C16 C18 internal olefin tested at least annually. Ratio of IO result compared with base fluid must be calculated at 1.0 or less.	Ratio = 0.8
PAH content of base fluid	High-performance liquid chromatography/UV-EPA method 1654 must give a PAH (as phenanthrene) content of less than 10 ppm.	Below 1 ppm limit of detection

Summary of permit requirements and performance of blended drilling fluid, base fluid against those permit mandated C16–18 Internal olefin standards.
From US-EPA (2001) NPDES General Permit for New and Existing Sources in the Offshore Subcategory of the Oil and Gas Extraction Category for the Western Portion of the Outer Continental Shelf of the Gulf of Mexico (GMG290000) 66 Fed. Reg. No. 243, p. 65209, December 18, 2001.

gases such as nitrogen to high-density muds and packer fluids. The application and requirements vary for each fluid.

Workover fluids are fluids used during the reworking of a well after its initial completion. They may be gases such as nitrogen or natural gas, brine waters, or muds. Workover fluids are used during operations such as well killing, cleaning out a well, drilling into a new production interval, and plugging back to complete a shallower interval.

Completion fluids are used during the process of establishing final contact between the productive formation and the wellbore. They may be a water-base mud, nitrogen, oil mud, solids-free brine, or acid-soluble system. The most significant requirement is that the fluid does not damage the producing formation and does not impair production performance.

Packer fluids are fluids placed in the annulus between the production tubing and casing. Packer fluids must provide the required hydrostatic pressure, must be nontoxic and noncorrosive, must not solidify or settle out of suspension over long periods of time, and must allow for minimal formation damage.

Various types of fluids may be used for completion and workover operations:

1. Oil fluids
   a. Crude
   b. Diesel
   c. Mineral oil
2. Clear water fluids
   a. Formation salt water
   b. Seawater
   c. Prepared salt water such as calcium chloride, potassium chloride or sodium chloride salt and zinc, calcium, or sodium-based bromides
3. Conventional water-base mud
4. Oil-base or invert emulsion muds

Completion or workover fluids may be categorized as

1. Water-base fluids containing oil-soluble organic particles
2. Acid-soluble and biodegradable
3. Water base with water-soluble solids
4. Oil-in-water emulsions
5. Oil-base fluids

Three types of completion or workover fluids are

1. Clear liquids (dense salt solutions)
2. Weighted suspensions containing calcium carbonate weighting material, a bridging agent to increase the density above that of saturated solutions
3. Water-in-oil emulsions made with emulsifiers for oil muds

Clear liquids have no suspended solids and can be referred to as solids-free fluids. Weighted suspensions are fluids with suspended solids for bridging or added density. These fluids can be referred to as solids-laden fluids.

For solids-free fluids, water may be used in conjunction with a defoamer, viscosifier, stabilized organic colloid, and usually a corrosion inhibitor. Solids-free completion and workover fluids have densities ranging from 7.0 to 19.2 pounds per gallon (ppg) (0.84 – 2.3 SG).

Solids-laden fluids may be composed of water, salt, a defoamer, suspension agent, stabilized organic colloid, pH stabilizer, and a weighting material/bridging agent.

1.4.1 Solids-Free Fluids

Brines used in completion and workover applications may be single-salt brines, two-salt brines, or brines containing three different salt compounds.

1.4.2 Single-Salt Brines

Single-salt brines are made with freshwater and one salt such as potassium chloride, sodium chloride, or calcium chloride. They are the simplest brines used in completion and workover fluids. Because they contain only one salt, their initial composition is easily understood. Their density is adjusted by adding either salt or water. Single-salt brines are available in densities of up to 11.6 ppg and are the least expensive brines used in completions.

Potassium chloride (KCl) brines are excellent completion fluids for water-sensitive formations when densities over (9.7 ppg)(1.16 SG) are not required. Corrosion rates are reasonably low and can be reduced even more by keeping the pH of the system between 7 and 10 and using corrosion inhibitors (1% by volume). Sodium chloride is one of the most used single-salt brines. Advantages of sodium chloride brines are low cost and wide availability. Densities up to 10.0 ppg are achievable for this single-salt brine. Calcium chloride ($CaCl_2$) brines are easily mixed at densities up to 11.6 ppg 1.39 SG. Sodium bromide brines can be used when the density of a calcium chloride brine is desired, but the presence of acid gas is possible. Sodium bromide has low corrosion rates even without the use of corrosion inhibitors. Although these brines are more expensive than $CaCl_2$ brines, they are useful in CO_2 environments.

1.4.3 Two-Salt Brines

The basic ingredient of calcium chloride/calcium bromide brines ($CaCl_2/CaBr_2$) is a calcium bromide solution that ranges in density from 14.1 to 14.3 ppg (1.72 SG); the pH range is 7.0 to 7.5. The density of $CaBr_2$ brine can be increased by adding calcium chloride pellets or flakes. However, a 1.81 S.G. $CaCl_2/CaBr_2$ solution crystallizes at approximately 65°F (18°C). $CaCl_2/CaBr_2$ brine can be diluted by adding a $CaCl_2$ brine weighing 11.6 ppg (1.39 SG). The corrosion rate for $CaCl_2/CaBr_2$ is no more than 5 mm per year on N-80 steel coupons at 300°F (149°C). If a corrosion inhibitor is desired, a corrosion inhibitor microbiostat is recommended.

The viscosity of $CaCl_2/CaBr_2$ brine can be increased by adding liquefied HEC viscosifier. Reduction in filtration may be obtained by the addition of $CaCO_3$ weighting material/bridging agent or by increasing the viscosity with polymeric materials.

There is not much of a crystallization problem with calcium chloride/calcium bromide brines at densities between 11.7 and 13.5 ppg 1.40 and

1.62 SG. However, the heavier $CaCl_2/CaBr_2$ brines require special formulation in cold weather applications.

1.4.4 Three-Salt Brines

Three-salt brines such as calcium chloride/calcium bromide/zinc bromide brines are composed of $CaCl_2$, $CaBr_2$, and $ZnBr_2$. At high temperatures, corrosion rates in brines containing $ZnBr_2$ are very high and can result in severe damage to equipment. For use at high temperatures, the brine should be treated with corrosion inhibitors. The corrosion rate of the treated brine is usually less than 3 mm per year.

1.4.5 Classification of Heavy Brines

Properties and Characteristics of Completion and Workover Fluids Although the properties required of a completion or workover fluid vary depending on the operation, formation protection should always be the primary concern.

Density The first function of a completion of workover fluid is to control formation pressure (Table 1.8). The density should be no higher than necessary to accomplish that function. Increased density can be obtained by using weighting materials such as calcium carbonate ($CaCO_3$), iron carbonate ($FeCO_3$), barite (BaSO4), or by using soluble salts such as NaCl, KCl, NaBr, $CaCl_2$, $CaBr_2$, or $ZnBr_2$. The Table 1.9 below shows the specific weight range and acid solubility of each type of solids-laden fluid.

Solids – Laden Fluids The density of a brine solution is a function of temperature. When measured at atmospheric pressure, brine densities decrease as temperature increases.

TABLE 1.8 Expansibility of Heavy Brine at 12,000 psi from 24° to 92°C or 76° to 198°F

Compressibility of Brine	Heavy Brine at 198°F Density SG – lb/gal	198°F at 2,000–12,000 psi (lb/gal/1,000 psi)
Nacl	1.14–9.49	0.019
$CaCl_2$	1.37–11.46	0.017
NaBr	1.50–12.48	0.021
$CaBr_2$	1.72–14.30	0.022
$ZnBr_2/CaBr_2/CaCl_2$	1.92–16.01	0.022
$ZnBr_2/CaBr_2$	2.31–19.27	0.031

TABLE 1.9 Solids-Laden Fluids

Weight Materials	Pounds/gallon	Acid Solubility
$CaCO_3$	10–14	98%
$FeCO_3$	10–18	90%
Barite	10–21	0%

Viscosity In many cases, the viscosity of the fluid must be increased to provide lifting capacity required to bring sand or cuttings to the surface at reasonable circulating rates. A popular viscosifier for completion and workover fluids is hydroxyethylcellulose (HEC). It is a highly refined, partially water soluble, and acid-soluble polymer with very little residue when acidized. Other materials used as viscosifiers include guar gums and biogums (xanthan). Although these materials are applicable in certain instances, they do not meet the acid or water-solubility standards of HEC. HEC is the most common viscosifier for all types of brines.

In some cases, solids must be suspended at low shear or static conditions. Available alternatives include clays and polymers. The most widely used suspension agent in completion and workover fluids is xanthan gum.

Suspension or Filtration Control: Filtration In most applications, some measure of filtration control is desirable. The standard approach to filtration control in completion and workover fluids is the use of property sized calcium carbonate particles for bridging in conjunction with colloidal size materials such as starch or CMC. The reason for the popularity of calcium carbonate is that it is acid soluble and can be removed. In some cases, oil-soluble resins are used as bridging agents, as are sized salts when used in saturated salt brines.

The seasonal ambient temperature must be considered when selecting a completion or workover fluid. If the temperature drops too low for the selected fluid, the fluid will crystallize or freeze. Each brine solution has a point at which crystallization or freezing occurs. Two definitions are important: FCTA, or fist crystal to appear, is the temperature at which the first crystal appears as a brine is cooled. LCTD, or last crystal to dissolve, is temperature at which the last salt crystal disappears as the brine is allowed to warm. Although this type of visual check may be somewhat inexact, it is an important part of the analysis of brines. Once the crystallization point of a fluid is determined, you can be reasonably sure that the fluid is safe at a temperature equal to or higher than the crystallization point. The FTCA and LCTD are run under normal-pressure conditions; pressure can greatly alter the formation of crystals in a brine, and more sophisticated tests are required to determine this value.

Special brine formulations are used to accommodate seasonal changes in temperature. Summer blends are fluids appropriate for use in warmer weather. Their crystallization points range from approximately 7–20°C$_2$ (45°F–68°F). Winter blends are used in colder weather or colder climates and have crystallization points ranging from approximately 20°F (−6°C) to below 0°F (−18°C). At times, a crystallization point between those of summer and winter blends is desirable. Special formulations are then used to prepare fluids that can be called fall, spring, or intermediate blends.

At first, it may seem practical to consistently formulate fluids having lower than necessary, and therefore safe, crystallization points. Although this approach may be easier, it is likely to be much more expensive. Generally, the lower the crystallization point, the more the fluid costs. If you provide a fluid having a crystallization point much lower than necessary, you are likely to be providing a fluid with a considerably higher cost than necessary. This is just one of numerous factors to consider when selecting a fluid that is both effective and economical.

Preparing Brines The typical blending procedure for NaCl and KCl brines is to begin with the required volume of water and then add sacked salt. Calcium chloride/calcium bromide brines and calcium chloride/calcium bromide/zinc bromide brines require special blending procedures.

Calcium Chloride/Calcium Bromide Solutions The ingredients in $CaCl_2/CaBr_2$ solutions must be added in a specific order. The necessary order of addition is as follows:

1. Start with the $CaBr_2$ brine.
2. Add sacked $CaCl_2$.
3. Allow approximately 45 minutes for most of the sacked $CaCl_2$ to dissolve.

Calcium Chloride/Calcium Bromide/Zinc Bromide Solutions For $CaCl_2/CaBr_2/ZnBr_2$ solutions using 15.0 ppg (1.80 SG) $CaCl_2/CaBr_2$ brine and 19.2 ppg (2.28 SG) $CaBr_2ZnBr_2$, the proper order of addition is as follows:

1. Start with the 15.0 ppg (1.80 SG) $CaCl_2/CaBr_2$ brine.
2. Add the 19.2 ppg (2.28 SG). $CaBr/ZnBr_2$ brine.

Rules of thumb for blending of brines

1. DO NOT CUT SACKS. Exception: $CaBr_2$ brines from 11.6 to 15.1 ppg.
2. An increase of $CaBr_2$ decreases the crystallization temperature for $CaBr_2$ brines.

3. An increase of $ZnBr_2$ decreases the crystallization temperature in any blend.
4. A decrease in crystallization temperature increases the cost of the fluid.
5. Do not mix fluids containing divalent ions (Ca^{2+}, Zn^{2+}) with fluids containing monovalent ions (Na^+, K^+), Precipitation may occur.
6. Do not increase the pH of $CaBr_2$ or $ZnBr_2$ fluids or precipitation may occur.
7. Do not add large volumes of water to $CaBr_2$ or $ZnBr_2$ brines or precipitation may occur.

Fluid Selection A good approach to selecting a fluid is to decide what functions the fluid is to perform and then select a base fluid and additives that will most effectively do the job. The first decision in selecting a fluid is identification of the required functions or properties. The next step is the selection of the type of fluid to be used. The properties or functions of the fluid dictate the type of fluid to be used. If the decision is made in reverse order, a poor performance is likely to result.

Completion and Workover Fluids Weighting Materials

Calcium Carbonate Calcium carbonate is available in five grades: 5, 50, 150, 600, and 2,300. At least 50% of the particles in each grade are large than the size (μm) indicated by the number. Other grind sizes can be made available.

Typical Physical Constants

Hardness (Mohr's scale)	3
Specific gravity	2.7
Bulk density, lb/ft^3, or ppg	168.3 or 22.5

Typical Chemical Composition

Total carbonates (Ca^{2+}, Mg^{2+})	98.0% (min)
Total impurities (Al_2O_3, Fe_2O_3, SiO_2, Mn)	2.0% (max)

$CaCO_3$ (5) (93% will pass through a 325 mesh) can be used alone or with ($FeCO_3$) to increase the densities of freshwater or brine fluids beyond their saturation limits. It may also be used to increase the density of oil base fluids (Table 1.10).

$CaCO_3$ (5) can be used instead of clays to provide wall cake buildup for acid-soluble fluids.

TABLE 1.10 Specific Gravity and Specific Weight of Common Materials

Material	Specific gravity	lb/gal	lb/bbl
Barite	4.2–4.3	35.0–35.8	1470–1504
Calcium carbonate	2.7	22.5	945
Cement	3.1–3.2	25.8–26.7	1085–1120
Clays and/or drilleld solids	2.4–2.7	20.0–22.5	840–945
Diesel oil	0.84	7.0	294
Dolomite	2.8–3.0	23.3–25.0	980–1050
Feldspar	2.4–2.7	20.0–22.5	840–945
Fresh water	1.0	8.33	350
Galena	6.5	54.1	2275
Gypsum	2.3	19.2	805
Halite (rock salt)	2.2	18.3	770
Iron	7.8	65.0	2730
Iron oxide (hematite)	5.1	42.5	1785
Lead	11.4	95.0	3990
Limestone	2.7–2.9	22.5–24.2	945–1015
Slate	2.7–2.8	22.5–23.3	945–980
Steel	7.0–8.0	58.3–66.6	2450–2800

It may also be used as an acid-soluble bridging agent for formations having pore sizes up to 15 μm.

$CaCO_3$ 50, 150, 600, and 2,300 grades are recommended for use as bridging agents for lost circulation problems, in squeeze mixtures, and in other similar applications. The particle size distribution is maintained in the slurry to provide effective bridging at the surface of the pay zone.

Acid solubility $CaCO_3$ is 98% soluble in 15% HCl solution. One gallon of 15% HCl dissolves 0.83 kg or 1.84 lb of $CaCO_3$.

Bridging agent Normal treatment is 2.27 to 5 kg (5 to 10 ppb) of the appropriate grade(s). From 5% to 10% of the material added should have particle size at least one third of the formation pore diameter.

Iron Carbonate Iron carbonate is used to achieve densities in excess of 14.0 ppg in (1.68 SG) solids-laden systems. The maximum density of a $CaCO_3$ fluid is approximately 14.0 ppg (1.68 SG) and the maximum density of iron carbonate fluids is 17.5–18.0 ppg (2.10–2.16 SG). For weighting fluids in the 13.0–16.5 ppg (1.56–1.98 SG) range, a blend is recommended.

The following precautions should be considered when using iron carbonate:

1. Iron carbonate is only 87% acid soluble, and after acidizing, 13% of the solids added may be left to plug the formation or may be flushed out, depending on the size and distribution of the formation flow channels.
2. *Mud acid, a combination of hydrofluoric acid and hydrochloric acid, should not be used with iron carbonate.* The hydrofluoric acid reacts with iron carbonate to produce insoluble salts of acidic and basic nature (iron fluoride and iron hydroxide). When using iron carbonate, use only hydrochloric acid.

1.5 SAFETY ASPECTS OF HANDLING BRINES

1.5.1 Potassium Chloride

Toxicity No published data indicate that potassium chloride is a hazardous material to handle. It is toxic only if ingested in very large amounts. It is considered a mild irritant to the eyes and skin. Inhalation of potassium chloride dust leaves a taste and causes mild irritation to mucous membranes in the nose and throat. Potassium is toxic to the Mysid shrimp used for aquatic toxicity testing in U.S. Federal waters. Potassium levels over 4% will likely fail the 30,000 ppm minimum required for discharge.

Safety precautions Prolonged contact with skin and eyes should be avoided. Inhalation of potassium chloride dust should be avoided as much as possible. Eye protection should be worn according to the degree of exposure, and dust masks should be used in severe dusting conditions. Personal protective equipment (PPE) should always be used when mixing or handling brines and all fluids used in drilling and completions operations.

First Aid Measures The following first aid measures should be used:

1. For contact with eyes, flush promptly with plenty of water for 15 minutes.
2. For contact with skin, flush with plenty of water to avoid irritation.
3. For ingestion, induce vomiting and get medical attention.
4. For inhalation, if illness occurs, remove the person to fresh air, keep him or her warm, and quiet, and get medical attention.

1.5.2 Sodium Chloride

Toxicity There are no published data indicating that salt is a hazardous material to handle. Sodium chloride is considered a mild irritant to the eyes and skin. Inhalation of dust leaves a taste and causes mild irritation to mucous membranes in the nose and throat.

Safety Precautions Prolonged contact with skin and eyes should be avoided. The inhalation of sodium chloride dust should be avoided as much as possible. Eye protection should be worn according to the degree of exposure, and dust masks may be needed in severe exposure.

First Aid Measures The following first aid measures should be used:

1. For contact with eyes, flush promptly with plenty of water for 15 minutes.
2. For contact with skin, flush with plenty of water to avoid irritation.

1.5.3 Calcium Chloride

Toxicity Three to five ounces of calcium chloride may be a lethal dose for a 45-kg (100-pound) person. However, calcium chloride is not likely to be absorbed through the skin in toxic amounts. Strong solutions are capable of causing severe irritation, superficial skin burns, and permanent eye damage. Normal solutions cause mild irritation to eyes and skin, and dust may be irritating.

Safety Precautions Contact with eyes and prolonged skin contact should be avoided. Clean, long-legged clothing must be worn. Hand and eye covering may be required, depending on the severity of possible exposure. For severe exposure, chemical goggles and a dust respirator should be worn. Cool water (27°C, 80°F or cooler) should always be used when dissolving calcium chloride. Because of an exothermic reaction, calcium chloride can burn bare hands if solids have been added. Barrier creams should always be used when handling brines such as calcium chloride, calcium bromide, and zinc bromide.

First Aid Measures The following first aid measures should be used:

1. For contact with eyes and skin, flush promptly with plenty of water for 15 minutes. Get medical attention in the event of contact with eyes. Remove contaminated clothing, and wash before reuse.
2. For inhalation, if illness occurs, remove the person to fresh air, keep him or her warm and quiet, and get medical attention.
3. For ingestion, induce vomiting, and get medical attention.

1.5.4 Calcium Bromide

Toxicity There are no published data indicating that calcium bromide is a hazardous material to handle. However, it is considered toxic when ingested in large amounts. It is also a mild irritant to the skin and eyes. Inhalation results in irritation of the mucous membranes in the nose and throat. Because of an exothermic reaction, calcium bromide fluid can burn bare hands when sacked $CaBr_2$ is added to the solution. Burns caused by these fluids are the result of a chemical reaction with moisture on the skin.

Safety Precautions Prolonged contact with the skin and eyes should be avoided. Clean, long-legged clothing and rubber boots should be worn. Eye protection should be worn and a dust respirator used for severe exposure. Contaminated clothing should be changed. Barrier creams should always be used when handling these brines.

First Aid Measures The following first aid measures should be used:

1. For contact with eyes and skin, flush promptly with plenty of water.
2. For inhalation, if illness occurs, remove the victim to fresh air, keep him or her warm and quiet, and get medical attention.
3. For ingestion, induce vomiting, and get medical attention.

Environmental Considerations Local regulations should be observed. Care should be taken to ensure that streams, ponds, lakes, or oceans are not polluted with calcium bromide.

1.5.5 Zinc Bromide

Toxicity There are no published data indicating that zinc bromide is a hazardous material to handle. It is considered toxic when ingested in large amounts. Zinc bromide is also a severe irritant to the skin and eyes. Inhalation results in irritation of mucous membranes in the nose and throat. Because of an exothermic reaction, zinc bromide fluid can burn bare hands if sacked materials have been added. Never expose eyes to zinc bromide; blindness can occur.

Safety Precautions Contact with skin and eyes should be avoided. Long-legged clothing and proper eye protection should be worn. Barrier creams should always be used when handling zinc bromide brines. Rubber boots and rubber protective clothing also is suggested. Contaminated clothing should be washed off or changed, because contact with the skin can cause burns.

First Aid Measures

1. For contact with eyes and skin, flush promptly with plenty of water. Wash skin with mild soap, and consider seeking medical attention.

2. For inhalation, if illness occurs, remove victim to fresh air, kept him or her warm and quiet, and get medical attention.
3. For ingestion, induce vomiting, and get medical attention.

Environmental Considerations Local regulations should be observed, and care should be taken to avoid polluting streams, lakes, ponds, or oceans. Regulations in the United States prohibit the discharge of zinc into federal waters. Zinc bromide fluids should be disposed of in the same matter as oil fluids.

Safety Rules of Thumb

1. Do not wear leather boots.
2. Wear eye goggles for $CaCl_2$, $CaBr_2$, and $ZnBr_2$.
3. Wear rubber gloves underneath regular gloves while tripping.
4. Wear slicker suits while tripping pipe in most brines.
5. Wash off $CaCl_2$ or $ZnBr_2$ spills within 15 minutes; reapply barrier creams.
6. Change clothes within 30 minutes for $CaCl_2$ and within 15 minutes for $CaBr_2$ or $ZnBr_2$ if a spill occurs.
7. Do not wear shoes or boots for more than 15 minutes if they have $CaCl_2$, $CaBr_2$, or $ZnBr_2$ spilled in them.
8. Use pipe wipers when tripping.

1.6 PREVENTING CONTAMINATION

1.6.1 Brine Filtration

Filtration is a critical step if a well is to produce at its full potential and remain on line for a longer period. Although filtering can be expensive and time consuming, the net production can be enough to pay the difference in only a matter of days.

Filtration can be defined as the removal of solids particles from a fluid. Because these particles are not uniform in size, various methods of removal must be used (Table 1.11).

Filtration has evolved from the surface filtering systems with low-flow volumes to highly sophisticated systems. Regardless of which system is used, a case for filtering fluid can be made for every well completed, every workover, and every secondary recovery project.

The purpose of filtering any fluid is to prevent the downhole contamination of the formation with undesirable solids present in the completion fluid. Contamination can impact production and shorten the productive life of the well. Contamination can occur during perforating, fracturing, acidizing, workover, water flooding, and gravel packing of a well.

TABLE 1.11 Drilling Fluids Contaminants Removal

Contaminant To Be Removed	Chemical Used To Remove Contaminant	Conversion Factor mg/L (Contaminant) × Factor = lb/bbl Chemical to Add
Ca^{2+}	Soda ash	0.000925
Ca^{2+}	Sodium bicarbonate	0.000734
Mg^{2+}	Caustic soda	0.00115
CO_3^{-2}	Lime	0.00043
HCO_3^{-1}	Lime	0.00043
H_2S	Lime	0.00076
H_2S	Zinc carbonate	0.00128
H_2S	Zinc oxide	0.000836

Any time a fluid containing solids is put into the wellbore, a chance of damaging the well exists.

Contaminants in fluids come in many sizes and forms. Cuttings from drilling operations, drilling mud, rust, scale, pipe dope, paraffin, undissolved polymer, and any other material on the casing or tubing string contributes to the solids in the fluid. At times, it is virtually impossible, because of particle size, to remove all of the solids from the fluid, but by filtering, the chance of success can be increased almost to 100%.

How clean does the fluid need to be? What size particle do we need to remove? Typically, the diameter of the grains of sand is three times the size of the pore throat, assuming the sand is perfectly round. Particles greater than one-third the diameter of the pore throat bridge instantly on the throat and do not penetrate the formation. These particles represent a problem, but one that can be remedied by hydraulic fracturing of the well and blowing the particles from the perforation tunnels, by perforation washing tools, or by acid. Particles less than one-tenth the diameter pass through the throat and through the formation without bridging or plugging. However, particles between one-third and one-tenth the pore throat diameter invade the formation and bridge in the pore throat deeper in the formation. These particles cause the serious problems because, with the pore throats plugged and no permeability, acid cannot be injected into the formation to clean the pore throats. Suggested guidelines for the degree of filtration are

Formation sand size (mesh)	Filtration level (μm)
11.84	40
5.41	80
2.49	2.09

In various stages of the completion process, we are faced with fluids contaminated by a high concentration of particles over a wide range of sizes.

To maintain production, the best filtering process should use a number of steps to remove contaminants, starting with the largest and working down to the smallest. This includes, in order of use, shale shaker or linear motion shaker, desilter, centrifuge, and cartridge filters.

In summary, successful completions primarily depend on following a set procedure without taking shortcuts and on good housekeeping practices. A key element in the entire process is using clean fluids, which is made possible in large part through filtration techniques.

1.6.2 Cartridge Filters

Each field, formation and well site has unique characteristics and conditions. These include reservoir rock permeabilities, pore sizes, connate fluid composition, downhole pressures, and so on. These conditions dictate the brine composition and level of clarity needed for a proper completion, which determine the level of filtration needed to achieve the fluid clarity level required.

Disposable cartridge filters are widely available around the world. They can be used alone, in combination (series), or in tandem with other types of filtration equipment. When very large particles or high solids concentrations are present, conventional solids control equipment should be used as prefilters if they are thoroughly washed and cleaned before use. After the filtration requirement is established, the goal becomes one of optimizing a filtration system design. This involves putting together a properly sized and operated system of prefilters and final filters to meet the filtration efficiency objective at the lowest operating cost. Cartridge filters are available in different configurations and various materials of construction. Filter media include yarns, felts, papers, resin-bonded fibers, woven-wire cloths, and sintered metallic and ceramic structures.

The cartridge is made of a perforated metal or plastic tube, layered with permeable material or wrapped with filament to form a permeable matrix around the tube. Coarser particles are stopped at or near the surface of the filter, and the finer particles are trapped within the matrix. Pleated outer surfaces are used to provide larger surface areas. Cartridge filters are rated by pore sizes such as 1, 2, 4, 10, 25, and 50 μm, which relate to the size of particles that the filter can remove. This rating is nominal or absolute, depending on how the cartridges are constructed. A nominally rated filter can be expected to remove approximately 90% of the particles that are larger than its nominal rating. Actually, solids larger than the rating pass through these filters, but the concentration of the larger particles is reduced. High flow rates and pressures cause their efficiencies to fall. They must be constantly monitored and changed when they begin to plug, or the fluid will begin to bypass the filters.

Absolute rated filters achieve a sharp cutoff at its rated size. They should remove all the particles larger than their rating and generally become plugged much faster than nominally rated cartridges. Cartridge filters are most often used downstream of other filters for final clarification.

1.6.3 Tubular Filters

Tubular filters consist of a fabric screen surrounding a perforated stainless steel tube. Dirty fluid flows from the outside, through the fabric, where solids are stopped, and the filtrate passes into the center tube. The fabric can remove particles down to 1 to 3 μm. Because the solids are trapped on the outside surface of fabric, the element is easy to backwash and clean. Backwashing is accomplished by changing the valving and forcing clean brine back through the filter in the opposite direction. In 8–15 seconds, the element can be filtering again.

CHAPTER

2

Drill String: Composition and Design

OUTLINE

The drill string is defined here as drill pipe with tool joints and drill collars. The stem consists of the drill string and other components of the drilling assembly that may include the following items:

- Kelly
- Subs
- Stabilizers
- Reamers
- Shock absorbing tools
- Drilling jars
- Junk baskets
- Directional tools
- Information gathering tools
- Nonmagnetic tools
- Nonmetallic tools
- Mud motors

The drill stem (1) transmits power by rotary motion from the surface to a rock bit, (2) conveys drilling fluid to the rock bit, (3) produces the weight on bit for efficient rock destruction by the bit, and (4) provides control of borehole direction. The drill pipe itself can be used for formation evaluation (drill stem testing [DST]), well stimulation (e.g., fracturing, acidizing), and fishing operations. The drill pipe is used to set tools in place that will remain in the hole. Therefore, the drill string is a fundamental part, perhaps on of the most important parts, of any drilling activity.

The schematic, typical arrangement of a drill stem is shown in Figure 2.1. The illustration includes a Kelly. Kellys are used on rigs with a rotary table to turn the pipe. Many rigs use top drives to supply rotational power to the pipe. Top drive rigs do not have Kellys.

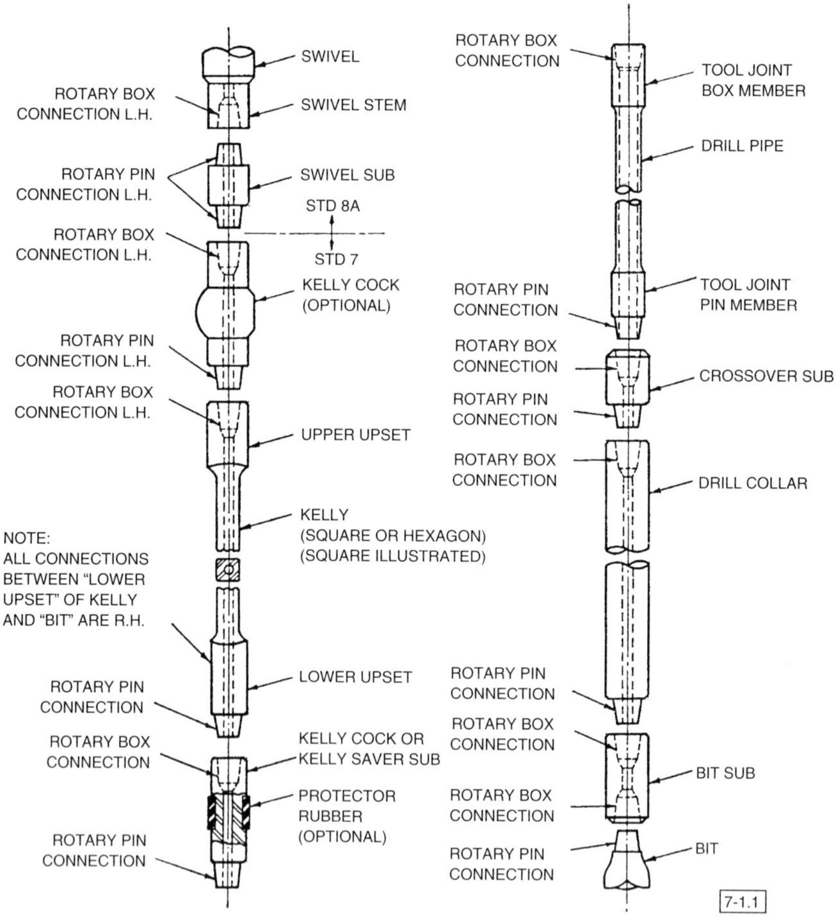

FIGURE 2.1 Typical drill-stem assembly [1].

2.1 DRILL COLLAR

The term *drill collar* derived from the short sub originally used to connect the bit to the drill pipe. A modern drill collar is about 30 ft long, and the total length of the string of drill collars may range from about 100 to 700 ft or longer. The purpose of drill collars is to furnish weight to the bit. However, the size and length of drill collars have an effect on bit performance, hold deviation, and drill pipe service life. Drill collars may be classified according to the shape of their cross-section as round drill collars (i.e., conventional drill collars), square drill collars, or spiral drill collars (i.e., drill collars with spiral grooves).

Square drill collars are used to increase the stiffness of the drill string and are used mainly for drilling in crooked hole areas. Square drill collars are rarely used. Spiral drill collars are used for drilling formations in which the differential pressure can cause sticking of drill collars. The spiral grooves on the drill collar surface reduce the area of contact between the drill collar and wall of the hole, which reduces the sticking force. Conventional drill collars and spiral drill collars are made with a uniform outside diameter (except for the spiral grooves) and with slip and elevator recesses. Slip and elevator recesses are grooves ranging from about $\frac{1}{4}$ in. deep on smaller collars to about $\frac{1}{2}$ in. deep on larger sizes. The grooves are about 18 in. long and placed on the box end of the collar. They provide a shoulder for the elevators and slips to engage. They reduce drill collar handling time while tripping by eliminating the need for lift subs and safety clamps. A lift sub is a tool about 2 ft long that makes up into the box of the drill collar, which has a groove for the elevator to engage. A safety clamp is a device clamped around the drill collar that provides a shoulder for the slips to engage.

The risk of drill collar failure is increased with slip and elevator grooves because they create stress risers where fatigue cracks can form. The slip and elevator grooves may be used together or separately.

Dimensions, physical properties, and unit weight of new, conventional drill collars are specified in Tables 2.1, 2.2, and 2.3, respectively. Technical data on square and spiral drill collars are available from manufacturers.

2.1.1 Selecting Drill Collar Size

Selection of the proper outside and inside diameter of drill collars is usually a difficult task. Perhaps the best way to select drill collar size is to study results obtained from offset wells previously drilled under similar conditions.

The most important factors in selecting drill collar size are

1. Bit size
2. Coupling diameter of the casing to be set in a hole

TABLE 2.1 Drill Collars [1]

Drill Collar Number	Outside Diameter (in.)	Bore (in.)	Length (ft)	Bevel Diameter (in.)	Ref. Bending Strength Ratio
NC23-31	$3\frac{1}{8}$	$1\frac{1}{4}$	30	3	2.57:1
NC26-35	$3\frac{1}{2}$	$1\frac{1}{2}$	30	$3\frac{17}{64}$	2.42:1
NC31-41	$4\frac{1}{8}$	2	30 or 31	$3\frac{61}{64}$	2.43:1
NC35-47	$4\frac{3}{4}$	2	30 or 31	$4\frac{33}{64}$	2.58:1
NC38-50	5	$2\frac{1}{4}$	30 or 31	$4\frac{49}{64}$	2.38:1
NC44-60	6	$2\frac{1}{4}$	30 or 31	$5\frac{11}{16}$	2.49:1
NC44-60	6	$2\frac{13}{16}$	30 or 31	$5\frac{11}{16}$	2.84:1
NC44-62	$6\frac{1}{4}$	$2\frac{1}{4}$	30 or 31	$5\frac{7}{8}$	2.91:1
NC46-62	$6\frac{1}{4}$	$2\frac{13}{16}$	30 or 31	$52\frac{9}{32}$	2.63:1
NC46-65	$6\frac{1}{2}$	$2\frac{1}{4}$	30 or 31	$6\frac{3}{32}$	2.76:1
NC46-65	$6\frac{1}{2}$	$2\frac{13}{16}$	30 or 31	$6\frac{3}{32}$	3.05:1
NC46-67	$6\frac{3}{4}$	$2\frac{1}{4}$	30 or 31	$6\frac{9}{32}$	3.18:1
NC50-70	7	$2\frac{1}{4}$	30 or 31	$6\frac{31}{64}$	2.54:1
NC50-70	7	$2\frac{13}{16}$	30 or 31	$6\frac{31}{64}$	2.73:1
NC50-72	$7\frac{1}{4}$	$2\frac{13}{16}$	30 or 31	$6\frac{43}{64}$	3.12:1
NC56-77	$7\frac{3}{4}$	$2\frac{13}{16}$	30 or 31	$7\frac{19}{64}$	2.70:1
NC56-80	8	$2\frac{13}{16}$	30 or 31	$7\frac{31}{64}$	3.02:1
$6\frac{5}{8}$ Reg	$8\frac{1}{4}$	$2\frac{13}{16}$	30 or 31	$7\frac{45}{64}$	2.93:1
NC61-90	9	$2\frac{13}{16}$	30 or 31	$8\frac{3}{8}$	3.17:1
$7\frac{5}{8}$ Reg	$9\frac{1}{2}$	3	30 or 31	$8\frac{13}{16}$	2.81:1
NC70-97	$9\frac{3}{4}$	3	30 or 31	$9\frac{5}{32}$	2.57:1
NC70-100	10	3	30 or 31	$9\frac{11}{32}$	2.81:1
$8\frac{5}{8}$ Reg	11	3	30 or 31	$10\frac{1}{2}$	2.84:1

TABLE 2.2 Physical Properties of New Drill Collars [1]

Drill Collar OD Range (in.)	Minimum Yield Strength (psi)	Minimum Tensile Strength (psi)	Elongation, Minimum
$3\frac{1}{8}$–$6\frac{7}{8}$	110,000	140,000	13%
7–10	100,000	135,000	13%

TABLE 2.3 Drill Collar Weight (Steel) (Pounds per Foot) [4]

(1)	(2)	(3)	(4)	(5)	(6)	(7)	(8)	(9)	(10)	(11)	(12)	(13)	(14)
Drill Collar OD,	Drill Collar ID, in.												
in.*	1	$1\frac{1}{4}$	$1\frac{1}{2}$	$1\frac{3}{4}$	2	$2\frac{1}{4}$	$2\frac{1}{2}$	$2\frac{13}{16}$	3	$3\frac{1}{4}$	$3\frac{1}{2}$	$3\frac{3}{4}$	4
$2\frac{7}{8}$	19	18	16										
3	21	20	18										
$3\frac{1}{8}$	22	22	20										
$3\frac{1}{4}$	26	24	22										
$3\frac{1}{2}$	30	29	27										
$3\frac{3}{4}$	35	33	32										
4	40	39	37	35	32	29							
$4\frac{1}{8}$	43	41	39	37	35	32							
$4\frac{1}{4}$	46	44	42	40	38	35							
$4\frac{1}{2}$	51	50	48	46	43	41							
$4\frac{3}{4}$			54	52	50	47	44						
5			61	69	56	53	50						
$5\frac{1}{4}$			68	65	63	60	57						
$5\frac{1}{2}$			75	73	70	67	64	60					
$5\frac{3}{4}$			82	80	78	75	72	67	64	60			
6			90	88	85	83	79	75	72	68			
$6\frac{1}{4}$			98	96	94	91	88	83	80	76	72		
$6\frac{1}{2}$			107	105	102	99	96	91	89	85	80		
$6\frac{3}{4}$			116	114	111	108	105	100	98	93	89		
7			125	123	120	117	114	110	107	103	98	93	84
$7\frac{1}{4}$			134	132	130	127	124	119	116	112	108	103	93
$7\frac{1}{2}$			144	142	139	137	133	129	126	122	117	113	102
$7\frac{3}{4}$			154	152	150	147	144	139	136	132	128	123	112
8			165	163	160	157	154	150	147	143	138	133	122
$8\frac{1}{4}$			176	174	171	168	165	160	158	154	149	144	133
$8\frac{1}{2}$			187	185	182	179	176	172	169	165	160	155	150
9			210	208	206	203	200	195	192	188	184	179	174
$9\frac{1}{2}$			234	232	230	227	224	220	216	212	209	206	198
$9\frac{3}{4}$			248	245	243	240	237	232	229	225	221	216	211
10			261	259	257	254	251	246	243	239	235	230	225
11			317	315	313	310	307	302	299	295	291	286	281
12			379	377	374	371	368	364	361	357	352	347	342

*See API Specification 7, Table 13 for API standard drill collar dimensions. For special configurations of drill collars, consult manufacture for reduction in weight.

3. Formation's tendency to produce sharp changes in hole deviation and direction
4. Hydraulic program
5. Possibility of washing over if the drill collar fails and is lost in the hole

To avoid an abrupt change in hole deviation (which may make it difficult or impossible to run casing) when drilling in crooked hole areas with an unstabilized bit and drill collars, the required outside diameter of the drill collar placed right above the bit can be found from the following formula [5]:

$$D_{dc} = 2 \cdot OD_{cc} - OD_{bit} \tag{2.1}$$

where OD_{bit} = outside diameter of bit size

OD_{cc} = outside diameter of casing coupling

OD_{dc} = outside diameter of drill collar

Example 2.1

The casing string for a certain well is to consist of $13\frac{3}{8}$ casing with a coupling outside diameter of 14.375 in. Determine the required outside diameter of the drill collar to avoid possible problems with running casing if the borehole diameter is assumed to be $17\frac{1}{2}$ in.

Solution

$$D_{dc} = 2 \cdot 14.375 - 17.5 = 11.15$$

Being aware of standard drill collar sizes (see Table 2.1), an 11-in. or 12-in. drill collar should be selected. To avoid such a large drill collar OD, a stabilizer or a proper-sized square drill collar (or a combination of the two) should be placed above the rock bit. If there is no tendency to cause an undersized hole, the largest drill collars that can be washed over are usually selected.

In general, if the optimal drilling programs require large drill collars, the operator should not hesitate to use them. Typical hole and drill collar sizes used in soft and hard formations are listed in Table 2.4.

2.1.2 Length of Drill Collars

The length of the drill collar string should be as short as possible but adequate to create the desired weight on bit. In vertical holes, ordinary drill pipe must never be used for exerting bit weight. In deviated holes, where the axial component of drill collar weight is sufficient for bit weight, heavy-weight pipe is often used in lieu of drill collars to reduce rotating torque. In highly deviated holes, where the axial component of the drill collar weight is below the needed bit weight, drill collars are not used, and heavy weight pipe is put high in the string in the vertical part of the hole. When this is done, bit weight is transmitted through the drill pipe.

TABLE 2.4 Popular Hole and Drill Collar Sizes [3]

$4\frac{3}{4}$	$2\frac{1}{8} \times 1\frac{1}{4}$ with	$3\frac{1}{2} \times 1\frac{1}{2}$ with
	$2\frac{7}{8}$ PAC or $2\frac{3}{8}$ API Reg	$2\frac{7}{8}$ PAC or $2\frac{3}{8}$ API Reg
$5\frac{7}{8}$–$6\frac{1}{8}$	$4\frac{1}{8} \times 2$ with NC31	$4\frac{3}{4} \times 2$ with NC31 or $3\frac{1}{2} \times$ H
$6\frac{1}{2}$–$6\frac{3}{4}$	$4\frac{3}{4} \times 2\frac{1}{4}$ with NC38	5 or $5\frac{1}{4} \times 2$ with NC38
$7\frac{5}{8}$–$7\frac{7}{8}$	$6 \times 2\frac{13}{16}$ with NC46	$6\frac{1}{4}$ or $6\frac{1}{2}$ OD $\times$ 2 or $2\frac{1}{4}$ with NC46
$8\frac{1}{2}$–$8\frac{3}{4}$	$6\frac{1}{4} \times 2\frac{13}{16}$ with NC46 or	
	$6\frac{1}{2} \times 2\frac{13}{16}$ with NC46 or NC50	$6\frac{3}{4}$ or $7 \times 2\frac{1}{4}$ with NC50
$9\frac{1}{2}$–$9\frac{7}{8}$	$7 \times 2\frac{13}{16}$ with NC50	$7 \times 2\frac{1}{4}$ with NC50
	$8 \times 2\frac{13}{16}$ with NC50	$8 \times 2\frac{13}{16}$ with $6\frac{5}{8}$ API Reg
$10\frac{5}{8}$–11	$7 \times 2\frac{13}{16}$ with NC50 or	$8 \times 2\frac{13}{16}$ with $6\frac{5}{8}$ API Reg
	$8 \times 2\frac{13}{16}$ with $6\frac{5}{8}$ API Reg	$9 \times 2\frac{13}{16}$ $7\frac{5}{8}$ API Reg
$12\frac{1}{4}$	$8 \times 2\frac{13}{16}$ with $6\frac{5}{8}$ API Reg	$8 \times 2\frac{13}{16}$ with $6\frac{5}{8}$ API Reg
		$9 \times 2\frac{13}{16}$ $7\frac{5}{8}$ API Reg
		$10 \times 2\frac{13}{16}$ or 3 with $7\frac{5}{8}$ API Reg
$17\frac{1}{2}$	$8 \times 2\frac{13}{16}$ with $6\frac{5}{8}$ API Reg	$8 \times 2\frac{13}{16}$ with $6\frac{5}{8}$ API Reg
		$9 \times 2\frac{13}{16}$ $7\frac{5}{8}$ API Reg
		$10 \times 2\frac{13}{16}$ or 3 with $7\frac{5}{8}$ API Reg
		11×3 with $8\frac{5}{8}$ API Reg
$18\frac{1}{2}$–26	Drill collar programs are the same as for the next reduced hole size.	

Extra care and planning must be done when apply bit weight through the pipe because buckling of the pipe can occur, which can lead to fatigue failures and accelerated wear of the pipe or tool joints.

The required length of drill collars can be obtained from the following formula:

$$L_{dc} = \frac{DF \cdot W}{W_{dc} \cdot K_b \cdot \cos\alpha} \tag{2.2}$$

where DF = is the design factor (DF = 1.2–1.3), and the weight on bit (lb) is determined by the following:

W_{dc} = unit weight of drill collar in air (lb/ft)

K_b = buoyancy factor

$K_b = 1 - \frac{\lambda_m}{\lambda_{st}}$

where λ_m is the drilling fluid density (lb/gal), λ_{st} is the drill collar density (lb/gal) (for steel, $\lambda_{st} = 65.5$ lb/gal), and α is the hole inclination from vertical (degrees).

When possible, the drill pipe should be in tension. The lower most collar has the maximum compressive load, which is transmitted to the bit, and the upper most collar has a tensile load. This means there is some point in the drill collars between the bit and the drill pipe that has a zero axial load. This is called the *neutral point*. The design factor (DF) is needed to place the neutral point below the top of the drill collar string. This will ensure that the pipe is not in compression because of axial vibration or bouncing of the bit and because of inaccurate handling of the brake by the driller.

The excess of drill collars also helps to prevent transverse movement of drill pipe due to the effect of centrifugal force. While the drill string rotates, a centrifugal force is generated that may produce a lateral movement of drill pipe, which causes bending stress and excessive torque. The centrifugal force also contributes to vibration of the drill pipe. Hence, some excess of drill collars is suggested. The magnitude of the design factor to control vibration can be determined by field experiments in any particular set of drilling conditions. Experimental determination of the design factor for preventing compressive loading on the pipe is more difficult. The result of running the pipe in compression can be a fatigue crack leading to a washout or parted pipe.

The pressure area method (PAM), occasionally used for evaluation of drill collar string length, is wrong because it does not consider the triaxial state of stresses that actually occur. Hydrostatic forces cannot cause any buckling of the drill string as long as the density of the string is greater than the density of the drilling fluid.

Example 2.2

Determine the required length of 7 in. × $2\frac{1}{4}$ in. drill collars with the following conditions:

Desired weight on bit: $W = 40{,}000$ lb
Drilling fluid density: $\lambda_m = 10$ lb/gal
Hole deviation from vertical: $\alpha = 20°$

Solution

From Table 2.3, the unit weight of a $7 \times 2\frac{1}{4}$ drill collar is 117 lb/ft. The buoyancy factor is

$$K_b = 1 - \frac{10}{65.5} = 0.847$$

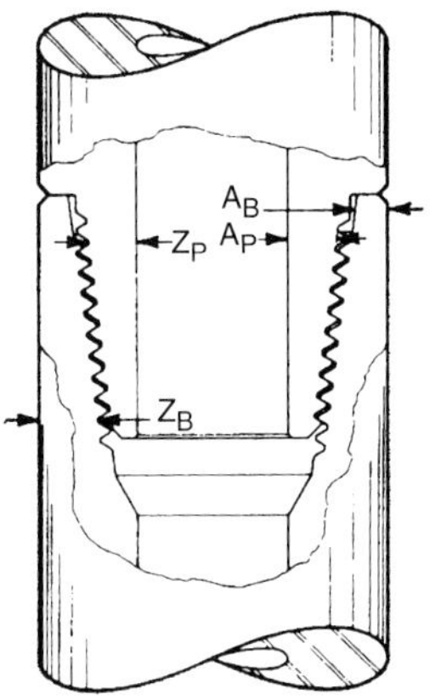

The section modulus, Z_b, of the box should be $2\frac{1}{2}$ times greater than the section modulus, Z_p, of the pin in a drill collar connection. On the right side of the connection are the spots at which the critical area of both the pin (A_p) and box (A_b) should be measured for calculating torsional strength.

FIGURE 2.2 The drill collar connection. (After C. E. Wilson and W. R. Garrett).

Applying Equation 2.2 gives

$$L_{dc} = \frac{(1.3)(40{,}000)}{(147)(0.847)(\cos 20)} = 444 \text{ ft}$$

The closest length, based on 30-ft collars, is 450 ft, which is 15 joints of drill collars.

2.1.3 Drill Collar Connections

It is current practice to select the rotary shoulder connection that provides a balanced bending fatigue resistance for the pin and the box. It has been determined empirically that the pin and boxes have approximately equally bending fatigue resistance if the section modulus of the box at its critical zone is 2.5 times the section modulus of the pin at its critical zone. This number is called the *bending strength ratio* (BSR). These critical zones are shown in Figure 2.2. Section modulus ratios from 2.25 to 2.75 are considered to be very good and satisfactory performance has been experienced with ratios from 2.0 to 2.3 [6].

The previous statements are valid if the connection is made up with the recommended makeup torque. A set of charts is available from the Drilco Division of Smith International, Inc. Some of these charts are presented in Figures 2.3 to 2.10. The charts are used as follows:

- The best group of connections are those that appear in the **shaded** sections of the charts.

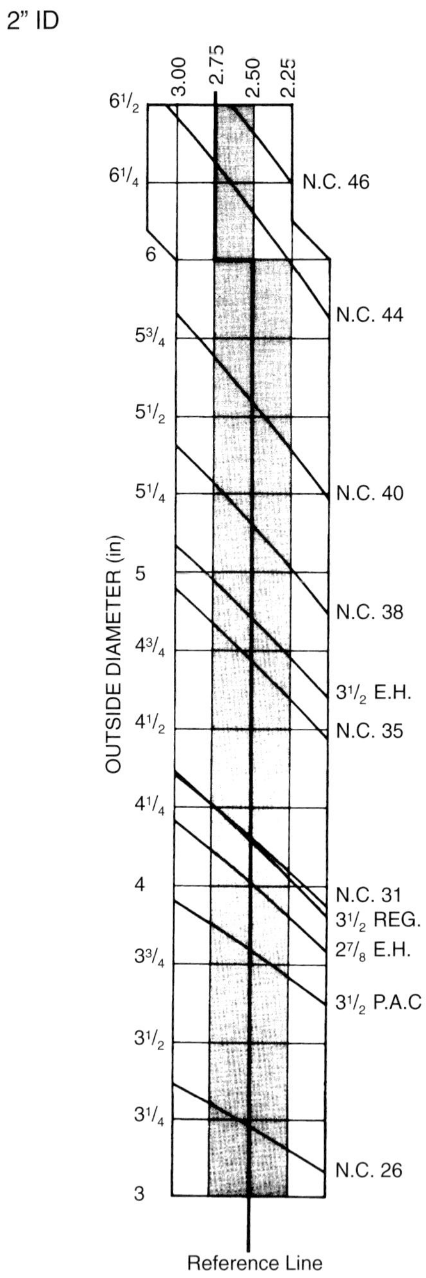

FIGURE 2.3 Practical chart for drill collar selection—2-in. ID. (From Drilco, Division of Smith International, Inc.).

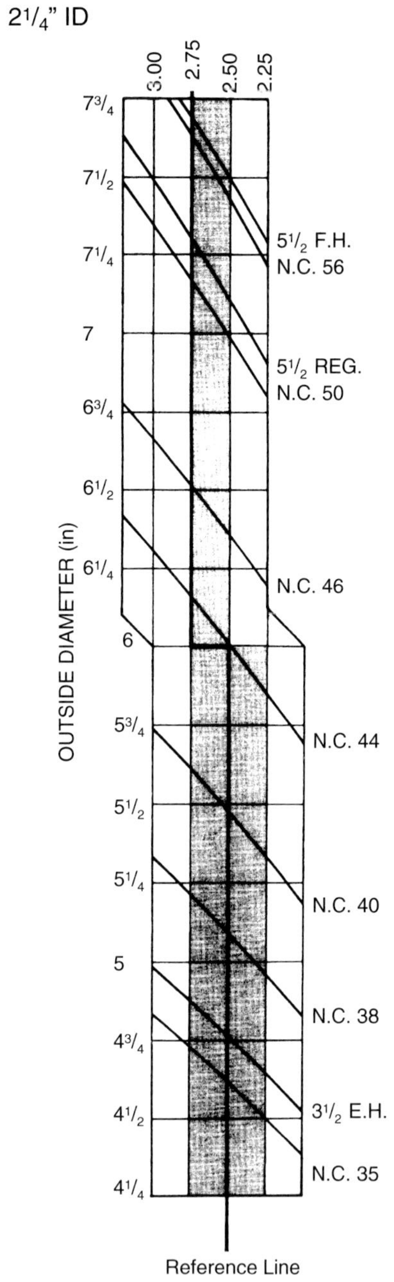

FIGURE 2.4 Practical chart for drill collar selection—2-$\frac{1}{4}$-in. ID. (From Drilco, Division of Smith International, Inc.).

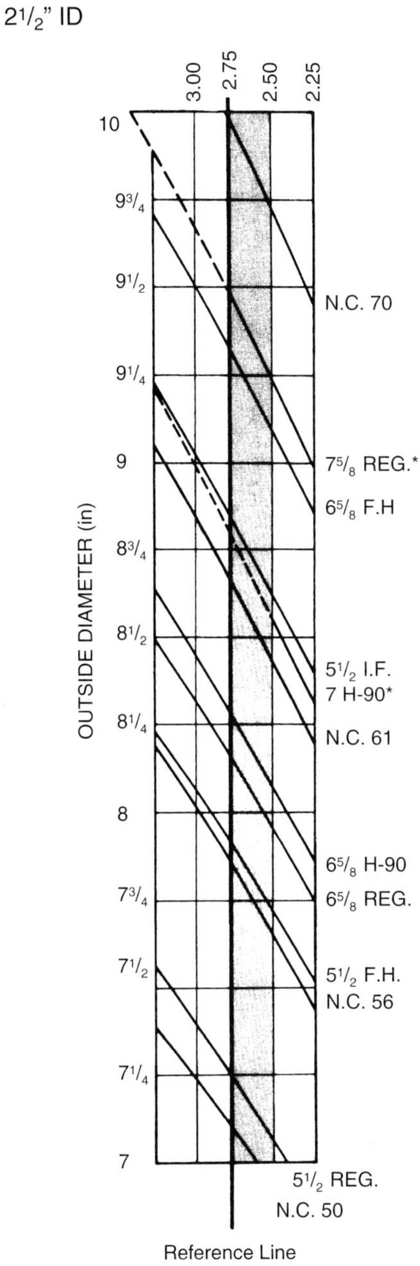

FIGURE 2.5 Practical chart for drill collar selection—2-$\frac{1}{2}$-in. ID. (From Drilco, Division of Smith International, Inc.).

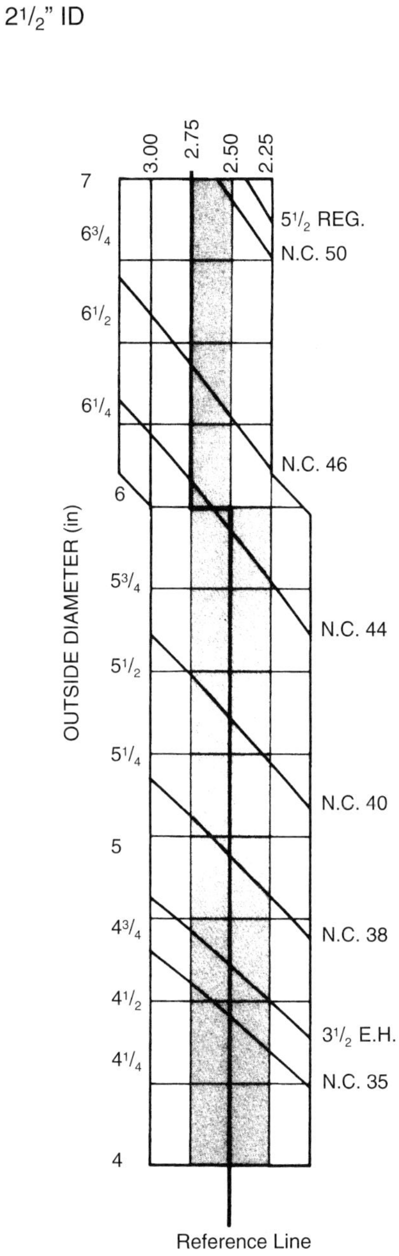

FIGURE 2.6 Practical chart for drill collar selection—2-$\frac{1}{4}$-in. ID. (From Drilco, Division of Smith International, Inc.).

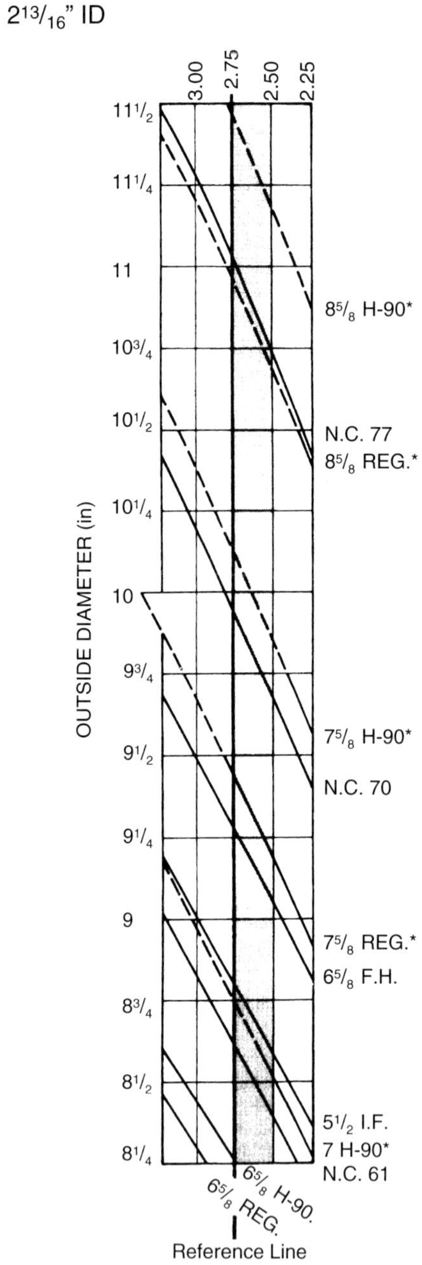

FIGURE 2.7 Practical chart for drill collar selection—2-$\frac{13}{16}$-in. ID. (From Drilco, Division of Smith International, Inc.).

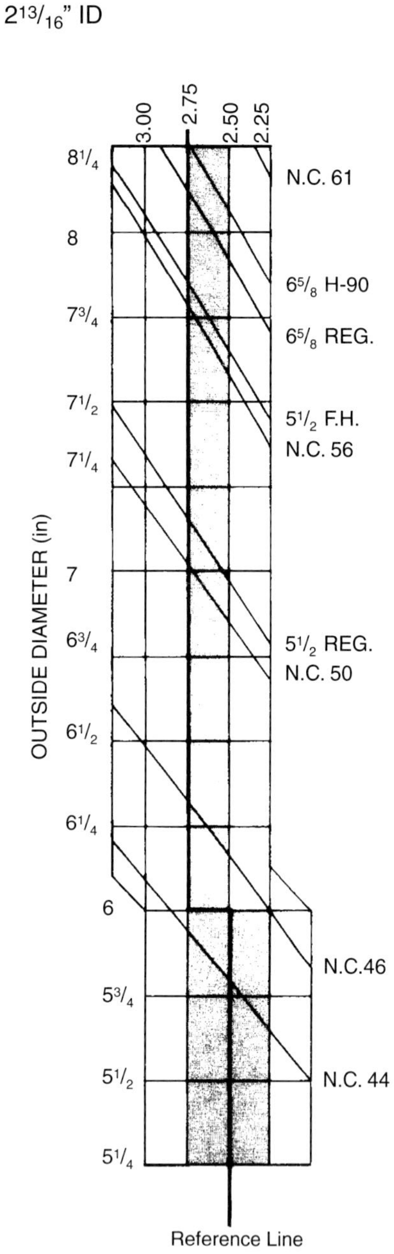

FIGURE 2.8 Practical chart for drill collar selection—2-$\frac{13}{16}$-in. ID. (From Drilco, Division of Smith International, Inc.).

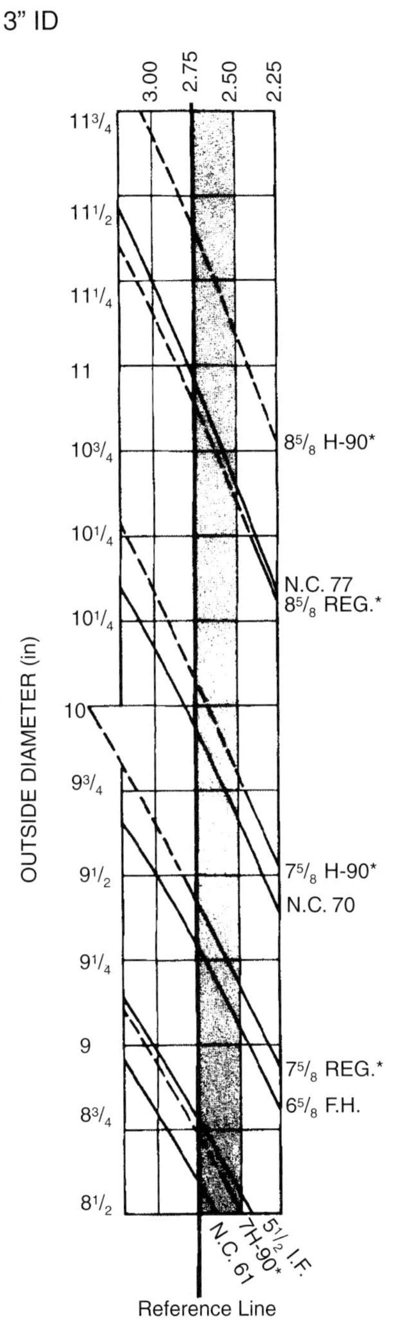

FIGURE 2.9 Practical chart for drill collar selection—3-in. ID. (From Drilco, Division of Smith International, Inc.).

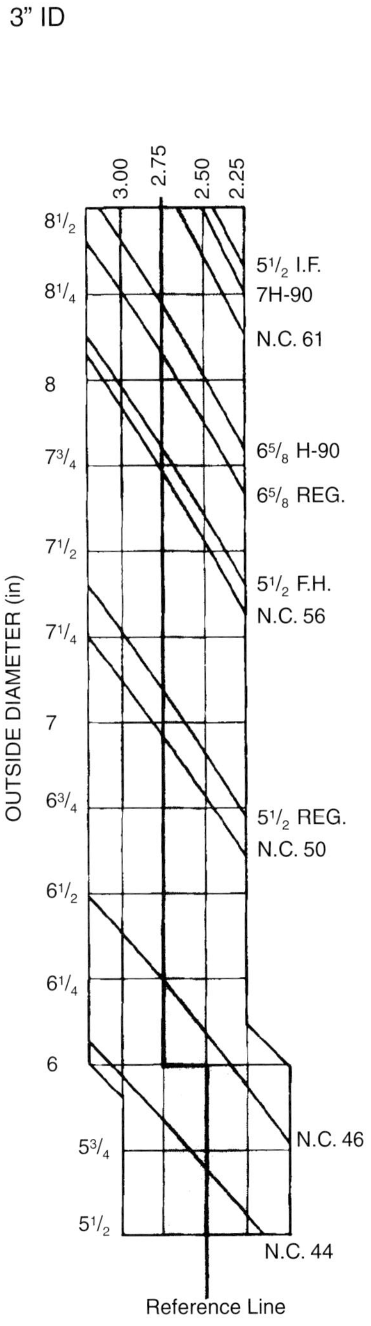

FIGURE 2.10 Practical chart for drill collar selection—3-in. ID. (From Drilco, Division of Smith International, Inc.).

- The second best group of connections are those that lie in the unshaded section to the left of the shaded section.
- The third best group of connections are those that lie in the unshaded section to right of the shaded section.
- The nearer the connection lies to the reference line, either to the right or the left, the more desirable is its selection.

Example 2.3

Select the best connection for $9\frac{3}{4}$ in. $\times 2\frac{9}{16}$ in. drill collars.

Solution

For average conditions, select in the following order:

- Best: NC70; shaded area and nearest reference line
- Second best: $7\frac{5}{8}$ REG low torque face; light area to left of reference line
- Third best: $7\frac{5}{8}$ H-90 low torque face; light area to right of reference line

In extremely abrasive or corrosive conditions, the following selections may be made. In this environment, box wear could be a factor, which means as the drill collars are used, the box diameter decreases and the BSR decreases. Corrosion decreased the fatigue resistance of material. The box is exposed to the corrosive environment, but the pin is not. A larger OD of the box increases its stiffness and decreases the bending stress. More bending is taken by the pin.

- Best: $7\frac{5}{8}$ REG low troque face; which has the strongest box; light area to left of reference line
- Second best: NC70, second strongest box
- Third best: $7\frac{5}{8}$ H-90, weakest box

2.1.4 Recommended Makeup Torque for Drill Collars

The rotary shouldered connections must be made up with a torque high enough to prevent the makeup shoulders from separating downhole. Shoulder separation can occur when rotating in doglegs, when drill collars are buckled by compressive loads, or when tensile loads, are applied to the drill collars. This is important because the makeup shoulder is the only seal in the connection. The threads themselves do not seal because they are designed with clearance between the crest of one thread and the root of its mating thread. The clearance acts as a channel for thread compound and small solid particles that could be in the drilling fluid. A second reason to prevent shoulder separation is because when the shoulders separate, the box no longer takes any bending load, and the total bending load is felt by the pin.

To keep the shoulders together, the shoulder load must be high enough to create a compressive stress at the shoulder face, capable of offsetting the bending that occurs due to drill collar buckling or rotating in doglegs. This compressive shoulder load is generated by makeup torque. Field observations indicate that the makeup torque should create an average axial stress of 62,500 psi in the pin or box, whichever is has the lower cross-sectional area. The cross-section of the pin is taken at $\frac{3}{4}$ in. from the makeup shoulder. The cross-sectional area of the box is taken at $\frac{3}{8}$ in. from the makeup shoulder. The formulas for calculating the cross-sectional areas and the makeup torque can be found in API Recommended Practice for Drill Stem Design and Operating Limits (API RP7G). Makeup torque creates a tensile stress in the pin and can reduce the fatigue life if makeup torque is too high.

The recommended makeup torque for drill collars is given in Table 2.5.

2.1.5 Drill Collar Buckling

In a straight vertical hole with no weight on the bit, a string of drill collars remains straight. As the weight on the bit is increased, compressive loads are induced into the collars, with the highest compressive load just above the bit and decreasing to zero at the neutral point. As weight is increased, the drill collars or drill pipe buckles and contacts the wall of the hole. If weight is increased further, the string buckles a second time and contacts the borehole at two points. With still further increased weight on the bit, the third and higher order of buckling occurs. The problem of drill collars buckling in vertical holes has been studied by A. Lubinski [8], and the weight on the bit that results in first- and second-order buckling can be calculated follows:

$$W_{crl} = 1.94(EIp^2)^{1/3} \tag{2.3}$$

$$W_{crll} = 3.75(EIp^2)^{1/3} \tag{2.4}$$

where E is the modulus of elasticity for drill collars ($lb/in.^2$). For steel, $E = 4320 \times 10^6\ lb/ft^2$, p is the unit weight of drill collar in drilling fluid (lb/ft), and I is the moment of inertial of drill collar cross-section with respect to its diameter (ft).

$$I = \frac{\pi}{64}(D^4 - d^4) \tag{2.5}$$

where D is the outside diameter of drill collar (ft) and d is the inside diameter of drill collar (in.).

Example 2.4

Find the magnitude of the weight on bit and corresponding length of $6\frac{3}{4}$ in. $\times\ 2\frac{1}{4}$ in. drill collars that result in second-order buckling. Mud weight = 12 lb/gal.

TABLE 2.5 Recommended Makeup Torque[1] for Rotary Shouldered Drill Collar Connections (See footnotes for use of this table.) [4]

(1)	(2)	(3)	(4)	(5)	(6)	(7)	(8)	(9)	(10)	(11)	(12)	(13)	(14)	(15)
Connection			Minimum Makeup Torque ft-lb[2]											
			Bore of Drill Collar, inches											
Size	Type	OD, in.	1	$1\frac{1}{4}$	$1\frac{1}{2}$	$1\frac{3}{4}$	2	$2\frac{1}{4}$	$2\frac{1}{2}$	$2\frac{13}{16}$	3	$3\frac{1}{4}$	$3\frac{1}{2}$	$3\frac{3}{4}$
API	NC23	3	*2,508	*2,508	*2,508									
		$3\frac{1}{8}$	*3,330	*3,330	2,647									
		$3\frac{1}{4}$	4,000	3,387	2,647									
$2\frac{3}{8}$	Regular	3		*2,241	*2,241	1,749								
		$3\frac{1}{8}$		*3,028	2,574	1,749								
		$3\frac{1}{4}$		3,285	2,574	1,749								
$2\frac{3}{8}$	PAC[3]	3		*3,797	*3,797	2,926								
		$3\frac{3}{8}$		*4,966	4,151	2,926								
		$3\frac{1}{4}$		5,206	4,151	2,926								
$2\frac{3}{8}$	API IF	$3\frac{1}{2}$		*4,606	*4,606	3,697								
API	NC26	$3\frac{3}{4}$		5,501	4,668	3,697								
$2\frac{7}{8}$	Regular	$3\frac{1}{2}$		*3,838	*3,838	*3,838								
		$3\frac{3}{4}$		5,766	4,951	4,002								
		$3\frac{7}{8}$		5,766	4,951	4,002								
$2\frac{7}{8}$	Slim hole													
$2\frac{7}{8}$	Extra hole	$3\frac{3}{4}$		*4,089	*4,089	*4,089								
$3\frac{1}{2}$	Double streamline	$3\frac{7}{8}$		*5,352	*5,352	*5,352								
$2\frac{7}{8}$	Mod. open	$4\frac{1}{8}$		*8,059	*8,059	7,433								
$2\frac{7}{8}$	API IF	$3\frac{7}{8}$		*4,640	*4,640	*4,640	*4,640							

API	NC31	$4\frac{1}{8}$	*7,390	*7,390	*7,390	6,853			
$3\frac{1}{2}$	Regular	$4\frac{1}{8}$	*6,466	*6,466	*6,466	*6,466	5,685		
		$4\frac{1}{4}$	*7,886	*7,886	*7,886	7,115	5,685		
		$4\frac{1}{2}$	10,471	9,514	8,394	7,115	5,685		
$3\frac{1}{2}$	Slim hole	$4\frac{1}{4}$	*8,858	*8,858	8,161	6,853	5,391		
		$4\frac{1}{2}$	10,286	9,307	8,161	6,853	5,391		
API	NC35	$4\frac{1}{2}$			*9,038	*9,038	*9,038	7,411	
		$4\frac{3}{4}$			12,273	10,826	9,202	7,411	
		5			12,273	10,826	9,202	7,411	
$3\frac{1}{2}$	Extra hole	$4\frac{1}{4}$			*5,161	*5,161	*5,161	*5,161	
4	Slim hole	$4\frac{1}{2}$			*8,479	*8,479	*8,479	8,311	
$3\frac{1}{2}$	Mod. open	$4\frac{3}{4}$			*12,074	11,803	10,144	8,311	
		5			13,283	11,803	10,144	8,311	
		$5\frac{1}{4}$			13,283	11,803	10,144	8,311	
$3\frac{1}{2}$	API IF	$4\frac{3}{4}$			*9,986	*9,986	*9,986	*9,986	8,315
API	NC38	5			*13,949	*13,949	12,907	10,977	8,315
$4\frac{1}{2}$	Slim hole	$5\frac{1}{4}$			16,207	14,643	12,907	10,977	8,315
		$5\frac{1}{2}$			16,207	14,643	12,907	10,977	8,315
$3\frac{1}{2}$	H-90[4]	$4\frac{3}{4}$			*8,786	*8,786	*8,786	*8,786	*8,786
		5			*12,794	*12,794	*12,794	*12,794	10,408
		$5\frac{1}{4}$			*17,094	16,929	15,137	13,151	10,408
		$5\frac{1}{2}$			18,522	16,929	15,137	13,151	10,408
4	Full hole	5			*10,910	*10,910	*10,910	*10,910	*10,910

(Continued)

TABLE 2.5 (Continued)

(1)	(2)	(3)	(4)	(5)	(6)	(7)	(8)	(9)	(10)	(11)	(12)	(13)	(14)	(15)
Connection			Minimum Makeup Torque ft-lb[2]											
			Bore of Drill Collar, inches											
Size	Type	OD, in.	1	$1\frac{1}{4}$	$1\frac{1}{2}$	$1\frac{3}{4}$	2	$2\frac{1}{4}$	$2\frac{1}{2}$	$2\frac{13}{16}$	3	$3\frac{1}{4}$	$3\frac{1}{2}$	$3\frac{3}{4}$
API	NC40	$5\frac{1}{4}$				*15,290	*15,290	*15,290	14,969	12,125				
4	Mod. open	$5\frac{1}{2}$				*19,985	18,886	17,028	14,969	12,125				
$4\frac{1}{2}$	Double streamline	$5\frac{3}{4}$				20,539	18,886	17,028	14,969	12,125				
		6				20,539	18,886	17,028	14,969	12,125				
4	H-90[4]	$5\frac{1}{4}$				*12,590	*12,590	*12,590	*12,590	*12,590				
		$5\frac{1}{2}$				*17,401	*17,401	*17,401	*17,401	16,536				
		$5\frac{3}{4}$				*22,531	*22,531	21,714	19,543	16,536				
		6				25,408	23,671	21,714	19,543	16,536				
		$6\frac{1}{4}$				25,408	23,671	21,714	19,543	16,536				
$4\frac{1}{2}$	API regular	$5\frac{1}{2}$					*15,576	*15,576	*15,576	*15,576	*15,576			
		$5\frac{3}{4}$					*20,609	*20,609	*20,609	19,601	16,629			
		6					25,407	23,686	21,749	19,601	16,629			
		$6\frac{1}{4}$					25,407	23,686	21,749	19,601	16,629			
API	NC44	$5\frac{3}{4}$					*20,895	*20,895	*20,895	*20,895	18,161			
		6					*26,453	25,510	23,493	21,157	18,161			
		$6\frac{1}{4}$					27,300	25,510	23,493	21,157	18,161			
		$6\frac{1}{2}$					27,300	25,510	23,493	21,157	18,161			
$4\frac{1}{2}$	API full hole	$5\frac{1}{2}$						*12,973	*12,973	*12,973	*12,973	*12,973		
		$5\frac{3}{4}$						*18,119	*18,119	*18,119	*18,119	17,900		

		6	∗23,605	∗23,605	23,028	19,921	17,900	
		$6\frac{1}{4}$	27,294	25,772	23,028	19,921	17,900	
		$6\frac{1}{2}$	27,294	25,772	23,028	19,921	17,900	
$4\frac{1}{2}$	Extra hole	$5\frac{3}{4}$		∗17,738	∗17,738	∗17,738	∗17,738	
API	NC46	6		∗23,422	∗23,422	22,426	20,311	
4	API IF	$6\frac{1}{4}$		28,021	25,676	22,426	20,311	
$4\frac{1}{2}$	Semi IF	$6\frac{1}{2}$		28,021	25,676	22,426	20,311	
5	Double Streamline	$6\frac{3}{4}$		28,021	25,676	22,426	20,311	
$4\frac{1}{2}$	Mod. open							
$4\frac{1}{2}$	H-90[4]	$5\frac{3}{4}$		∗18,019	∗18,019	∗18,019	∗18,019	
		6		∗23,681	∗23,681	23,159	21,051	
		$6\frac{1}{4}$		28,732	26,397	23,159	21,051	
		$6\frac{1}{2}$		28,732	26,397	23,159	21,051	
		$6\frac{3}{4}$		28,732	26,397	23,159	21,051	
5	H-90[4]	$6\frac{1}{4}$		∗25,360	∗25,360	∗25,360	∗25,360	23,988
		$6\frac{1}{2}$		∗31,895	∗31,895	29,400	27,167	23,988
		$6\frac{3}{4}$		35,292	32,825	29,400	27,167	23,988
		7		35,292	32,825	29,400	27,167	23,988
$4\frac{1}{2}$	API IF	$6\frac{1}{4}$		∗23,004	∗23,004	∗23,004	∗23,004	∗23,004
API	NC50	$6\frac{1}{2}$		∗29,679	∗29,679	∗29,679	∗29,679	26,675
5	Extra hole	$6\frac{3}{4}$		∗36,742	35,824	32,277	29,966	26,675
5	Mod. open	7		38,379	35,824	32,277	29,966	26,675
$5\frac{1}{2}$	Double streamline	$7\frac{1}{4}$		38,379	35,824	32,277	29,966	26,675

(Continued)

TABLE 2.5 (*Continued*)

(1)	(2)	(3)	(4)	(5)	(6)	(7)	(8)	(9)	(10)	(11)	(12)	(13)	(14)	(15)
Connection			Minimum Makeup Torque ft-lb^2											
			Bore of Drill Collar, inches											
Size	Type	OD, in.	1	$1\frac{1}{4}$	$1\frac{1}{2}$	$1\frac{3}{4}$	2	$2\frac{1}{4}$	$2\frac{1}{2}$	$2\frac{13}{16}$	3	$3\frac{1}{4}$	$3\frac{1}{2}$	$3\frac{3}{4}$
5	Semi-IF	$7\frac{1}{2}$							38,379	35,824	32,277	29,966	26,675	
$5\frac{1}{2}$	H-90[4]	$6\frac{3}{4}$							*34,508	*34,508	*34,508	34,142	30,781	
		7							*41,993	40,117	36,501	34,142	30,781	
		$7\frac{1}{4}$							42,719	40,117	36,501	34,142	30,781	
		$7\frac{1}{2}$							42,719	40,117	36,501	34,142	30,781	
$5\frac{1}{2}$	API regular	$6\frac{3}{4}$							*31,941	*31,941	*31,941	*31,941	30,495	
		7							*39,419	*39,419	36,235	33,868	30,495	
		$7\frac{1}{4}$						42,481	39,866	36,235	33,868	30,495		
		$7\frac{1}{2}$						42,481	39,866	36,235	33,868	30,495		
$5\frac{1}{2}$	API full hole	7						*32,762	*32,762	*32,762	*32,762	*32,762		
		$7\frac{1}{4}$						*40,998	*40,998	*40,998	*40,998	*40,998		
		$7\frac{1}{2}$						*49,661	*49,661	47,756	45,190	41,533		
		$7\frac{3}{4}$						54,515	51,687	47,756	45,190	41,533		
API	NC56	$7\frac{1}{4}$							*40,498	*40,498	*40,498	*40,498		
		$7\frac{1}{2}$							*49,060	48,221	45,680	42,058		
		$7\frac{3}{4}$							52,115	48,221	45,680	42,058		
		8							52,115	48,221	45,680	42,058		
$6\frac{5}{8}$	API regular	$7\frac{1}{2}$							*46,399	*46,399	*46,399	*46,399		

		$7\frac{3}{4}$	*55,627	53,346	50,704	46,936		
		8	57,393	53,346	50,704	46,936		
		$8\frac{1}{4}$	57,393	53,346	50,704	46,936		
$6\frac{5}{8}$	H-90[4]	$7\frac{1}{2}$	*46,509	*46,509	*46,509	*46,509		
		$7\frac{3}{4}$	*55,708	*55,708	53,629	49,855		
		8	60,321	56,273	53,629	49,855		
		$8\frac{1}{4}$	60,321	56,273	53,629	49,855		
API	NC61	8	*55,131	*55,131	*55,131	*55,131		
		$8\frac{1}{4}$	*65,438	*65,438	*65,438	61,624		
		$8\frac{1}{2}$	72,670	68,398	65,607	61,624		
		$8\frac{3}{4}$	72,670	68,398	65,607	61,624		
		9	72,670	68,398	65,607	61,624		
$5\frac{1}{2}$	API IF	8	*56,641	*56,641	*56,641	*56,641	*56,641	
		$8\frac{1}{4}$	*67,133	*67,133	*67,133	63,381	59,027	
		$8\frac{1}{2}$	74,626	70,277	67,436	63,381	59,027	
		$8\frac{3}{4}$	74,626	70,277	67,436	63,381	59,027	
		9	74,626	70,277	67,436	63,381	59,027	
		$9\frac{1}{4}$	74,626	70,277	67,436	63,381	59,027	
$6\frac{5}{8}$	API full hole	$8\frac{1}{2}$	*67,789	*67,789	*67,789	*67,789	*67,789	67,184
		$8\frac{3}{4}$	*79,554	*79,554	*79,554	76,706	72,102	67,184
		9	88,582	83,992	80,991	76,706	72,102	67,184
		$9\frac{1}{4}$	88,582	83,992	80,991	76,706	72,102	67,184
		$9\frac{1}{2}$	88,582	83,992	80,991	76,706	72,102	67,184

(Continued)

TABLE 2.5 (*Continued*)

(1)	(2)	(3)	(4)	(5)	(6)	(7)	(8)	(9)	(10)	(11)	(12)	(13)	(14)	(15)
Connection			Minimum Makeup Torque ft-lb[2]											
			Bore of Drill Collar, inches											
Size	Type	OD, in.	1	$1\frac{1}{4}$	$1\frac{1}{2}$	$1\frac{3}{4}$	2	$2\frac{1}{4}$	$2\frac{1}{2}$	$2\frac{13}{16}$	3	$3\frac{1}{4}$	$3\frac{1}{2}$	$3\frac{3}{4}$
API	NC70	9							*75,781	*75,781	*75,781	*75,781	*75,781	*75,781
		$9\frac{1}{4}$							*88,802	*88,802	*88,802	*88,802	*88,802	*88,802
		$9\frac{1}{2}$							*102,354	*102,354	*102,354	101,107	96,214	90,984
		$9\frac{3}{4}$							113,710	108,841	105,657	101,107	96,214	90,984
		10							113,710	108,841	105,657	101,107	96,214	90,984
		$10\frac{1}{4}$							113,710	108,841	105,657	101,107	96,214	90,984
API	NC77	10							*108,194	*108,194	*108,194	*108,194	*108,194	*108,194
		$10\frac{1}{4}$							*124,051	*124,051	*124,051	*124,051	*124,051	*124,051
		$10\frac{1}{2}$							*140,491	*140,491	*140,491	140,488	135,119	129,375
		$10\frac{3}{4}$							154,297	148,965	145,476	140,488	135,119	129,375
		11							154,297	148,965	145,476	140,488	135,119	129,375
7	H-90[4]	8							*53,454	*53,454	*53,454	*53,454	*53,454	*53,454
		$8\frac{1}{4}$							*63,738	*63,738	*63,738	*63,738	60,971	56,382
		$8\frac{1}{2}$							*74,478	72,066	69,265	65,267	60,971	56,382
$7\frac{5}{8}$	API regular	$8\frac{1}{2}$							*60,402	*60,402	*60,402	*60,402	*60,402	*60,402
		$8\frac{3}{4}$							*72,169	*72,169	*72,169	*72,169	*72,169	*72,169
$7\frac{5}{8}$	API regular	9							*84,442	*84,442	*84,442	84,221	79,536	74,529
		$9\frac{1}{4}$							96,301	91,633	88,580	84,221	79,536	74,529

		$9\frac{1}{2}$	96,301	91,633	83,580	84,221	79,536	74,529
$7\frac{5}{8}$	H-90[4]	9	*73,017	*73,017	*73,017	*73,017	*73,017	*73,017
		$9\frac{1}{4}$	*86,006	*86,006	*86,006	*86,006	*86,006	*86,006
		$9\frac{1}{2}$	*99,508	*99,508	*99,508	*99,508	*99,508	96,285
$8\frac{5}{8}$	API regular	10	*109,345	*109,345	*109,345	*109,345	*109,345	*109,345
		$10\frac{1}{4}$	*125,263	*125,263	*125,263	*125,263	*125,263	125,034
		$10\frac{1}{2}$	*141,767	*141,767	141,134	136,146	130,077	125,034
$8\frac{5}{8}$	H-90[4]	$10\frac{1}{4}$	*113,482	*113,482	*113,482	*113,482	*113,482	*113,482
		$10\frac{1}{2}$	*130,063	*130,063	*130,063	*130,063	*130,063	*130,063
7	H-90[4]	$8\frac{3}{4}$		*68,061	*68,061	67,257	62,845	58,131
	(with low torque face)	9		74,235	71,361	67,257	62,845	58,131
$7\frac{5}{8}$	API regular	$9\frac{1}{4}$			*73,099	*73,099	*73,099	*73,099
	(with low torque face)	$9\frac{1}{2}$			*86,463	*86,463	82,457	77,289
		$9\frac{3}{4}$			91,789	87,292	82,457	77,289
		10			91,789	87,292	82,457	77,289
$7\frac{5}{8}$	H-90[4]	$9\frac{3}{4}$		*91,667	*91,667	*91,667	*91,667	*91,667
	(with low torque face)	10		*106,260	*106,260	*106,260	104,171	98,804
		$10\frac{1}{4}$		117,112	113,851	109,188	104,171	98,804
		$10\frac{1}{2}$		117,112	113,851	109,188	104,171	98,804
$8\frac{5}{8}$	API regular	$10\frac{3}{4}$			*112,883	*112,883	*112,883	*112,883
	(with low torque face)	11			*130,672	*130,672	*130,672	*130,672

(Continued)

TABLE 2.5 (*Continued*)

(1)	(2)	(3)	(4)	(5)	(6)	(7)	(8)	(9)	(10)	(11)	(12)	(13)	(14)	(15)
Connection			Minimum Makeup Torque ft-lb[2]											
			Bore of Drill Collar, inches											
Size	Type	OD, in.	1	$1\frac{1}{4}$	$1\frac{1}{2}$	$1\frac{3}{4}$	2	$2\frac{1}{4}$	$2\frac{1}{2}$	$2\frac{13}{16}$	3	$3\frac{1}{4}$	$3\frac{1}{2}$	$3\frac{3}{4}$
		$11\frac{1}{4}$									147,616	142,430	136,846	130,871
$8\frac{5}{8}$	H-90[4]	$10\frac{3}{4}$									*92,960	*92,960	*92,960	*92,960
	(with low torque face)	11									*110,781	*110,781	*110,781	*110,781
		$11\frac{1}{4}$									*129,203	*129,203	*129,203	*129,203

[1]Torque figures preceded by an asterisk (*) indicate that the weaker member for the corresponding outside diameter (OD) and bore is the BOX; for all other torque values, the weaker member is the PIN.

[2]In each connection size and type group, torque values apply to all connection types in the group, when used with the same drill collar outside diameter and bore, i.e., $2\frac{3}{8}$ API IF, API NC 26, and $2\frac{7}{8}$ Slim Hole connections used with $3\frac{1}{2} \times 1\frac{1}{4}$ drill collars all have the same minimum make-up torque of 4600 ft. lb., and the BOX is the weaker member.

[3]Stress-relief features are disregarded for make-up torque.

[1]Basis of calculations for recommended make-up torque assumed the use of a thread compound containing 40–60% by weight of finely powdered metallic zinc or 60% by weight of finely powdered metallic lead, with not more than 0.3% total active sulfur (reference the caution regarding the use of hazardous materials in Appendix G of Specification 7) applied thoroughly to all threads and shoulders and using the modified Screw Jack formula in A.8, and a unit stress of 62,500 psi in the box or pin, whichever is weaker.

[2]Normal torque range is tabulated value plus 10 percent. Higher torque values may be used under extreme conditions.

[3]Makeup torque for 2? PAC connection is based on 87,500 psi stress and other factors listed in footnote 1.

[4]Makeup torque for H-90 connection is based on 56,200 psi stress and other factors listed in footnote 1.

Solution

Moment of inertia:

$$I = \frac{\pi}{64}\left[\left(\frac{6.75}{12}\right)^4 - \left(\frac{2.25}{12}\right)^4\right] = 4.853 \times 10^{-3}\,\text{ft}^4$$

Unit weight of drill collar in air = 108 lb/ft.
Unit weight of drill collar in drilling fluid:

$$p = 108\left(1 - \frac{12}{65.4}\right) = 88.18\ \text{lb/ft}$$

For weight on the bit that results in second-order buckling, use Equation 2.4:

$$W_{crll} = 3.75\left[\left(4320 \times 10^6\right)\left(48.53 \times 10^{-3}\right)\left(88.18\right)^2\right]^{\frac{1}{3}}$$
$$= 20{,}485\ \text{lb}$$

Corresponding length of drill collars:

$$L_{dc} = \frac{20{,}485}{88.18} = 232\,\text{ft}$$

L_{dc} is the distance in which the drill collars are in compression or subject to buckling. It is the distance from the bit to the neutral point.

Lubinski also found [8] that do drill a vertical hole in homogeneous formations, it is best to carry less weight on the bit than the critical value of the first order at which the drill string buckles. However, if such weight is not sufficient, it is advisable to avoid the weight that falls between the first- and second-order buckling and to carry a weight close to the critical value of the third order.

In many instances, the previous statement holds true if formations being drilled are horizontal. When drilling in dipping formations, a proper drill collar stabilization is required for vertical or nearly vertical hole drilling. In an inclined hole, a critical value of weight on the bit that produces buckling may be calculated from the formula given by R. Dawson and P. R. Palsy [7].

$$W_{crit} = 2\left(\frac{EIp\sin\alpha}{r}\right) \tag{2.6}$$

where α is the hole inclination measured from vertical (degrees), r is the radial clearance between drill collar and borehole wall (ft), and E, I, and p are as defined in Equations 2.3, 2.4, and 2.5.

It can be seen by this equation that the critical weight is very high in highly deviated holes. As the hole angle increases, the resistant to buckling increases the drill collar is supported by the borehole wall and lateral component of the buckling forces must overcome the gravity forces.

This explains why heavy-weight drill pipe is successfully used for creating weight on the bit in highly deviated holes. However, in drilling a vertical or nearly vertical hole, a drill pipe must never be run in effective compression; the neutral point must always reside in the drill collar string.

2.1.6 Rig Maintenance of Drill Collars

It is recommended practice to break a different joint on each trip, giving the crew an opportunity to look at each pin and box every third trip. Inspect the shoulders for signs of loose connections, galls, and possible washouts.

Thread protectors should be used on pins and boxes when picking up or laying down the drill collars.

Periodically, based on drilling conditions and experience, a magnetic particle inspection should be performed using a wet fluorescent and black light method. Before storing, the drill collars should be cleaned. If necessary, reface the shoulders with a shoulder refacing tool, and remove the fins on the shoulders by beveling. A good rust-preventative or drill collar compound should be applied to the connections liberally, and thread protectors should be installed.

2.2 DRILL PIPE

The major portion of drill string is composed of drill pipe. Drill pipe consists of three components: a tube with a pin tool joint welded to one end and a box tool joint welded to the other. Figure 2.11 shows a sectioned view of a drill pipe assembly. Before the tool joints are welded to the tube, the tube is upset, or forged, on each end to increase the wall thickness. After upsetting, the tube is heat treated to the proper grade strength. All tool joints are heat treated to the same material yield strength (120,000 psi), regardless of the grade of pipe to which they are attached. Most drill pipe is made from material similar to AISI 4125/30 steel seamless tube. Most tool joints are made from material similar to AISI 4140 steel forgings, tubing, or bars stock.

Drill pipe dimensional and metallurgical specifications are defined by the American Petroleum Institute (API) and published in *API Spec*

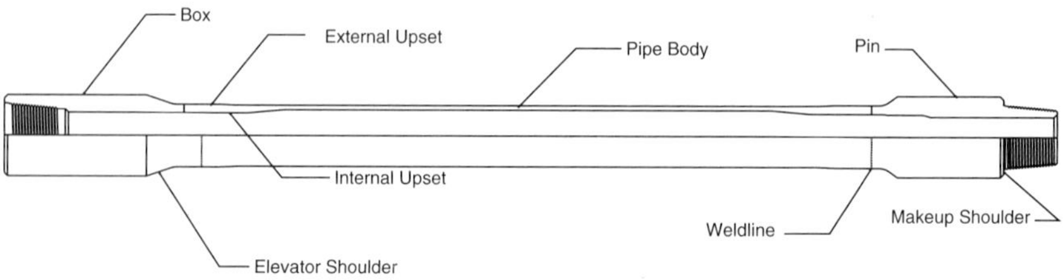

FIGURE 2.11 Section view of a drill pipe assembly.

7 Specifications for Drill Stem Elements and *API Spec 5D Specifications for Drill Pipe*. Performance characteristics, guidelines for drill pipe use and inspection standards are in *API RP7G Recommended Practice for Drill Stem Design and Operating Limits*. Drill pipe specifications and performance characteristics can also be found in ISO 10407-1, ISO 10407-2, ISO 10424-1, ISO 10424-2, and ISO 11961.

A list of drill pipe assemblies is shown in Table 2.6. This list is not all inclusive; it contains drill pipe assemblies with various tool joints but not all OD and ID combinations of tool joints.

The following items are required to completely identify a length of drill pipe:

Pipe size is the pipe OD (in., mm). Drill pipe sizes and wall thicknesses are shown in Table 2.8. The API specification for the drill pipe tube is API 5D.

Pipe weight (lb/ft, kg/m) is for the tube only exclusive of tool joints and upset ends and is used to specify wall thickness. Except in a few cases, the tabulated pipe weight is not the calculated pipe weight.

Pipe grade is the pipe yield strength. The API grades are listed in Tables 2.7 and 2.8. Drill pipe manufacturers may offer higher strength grades or grades designed for specific applications such as drilling in H_2S environments.

Pipe upset The drill pipe tubes are upset on each end to increase the wall thickness. The thicker wall is needed to compensate for loss of material strength during the welding process. There are three configurations of upsets:

- Internal upset (IU), in which the wall thickness is increased by decreasing the ID. This allows smaller OD tool joints to be welded to the pipe. This pipe is sometimes referred to as a *slim hole pipe* and is used in smaller-diameter holes.
- External upset (EU), in which the wall thickness is increased by increasing the OD. This allows larger tool joints to be welded to the pipe. The larger tool joints provide more torsional strength and create a lower pressure drop than those used on IU pipe.
- Internal external upset (IEU), in which the wall thickness is increased by increasing the OD and decreasing the ID. This is the most common upset type on pipe larger than 4 in.

Tool joint type. Table 2.9 is a tool joint interchangeability chart which shows a number of tool joint types. The API tool joint types are printed in bold. These include API Reg (used mostly for drill collars, bits, subs and other bottom-hole assembly components), NC numbered connections, and $5\frac{1}{2}$ and $6\frac{5}{8}$ full hole (FH). Before the NC-type connections were established, tool joint manufacturers often produced interchangeable tool joints with different names. The NC connections were established to reduce the

TABLE 2.6 Drill Pipe Assembly Properties

	Pipe Data								
Size OD	Nominal Weight	Grade and Upset Type	Torsional Yield Strength	Tensile Yield Strength	Wall Thickness	Nominal ID	Pipe Body Section Area	Pipe Body Moment of Inertia	Pipe Body Polar Moment of Inertia
in.	lb/ft		ft-lb	lb	in.	in.	sq in.	cu in.	cu in.
$2\frac{3}{8}$	4.85	E-75 EU	4,800	97,800	0.190	1.995	1.304	0.784	1.568
	4.85	E-75 EU	4,800	97,800	0.190	1.995	1.304	0.784	1.568
	4.85	E-75 EU	4,800	97,800	0.190	1.995	1.304	0.784	1.563
	4.85	E-75 EU	4,800	97,800	0.190	1.995	1.304	0.784	1.563
$2\frac{3}{8}$	4.85	X-95 EU	6,000	123,900	0.190	1.995	1.304	0.784	1.568
	4.85	X-95 EU	6,000	123,900	0.190	1.995	1.304	0.784	1.568
	4.85	X-95 EU	6,000	123,900	0.190	1.995	1.304	0.784	1.568
	4.85	X-95 EU	6,000	123,900	0.190	1.995	1.304	0.784	1.568
$2\frac{3}{8}$	4.85	G-105 EU	6,700	136,900	0.190	1.995	1.304	0.784	1.568
	4.85	G-105 EU	6,700	136,900	0.190	1.995	1.304	0.784	1.568
	4.85	G-105 EU	6,700	136,900	0.190	1.995	1.304	0.784	1.568
	4.85	G-105 EU	6,700	136,900	0.190	1.995	1.304	0.784	1.568
$2\frac{3}{8}$	4.85	S-135 EU	8,600	176,100	0.190	1.995	1.304	0.784	1.568
	4.85	S-135 EU	8,600	176,100	0.190	1.995	1.304	0.784	1.568
	4.85	S-135 EU	8,600	176,100	0.190	1.995	1.304	0.784	1.568
	4.85	S-135 EU	8,600	176,100	0.190	1.995	1.304	0.784	1.568
$2\frac{3}{8}$	4.85	Z-140 EU	8,900	182,600	0.190	1.995	1.304	0.784	1.568
	4.85	Z-140 EU	8,900	182,600	0.190	1.995	1.304	0.784	1.568
	4.85	Z-140 EU	8,900	182,600	0.190	1.995	1.304	0.784	1.568

$2\frac{3}{8}$	4.85	V-150 EU	9,500	195,600	0.190	1.995	1.304	0.784	1.568
	4.85	V-150 EU	9,500	195,600	0.190	1.995	1.304	0.784	1.568
	4.85	V-150 EU	9,500	195,600	0.190	1.995	1.304	0.784	1.568
$2\frac{3}{8}$	6.65	E-75 EU	6,300	138,200	0.280	1.815	1.843	1.029	2.058
	6.65	E-75 EU	6,300	138,200	0.280	1.815	1.843	1.029	2.058
	6.65	E-75 EU	6,300	138,200	0.280	1.815	1.843	1.029	2.058
	6.65	E-75 EU	6,300	138,200	0.280	1.815	1.843	1.029	2.058
$2\frac{3}{8}$	6.65	X-95 EU	7,900	175,100	0.280	1.815	1.843	1.029	2.058
	6.65	X-95 EU	7,900	175,100	0.280	1.815	1.843	1.029	2.058
	6.65	X-95 EU	7,900	175,100	0.280	1.815	1.843	1.029	2.058
	6.65	X-95 EU	7,900	175,100	0.280	1.815	1.843	1.029	2.058
$2\frac{3}{8}$	6.65	G-105 EU	8,800	193,500	0.280	1.815	1.843	1.029	2.058
	6.65	G-105 EU	8,800	193,500	0.280	1.815	1.843	1.029	2.058
	6.65	G-105 EU	8,800	193,500	0.280	1.815	1.843	1.029	2.058
	6.65	G-105 EU	8,800	193,500	0.280	1.815	1.843	1.029	2.058
$2\frac{3}{8}$	6.65	S-135 EU	11,300	248,800	0.280	1.815	1.843	1.029	2.058
	6.65	S-135 EU	11,300	248,800	0.280	1.815	1.843	1.029	2.058
	6.65	S-135 EU	11,300	248,800	0.280	1.815	1.843	1.029	2.058
	6.65	S-135 EU	11,300	248,800	0.280	1.815	1.843	1.029	2.058
	6.65	S-135 EU	11,300	248,800	0.280	1.815	1.843	1.029	2.058
$2\frac{3}{8}$	6.65	Z-140 EU	11,700	258,000	0.280	1.815	1.843	1.029	2.058
	6.65	Z-140 EU	11,700	258,000	0.280	1.815	1.843	1.029	2.058
	6.65	Z-140 EU	11,700	258,000	0.280	1.815	1.843	1.029	2.058
	6.65	Z-140 EU	11,700	258,000	0.280	1.815	1.843	1.029	2.058
$2\frac{3}{8}$	6.65	V-150 EU	12,500	276,400	0.280	1.815	1.843	1.029	2.058
	6.65	V-150 EU	12,500	276,400	0.280	1.815	1.843	1.029	2.058
	6.65	V-150 EU	12,500	276,400	0.280	1.815	1.843	1.029	2.058

(*Continued*)

TABLE 2.6 (*Continued*)

Pipe Data									
Size OD	Nominal Weight	Grade and Upset Type	Torsional Yield Strength	Tensile Yield Strength	Wall Thickness	Nominal ID	Pipe Body Section Area	Pipe Body Moment of Inertia	Pipe Body Polar Moment of Inertia
in.	lb/ft		ft-lb	lb	in.	in.	sq in.	cu in.	cu in.
	6.65	V-150 EU	12,500	276,400	0.280	1.815	1.843	1.029	2.058
$2\frac{7}{8}$	6.85	E-75 IU	8,100	135,900	0.217	2.441	1.812	1.611	3.222
	6.85	E-75 IU	8,100	135,900	0.217	2.441	1.812	1.611	3.222
	6.85	E-75 EU	8,100	135,900	0.217	2.441	1.812	1.611	3.222
	6.85	E-75 IU	8,100	135,900	0.217	2.441	1.812	1.611	3.222
	6.85	E-75 EU	8,100	135,900	0.217	2.441	1.812	1.611	3.222
	6.65	E-75 IU	8,100	135,900	0.217	2.441	1.812	1.611	3.222
	6.85	E-75 EU	8,100	135,900	0.217	2.441	1.812	1.611	3.222
	6.85	E-75 EU	8,100	135,900	0.217	2.441	1.812	1.611	3.222
$2\frac{7}{8}$	6.85	X-95 IU	10,200	172,100	0.217	2.441	1.812	1.611	3.222
	6.85	X-95 IU	10,200	172,100	0.217	2.441	1.812	1.611	3.222
	6.85	X-95 EU	10,200	172,100	0.217	2.441	1.812	1.611	3.222
	6.85	X-95 IU	10,200	172,100	0.217	2.441	1.812	1.611	3.222
	6.85	X-95 EU	10,200	172,100	0.217	2.441	1.812	1.611	3.222
	6.65	X-95 IU	10,200	172,100	0.217	2.441	1.812	1.611	3.222
	6.85	X-95 EU	10,200	172,100	0.217	2.441	1.812	1.611	3.222
	6.85	X-95 EU	10,200	172,100	0.217	2.441	1.812	1.611	3.222
$2\frac{7}{8}$	6.85	G-105 IU	11,300	190,300	0.217	2.441	1.812	1.611	3.222
	6.85	G-105 IU	11,300	190,300	0.217	2.441	1.812	1.611	3.222
	6.85	G-105 EU	11,300	190,300	0.217	2.441	1.812	1.611	3.222

	6.85	G-105 IU	11,300	190,300	0.217	2.441	1.812	1.611	3.222
	6.85	G-105 EU	11,300	190,300	0.217	2.441	1.812	1.611	3.222
	6.65	G-105 IU	11,300	190,300	0.217	2.441	1.812	1.611	3.222
	6.85	G-105 EU	11,300	190,300	0.217	2.441	1.812	1.611	3.222
	6.85	G-105 EU	11,300	190,300	0.217	2.441	1.812	1.611	3.222
$2\frac{7}{8}$	6.85	S-135 IU	14,500	244,600	0.217	2.441	1.812	1.611	3.222
	6.85	S-135 IU	14,500	244,600	0.217	2.441	1.812	1.611	3.222
	6.85	S-135 EU	14,500	244,600	0.217	2.441	1.812	1.611	3.222
	6.85	S-135 IU	14,500	244,600	0.217	2.441	1.812	1.611	3.222
	6.85	S-135 EU	14,500	244,600	0.217	2.441	1.812	1.611	3.222
	6.65	S-135 IU	14,500	244,600	0.217	2.441	1.812	1.611	3.222
	6.85	S-135 EU	14,500	244,600	0.217	2.441	1.812	1.611	3.222
	6.85	S-135 EU	14,500	244,600	0.217	2.441	1.812	1.611	3.222
$2\frac{7}{8}$	6.85	Z-140 IU	15,100	253,700	0.217	2.441	1.812	1.611	3.222
	6.65	Z-140 IU	15,100	253,700	0.217	2.441	1.812	1.611	3.222
	6.85	Z-140 EU	15,100	253,700	0.217	2.441	1.812	1.611	3.222
	6.85	Z-140 EU	15,100	253,700	0.217	2.441	1.812	1.611	3.222
$2\frac{7}{8}$	6.85	V-150 IU	16,200	271,800	0.217	2.441	1.812	1.611	3.222
	6.65	V-150 IU	16,200	271,800	0.217	2.441	1.812	1.611	3.222
	6.85	V-150 EU	16,200	271,800	0.217	2.441	1.812	1.611	3.222
	6.85	V-150 EU	16,200	271,800	0.217	2.441	1.812	1.611	3.222
$2\frac{7}{8}$	10.4	E-75 EU	11,600	214,300	0.362	2.151	2.858	2.303	4.606
	10.4	E-75 EU	11,600	214,300	0.362	2.151	2.858	2.303	4.606
	10.4	E-75 IU	11,600	214,300	0.362	2.151	2.858	2.303	4.606
	10.4	E-75 EU	11,600	214,300	0.362	2.151	2.858	2.303	4.606
	10.4	E-75 EU	11,600	214,300	0.362	2.151	2.858	2.303	4.606
	10.4	E-75 IU	11,600	214,300	0.362	2.151	2.858	2.303	4.606
	10.4	E-75 IU	11,600	214,300	0.362	2.151	2.858	2.303	4.606

(Continued)

TABLE 2.6 (Continued)

Pipe Data									
Size OD	Nominal Weight	Grade and Upset Type	Torsional Yield Strength	Tensile Yield Strength	Wall Thickness	Nominal ID	Pipe Body Section Area	Pipe Body Moment of Inertia	Pipe Body Polar Moment of Inertia
in.	lb/ft		ft-lb	lb	in.	in.	sq in.	cu in.	cu in.
	10.4	E-75 EU	11,600	214,300	0.362	2.151	2.858	2.303	4.606
	10.4	E-75 IU	11,600	214,300	0.362	2.151	2.858	2.303	4.606
	10.4	E-75 EU	11,600	214,300	0.362	2.151	2.858	2.303	4.605
$2\frac{7}{8}$	10.4	X-95 EU	14,600	271,500	0.362	2.151	2.858	2.303	4.605
	10.4	X-95 IU	14,600	271,500	0.362	2.151	2.858	2.303	4.606
	10.4	X-95 IU	14,600	271,500	0.362	2.151	2.858	2.303	4.606
	10.4	X-95 EU	14,600	271,500	0.362	2.151	2.858	2.303	4.606
	10.4	X-95 EU	14,600	271,500	0.362	2.151	2.858	2.303	4.606
	10.4	X-95 IU	14,600	271,500	0.362	2.151	2.858	2.303	4.606
	10.4	X-95 IU	14,600	271,500	0.362	2.151	2.858	2.303	4.606
	10.4	X-95 EU	14,600	271,500	0.362	2.151	2.858	2.303	4.606
	10.4	X-95 IU	14,600	271,500	0.362	2.151	2.858	2.303	4.606
	10.4	X-95 EU	14,600	271,500	0.362	2.151	2.858	2.303	4.606
$2\frac{7}{8}$	10.4	G-105 EU	16,200	300,100	0.362	2.151	2.858	2.303	4.606
	10.4	G-105 IU	16,200	300,100	0.362	2.151	2.858	2.303	4.606
	10.4	G-105 IU	16,200	300,100	0.362	2.151	2.858	2.303	4.606
	10.4	G-105 EU	16,200	300,100	0.362	2.151	2.858	2.303	4.606
	10.4	G-105 EU	16,200	300,100	0.362	2.151	2.858	2.303	4.606
	10.4	G-105 IU	16,200	300,100	0.362	2.151	2.858	2.303	4.606
	10.4	G-105 IU	16,200	300,100	0.362	2.151	2.858	2.303	4.606

	10.4	G-105 EU	16,200	300,100	0.362	2.151	2.858	2.303	4.606
	10.4	G-105 IU	16,200	300,100	0.362	2.151	2.858	2.303	4.606
	10.4	G-150 EU	16,200	300,100	0.362	2.151	2.858	2.303	4.606
$2\frac{7}{8}$	10.4	S-135 EU	20,800	385,800	0.362	2.151	2.858	2.303	4.606
	10.4	S-135 IU	20,800	385,800	0.362	2.151	2.858	2.303	4.606
	10.4	S-135 IU	20,800	385,800	0.362	2.151	2.858	2.303	4.606
	10.4	S-135 EU	20,800	385,800	0.362	2.151	2.858	2.303	4.606
	10.4	S-135 EU	20,800	385,800	0.362	2.151	2.858	2.303	4.606
	10.4	S-135 IU	20,800	385,800	0.362	2.151	2.858	2.303	4.606
	10.4	S-135 IU	20,800	385,800	0.362	2.151	2.858	2.303	4.606
	10.4	S-135 EU	20,800	385,800	0.362	2.151	2.858	2.303	4.606
	10.4	S-135 IU	20,800	385,800	0.362	2.151	2.858	2.303	4.606
	10.4	S-135 EU	20,800	385,800	0.362	2.151	2.858	2.303	4.606
	10.4	S-135 EU	20,800	385,800	0.362	2.151	2.858	2.303	4.606
$2\frac{7}{8}$	10.4	Z-140 IU	21,600	400,100	0.362	2.151	2.858	2.303	4.606
	10.4	Z-140 EU	21,600	400,100	0.362	2.151	2.858	2.303	4.606
	10.4	Z-140 IU	21,600	400,100	0.362	2.151	2.858	2.303	4.606
	10.4	Z-140 EU	21,600	400,100	0.362	2.151	2.858	2.303	4.606
	10.4	Z-140 EU	21,600	400,100	0.362	2.151	2.858	2.303	4.606
$2\frac{7}{8}$	10.4	V-150 IU	23,100	428,700	0.362	2.151	2.858	2.303	4.606
	10.4	V-150 EU	23,100	428,700	0.362	2.151	2.858	2.303	4.606
	10.4	V-150 IU	23,100	428,700	0.362	2.151	2.858	2.303	4.606
	10.4	V-150 EU	23,100	428,700	0.362	2.151	2.858	2.303	4.606
	10.4	V-150 EU	23,100	428,700	0.362	2.151	2.858	2.303	4.606
$3\frac{1}{2}$	9.50	E-75 EU	14,100	194,300	0.254	2.992	2.590	3.432	6.865
	9.50	E-75 IU	14,100	194,300	0.254	2.992	2.590	3.432	6.865
	9.50	E-75 IU	14,100	194,300	0.254	2.992	2.590	3.432	6.865
	9.50	E-75 EU	14,100	194,300	0.254	2.992	2.590	3.432	6.865
	9.50	E-75 EU	14,100	194,300	0.254	2.992	2.590	3.432	6.865

(*Continued*)

TABLE 2.6 (*Continued*)

			Pipe Data						
Size OD	Nominal Weight	Grade and Upset Type	Torsional Yield Strength	Tensile Yield Strength	Wall Thickness	Nominal ID	Pipe Body Section Area	Pipe Body Moment of Inertia	Pipe Body Polar Moment of Inertia
in.	lb/ft		ft-lb	lb	in.	in.	sq in.	cu in.	cu in.
	9.50	E-75 IU	14,100	194,300	0.254	2.992	2.590	3.432	6.865
	9.50	E-75 EU	14,100	194,300	0.254	2.992	2.590	3.432	6.865
$3\frac{1}{2}$	9.50	X-95 EU	17,900	246,100	0.254	2.992	2.590	3.432	6.865
	9.50	X-95 IU	17,900	246,100	0.254	2.992	2.590	3.432	6.865
	9.50	X-95 IU	17,900	246,100	0.254	2.992	2.590	3.432	6.865
	9.50	X-95 EU	17,900	246,100	0.254	2.992	2.590	3.432	6.865
	9.50	X-95 EU	17,900	246,100	0.254	2.992	2.590	3.432	6.865
	9.50	X-95 IU	17,900	246,100	0.254	2.992	2.590	3.432	6.865
	9.50	X-95 EU	17,900	246,100	0.254	2.992	2.590	3.432	6.865
$3\frac{1}{2}$	9.50	G-105 EU	19,800	272,000	0.254	2.992	2.590	3.432	6.865
	9.50	G-105 IU	19,800	272,000	0.254	2.992	2.590	3.432	6.865
	9.50	G-105 IU	19,800	272,000	0.254	2.992	2.590	3.432	6.865
	9.50	G-105 EU	19,800	272,000	0.254	2.992	2.590	3.432	6.865
	9.50	G-105 EU	19,800	272,000	0.254	2.992	2.590	3.432	6.865
	9.50	G-105 IU	19,800	272,000	0.254	2.992	2.590	3.432	6.865
	9.50	G-105 EU	19,800	272,000	0.254	2.992	2.590	3.432	6.865
$3\frac{1}{2}$	9.50	S-135 EU	25,500	349,700	0.254	2.992	2.590	3.432	6.865
	9.50	S-135 IU	25,500	349,700	0.254	2.992	2.590	3.432	6.865
	9.50	S-135 IU	25,500	349,700	0.254	2.992	2.590	3.432	6.865
	9.50	S-135 EU	25,500	349,700	0.254	2.992	2.590	3.432	6.865

	9.50	S-135 EU	25,500	349,700	0.254	2.992	2.590	3.432	6.865
	9.50	S-135 IU	25,500	349,700	0.254	2.992	2.590	3.432	6.865
	9.50	S-135 EU	25,500	349,700	0.254	2.992	2.590	3.432	6.865
$3\frac{1}{2}$	9.50	Z-140 IU	26,400	362,600	0.254	2.992	2.590	3.432	6.865
	9.50	Z-140 EU	26,400	362,600	0.254	2.992	2.590	3.432	6.865
	9.50	Z-140 IU	26,400	362,600	0.254	2.992	2.590	3.432	6.865
	9.50	Z-140 EU	26,400	362,600	0.254	2.992	2.590	3.432	6.865
$3\frac{1}{2}$	9.50	V-150 IU	28,300	388,500	0.254	2.992	2.590	3.432	6.865
	9.50	V-150 EU	28,300	388,500	0.254	2.992	2.590	3.432	6.865
	9.50	V-150 IU	28,300	388,500	0.254	2.992	2.590	3.432	6.865
	9.50	V-150 EU	28,300	388,500	0.254	2.992	2.590	3.432	6.865
$3\frac{1}{2}$	13.30	E-75 EU	18,600	271,600	0.368	2.764	3.621	4.501	9.002
	13.30	E-75 IU	18,600	271,600	0.368	2.764	3.621	4.501	9.002
	13.30	E-75 IU	18,600	271,600	0.368	2.764	3.621	4.501	9.002
	13.30	E-75 EU	18,600	271,600	0.368	2.764	3.621	4.501	9.002
	13.30	E-75 EU	18,600	271,600	0.368	2.764	3.621	4.501	9.002
	13.30	E-75 IU	18,600	271,600	0.368	2.764	3.621	4.501	9.002
	13.30	E-75 EU	18,600	271,600	0.368	2.764	3.621	4.501	9.002
$3\frac{1}{2}$	13.30	X-95 EU	23,500	344,000	0.368	2.764	3.621	4.501	9.002
	13.30	X-95 IU	23,500	344,000	0.368	2.764	3.621	4.501	9.002
	13.30	X-95 IU	23,500	344,000	0.368	2.764	3.621	4.501	9.002
	13.30	X-95 EU	23,500	344,000	0.368	2.764	3.621	4.501	9.002
	13.30	X-95 EU	23,500	344,000	0.368	2.764	3.621	4.501	9.002
	13.30	X-95 IU	23,500	344,000	0.368	2.764	3.621	4.501	9.002
	13.30	X-95 EU	23,500	344,000	0.368	2.764	3.621	4.501	9.002
$3\frac{1}{2}$	13.30	G-105 EU	26,000	380,200	0.368	2.764	3.621	4.501	9.002
	13.30	G-105 IU	26,000	380,200	0.368	2.764	3.621	4.501	9.002
	13.30	G-105 IU	26,000	380,200	0.368	2.764	3.621	4.501	9.002
	13.30	G-105 EU	26,000	380,200	0.368	2.764	3.621	4.501	9.002

(Continued)

TABLE 2.6 *(Continued)*

Pipe Data									
Size OD	Nominal Weight	Grade and Upset Type	Torsional Yield Strength	Tensile Yield Strength	Wall Thickness	Nominal ID	Pipe Body Section Area	Pipe Body Moment of Inertia	Pipe Body Polar Moment of Inertia
in.	lb/ft		ft-lb	lb	in.	in.	sq in.	cu in.	cu in.
	13.30	G-105 EU	26,000	380,200	0.368	2.764	3.621	4.501	9.002
	13.30	G-105 IU	26,000	380,200	0.368	2.764	3.621	4.501	9.002
	13.30	G-105 EU	26,000	380,200	0.368	2.764	3.621	4.501	9.002
$3\frac{1}{2}$	13.30	S-135 EU	33,400	488,800	0.368	2.764	3.621	4.501	9.002
	13.30	S-135 IU	33,400	488,800	0.368	2.764	3.621	4.501	9.002
	13.30	S-135 IU	33,400	488,800	0.368	2.764	3.621	4.501	9.002
	13.30	S-135 EU	33,400	488,800	0.368	2.764	3.621	4.501	9.002
	13.30	S-135 EU	33,400	488,800	0.368	2.764	3.621	4.501	9.002
	13.30	S-135 IU	33,400	488,800	0.368	2.764	3.621	4.501	9.002
	13.30	S-135 EU	33,400	488,800	0.368	2.764	3.621	4.501	9.002
	13.30	S-135 EU	33,400	488,800	0.368	2.764	3.621	4.501	9.002
$3\frac{1}{2}$	13.30	Z-140 IU	34,600	506,900	0.368	2.764	3.621	4.501	9.002
	13.30	Z-140 EU	34,600	506,900	0.368	2.764	3.621	4.501	9.002
	13.30	Z-140 IU	34,600	506,900	0.368	2.764	3.621	4.501	9.002
	13.30	Z-140 EU	34,600	506,900	0.368	2.764	3.621	4.501	9.002
	13.30	Z-140 EU	34,600	506,900	0.368	2.764	3.621	4.501	9.002
$3\frac{1}{2}$	13.30	V-150 IU	37,100	543,100	0.368	2.764	3.621	4.501	9.002
	13.30	V-150 EU	37,100	543,100	0.368	2.764	3.621	4.501	9.002
	13.30	V-150 IU	37,100	543,100	0.368	2.764	3.621	4.501	9.002
	13.30	V-150 EU	37,100	543,100	0.368	2.764	3.621	4.501	9.002

	13.30	V-150 EU	37,100	543,100	0.368	2.764	3.621	4.501	9.002
$3\frac{1}{2}$	15.50	E-75 EU	21,100	322,800	0.499	2.602	4.304	5.116	10.232
	15.50	E-75 EU	21,100	322,800	0.499	2.602	4.304	5.116	10.232
	15.50	E-75 EU	21,100	322,800	0.499	2.602	4.304	5.116	10.232
$3\frac{1}{2}$	15.50	X-95 EU	26,700	408,800	0.499	2.602	4.304	5.116	10.232
	15.50	X-95 EU	26,700	408,800	0.499	2.602	4.304	5.116	10.232
	15.50	X-95 EU	26,700	408,800	0.499	2.602	4.304	5.116	10.232
$3\frac{1}{2}$	15.50	G-105 EU	29,500	451,900	0.499	2.602	4.304	5.116	10.232
	15.50	G-105 EU	29,500	451,900	0.499	2.602	4.304	5.116	10.232
	15.50	G-105 EU	29,500	451,900	0.499	2.602	4.304	5.116	10.232
	15.50	G-105 EU	29,500	451,900	0.499	2.602	4.304	5.116	10.232
$3\frac{1}{2}$	15.50	S-135 EU	38,000	581,000	0.499	2.602	4.304	5.116	10.232
	15.50	S-135 EU	38,000	581,000	0.499	2.602	4.304	5.116	10.232
	15.50	S-135 EU	38,000	581,000	0.499	2.602	4.304	5.116	10.232
	15.50	S-135 EU	38,000	581,000	0.499	2.602	4.304	5.116	10.232
	15.50	S-135 EU	38,000	581,000	0.499	2.602	4.304	5.116	10.232
	15.50	S-135 EU	38,000	581,000	0.499	2.602	4.304	5.116	10.232
$3\frac{1}{2}$	15.50	Z-140 EU	39,400	602,500	0.499	2.602	4.304	5.116	10.232
	15.50	Z-140 EU	39,400	602,500	0.499	2.602	4.304	5.116	10.232
	15.50	Z-140 EU	39,400	602,500	0.499	2.602	4.304	5.116	10.232
	15.50	Z-140 EU	39,400	602,500	0.499	2.602	4.304	5.116	10.232
$3\frac{1}{2}$	15.50	V-150 EU	42,200	645,500	0.499	2.602	4.304	5.116	10.232
	15.50	V-150 EU	42,200	645,500	0.499	2.602	4.304	5.116	10.232
	15.50	V-150 EU	42,200	645,500	0.499	2.602	4.304	5.116	10.232
	15.50	V-150 EU	42,200	645,500	0.499	2.602	4.304	5.116	10.232
4	11.85	E-75 IU	19,500	230,800	0.262	3.476	0.3077	5.400	10.800
	11.85	E-75 IU	19,500	230,800	0.262	3.476	0.3077	5.400	10.800
	11.85	E-75 IU	19,500	230,800	0.262	3.476	0.3077	5.400	10.800
	11.85	E-75 IU	19,500	230,800	0.262	3.476	0.3077	5.400	10.800

(Continued)

TABLE 2.6 (*Continued*)

Pipe Data									
Size OD	Nominal Weight	Grade and Upset Type	Torsional Yield Strength	Tensile Yield Strength	Wall Thickness	Nominal ID	Pipe Body Section Area	Pipe Body Moment of Inertia	Pipe Body Polar Moment of Inertia
in.	lb/ft		ft-lb	lb	in.	in.	sq in.	cu in.	cu in.
	11.85	E-75 IU	19,500	230,800	0.262	3.476	0.3077	5.400	10.800
4	11.85	X-95 IU	24,700	292,300	0.262	3.476	0.3077	5.400	10.800
	11.85	X-95 IU	24,700	292,300	0.262	3.476	0.3077	5.400	10.800
	11.85	X-95 IU	24,700	292,300	0.262	3.476	0.3077	5.400	10.800
	11.85	X-95 IU	24,700	292,300	0.262	3.476	0.3077	5.400	10.800
	11.85	X-95 IU	24,700	292,300	0.262	3.476	0.3077	5.400	10.800
4	11.85	G-105 IU	27,300	323,100	0.262	3.476	0.3077	5.400	10.800
	11.85	G-105 IU	27,300	323,100	0.262	3.476	0.3077	5.400	10.800
	11.85	G-105 IU	27,300	323,100	0.262	3.476	0.3077	5.400	10.800
	11.85	G-105 IU	27,300	323,100	0.262	3.476	0.3077	5.400	10.800
	11.85	G-105 IU	27,300	323,100	0.262	3.476	0.3077	5.400	10.800
4	11.85	S-135 IU	35,100	415,400	0.262	3.476	0.3077	5.400	10.800
	11.85	S-135 IU	35,100	415,400	0.262	3.476	0.3077	5.400	10.800
	11.85	S-135 IU	35,100	415,400	0.262	3.476	0.3077	5.400	10.800
	11.85	S-135 IU	35,100	415,400	0.262	3.476	0.3077	5.400	10.800
	11.85	S-135 IU	35,100	415,400	0.262	3.476	0.3077	5.400	10.800
4	11.85	Z-140 IU	36,400	430,700	0.262	3.476	0.3077	5.400	10.800
	11.85	Z-140 IU	36,400	430,700	0.262	3.476	0.3077	5.400	10.800
	11.85	Z-140 IU	36,400	430,700	0.262	3.476	0.3077	5.400	10.800
4	11.85	V-150 IU	38,900	461,500	0.262	3.476	0.3077	5.400	10.800

	11.85	V-150 IU	38,900	461,500	0.262	3.476	0.3077	5.400	10.800
	11.85	V-150 IU	38,900	461,500	0.262	3.476	0.3077	5.400	10.800
4	14.00	E-75 IU	23,300	285,400	0.330	3.340	3.805	6.458	12.915
	14.00	E-75 IU	23,300	285,400	0.330	3.340	3.805	6.458	12.915
	14.00	E-75 IU	23,300	285,400	0.330	3.340	3.805	6.458	12.915
	14.00	E-75 IU	23,300	285,400	0.330	3.340	3.805	6.458	12.915
	14.00	E-75 EU	23,300	285,400	0.330	3.340	3.805	6.458	12.915
	14.00	E-75 IU	23,300	285,400	0.330	3.340	3.805	6.458	12.915
	14.00	E-75 IU	23,300	285,400	0.330	3.340	3.805	6.458	12.915
4	14.00	X-95 IU	29,500	361,500	0.330	3.340	3.805	6.458	12.915
	14.00	X-95 IU	29,500	361,500	0.330	3.340	3.805	6.458	12.915
	14.00	X-95 IU	29,500	361,500	0.330	3.340	3.805	6.458	12.915
	14.00	X-95 IU	29,500	361,500	0.330	3.340	3.805	6.458	12.915
	14.00	X-95 EU	29,500	361,500	0.330	3.340	3.805	6.458	12.915
	14.00	X-95 IU	29,500	361,500	0.330	3.340	3.805	6.458	12.915
	14.00	X-95 IU	29,500	361,500	0.330	3.340	3.805	6.458	12.915
4	14.00	G-105 IU	32,600	399,500	0.330	3.340	3.805	6.458	12.915
	14.00	G-105 IU	32,600	399,500	0.330	3.340	3.805	6.458	12.915
	14.00	G-105 IU	32,600	399,500	0.330	3.340	3.805	6.458	12.915
	14.00	G-105 IU	32,600	399,500	0.330	3.340	3.805	6.458	12.915
	14.00	G-105 EU	32,600	399,500	0.330	3.340	3.805	6.458	12.915
	14.00	G-105 IU	32,600	399,500	0.330	3.340	3.805	6.458	12.915
	14.00	G-105 IU	32,600	399,500	0.330	3.340	3.805	6.458	12.915
4	14.00	S-135 IU	41,900	513,600	0.330	3.340	3.805	6.458	12.915
	14.00	S-135 IU	41,900	513,600	0.330	3.340	3.805	6.458	12.915
	14.00	S-135 IU	41,900	513,600	0.330	3.340	3.805	6.458	12.915
	14.00	S-135 IU	41,900	513,600	0.330	3.340	3.805	6.458	12.915
	14.00	S-135 EU	41,900	513,600	0.330	3.340	3.805	6.458	12.915
	14.00	S-135 IU	41,900	513,600	0.330	3.340	3.805	6.458	12.915
	14.00	S-135 IU	41,900	513,600	0.330	3.340	3.805	6.458	12.915

(Continued)

TABLE 2.6 (*Continued*)

Pipe Data									
Size OD	Nominal Weight	Grade and Upset Type	Torsional Yield Strength	Tensile Yield Strength	Wall Thickness	Nominal ID	Pipe Body Section Area	Pipe Body Moment of Inertia	Pipe Body Polar Moment of Inertia
in.	lb/ft		ft-lb	lb	in.	in.	sq in.	cu in.	cu in.
4	14.00	Z-140 IU	43,500	532,700	0.330	3.340	3.805	6.458	12.915
	14.00	Z-140 IU	43,500	532,700	0.330	3.340	3.805	6.458	12.915
	14.00	Z-140 IU	43,500	532,700	0.330	3.340	3.805	6.458	12.915
	14.00	Z-140 IU	43,500	532,700	0.330	3.340	3.805	6.458	12.915
	14.00	Z-140 IU	43,500	532,700	0.330	3.340	3.805	6.458	12.915
4	14.00	V-150 IU	46,600	570,700	0.330	3.340	3.805	6.458	12.915
	14.00	V-150 IU	46,600	570,700	0.330	3.340	3.805	6.458	12.915
	14.00	V-150 IU	46,600	570,700	0.330	3.340	3.805	6.458	12.915
	14.00	V-150 IU	46,600	570,700	0.330	3.340	3.805	6.458	12.915
	14.00	V-150 IU	46,600	570,700	0.330	3.340	3.805	6.458	12.915
4	15.70	E-75 IU	25,800	324,100	0.380	3.240	4.322	7.157	14.314
	15.70	E-75 IU	25,800	324,100	0.380	3.240	4.322	7.157	14.314
	15.70	E-75 IU	25,800	324,100	0.380	3.240	4.322	7.157	14.314
	15.70	E-75 EU	25,800	324,100	0.380	3.240	4.322	7.157	14.314
	15.70	E-75 IU	25,800	324,100	0.380	3.240	4.322	7.157	14.314
	15.70	E-75 IU	25,800	324,100	0.380	3.240	4.322	7.157	14.314
4	15.70	X-95 IU	32,700	410,500	0.380	3.240	4.322	7.157	14.314
	15.70	X-95 IU	32,700	410,500	0.380	3.240	4.322	7.157	14.314
	15.70	X-95 IU	32,700	410,500	0.380	3.240	4.322	7.157	14.314
	15.70	X-95 EU	32,700	410,500	0.380	3.240	4.322	7.157	14.314

	15.70	X-95 IU	32,700	410,500	0.380	3.240	4.322	7.157	14.314
	15.70	X-95 IU	32,700	410,500	0.380	3.240	4.322	7.157	14.314
4	15.70	G-105 IU	36,100	453,800	0.380	3.240	4.322	7.157	14.314
	15.70	G-105 IU	36,100	453,800	0.380	3.240	4.322	7.157	14.314
	15.70	G-105 IU	36,100	453,800	0.380	3.240	4.322	7.157	14.314
	15.70	G-105 EU	36,100	453,800	0.380	3.240	4.322	7.157	14.314
	15.70	G-105 IU	36,100	453,800	0.380	3.240	4.322	7.157	14.314
	15.70	G-105 IU	36,100	453,800	0.380	3.240	4.322	7.157	14.314
4	15.70	S-135 IU	46,500	583,400	0.380	3.240	4.322	7.157	14.314
	15.70	S-135 IU	46,500	583,400	0.380	3.240	4.322	7.157	14.314
	15.70	S-135 IU	46,500	583,400	0.380	3.240	4.322	7.157	14.314
	15.70	S-135 EU	46,500	583,400	0.380	3.240	4.322	7.157	14.314
	15.70	S-135 IU	46,500	583,400	0.380	3.240	4.322	7.157	14.314
	15.70	S-135 IU	46,500	583,400	0.380	3.240	4.322	7.157	14.314
	15.70	S-135 IU	46,500	583,400	0.380	3.240	4.322	7.157	14.314
4	15.70	Z-140 IU	48,200	605,000	0.380	3.240	4.322	7.157	14.314
	15.70	Z-140 IU	48,200	605,000	0.380	3.240	4.322	7.157	14.314
	15.70	Z-140 IU	48,200	605,000	0.380	3.240	4.322	7.157	14.314
	15.70	Z-140 IU	48,200	605,000	0.380	3.240	4.322	7.157	14.314
4	15.70	V-150 IU	51,600	648,200	0.380	3.240	4.322	7.157	14.314
	15.70	V-150 IU	51,600	648,200	0.380	3.240	4.322	7.157	14.314
	15.70	V-150 IU	51,600	648,200	0.380	3.240	4.322	7.157	14.314
	15.70	V-150 IU	51,600	648,200	0.380	3.240	4.322	7.157	14.314
$4\frac{1}{2}$	16.60	E-75 IEU	30,800	330,600	0.337	3.826	4.407	9.610	19.221
	16.60	E-75 EU	30,800	330,600	0.337	3.826	4.407	9.610	19.221
	16.60	E-75 IEU	30,800	330,600	0.337	3.826	4.407	9.610	19.221
	16.60	E-75 IEU	30,800	330,600	0.337	3.826	4.407	9.610	19.221
	16.60	E-75 IEU	30,800	330,600	0.337	3.826	4.407	9.610	19.221

(Continued)

TABLE 2.6 (*Continued*)

Pipe Data									
Size OD	Nominal Weight	Grade and Upset Type	Torsional Yield Strength	Tensile Yield Strength	Wall Thickness	Nominal ID	Pipe Body Section Area	Pipe Body Moment of Inertia	Pipe Body Polar Moment of Inertia
in.	lb/ft		ft-lb	lb	in.	in.	sq in.	cu in.	cu in.
	16.60	E-75 EU	30,800	330,600	0.337	3.826	4.407	9.610	19.221
	16.60	E-75 EU	30,800	330,600	0.337	3.826	4.407	9.610	19.221
	16.60	E-75 IEU	30,800	330,600	0.337	3.826	4.407	9.610	19.221
	16.60	E-75 IEU	30,800	330,600	0.337	3.826	4.407	9.610	19.221
	16.60	E-75 EU	30,800	330,600	0.337	3.826	4.407	9.610	19.221
$4\frac{1}{2}$	16.60	X-95 IEU	39,000	418,700	0.337	3.826	4.407	9.610	19.221
	16.60	X-95 EU	39,000	418,700	0.337	3.826	4.407	9.610	19.221
	16.60	X-95 IEU	39,000	418,700	0.337	3.826	4.407	9.610	19.221
	16.60	X-95 IEU	39,000	418,700	0.337	3.826	4.407	9.610	19.221
	16.60	X-95 IEU	39,000	418,700	0.337	3.826	4.407	9.610	19.221
	16.60	X-95 EU	39,000	418,700	0.337	3.826	4.407	9.610	19.221
	16.60	X-95 EU	39,000	418,700	0.337	3.826	4.407	9.610	19.221
	16.60	X-95 IEU	39,000	418,700	0.337	3.826	4.407	9.610	19.221
	16.60	X-95 IEU	39,000	418,700	0.337	3.826	4.407	9.610	19.221
	16.60	X-95 EU	39,000	418,700	0.337	3.826	4.407	9.610	19.221
$4\frac{1}{2}$	16.60	G-105 IEU	43,100	462,800	0.337	3.826	4.407	9.610	19.221
	16.60	G-105 EU	43,100	462,800	0.337	3.826	4.407	9.610	19.221
	16.60	G-105 IEU	43,100	462,800	0.337	3.826	4.407	9.610	19.221
	16.60	G-105 IEU	43,100	462,800	0.337	3.826	4.407	9.610	19.221
	16.60	G-105 IEU	43,100	462,800	0.337	3.826	4.407	9.610	19.221
	16.60	G-105 EU	43,100	462,800	0.337	3.826	4.407	9.610	19.221

	16.60	G-105 EU	43,100	462,800	0.337	3.826	4.407	9.610	19.221
	16.60	G-105 IEU	43,100	462,800	0.337	3.826	4.407	9.610	19.221
	16.60	G-105 IEU	43,100	462,800	0.337	3.826	4.407	9.610	19.221
	16.60	G-105 EU	43,100	462,800	0.337	3.826	4.407	9.610	19.221
$4\frac{1}{2}$	16.60	S-135 IEU	55,500	595,000	0.337	3.826	4.407	9.610	19.221
	16.60	S-135 EU	55,500	595,000	0.337	3.826	4.407	9.610	19.221
	16.60	S-135 IEU	55,500	595,000	0.337	3.826	4.407	9.610	19.221
	16.60	S-135 IEU	55,500	595,000	0.337	3.826	4.407	9.610	19.221
	16.60	S-135 IEU	55,500	595,000	0.337	3.826	4.407	9.610	19.221
	16.60	S-135 EU	55,500	595,000	0.337	3.826	4.407	9.610	19.221
	16.60	S-135 EU	55,500	595,000	0.337	3.826	4.407	9.610	19.221
	16.60	S-135 IEU	55,500	595,000	0.337	3.826	4.407	9.610	19.221
	16.60	S-135 IEU	55,500	595,000	0.337	3.826	4.407	9.610	19.221
	16.60	S-135 EU	55,500	595,000	0.337	3.826	4.407	9.610	19.221
	16.60	S-135 IEU	55,500	595,000	0.337	3.826	4.407	9.610	19.221
$4\frac{1}{2}$	16.60	Z-140 IEU	57,500	617,000	0.337	3.826	4.407	9.610	19.221
	16.60	Z-140 EU	57,500	617,000	0.337	3.826	4.407	9.610	19.221
	16.60	Z-140 IEU	57,500	617,000	0.337	3.826	4.407	9.610	19.221
	16.60	Z-140 IEU	57,500	617,000	0.337	3.826	4.407	9.610	19.221
	16.60	Z-140 EU	57,500	617,000	0.337	3.826	4.407	9.610	19.221
	16.60	Z-140 IEU	57,500	617,000	0.337	3.826	4.407	9.610	19.221
$4\frac{1}{2}$	16.60	V-150 IEU	61,600	661,100	0.337	3.826	4.407	9.610	19.221
	16.60	V-150 EU	61,600	661,100	0.337	3.826	4.407	9.610	19.221
	16.60	V-150 IEU	61,600	661,100	0.337	3.826	4.407	9.610	19.221
	16.60	V-150 IEU	61,600	661,100	0.337	3.826	4.407	9.610	19.221
	16.60	V-150 EU	61,600	661,100	0.337	3.826	4.407	9.610	19.221
	16.60	V-150 IEU	61,600	661,100	0.337	3.826	4.407	9.610	19.221
$4\frac{1}{2}$	20.00	E-75 IEU	36,900	412,400	0.430	3.640	5.498	11.512	23.023
	20.00	E-75 EU	36,900	412,400	0.430	3.640	5.498	11.512	23.023

(Continued)

TABLE 2.6 *(Continued)*

Pipe Data									
Size OD	Nominal Weight	Grade and Upset Type	Torsional Yield Strength	Tensile Yield Strength	Wall Thickness	Nominal ID	Pipe Body Section Area	Pipe Body Moment of Inertia	Pipe Body Polar Moment of Inertia
in.	lb/ft		ft-lb	lb	in.	in.	sq in.	cu in.	cu in.
	20.00	E-75 IEU	36,900	412,400	0.430	3.640	5.498	11.512	23.023
	20.00	E-75 IEU	36,900	412,400	0.430	3.640	5.498	11.512	23.023
	20.00	E-75 EU	36,900	412,400	0.430	3.640	5.498	11.512	23.023
	20.00	E-75 EU	36,900	412,400	0.430	3.640	5.498	11.512	23.023
	20.00	E-75 IEU	36,900	412,400	0.430	3.640	5.498	11.512	23.023
	20.00	E-75 EU	36,900	412,400	0.430	3.640	5.498	11.512	23.023
$4\frac{1}{2}$	20.00	X-95 IEU	46,700	522,300	0.430	3.640	5.498	11.512	23.023
	20.00	X-95 EU	46,700	522,300	0.430	3.640	5.498	11.512	23.023
	20.00	X-95 IEU	46,700	522,300	0.430	3.640	5.498	11.512	23.023
	20.00	X-95 IEU	46,700	522,300	0.430	3.640	5.498	11.512	23.023
	20.00	X-95 EU	46,700	522,300	0.430	3.640	5.498	11.512	23.023
	20.00	X-95 EU	46,700	522,300	0.430	3.640	5.498	11.512	23.023
	20.00	X-95 IEU	46,700	522,300	0.430	3.640	5.498	11.512	23.023
	20.00	X-95 EU	46,700	522,300	0.430	3.640	5.498	11.512	23.023
$4\frac{1}{2}$	20.00	G-105 IEU	51,700	577,300	0.430	3.640	5.498	11.512	23.023
	20.00	G-105 EU	51,700	577,300	0.430	3.640	5.498	11.512	23.023
	20.00	G-105 IEU	51,700	577,300	0.430	3.640	5.498	11.512	23.023
	20.00	G-105 IEU	51,700	577,300	0.430	3.640	5.498	11.512	23.023
	20.00	G-105 EU	51,700	577,300	0.430	3.640	5.498	11.512	23.023
	20.00	G-105 EU	51,700	577,300	0.430	3.640	5.498	11.512	23.023
	20.00	G-105 IEU	51,700	577,300	0.430	3.640	5.498	11.512	23.023
	20.00	G-105 EU	51,700	577,300	0.430	3.640	5.498	11.512	23.023

4½	20.00	S-135 IEU	66,400	742,200	0.430	3.640	5.498	11.512	23.023
	20.00	S-135 EU	66,400	742,200	0.430	3.640	5.498	11.512	23.023
	20.00	S-135 IEU	66,400	742,200	0.430	3.640	5.498	11.512	23.023
	20.00	S-135 IEU	66,400	742,200	0.430	3.640	5.498	11.512	23.023
	20.00	S-135 EU	66,400	742,200	0.430	3.640	5.498	11.512	23.023
	20.00	S-135 EU	66,400	742,200	0.430	3.640	5.498	11.512	23.023
	20.00	S-135 IEU	66,400	742,200	0.430	3.640	5.498	11.512	23.023
	20.00	S-135 EU	66,400	742,200	0.430	3.640	5.498	11.512	23.023
	20.00	S-135 IEU	66,400	742,200	0.430	3.640	5.498	11.512	23.023
4½	20.00	Z-140 IEU	68,900	769,700	0.430	3.640	5.498	11.512	23.023
	20.00	Z-140 EU	68,900	769,700	0.430	3.640	5.498	11.512	23.023
	20.00	Z-140 IEU	68,900	769,700	0.430	3.640	5.498	11.512	23.023
	20.00	Z-140 EU	68,900	769,700	0.430	3.640	5.498	11.512	23.023
	20.00	Z-140 IEU	68,900	769,700	0.430	3.640	5.498	11.512	23.023
4½	20.00	V-150 IEU	73,800	824,700	0.430	3.640	5.498	11.512	23.023
	20.00	V-150 EU	73,800	824,700	0.430	3.640	5.498	11.512	23.023
	20.00	V-150 IEU	73,800	824,700	0.430	3.640	5.498	11.512	23.023
	20.00	V-150 EU	73,800	824,700	0.430	3.640	5.498	11.512	23.023
	20.00	V-150 IEU	73,800	824,700	0.430	3.640	5.498	11.512	23.023
5	19.50	E-75 IEU	41,200	395,600	0.362	4.276	5.275	14.269	28.538
	19.50	E-75 IEU	41,200	395,600	0.362	4.276	5.275	14.269	28.538
	19.50	E-75 IEU	41,200	395,600	0.362	4.276	5.275	14.269	28.538
	19.50	E-75 IEU	41,200	395,600	0.362	4.276	5.275	14.269	28.538
	19.50	E-75 IEU	41,200	395,600	0.362	4.276	5.275	14.269	28.538
5	19.50	X-95 IEU	52,100	501,100	0.362	4.276	5.275	14.269	28.538
	19.50	X-95 IEU	52,100	501,100	0.362	4.276	5.275	14.269	28.538
	19.50	X-95 IEU	52,100	501,100	0.362	4.276	5.275	14.269	28.538
	19.50	X-95 IEU	52,100	501,100	0.362	4.276	5.275	14.269	28.538
	19.50	X-95 IEU	52,100	501,100	0.362	4.276	5.275	14.269	28.538

(*Continued*)

TABLE 2.6 (*Continued*)

Pipe Data									
Size OD	Nominal Weight	Grade and Upset Type	Torsional Yield Strength	Tensile Yield Strength	Wall Thickness	Nominal ID	Pipe Body Section Area	Pipe Body Moment of Inertia	Pipe Body Polar Moment of Inertia
in.	lb/ft		ft-lb	lb	in.	in.	sq in.	cu in.	cu in.
5	19.50	G-105 IEU	57,600	553,800	0.362	4.276	5.275	14.269	28.538
	19.50	G-105 IEU	57,600	553,800	0.362	4.276	5.275	14.269	28.538
	19.50	G-105 IEU	57,600	553,800	0.362	4.276	5.275	14.269	28.538
	19.50	G-105 IEU	57,600	553,800	0.362	4.276	5.275	14.269	28.538
	19.50	G-105 IEU	57,600	553,800	0.362	4.276	5.275	14.269	28.538
	19.50	G-105 IEU	57,600	553,800	0.362	4.276	5.275	14.269	28.538
5	19.50	S-135 IEU	74,100	712,100	0.362	4.276	5.275	14.269	28.538
	19.50	S-135 IEU	74,100	712,100	0.362	4.276	5.275	14.269	28.538
	19.50	S-135 IEU	74,100	712,100	0.362	4.276	5.275	14.269	28.538
	19.50	S-135 IEU	74,100	712,100	0.362	4.276	5.275	14.269	28.538
	19.50	S-135 IEU	74,100	712,100	0.362	4.276	5.275	14.269	28.538
	19.50	S-135 IEU	74,100	712,100	0.362	4.276	5.275	14.269	28.538
5	19.50	Z-140 IEU	76,800	738,400	0.362	4.276	5.275	14.269	28.538
	19.50	Z-140 IEU	76,800	738,400	0.362	4.276	5.275	14.269	28.538
	19.50	Z-140 IEU	76,800	738,400	0.362	4.276	5.275	14.269	28.538
	19.50	Z-140 IEU	76,800	738,400	0.362	4.276	5.275	14.269	28.538
5	19.50	V-150 IEU	82,300	791,200	0.362	4.276	5.275	14.269	28.538
	19.50	V-150 IEU	82,300	791,200	0.362	4.276	5.275	14.269	28.538
	19.50	V-150 IEU	82,300	791,200	0.362	4.276	5.275	14.269	28.538
	19.50	V-150 IEU	82,300	791,200	0.362	4.276	5.275	14.269	28.538

5	25.60	E-75 IEU	52,300	530,100	0.500	4.000	7.069	18.113	36.226
	25.60	E-75 IEU	52,300	530,100	0.500	4.000	7.069	18.113	36.226
	25.60	E-75 IEU	52,300	530,100	0.500	4.000	7.069	18.113	36.226
	25.60	E-75 IEU	52,300	530,100	0.500	4.000	7.069	18.113	36.226
5	25.60	X-95 IEU	66,200	671,500	0.500	4.000	7.069	18.113	36.226
	25.60	X-95 IEU	66,200	671,500	0.500	4.000	7.069	18.113	36.226
	25.60	X-95 IEU	66,200	671,500	0.500	4.000	7.069	18.113	36.226
	25.60	X-95 IEU	66,200	671,500	0.500	4.000	7.069	18.113	36.226
5	25.60	G-105 IEU	73,200	742,200	0.500	4.000	7.069	18.113	36.226
	25.60	G-105 IEU	73,200	742,200	0.500	4.000	7.069	18.113	36.226
	25.60	G-105 IEU	73,200	742,200	0.500	4.000	7.069	18.113	36.226
	25.60	G-105 IEU	73,200	742,200	0.500	4.000	7.069	18.113	36.226
	25.60	G-105 IEU	73,200	742,200	0.500	4.000	7.069	18.113	36.226
5	25.60	S-135 IEU	94,100	954,300	0.500	4.000	7.069	18.113	36.226
	25.60	S-135 IEU	94,100	954,300	0.500	4.000	7.069	18.113	36.226
	25.60	S-135 IEU	94,100	954,300	0.500	4.000	7.069	18.113	36.226
	25.60	S-135 IEU	94,100	954,300	0.500	4.000	7.069	18.113	36.226
	25.60	S-135 IEU	94,100	954,300	0.500	4.000	7.069	18.113	36.226
5	25.60	Z-140 IEU	97,500	989,600	0.500	4.000	7.069	18.113	36.226
	25.60	Z-140 IEU	97,500	989,600	0.500	4.000	7.069	18.113	36.226
	25.60	Z-140 IEU	97,500	989,600	0.500	4.000	7.069	18.113	36.226
5	25.60	V-150 IEU	104,500	1,060,300	0.500	4.000	7.069	18.113	36.226
	25.60	V-150 IEU	104,500	1,060,300	0.500	4.000	7.069	18.113	36.226
	25.60	V-150 IEU	104,500	1,060,300	0.500	4.000	7.069	18.113	36.226
$5\frac{1}{2}$	21.90	E-75 IEU	50,700	437,100	0.361	4.778	5.828	19.335	38.670
	21.90	E-75 IEU	50,700	437,100	0.361	4.778	5.828	19.335	38.670
	21.90	E-75 IEU	50,700	437,100	0.361	4.778	5.828	19.335	38.670
	21.90	E-75 IEU	50,700	437,100	0.361	4.778	5.828	19.335	38.670

(Continued)

TABLE 2.6 *(Continued)*

Pipe Data									
Size OD	Nominal Weight	Grade and Upset Type	Torsional Yield Strength	Tensile Yield Strength	Wall Thickness	Nominal ID	Pipe Body Section Area	Pipe Body Moment of Inertia	Pipe Body Polar Moment of Inertia
in.	lb/ft		ft-lb	lb	in.	in.	sq in.	cu in.	cu in.
$5\frac{1}{2}$	21.90	X-95 IEU	64,200	553,700	0.361	4.778	5.828	19.335	38.670
	21.90	X-95 IEU	64,200	553,700	0.361	4.778	5.828	19.335	38.670
	21.90	X-95 IEU	64,200	553,700	0.361	4.778	5.828	19.335	38.670
	21.90	X-95 IEU	64,200	553,700	0.361	4.778	5.828	19.335	38.670
$5\frac{1}{2}$	21.90	G-105 IEU	71,000	612,000	0.361	4.778	5.828	19.335	38.670
	21.90	G-105 IEU	71,000	612,000	0.361	4.778	5.828	19.335	38.670
	21.90	G-105 IEU	71,000	612,000	0.361	4.778	5.828	19.335	38.670
	21.90	G-105 IEU	71,000	612,000	0.361	4.778	5.828	19.335	38.670
	21.90	G-105 IEU	71,000	612,000	0.361	4.778	5.828	19.335	38.670
$5\frac{1}{2}$	21.90	S-135 IEU	91,300	786,800	0.361	4.778	5.828	19.335	38.670
	21.90	S-135 IEU	91,300	786,800	0.361	4.778	5.828	19.335	38.670
	21.90	S-135 IEU	91,300	786,800	0.361	4.778	5.828	19.335	38.670
	21.90	S-135 IEU	91,300	786,800	0.361	4.778	5.828	19.335	38.670
	21.90	S-135 IEU	91,300	786,800	0.361	4.778	5.828	19.335	38.670
$5\frac{1}{2}$	21.90	Z-140 IEU	94,700	816,000	0.361	4.778	5.828	19.335	38.670
	21.90	Z-140 IEU	94,700	816,000	0.361	4.778	5.828	19.335	38.670
	21.90	Z-140 IEU	94,700	816,000	0.361	4.778	5.828	19.335	38.670
	21.90	Z-140 IEU	94,700	816,000	0.361	4.778	5.828	19.335	38.670
	21.90	Z-140 IEU	94,700	816,000	0.361	4.778	5.828	19.335	38.670
$5\frac{1}{2}$	21.90	V-150 IEU	101,400	874,200	0.361	4.778	5.828	19.335	38.670
	21.90	V-150 IEU	101,400	874,200	0.361	4.778	5.828	19.335	38.670
	21.90	V-150 IEU	101,400	874,200	0.361	4.778	5.828	19.335	38.670

	21.90	V-150 IEU	101,400	874,200	0.361	4.778	5.828	19.335	38.670
	21.90	V-150 IEU	101,400	874,200	0.361	4.778	5.828	19.335	38.670
$5\frac{1}{2}$	24.70	E-75 IEU	56,600	497,200	0.415	4.670	6.630	21.571	43.141
	24.70	E-75 IEU	56,600	497,200	0.415	4.670	6.630	21.571	43.141
	24.70	E-75 IEU	56,600	497,200	0.415	4.670	6.630	21.571	43.141
	24.70	E-75 IEU	56,600	497,200	0.415	4.670	6.630	21.571	43.141
	24.70	E-75 IEU	56,600	497,200	0.415	4.670	6.630	21.571	43.141
$5\frac{1}{2}$	24.70	X-95 IEU	71,700	629,800	0.415	4.670	6.630	21.571	43.141
	24.70	X-95 IEU	71,700	629,800	0.415	4.670	6.630	21.571	43.141
	24.70	X-95 IEU	71,700	629,800	0.415	4.670	6.630	21.571	43.141
	24.70	X-95 IEU	71,700	629,800	0.415	4.670	6.630	21.571	43.141
$5\frac{1}{2}$	24.70	G-105 IEU	79,200	696,100	0.415	4.670	6.630	21.571	43.141
	24.70	G-105 IEU	79,200	696,100	0.415	4.670	6.630	21.571	43.141
	24.70	G-105 IEU	79,200	696,100	0.415	4.670	6.630	21.571	43.141
	24.70	G-105 IEU	79,200	696,100	0.415	4.670	6.630	21.571	43.141
	24.70	G-105 IEU	79,200	696,100	0.415	4.670	6.630	21.571	43.141
$5\frac{1}{2}$	24.70	S-135 IEU	101,800	895,000	0.415	4.670	6.630	21.571	43.141
	24.70	S-135 IEU	101,800	895,000	0.415	4.670	6.630	21.571	43.141
	24.70	S-135 IEU	101,800	895,000	0.415	4.670	6.630	21.571	43.141
	24.70	S-135 IEU	101,800	895,000	0.415	4.670	6.630	21.571	43.141
	24.70	S-135 IEU	101,800	895,000	0.415	4.670	6.630	21.571	43.141
$5\frac{1}{2}$	24.70	Z-140 IEU	105,600	928,100	0.415	4.670	6.630	21.571	43.141
	24.70	Z-140 IEU	105,600	928,100	0.415	4.670	6.630	21.571	43.141
	24.70	Z-140 IEU	105,600	928,100	0.415	4.670	6.630	21.571	43.141
	24.70	Z-140 IEU	105,600	928,100	0.415	4.670	6.630	21.571	43.141
	24.70	Z-140 IEU	105,600	928,100	0.415	4.670	6.630	21.571	43.141
$5\frac{1}{2}$	24.70	V-150 IEU	113,100	994,400	0.415	4.670	6.630	21.571	43.141
	24.70	V-150 IEU	113,100	994,400	0.415	4.670	6.630	21.571	43.141
	24.70	V-150 IEU	113,100	994,400	0.415	4.670	6.630	21.571	43.141

(Continued)

TABLE 2.6 *(Continued)*

Pipe Data									
Size OD	Nominal Weight	Grade and Upset Type	Torsional Yield Strength	Tensile Yield Strength	Wall Thickness	Nominal ID	Pipe Body Section Area	Pipe Body Moment of Inertia	Pipe Body Polar Moment of Inertia
in.	lb/ft		ft-lb	lb	in.	in.	sq in.	cu in.	cu in.
	24.70	V-150 IEU	113,100	994,400	0.415	4.670	6.630	21.571	43.141
	24.70	V-150 IEU	113,100	994,400	0.415	4.670	6.630	21.571	43.141
$5\frac{7}{8}$	23.40	E-75 IEU	58,600	469,000	0.361	5.153	6.254	23.868	47.737
$5\frac{7}{8}$	23.40	X-95 IEU	74,200	594,100	0.361	5.153	6.254	23.868	47.737
$5\frac{7}{8}$	23.40	G-105 IEU	82,000	656,600	0.361	5.153	6.254	23.868	47.737
$5\frac{7}{8}$	23.40	S-135 IEU	105,500	844,200	0.361	5.153	6.254	23.868	47.737
$5\frac{7}{8}$	23.40	Z-140 IEU	109,400	875,500	0.361	5.153	6.254	23.868	47.737
$5\frac{7}{8}$	23.40	V-150 IEU	117,200	938,000	0.361	5.153	6.254	23.868	47.737
$5\frac{7}{8}$	26.40	E-75 IEU	65,500	533,900	0.415	5.045	7.119	26.680	53.360
$5\frac{7}{8}$	26.40	X-95 IEU	83,000	676,300	0.415	5.045	7.119	26.680	53.360
$5\frac{7}{8}$	26.40	G-105 IEU	91,700	747,400	0.415	5.045	7.119	26.680	53.360
$5\frac{7}{8}$	26.40	S-135 IEU	117,900	961,000	0.415	5.045	7.119	26.680	53.360
$5\frac{7}{8}$	26.40	Z-140 IEU	122,300	996,600	0.415	5.045	7.119	26.680	53.360
$5\frac{7}{8}$	26.40	V-150 IEU	131,000	1,067,800	0.415	5.045	7.119	26.680	53.360
$6\frac{5}{8}$	25.20	E-75 IEU	70,600	489,500	0.330	5.965	6.526	32.416	64.831
	25.20	E-75 IEU	70,600	489,500	0.330	5.965	6.526	32.416	64.831
	25.20	E-75 IEU	70,600	489,500	0.330	5.965	6.526	32.416	64.831

$6\frac{5}{8}$	25.20	X-95 IEU	89,400	620,000	0.330	5.965	6.526	32.416	64.831
	25.20	X-95 IEU	89,400	620,000	0.330	5.965	6.526	32.416	64.831
	25.20	X-95 IEU	89,400	620,000	0.330	5.965	6.526	32.416	64.831
$6\frac{5}{8}$	25.20	G-105 IEU	98,800	685,200	0.330	5.965	6.526	32.416	64.831
	25.20	G-105 IEU	98,800	685,200	0.330	5.965	6.526	32.416	64.831
	25.20	G-105 IEU	98,800	685,200	0.330	5.965	6.526	32.416	64.831
$6\frac{5}{8}$	25.20	S-135 IEU	127,000	881,000	0.330	5.965	6.526	32.416	64.831
	25.20	S-135 IEU	127,000	881,000	0.330	5.965	6.526	32.416	64.831
	25.20	S-135 IEU	127,000	881,000	0.330	5.965	6.526	32.416	64.831
	25.20	S-135 IEU	127,000	881,000	0.330	5.965	6.526	32.416	64.831
$6\frac{5}{8}$	25.20	Z-140 IEU	131,700	913,700	0.330	5.965	6.526	32.416	64.831
	25.20	Z-140 IEU	131,700	913,700	0.330	5.965	6.526	32.416	64.831
	25.20	Z-140 IEU	131,700	913,700	0.330	5.965	6.526	32.416	64.831
	25.20	Z-140 IEU	131,700	913,700	0.330	5.965	6.526	32.416	64.831
$6\frac{5}{8}$	25.20	V-150 IEU	141,200	978,900	0.330	5.965	6.526	32.416	64.831
	25.20	V-150 IEU	141,200	978,900	0.330	5.965	6.526	32.416	64.831
	25.20	V-150 IEU	141,200	978,900	0.330	5.965	6.526	32.416	64.831
	25.20	V-150 IEU	141,200	978,900	0.330	5.965	6.526	32.416	64.831
$6\frac{5}{8}$	27.70	E-75 IEU	76,300	534,200	0.362	5.901	7.123	35.040	70.080
	27.70	E-75 IEU	76,300	534,200	0.362	5.901	7.123	35.040	70.080
	27.70	E-75 IEU	76,300	534,200	0.362	5.901	7.123	35.040	70.080
$6\frac{5}{8}$	27.70	X-95 IEU	96,600	676,700	0.362	5.901	7.123	35.040	70.080
	27.70	X-95 IEU	96,600	676,700	0.362	5.901	7.123	35.040	70.080
	27.70	X-95 IEU	96,600	676,700	0.362	5.901	7.123	35.040	70.080
$6\frac{5}{8}$	27.70	G-105 IEU	106,800	747,900	0.362	5.901	7.123	35.040	70.080
	27.70	G-105 IEU	106,800	747,900	0.362	5.901	7.123	35.040	70.080
	27.70	G-105 IEU	106,800	747,900	0.362	5.901	7.123	35.040	70.080
$6\frac{5}{8}$	27.70	S-135 IEU	137,300	961,600	0.362	5.901	7.123	35.040	70.080
	27.70	S-135 IEU	137,300	961,600	0.362	5.901	7.123	35.040	70.080

(Continued)

TABLE 2.6 (*Continued*)

Pipe Data

Size OD	Nominal Weight	Grade and Upset Type	Torsional Yield Strength	Tensile Yield Strength	Wall Thickness	Nominal ID	Pipe Body Section Area	Pipe Body Moment of Inertia	Pipe Body Polar Moment of Inertia
in.	lb/ft		ft-lb	lb	in.	in.	sq in.	cu in.	cu in.
	27.70	S-135 IEU	137,300	961,600	0.362	5.901	7.123	35.040	70.080
	27.70	S-135 IEU	137,300	961,600	0.362	5.901	7.123	35.040	70.080
$6\frac{5}{8}$	27.70	Z-140 IEU	142,400	997,200	0.362	5.901	7.123	35.040	70.080
	27.70	Z-140 IEU	142,400	997,200	0.362	5.901	7.123	35.040	70.080
	27.70	Z-140 IEU	142,400	997,200	0.362	5.901	7.123	35.040	70.080
	27.70	Z-140 IEU	142,400	997,200	0.362	5.901	7.123	35.040	70.080
$6\frac{5}{8}$	27.70	V-150 IEU	152,600	1,068,400	0.362	5.901	7.123	35.040	70.080
	27.70	V-150 IEU	152,600	1,068,400	0.362	5.901	7.123	35.040	70.080
	27.70	V-150 IEU	152,600	1,068,400	0.362	5.901	7.123	35.040	70.080
	27.70	V-150 IEU	152,600	1,068,400	0.362	5.901	7.123	35.040	70.080

Tool Joint Data

Internal Pressure	Collapse Pressure	Connection Type	Outside Diameter	Inside Diameter	Torsional Yield Strength	Tensile Yield Strength	Make-up Torque	Torsional Ratio Tool Joint to Pipe	* Pin Tong Space
psi	psi		in.	in.	ft-lb	lb	ft-lb		in.
10,500	11,040	NC26	$3\frac{3}{8}$	$1\frac{3}{4}$	6,900	313,700	3,900	1.44	9
10,500	11,040	$2\frac{3}{8}$ OH	$3\frac{3}{8}$	$1\frac{3}{4}$	6,600	294,600	3,700	1.38	9

10,500	11,040	$2\frac{3}{8}$ WO	$3\frac{3}{8}$	$1\frac{7}{8}$	5,400	241,300	3,000	1.13	9
10,500	11,040	$2\frac{3}{8}$ SLH90	$3\frac{3}{8}$	2	5,200	202,900	2,700	1.08	9
13,300	13,984	NC26	$3\frac{3}{8}$	$1\frac{3}{4}$	6,900	313,700	3,900	1.15	9
13,300	13,984	$2\frac{3}{8}$ OH	$3\frac{3}{8}$	$1\frac{3}{4}$	6,600	294,600	3,700	1.10	9
13,300	13,984	$2\frac{3}{8}$ WO	$3\frac{3}{8}$	$1\frac{7}{8}$	5,400	241,300	3,000	0.90	9
13,300	13,984	$2\frac{3}{8}$ SLH90	$3\frac{3}{8}$	$1\frac{7}{8}$	6,400	248,500	3,400	1.07	9
14,700	15,456	NC26	$3\frac{3}{8}$	$1\frac{3}{4}$	6,900	313,700	3,900	1.03	9
14,700	15,456	$2\frac{3}{8}$ OH	$3\frac{3}{8}$	$1\frac{3}{4}$	6,600	294,600	3,700	0.99	9
14,700	15,456	$2\frac{3}{8}$ WO	$3\frac{3}{8}$	$1\frac{7}{8}$	5,400	241,300	3,000	0.81	9
14,700	15,456	$2\frac{3}{8}$ SLH90	$3\frac{3}{8}$	$1\frac{7}{8}$	6,400	248,500	3,400	0.96	9
18,900	19,035	NC26	$3\frac{3}{8}$	$1\frac{3}{4}$	6,900	313,700	3,900	0.80	9
18,900	19,035	$2\frac{3}{8}$ OH	$3\frac{3}{8}$	$1\frac{1}{2}$	8,300	371,200	4,700	0.97	9
18,900	19,035	$2\frac{3}{8}$ WO	$3\frac{3}{8}$	$1\frac{5}{8}$	7,300	323,800	4,000	0.85	9
18,900	19,035	$2\frac{3}{8}$ SLH90	$3\frac{3}{8}$	$1\frac{3}{4}$	7,500	291,200	4,000	0.87	9
19,600	19,588	XT24	$3\frac{1}{8}$	$1\frac{1}{2}$	10,000	261,500	6,000	1.12	10
19,600	19,588	XT26	$3\frac{3}{8}$	$1\frac{3}{4}$	11,600	290,900	7,000	1.30	10
19,600	19,588	HT26	$3\frac{3}{8}$	$1\frac{3}{4}$	8,700	313,700	5,200	0.98	9
21,000	20,661	XT24	$3\frac{1}{8}$	$1\frac{1}{2}$	10,000	261,500	6,000	1.05	10
21,000	20,661	XT26	$3\frac{3}{8}$	$1\frac{3}{4}$	11,600	290,900	7,000	1.22	10
21,000	20,661	HT26	$3\frac{3}{8}$	$1\frac{3}{4}$	8,700	313,700	5,200	0.92	9
15,474	15,599	NC26	$3\frac{3}{8}$	$1\frac{3}{4}$	6,900	313,700	3,900	1.10	9
15,474	15,599	HT26	$3\frac{3}{8}$	$1\frac{3}{4}$	8,700	313,700	5,200	1.38	9
15,474	15,599	$2\frac{3}{8}$ OH	$3\frac{1}{4}$	$1\frac{3}{4}$	6,500	294,600	3,700	1.03	9
15,474	15,599	$2\frac{3}{8}$ SLH90	$3\frac{1}{4}$	$1\frac{13}{16}$	6,900	270,200	3,700	1.10	9

(Continued)

TABLE 2.6 *(Continued)*

Tool Joint Data									
Internal Pressure	Collapse Pressure	Connection Type	Outside Diameter	Inside Diameter	Torsional Yield Strength	Tensile Yield Strength	Make-up Torque	Torsional Ratio Tool Joint to Pipe	* Pin Tong Space
psi	psi		in.	in.	ft-lb	lb	ft-lb		in.
19,600	19,759	NC26	$3\frac{3}{8}$	$3\frac{1}{4}$	6,900	313,700	3,900	0.87	9
19,600	19,759	HT26	$3\frac{3}{8}$	$3\frac{1}{4}$	8,700	313,700	5,200	1.10	9
19,600	19,759	$2\frac{3}{8}$ OH	$3\frac{1}{4}$	$3\frac{1}{4}$	6,500	294,600	3,700	0.82	9
19,600	19,759	$2\frac{3}{8}$ SLH90	$3\frac{1}{4}$	$1\frac{13}{16}$	6,900	270,200	3,700	0.87	9
21,663	21,839	NC26	$3\frac{3}{8}$	$1\frac{3}{4}$	6,900	313,700	3,900	0.78	9
21,663	21,839	HT26	$3\frac{3}{8}$	$1\frac{3}{4}$	8,700	313,700	5,200	0.99	9
21,663	21,839	$2\frac{3}{8}$ OH	$3\frac{1}{4}$	$1\frac{3}{4}$	6,500	294,600	3,700	0.74	9
21,663	21,839	$2\frac{3}{8}$ SLH90	$3\frac{1}{4}$	$1\frac{13}{16}$	6,900	270,200	3,700	0.78	9
27,853	28,079	NC26	$3\frac{5}{8}$	$1\frac{1}{2}$	9,000	390,300	4,900	0.80	9
27,853	28,079	HT26	$3\frac{3}{8}$	$1\frac{3}{4}$	8,700	313,700	5,200	0.77	9
27,853	28,079	$2\frac{3}{8}$ OH	$3\frac{1}{4}$	$1\frac{3}{4}$	6,500	294,600	3,700	0.58	9
27,853	28,079	$2\frac{3}{8}$ SLH90	$3\frac{1}{4}$	$1\frac{13}{16}$	6,900	270,200	3,700	0.61	9
27,853	28,079	GPDS26	$3\frac{1}{2}$	$1\frac{3}{4}$	8,800	313,700	5,300	0.78	9
28,884	29,119	XT24	$3\frac{1}{8}$	$1\frac{1}{2}$	10,000	261,500	6,000	0.85	10
28,884	29,119	XT26	$3\frac{3}{8}$	$1\frac{3}{4}$	11,600	290,900	7,000	0.99	10
28,884	29,119	HT26	$3\frac{3}{8}$	$1\frac{3}{4}$	8,700	313,700	5,200	0.74	9
28,884	29,119	GPDS26	$3\frac{1}{2}$	$1\frac{11}{16}$	9,700	333,900	5,800	0.83	9

30,947	31,199	XT24	$3\frac{1}{8}$	$1\frac{3}{8}$	10,900	295,400	6,500	0.87	10
30,947	31,199	XT26	$3\frac{3}{8}$	$1\frac{3}{4}$	11,600	290,900	7,000	0.93	10
30,947	31,199	HT26	$3\frac{3}{8}$	$1\frac{3}{4}$	8,700	313,700	5,200	0.70	9
30,947	31,199	GPDS26	$3\frac{1}{2}$	$1\frac{5}{8}$	10,500	353,400	6,300	0.84	9
9,907	10,467	NC26	$3\frac{3}{8}$	$1\frac{3}{4}$	6,900	313,700	3,900	0.85	9
9,907	10,467	HT26	$3\frac{3}{8}$	$1\frac{3}{4}$	8,700	313,700	5,200	1.07	9
9,907	10,467	NC31	$4\frac{1}{8}$	$2\frac{5}{32}$	11,500	434,500	6,200	1.42	9
9,907	10,467	$2\frac{7}{8}$ PAC	$3\frac{1}{8}$	$1\frac{1}{2}$	5,700	273,000	3,200	0.70	9
9,907	10,467	$2\frac{7}{8}$ OH	$3\frac{3}{4}$	$2\frac{3}{8}$	6,300	252,100	3,500	0.78	9
9,907	10,467	XT 26	$3\frac{3}{8}$	$1\frac{3}{4}$	11,600	290,900	7,000	1.43	10
9,907	10,467	HT31	4	$2\frac{5}{32}$	14,900	434,600	8,900	1.84	9
9,907	10,467	XT31	4	$2\frac{3}{8}$	12,900	309,100	7,700	1.59	10
12,548	12,940	NC26	$3\frac{1}{2}$	$1\frac{1}{2}$	8,800	390,300	4,900	0.86	9
12,548	12,940	HT26	$3\frac{3}{8}$	$1\frac{3}{4}$	8,700	313,700	5,200	0.85	9
12,548	12,940	NC31	$4\frac{1}{8}$	$2\frac{5}{32}$	11,500	434,500	6,200	1.13	9
12,548	12,940	$2\frac{7}{8}$ PAC	$3\frac{1}{8}$	$1\frac{1}{2}$	5,700	273,000	3,200	0.56	9
12,548	12,940	$2\frac{7}{8}$ OH	$3\frac{3}{4}$	$2\frac{3}{8}$	6,300	252,100	3,500	0.62	9
12,548	12,940	XT26	$3\frac{3}{8}$	$1\frac{3}{4}$	11,600	290,900	7,000	1.14	10
12,548	12,940	HT31	4	$2\frac{5}{32}$	14,900	434,500	8,900	1.46	9
12,548	12,940	XT31	4	$2\frac{3}{8}$	12,900	309,100	7,700	1.26	10
13,869	14,020	NC26	$3\frac{5}{8}$	$1\frac{3}{4}$	7,200	313,700	3,900	0.64	9
13,869	14,020	HT26	$3\frac{3}{8}$	$1\frac{3}{4}$	8,700	313,700	5,200	0.77	9

(*Continued*)

TABLE 2.6 *(Continued)*

		Tool Joint Data							
Internal Pressure	Collapse Pressure	Connection Type	Outside Diameter	Inside Diameter	Torsional Yield Strength	Tensile Yield Strength	Make-up Torque	Torsional Ratio Tool Joint to Pipe	* Pin Tong Space
psi	psi		in.	in.	ft-lb	lb	ft-lb		in.
13,869	14,020	NC31	$4\frac{1}{8}$	$2\frac{5}{32}$	11,500	434,500	6,200	1.02	9
13,869	14,020	$2\frac{7}{8}$ PAC	$3\frac{1}{8}$	$1\frac{7}{8}$	5,700	273,000	3,200	0.50	9
13,869	14,020	$2\frac{7}{8}$ OH	$3\frac{7}{8}$	$2\frac{5}{32}$	8,800	345,500	4,800	0.78	9
13,869	14,020	XT26	$3\frac{3}{8}$	$1\frac{3}{4}$	11,600	290,900	7,000	1.03	10
13,869	14,020	HT31	4	$2\frac{5}{32}$	14,900	434,500	8,900	1.32	9
13,869	14,020	XT31	4	$2\frac{3}{8}$	12,900	309,100	7,700	1.14	10
17,832	17,034	NC26	$3\frac{5}{8}$	$1\frac{1}{2}$	9,000	390,300	4,900	0.62	9
17,832	17,034	HT26	$3\frac{1}{2}$	$1\frac{1}{2}$	12,100	390,300	7,300	0.83	9
17,832	17,034	NC31	$4\frac{1}{8}$	$2\frac{1}{8}$	11,900	447,100	6,400	0.82	9
17,832	17,034	$2\frac{7}{8}$ PAC	$3\frac{1}{8}$	$1\frac{1}{2}$	5,700	273,000	3,200	0.39	9
17,832	17,034	$2\frac{7}{8}$ OH	$3\frac{7}{8}$	$2\frac{5}{32}$	8,800	345,500	4,800	0.61	9
17,832	17,034	XT26	$3\frac{3}{8}$	$1\frac{3}{4}$	11,600	290,900	7,000	0.80	10
17,832	17,034	HT31	4	$2\frac{5}{32}$	14,900	434,500	8,900	1.03	9
17,832	17,034	XT31	4	$2\frac{3}{8}$	12,900	309,100	7,700	0.89	10
18,492	17,500	HT26	$3\frac{1}{2}$	$1\frac{1}{2}$	12,100	390,300	7,300	0.80	9
18,492	17,500	XT26	$3\frac{3}{8}$	$1\frac{3}{4}$	11,600	290,900	7,000	0.77	10
18,492	17,500	HT31	4	$2\frac{5}{32}$	14,900	434,500	8,900	0.99	9
18,492	17,500	XT31	4	$2\frac{3}{8}$	12,900	309,100	7,700	0.85	10

19,813	18,398	HT26	$3\frac{1}{2}$	$1\frac{1}{2}$	12,100	390,300	7,300	0.75	9
19,813	18,398	XT26	$3\frac{3}{8}$	$1\frac{3}{4}$	11,600	290,900	7,000	0.72	10
19,813	18,398	HT31	4	$2\frac{5}{32}$	14,900	434,500	8,900	0.92	9
19,813	18,398	XT31	4	$2\frac{3}{8}$	12,900	309,100	7,700	0.80	10
16,526	16,509	NC31	$4\frac{1}{8}$	$2\frac{5}{32}$	11,500	434,500	6,200	0.99	9
16,526	16,509	NC26	$3\frac{1}{2}$	$1\frac{1}{2}$	8,800	390,300	4,900	0.76	9
16,526	16,509	$2\frac{7}{8}$ PAC	$3\frac{1}{8}$	$1\frac{1}{2}$	5,700	273,000	3,200	0.49	9
16,526	16,509	$2\frac{7}{8}$ OH	$3\frac{7}{8}$	$2\frac{5}{32}$	8,800	345,500	4,800	0.76	9
16,526	16,509	$2\frac{7}{8}$ SLH90	$4\frac{1}{8}$	$2\frac{5}{32}$	11,500	382,800	5,900	0.99	9
16,526	16,509	2-78 HTPAC	$3\frac{1}{8}$	$1\frac{1}{2}$	8,500	273,000	5,100	0.73	9
16,526	16,509	HT26	$3\frac{1}{2}$	$1\frac{1}{2}$	12,100	390,300	7,300	1.04	9
16,526	16,509	HT31	$4\frac{1}{8}$	$2\frac{5}{32}$	16,000	434,500	9,600	1.38	9
16,526	16,509	XT26	$3\frac{1}{2}$	$1\frac{1}{2}$	14,900	367,400	8,900	1.28	10
16,526	16,509	XT31	4	$2\frac{5}{32}$	17,700	402,500	10,600	1.53	10
20,933	20,911	NC31	$4\frac{1}{8}$	2	13,200	495,700	7,100	0.90	9
20,933	20,911	NC26	$3\frac{1}{2}$	$1\frac{1}{2}$	8,800	390,300	4,900	0.60	9
20,933	20,911	$2\frac{7}{8}$ PAC	$3\frac{1}{8}$	$1\frac{1}{2}$	5,700	273,000	3,200	0.39	9
20,933	20,911	$2\frac{7}{8}$ OH	$3\frac{7}{8}$	$2\frac{5}{32}$	8,800	345,500	4,800	0.60	9
20,933	20,911	$2\frac{7}{8}$ SLH90	$4\frac{1}{8}$	$2\frac{5}{32}$	11,500	382,800	5,900	0.79	9
20,933	20,911	2-78 HTPAC	$3\frac{1}{8}$	$1\frac{1}{2}$	8,500	273,000	5,100	0.58	9
20,933	20,911	HT26	$3\frac{1}{2}$	$1\frac{1}{2}$	12,100	390,300	7,300	0.83	9
20,933	20,911	HT31	$4\frac{1}{8}$	$2\frac{5}{32}$	16,000	434,500	9,600	1.10	9
20,933	20,911	XT26	$3\frac{1}{2}$	$1\frac{1}{2}$	14,900	367,400	8,900	1.02	10
20,933	20,911	XT31	4	$2\frac{5}{32}$	17,700	402,500	10,600	1.21	10

(Continued)

TABLE 2.6 (*Continued*)

Tool Joint Data									
Internal Pressure	Collapse Pressure	Connection Type	Outside Diameter	Inside Diameter	Torsional Yield Strength	Tensile Yield Strength	Make-up Torque	Torsional Ratio Tool Joint to Pipe	* Pin Tong Space
psi	psi		in.	in.	ft-lb	lb	ft-lb		in.
23,137	23,112	NC31	$4\frac{1}{8}$	2	13,200	495,700	7,100	0.81	9
23,137	23,112	NC26	$3\frac{1}{2}$	$1\frac{1}{2}$	8,800	390,300	4,900	0.54	9
23,137	23,112	$2\frac{7}{8}$ PAC	$3\frac{1}{8}$	$1\frac{3}{8}$	6,500	306,900	3,600	0.40	9
23,137	23,112	$2\frac{7}{8}$ OH	$3\frac{7}{8}$	$2\frac{5}{32}$	8,800	345,500	4,800	0.54	9
23,137	23,112	$2\frac{7}{8}$ SLH90	$4\frac{1}{8}$	2	13,300	444,000	6,900	0.82	9
23,137	23,112	$2\frac{7}{8}$ HTPAC	$3\frac{1}{8}$	$1\frac{3}{8}$	9,800	306,900	5,900	0.60	9
23,137	23,112	HT26	$3\frac{5}{8}$	$1\frac{1}{2}$	13,100	390,300	7,900	0.81	9
23,137	23,112	HT31	$4\frac{1}{8}$	$2\frac{5}{32}$	16,300	434,500	9,600	0.99	9
23,137	23,112	XT26	$3\frac{1}{2}$	$1\frac{1}{2}$	14,900	367,400	8,900	0.92	10
23,137	23,112	XT31	4	$2\frac{5}{32}$	17,700	402,500	10,600	1.09	10
29,747	29,716	NC31	$4\frac{3}{8}$	$1\frac{5}{8}$	16,900	623,800	9,000	0.81	9
29,747	29,716	NC26	$3\frac{5}{8}$	$1\frac{1}{2}$	9,000	390,300	4,900	0.43	9
29,747	29,716	$2\frac{7}{8}$ PAC	$3\frac{1}{8}$	$1\frac{1}{2}$	5,700	273,000	3,200	0.27	9
29,747	29,716	$2\frac{7}{8}$ OH	$3\frac{7}{8}$	2	10,400	406,700	5,700	0.50	9
29,747	29,716	$2\frac{7}{8}$ SLH90	$4\frac{1}{8}$	2	13,300	444,000	6,900	0.64	9
29,747	29,716	$2\frac{7}{8}$ HTPAC	$3\frac{1}{8}$	$1\frac{1}{2}$	8,500	273,000	5,100	0.41	9
29,747	29,716	HT26	$3\frac{5}{8}$	$1\frac{1}{2}$	13,100	390,300	7,900	0.63	9

29,747	29,716	HT31	$4\frac{1}{8}$	2	18,900	495,700	11,300	0.91	9
29,747	29,716	XT26	$3\frac{1}{2}$	$1\frac{3}{8}$	16,000	401,300	9,600	0.77	10
29,747	29,716	XT31	4	$2\frac{5}{32}$	17,700	402,500	10,600	0.85	10
29,747	29,716	GPDS31	$4\frac{1}{8}$	2	17,200	495,700	10,300	0.83	9
30,849	30,817	HT26	$3\frac{5}{8}$	$1\frac{3}{8}$	14,500	424,100	8,700	0.67	9
30,849	30,817	HT31	$4\frac{1}{8}$	2	18,900	495,700	11,300	0.88	9
30,849	30,817	XT26	$3\frac{1}{2}$	$1\frac{1}{4}$	16,500	432,200	9,900	0.76	10
30,849	30,817	XT31	4	$2\frac{5}{32}$	17,700	402,500	10,600	0.82	10
30,849	30,817	GPDS31	$4\frac{1}{8}$	2	17,200	495,700	10,300	0.80	9
33,052	33,018	HT26	$3\frac{5}{8}$	$1\frac{1}{4}$	15,300	455,100	9,200	0.66	9
33,052	33,018	HT31	$4\frac{1}{8}$	2	18,900	495,700	11,300	0.82	9
33,052	33,018	XT26	$3\frac{1}{2}$	$1\frac{1}{4}$	16,500	432,200	9,900	0.71	10
33,052	33,018	XT31	4	2	20,100	463,700	12,100	0.87	10
33,052	33,018	GPDS31	$4\frac{1}{8}$	$1\frac{7}{8}$	18,200	541,400	10,900	0.79	9
9,525	10,001	NC38	$4\frac{3}{4}$	$2\frac{11}{16}$	18,100	587,300	9,700	1.28	10
9,525	10,001	NC31	$4\frac{1}{8}$	$2\frac{1}{8}$	11,900	447,100	6,400	0.84	9
9,525	10,001	HT31	$4\frac{1}{8}$	$2\frac{1}{8}$	16,600	447,100	10,000	1.18	9
9,525	10,001	HT38	$4\frac{3}{4}$	$2\frac{11}{16}$	25,300	587,300	15,200	1.79	10
9,525	10,001	$3\frac{1}{2}$ SLH90	$4\frac{3}{4}$	$2\frac{11}{16}$	18,688	534,200	11,100	1.33	10
9,525	10,001	XT31	4	$2\frac{1}{8}$	18,300	415,100	11,000	1.30	10
9,525	10,001	XT38	$4\frac{5}{8}$	$2\frac{13}{16}$	24,000	473,000	14,400	1.70	10

(Continued)

TABLE 2.6 (*Continued*)

Tool Joint Data									
Internal Pressure	Collapse Pressure	Connection Type	Outside Diameter	Inside Diameter	Torsional Yield Strength	Tensile Yield Strength	Make-up Torque	Torsional Ratio Tool Joint to Pipe	* Pin Tong Space
psi	psi		in.	in.	ft-lb	lb	ft-lb		in.
12,065	12,077	NC38	$4\frac{3}{4}$	$2\frac{11}{16}$	18,100	587,300	9,700	1.01	10
12,065	12,077	NC31	$4\frac{1}{8}$	2	13,200	495,700	7,100	0.74	9
12,065	12,077	HT31	$4\frac{1}{8}$	$2\frac{1}{8}$	16,600	447,100	10,000	0.93	9
12,065	12,077	HT38	$4\frac{3}{4}$	$2\frac{11}{16}$	25,300	587,300	15,200	1.41	10
12,065	12,077	$3\frac{1}{2}$ SLH90	$4\frac{3}{4}$	$2\frac{11}{16}$	18,700	534,200	11,100	1.04	10
12,065	12,077	XT31	4	$2\frac{1}{8}$	18,300	415,100	11,000	1.02	10
12,065	12,077	XT38	$4\frac{5}{8}$	$2\frac{13}{16}$	24,000	473,000	14,400	1.34	10
13,335	13,055	NC38	$4\frac{3}{4}$	$2\frac{11}{16}$	18,100	587,300	9,700	0.91	10
13,335	13,055	NC31	$4\frac{1}{8}$	2	13,200	495,700	7,100	0.67	9
13,335	13,055	HT31	$4\frac{1}{8}$	2	18,900	495,700	11,300	0.95	9
13,335	13,055	HT38	$4\frac{3}{4}$	$2\frac{11}{16}$	25,300	587,300	15,200	1.28	10
13,335	13,055	$3\frac{1}{2}$ SLH90	$4\frac{3}{4}$	$2\frac{11}{16}$	18,700	534,200	11,100	0.94	10
13,335	13,055	XT31	4	$2\frac{1}{8}$	18,300	415,100	11,000	0.92	10
13,335	13,055	XT38	$4\frac{5}{8}$	$2\frac{13}{16}$	24,000	473,000	14,400	1.21	10
17,145	15,748	NC38	5	$2\frac{9}{16}$	20,300	649,200	10,700	0.80	10
17,145	15,748	NC31	$4\frac{1}{8}$	2	13,200	495,700	7,100	0.52	9
17,145	15,748	HT31	$4\frac{1}{8}$	2	18,900	495,700	11,300	0.74	9
17,145	15,748	HT38	$4\frac{3}{4}$	$2\frac{11}{16}$	25,300	587,300	15,200	0.99	10

17,145	15,748	$3\frac{1}{2}$ SLH90	$4\frac{3}{4}$	$2\frac{9}{16}$	20,900	596,100	12,400	0.82	10
17,145	15,748	XT31	4	2	20,100	463,700	12,100	0.79	10
17,145	15,748	XT38	$4\frac{5}{8}$	$2\frac{13}{16}$	24,000	473,000	14,400	0.94	10
17,780	16,158	HT31	$4\frac{1}{8}$	2	18,900	495,700	11,300	0.72	9
17,780	16,158	HT38	$4\frac{3}{4}$	$2\frac{11}{16}$	25,300	587,300	15,200	0.96	10
17,780	16,158	XT31	4	2	20,100	463,700	12,100	0.79	10
17,780	16,158	XT38	$4\frac{5}{8}$	$2\frac{13}{16}$	24,000	473,000	14,400	0.91	10
19,050	16,943	HT31	$4\frac{1}{4}$	$1\frac{3}{4}$	23,400	584,100	14,000	0.83	9
19,050	16,943	HT38	$4\frac{3}{4}$	$2\frac{11}{16}$	25,300	587,300	15,200	0.89	10
19,050	16,943	XT31	4	2	20,100	463,700	12,100	0.71	10
19,050	16,943	XT38	$4\frac{5}{8}$	$2\frac{13}{16}$	24,000	473,000	14,400	0.85	10
13,800	14,113	NC38	$4\frac{3}{4}$	$2\frac{11}{16}$	18,100	587,300	9,700	0.97	10
13,800	14,113	NC31	$4\frac{1}{8}$	2	13,200	495,700	7,100	0.71	9
13,800	14,113	HT31	$4\frac{1}{8}$	$2\frac{1}{8}$	16,600	447,100	10,000	0.89	9
13,800	14,113	HT38	$4\frac{3}{4}$	$2\frac{11}{16}$	25,300	587,300	15,200	1.36	10
13,800	14,113	$3\frac{1}{2}$ SLH90	$4\frac{3}{4}$	$2\frac{13}{16}$	16,400	469,400	9,800	0.88	10
13,800	14,113	XT31	4	$2\frac{1}{8}$	18,300	415,100	11,000	0.98	10
13,800	14,113	XT38	$4\frac{3}{4}$	$2\frac{11}{16}$	27,900	537,800	16,700	1.50	10
17,480	17,877	NC38	5	$2\frac{9}{16}$	20,300	649,200	10,700	0.86	10
17,480	17,877	NC31	$4\frac{1}{8}$	2	13,200	495,700	7,100	0.56	9
17,480	17,877	HT31	$4\frac{1}{8}$	2	18,900	495,700	11,300	0.80	9
17,480	17,877	HT38	$4\frac{3}{4}$	$2\frac{11}{16}$	25,300	587,300	15,200	1.08	10
17,480	17,877	$3\frac{1}{2}$ SLH90	$4\frac{3}{4}$	$2\frac{11}{16}$	18,700	534,200	11,100	0.80	10

(Continued)

TABLE 2.6 (*Continued*)

Tool Joint Data									
Internal Pressure	Collapse Pressure	Connection Type	Outside Diameter	Inside Diameter	Torsional Yield Strength	Tensile Yield Strength	Make-up Torque	Torsional Ratio Tool Joint to Pipe	* Pin Tong Space
psi	psi		in.	in.	ft-lb	lb	ft-lb		in.
17,480	17,877	XT31	4	$2\frac{1}{8}$	18,300	415,100	11,000	0.78	10
17,480	17,877	XT38	$4\frac{3}{4}$	$2\frac{11}{16}$	27,900	537,800	16,700	1.19	10
19,320	19,758	NC38	5	$2\frac{7}{16}$	22,200	708,100	11,700	0.85	10
19,320	19,758	NC31	$4\frac{1}{8}$	2	13,200	495,700	7,100	0.51	9
19,320	19,758	HT31	$4\frac{1}{8}$	2	18,900	495,700	11,300	0.73	9
19,320	19,758	HT38	$4\frac{3}{4}$	$2\frac{11}{16}$	25,300	587,300	15,200	0.97	10
19,320	19,758	$3\frac{1}{2}$ SLH90	$4\frac{3}{4}$	$2\frac{9}{16}$	20,900	596,100	12,400	0.80	10
19,320	19,758	XT31	$4\frac{1}{8}$	2	20,900	463,700	12,500	0.80	10
19,320	19,758	XT38	$4\frac{3}{4}$	$2\frac{11}{16}$	27,900	537,800	16,700	1.07	10
24,840	25,404	NC38	5	$2\frac{1}{8}$	26,500	842,400	14,000	0.79	10
24,840	25,404	NC31	$4\frac{1}{8}$	2	13,200	495,700	7,100	0.40	9
24,840	25,404	HT31	$4\frac{1}{8}$	2	18,900	495,700	11,300	0.57	9
24,840	25,404	HT38	$4\frac{3}{4}$	$2\frac{9}{16}$	26,900	649,200	16,100	0.81	10
24,840	25,404	$3\frac{1}{2}$ SLH90	$4\frac{3}{4}$	$2\frac{9}{16}$	20,900	596,100	12,400	0.63	10
24,840	25,404	XT31	$4\frac{1}{8}$	$1\frac{7}{8}$	23,100	509,400	13,900	0.69	10
24,840	25,404	XT38	$4\frac{3}{4}$	$2\frac{11}{16}$	27,900	537,800	16,700	0.84	10
24,840	25,404	GPDS38	$4\frac{7}{8}$	$2\frac{9}{16}$	25,700	649,200	15,400	0.77	10

25,760	26,345	HT31	$4\frac{1}{8}$	$1\frac{7}{8}$	19,900	541,400	11,900	0.58	9
25,760	26,345	HT38	$4\frac{3}{4}$	$2\frac{11}{16}$	25,300	587,300	15,200	0.73	10
25,760	26,345	HT31	$4\frac{1}{8}$	$1\frac{3}{4}$	24,800	552,100	14,900	0.72	10
25,760	26,345	XT38	$4\frac{5}{8}$	$2\frac{11}{16}$	27,500	537,800	16,500	0.79	10
25,760	26,345	GPDS38	$4\frac{7}{8}$	$2\frac{7}{8}$	27,400	679,000	16,400	0.79	10
27,600	28,226	HT31	$4\frac{1}{4}$	$1\frac{3}{4}$	23,400	584,100	14,000	0.63	9
27,600	28,226	HT38	$4\frac{3}{4}$	$2\frac{11}{16}$	25,300	587,300	15,200	0.68	10
27,600	28,226	XT31	$4\frac{1}{8}$	$1\frac{3}{4}$	24,800	552,100	14,900	0.67	10
27,600	28,226	XT38	$4\frac{3}{4}$	$2\frac{9}{16}$	31,500	599,600	18,900	0.85	10
27,600	28,226	GPDS38	5	$2\frac{7}{16}$	29,200	708,100	17,500	0.79	10
16,838	16,774	NC38	$4\frac{3}{4}$	$2\frac{9}{16}$	19,200	649,200	10,700	0.91	10
16,838	16,774	HT38	$4\frac{3}{4}$	$2\frac{11}{16}$	25,300	587,300	15,200	1.20	10
16,838	16,774	XT38	$4\frac{3}{4}$	$2\frac{9}{16}$	31,500	599,600	18,900	1.49	10
21,328	21,247	NC38	5	$2\frac{7}{16}$	22,200	708,100	11,700	0.83	10
21,328	21,247	HT38	$4\frac{3}{4}$	$2\frac{11}{16}$	25,300	587,300	15,200	0.95	10
21,328	21,247	XT38	$4\frac{3}{4}$	$2\frac{9}{16}$	31,500	599,600	18,900	1.18	10
23,573	23,484	NC38	5	$2\frac{1}{8}$	26,500	842,400	14,000	0.90	10
23,573	23,484	HT38	$4\frac{3}{4}$	$2\frac{9}{16}$	26,900	649,200	16,100	0.91	10
23,573	23,484	NC40	$5\frac{1}{4}$	$2\frac{9}{16}$	27,800	838,300	14,600	0.94	9
23,573	23,484	XT38	$4\frac{3}{4}$	$2\frac{9}{16}$	31,500	599,600	18,900	1.07	10
30,308	30,194	NC38	5	$2\frac{1}{8}$	26,500	842,400	14,000	0.70	10
30,308	30,194	HT38	$4\frac{3}{4}$	$2\frac{7}{16}$	28,400	708,100	17,000	0.75	10
30,308	30,194	NC40	$5\frac{1}{2}$	$2\frac{1}{4}$	32,900	980,000	17,100	0.87	10

(Continued)

TABLE 2.6 (*Continued*)

Tool Joint Data									
Internal Pressure	Collapse Pressure	Connection Type	Outside Diameter	Inside Diameter	Torsional Yield Strength	Tensile Yield Strength	Make-up Torque	Torsional Ratio Tool Joint to Pipe	* Pin Tong Space
psi	psi		in.	in.	ft-lb	lb	ft-lb		in.
30,308	30,194	XT38	$4\frac{3}{4}$	$2\frac{9}{16}$	31,500	599,600	18,900	0.83	10
30,308	30,194	XT39	5	$2\frac{9}{16}$	40,800	729,700	24,500	1.07	10
30,308	30,194	GPDS38	5	$2\frac{3}{8}$	30,800	736,400	18,500	0.81	10
31,430	31,312	HT38	$4\frac{3}{4}$	$2\frac{7}{16}$	28,400	708,100	17,000	0.72	10
31,430	31,312	XT38	$4\frac{3}{4}$	$2\frac{7}{16}$	34,400	658,500	20,600	0.87	10
31,430	31,312	XT39	5	$2\frac{9}{16}$	40,800	729,700	24,500	1.04	10
31,430	31,312	GPDS38	5	$2\frac{3}{8}$	30,800	736,400	18,500	0.78	10
33,675	33,549	HT38	5	$2\frac{7}{16}$	33,000	708,100	19,800	0.78	10
33,675	33,549	XT38	$4\frac{3}{4}$	$2\frac{7}{16}$	34,400	658,500	20,600	0.82	10
33,675	33,549	XT39	5	$2\frac{9}{16}$	40,800	729,700	24,500	0.97	10
33,675	33,549	GPDS38	5	$2\frac{1}{4}$	33,900	790,900	20,300	0.80	10
8,597	8,381	NC40	$5\frac{1}{4}$	$2\frac{13}{16}$	23,500	711,600	12,400	1.21	9
8,597	8,381	4 SH	$4\frac{3}{4}$	$2\frac{9}{16}$	15,300	512,000	8,100	0.78	9
8,597	8,381	HT38	$4\frac{3}{4}$	$2\frac{11}{16}$	25,300	587,300	15,200	1.30	10
8,597	8,381	XT38	$4\frac{3}{4}$	$2\frac{11}{16}$	27,900	537,800	16,700	1.43	10
8,597	8,381	XT39	5	$2\frac{7}{8}$	31,000	569,500	18,600	1.59	10

10,889	9,978	NC40	$5\frac{1}{4}$	$2\frac{13}{16}$	23,500	711,600	12,400	0.95	9
10,889	9,978	4 SH	$4\frac{3}{4}$	$2\frac{9}{16}$	15,300	512,000	8,100	0.62	9
10,889	9,978	HT38	$4\frac{3}{4}$	$2\frac{11}{16}$	25,300	587,300	15,200	1.02	10
10,889	9,978	XT38	$4\frac{3}{4}$	$2\frac{11}{16}$	27,900	537,800	16,700	1.13	10
10,889	9,978	XT39	5	$2\frac{7}{8}$	31,000	569,500	18,600	1.26	10
12,036	10,708	NC40	$5\frac{1}{4}$	$2\frac{13}{16}$	23,500	711,600	12,400	0.86	9
12,036	10,708	4 SH	$4\frac{3}{4}$	$2\frac{9}{16}$	15,300	512,000	8,100	0.56	9
12,036	10,708	HT38	$4\frac{3}{4}$	$2\frac{9}{16}$	26,900	649,200	16,100	0.99	10
12,036	10,708	XT38	$4\frac{3}{4}$	$2\frac{11}{16}$	27,900	537,800	16,700	1.02	10
12,036	10,708	XT39	5	$2\frac{7}{8}$	31,000	569,500	18,600	1.14	10
15,474	12,618	NC40	$5\frac{1}{2}$	$2\frac{9}{16}$	28,100	838,300	14,600	0.80	9
15,474	12,618	4 SH	$4\frac{3}{4}$	$2\frac{9}{16}$	15,300	512,000	8,100	0.44	9
15,474	12,618	HT38	$4\frac{3}{4}$	$2\frac{7}{16}$	28,400	708,100	17,000	0.81	10
15,474	12,618	XT38	$4\frac{3}{4}$	$2\frac{11}{16}$	27,900	537,800	16,700	0.79	10
15,474	12,618	XT39	5	$2\frac{7}{8}$	31,000	569,500	18,600	0.88	10
16,048	12,894	HT38	$4\frac{3}{4}$	$2\frac{7}{16}$	28,400	708,100	17,000	0.78	10
16,048	12,894	XT38	$4\frac{3}{4}$	$2\frac{11}{16}$	27,000	537,800	16,700	0.77	10
16,048	12,894	XT39	5	$2\frac{7}{8}$	31,000	569,500	18,600	0.85	10
17,194	13,404	HT38	5	$2\frac{7}{16}$	33,000	708,100	19,800	0.85	10
17,194	13,404	XT38	$4\frac{3}{4}$	$2\frac{9}{16}$	31,500	599,600	18,900	0.81	10
17,194	13,404	XT39	5	$2\frac{7}{8}$	31,000	569,500	18,600	0.80	10
10,828	11,354	NC40	$5\frac{1}{4}$	$2\frac{13}{16}$	23,500	711,600	12,400	1.01	9
10,828	11,354	HT38	$4\frac{3}{4}$	$2\frac{11}{16}$	25,300	587,300	15,200	1.09	10
10,828	11,354	4 SH	$4\frac{3}{4}$	$2\frac{7}{16}$	17,100	570,900	9,100	0.73	9

(Continued)

TABLE 2.6 *(Continued)*

		Tool Joint Data							
Internal Pressure	Collapse Pressure	Connection Type	Outside Diameter	Inside Diameter	Torsional Yield Strength	Tensile Yield Strength	Make-up Torque	Torsional Ratio Tool Joint to Pipe	* Pin Tong Space
psi	psi		in.	in.	ft-lb	lb	ft-lb		in.
10,828	11,354	HT40	$5\frac{1}{4}$	$2\frac{13}{16}$	31,900	711,600	19,100	1.37	9
10,828	11,354	NC46	6	$3\frac{1}{4}$	33,600	901,200	17,600	1.44	9
10,828	11,354	XT38	$4\frac{3}{4}$	$2\frac{11}{16}$	27,900	537,800	16,700	1.20	10
10,828	11,354	XT39	5	$2\frac{13}{16}$	33,100	603,000	19,900	1.42	10
13,716	14,382	NC40	$5\frac{1}{4}$	$2\frac{13}{16}$	23,500	711,600	12,400	0.80	9
13,716	14,382	HT38	$4\frac{3}{4}$	$2\frac{11}{16}$	25,300	587,300	15,200	0.86	10
13,716	14,382	4 SH	$4\frac{3}{4}$	$2\frac{7}{16}$	17,100	570,900	9,100	0.58	9
13,716	14,382	HT40	$5\frac{1}{4}$	$2\frac{13}{16}$	31,900	711,600	19,100	1.08	9
13,716	14,382	NC46	6	$3\frac{1}{4}$	33,600	901,200	17,600	1.14	9
13,716	14,382	XT38	$4\frac{3}{4}$	$2\frac{11}{16}$	27,900	537,800	16,700	0.95	10
13,716	14,382	XT39	5	$2\frac{13}{16}$	33,100	603,000	19,900	1.12	10
15,159	15,896	NC40	$5\frac{1}{4}$	$2\frac{13}{16}$	23,500	711,600	12,400	0.72	9
15,159	15,896	HT38	5	$2\frac{9}{16}$	29,600	649,200	17,800	0.91	10
15,159	15,896	4 SH	$4\frac{3}{4}$	$2\frac{7}{16}$	17,100	570,900	9,100	0.52	9
15,159	15,896	HT40	$5\frac{1}{4}$	$2\frac{13}{16}$	31,900	711,600	19,100	0.98	9
15,159	15,896	NC46	6	$3\frac{1}{4}$	33,600	901,200	17,600	1.03	9
15,159	15,896	XT38	$4\frac{3}{4}$	$2\frac{11}{16}$	27,900	537,800	16,700	0.86	10

15,159	15,896	XT39	5	$2\frac{13}{16}$	33,100	603,000	19,900	1.02	10
19,491	20,141	NC40	$5\frac{1}{2}$	$2\frac{9}{16}$	28,100	838,300	14,600	0.67	9
19,491	20,141	HT38	5	$2\frac{7}{16}$	33,000	708,100	19,800	0.79	10
19,491	20,141	4 SH	$4\frac{3}{4}$	$2\frac{7}{16}$	17,100	570,900	9,100	0.41	9
19,491	20,141	HT40	$5\frac{1}{4}$	$2\frac{11}{16}$	35,900	776,400	21,500	0.86	9
19,491	20,141	NC46	6	3	39,200	1,048,400	20,500	0.94	9
19,491	20,141	XT38	5	$2\frac{9}{16}$	31,800	599,600	19,100	0.76	10
19,491	20,141	XT39	5	$2\frac{13}{16}$	33,100	603,000	19,900	0.79	10
19,491	20,141	GPDS40	$5\frac{1}{4}$	$2\frac{11}{16}$	32,700	776,400	19,600	0.78	9
20,213	20,742	HT38	5	$2\frac{7}{16}$	33,000	708,100	19,800	0.76	10
20,213	20,742	HT40	$5\frac{1}{4}$	$2\frac{11}{16}$	35,900	776,400	21,500	0.83	9
20,213	20,742	XT38	5	$2\frac{9}{16}$	31,800	599,600	19,100	0.73	10
20,213	20,742	XT39	5	$2\frac{13}{16}$	33,100	603,000	19,900	0.76	10
20,213	20,742	GPDS40	$5\frac{1}{4}$	$2\frac{5}{8}$	34,600	807,700	20,800	0.80	9
21,656	21,912	HT38	5	$2\frac{7}{16}$	33,000	708,100	19,800	0.71	10
21,656	21,912	HT40	$5\frac{1}{4}$	$2\frac{11}{16}$	35,900	776,400	21,500	0.77	9
21,656	21,912	XT38	5	$2\frac{7}{16}$	35,200	658,500	21,100	0.76	10
21,656	21,912	XT39	5	$2\frac{13}{16}$	33,100	603,000	19,900	0.71	10
21,656	21,912	GPDS40	$5\frac{1}{4}$	$2\frac{1}{2}$	37,300	868,100	22,400	0.80	9
12,469	12,896	NC40	$5\frac{1}{4}$	$2\frac{13}{16}$	23,500	711,600	12,400	0.91	9
12,469	12,896	HT40	$5\frac{1}{4}$	$2\frac{13}{16}$	31,900	711,600	19,100	1.24	9
12,469	12,896	4 H90	$5\frac{1}{2}$	$2\frac{13}{16}$	35,400	913,700	20,400	1.37	9
12,469	12,896	NC46	6	$3\frac{1}{4}$	33,600	901,200	17,600	1.30	9

(Continued)

TABLE 2.6 *(Continued)*

		Tool Joint Data							
Internal Pressure	Collapse Pressure	Connection Type	Outside Diameter	Inside Diameter	Torsional Yield Strength	Tensile Yield Strength	Make-up Torque	Torsional Ratio Tool Joint to Pipe	* Pin Tong Space
psi	psi		in.	in.	ft-lb	lb	ft-lb		in.
12,469	12,896	XT39	5	$2\frac{13}{16}$	33,100	603,000	19,900	1.28	10
12,469	12,896	XT40	$5\frac{1}{4}$	$2\frac{13}{16}$	44,100	751,600	26,500	1.71	10
15,794	16,335	NC40	$5\frac{1}{4}$	$2\frac{9}{16}$	27,800	838,300	14,600	0.85	9
15,794	16,335	HT40	$5\frac{1}{4}$	$2\frac{13}{16}$	31,900	711,600	19,100	0.98	9
15,794	16,335	4 H90	$5\frac{1}{2}$	$2\frac{13}{16}$	35,400	913,700	20,400	1.08	9
15,794	16,335	NC46	6	$3\frac{1}{4}$	33,600	901,200	17,600	1.03	9
15,794	16,335	XT39	5	$2\frac{13}{16}$	33,100	603,000	19,900	1.01	10
15,794	16,335	XT40	$5\frac{1}{4}$	$2\frac{13}{16}$	44,100	751,600	26,500	1.35	10
17,456	18,055	NC40	$5\frac{1}{2}$	$2\frac{7}{16}$	30,100	897,200	15,600	0.83	9
17,456	18,055	HT40	$5\frac{1}{4}$	$2\frac{13}{16}$	31,900	711,600	19,100	0.88	9
17,456	18,055	4 H90	$5\frac{1}{2}$	$2\frac{13}{16}$	35,400	913,700	20,400	0.98	9
17,456	18,055	NC46	6	$3\frac{1}{4}$	33,600	901,200	17,600	0.93	9
17,456	18,055	XT39	5	$2\frac{13}{16}$	33,100	603,000	19,900	0.92	10
17,456	18,055	XT40	$5\frac{1}{4}$	$2\frac{13}{16}$	44,100	751,600	26,500	1.22	10
22,444	23,213	NC40	$5\frac{1}{2}$	2	36,400	1,080,100	18,900	0.78	9
22,444	23,213	HT40	$5\frac{1}{2}$	$2\frac{9}{16}$	39,900	838,300	23,900	0.86	9

22,444	23,213	4 H90	$5\frac{3}{4}$	$2\frac{11}{16}$	38,400	978,500	21,800	0.83	9
22,444	23,213	NC46	6	3	39,200	1,048,400	20,500	0.84	9
22,444	23,213	XT39	5	$2\frac{11}{16}$	37,000	667,800	22,200	0.80	10
22,444	23,213	XT40	$5\frac{1}{4}$	$2\frac{13}{16}$	44,100	751,600	26,500	0.95	10
22,444	23,213	GPDS40	$5\frac{1}{4}$	$2\frac{9}{16}$	36,400	838,300	21,800	0.78	9
23,275	24,073	HT40	$5\frac{1}{2}$	$2\frac{9}{16}$	39,900	838,300	23,900	0.83	9
23,275	24,073	XT39	5	$2\frac{11}{16}$	37,000	667,800	22,200	0.77	10
23,275	24,073	XT40	$5\frac{1}{4}$	$2\frac{13}{16}$	44,100	751,600	26,500	0.91	10
23,275	24,073	GPDS40	$5\frac{3}{8}$	$2\frac{1}{2}$	38,400	868,100	23,000	0.80	9
24,938	25,793	HT40	$5\frac{1}{2}$	$2\frac{9}{16}$	39,900	838,300	23,900	0.77	9
24,938	25,793	XT39	5	$2\frac{9}{16}$	40,800	729,700	24,500	0.79	10
24,938	25,793	XT40	$5\frac{1}{4}$	$2\frac{13}{16}$	44,100	751,600	26,500	0.85	10
24,938	25,793	GPDS40	$5\frac{1}{2}$	$2\frac{7}{16}$	40,300	897,200	24,200	0.78	9
9,829	10,392	NC46	$6\frac{1}{4}$	$3\frac{1}{4}$	34,000	901,200	17,600	1.10	9
9,829	10,392	$4\frac{1}{2}$ OH	$5\frac{7}{8}$	$3\frac{3}{4}$	27,300	714,000	14,600	0.89	9
9,829	10,392	$4\frac{1}{2}$ FH	6	3	34,800	976,200	17,600	1.13	9
9,829	10,392	$4\frac{1}{2}$ H90	6	$3\frac{1}{4}$	39,000	938,400	18,800	1.27	9
9,829	10,392	HT46	$6\frac{1}{4}$	$3\frac{1}{4}$	47,600	901,200	28,600	1.55	9
9,829	10,392	NC50	$6\frac{3}{8}$	$3\frac{3}{4}$	37,700	939,100	19,800	1.22	9
9,829	10,392	HT50	$6\frac{1}{4}$	$3\frac{3}{4}$	52,700	939,100	31,600	1.71	9
9,829	10,392	XT40	$5\frac{1}{4}$	3	37,500	648,900	22,500	1.22	10
9,829	10,392	XT46	6	$3\frac{1}{2}$	58,100	910,300	34,900	1.89	10
9,829	10,392	XT50	$6\frac{3}{8}$	$3\frac{3}{4}$	75,200	1,085,500	45,100	2.44	10
12,450	12,765	NC46	$6\frac{1}{4}$	$3\frac{1}{4}$	34,000	901,200	17,600	0.87	9

(*Continued*)

TABLE 2.6 (Continued)

Tool Joint Data									
Internal Pressure	Collapse Pressure	Connection Type	Outside Diameter	Inside Diameter	Torsional Yield Strength	Tensile Yield Strength	Make-up Torque	Torsional Ratio Tool Joint to Pipe	* Pin Tong Space
psi	psi		in.	in.	ft-lb	lb	ft-lb		in.
12,450	12,765	$4\frac{1}{2}$ OH	$5\frac{7}{8}$	$3\frac{1}{2}$	33,900	884,800	18,200	0.87	9
12,450	12,765	$4\frac{1}{2}$ FH	6	3	34,800	976,200	17,600	0.89	9
12,450	12,765	$4\frac{1}{2}$ H90	6	$3\frac{1}{4}$	39,000	938,400	18,800	1.00	9
12,450	12,765	HT46	$6\frac{1}{4}$	$3\frac{1}{4}$	47,600	901,200	28,600	1.22	9
12,450	12,765	NC50	$6\frac{3}{8}$	$3\frac{3}{4}$	37,700	939,100	19,800	0.97	9
12,450	12,765	HT50	$6\frac{1}{4}$	$3\frac{3}{4}$	52,700	939,100	31,600	1.35	9
12,450	12,765	XT40	$5\frac{1}{4}$	3	37,500	648,900	22,500	0.96	10
12,450	12,765	XT46	6	$3\frac{1}{2}$	58,100	910,300	34,900	1.49	10
12,450	12,765	XT50	$6\frac{3}{8}$	$3\frac{3}{4}$	75,200	1,085,500	45,100	1.93	10
13,761	13,825	NC46	$6\frac{1}{4}$	3	39,700	1,048,400	20,500	0.92	9
13,761	13,825	$4\frac{1}{2}$ OH	6	$3\frac{1}{4}$	40,300	1,043,800	21,500	0.94	9
13,761	13,825	$4\frac{1}{2}$ FH	$6\frac{1}{4}$	$2\frac{3}{4}$	40,200	1,111,600	20,100	0.93	9
13,761	13,825	$4\frac{1}{2}$ H90	6	$3\frac{1}{4}$	39,000	938,400	18,800	0.90	9
13,761	13,825	HT46	$6\frac{1}{4}$	$3\frac{1}{4}$	47,600	901,200	28,600	1.10	9
13,761	13,825	NC50	$6\frac{3}{8}$	$3\frac{3}{4}$	37,700	939,100	19,800	0.87	9
13,761	13,825	HT50	$6\frac{1}{4}$	$3\frac{3}{4}$	52,700	939,100	31,600	1.22	9
13,761	13,825	XT40	$5\frac{1}{4}$	3	37,500	648,900	22,500	0.87	10

13,761	13,825	XT46	6	$3\frac{1}{2}$	58,100	910,300	34,900	1.35	10
13,761	13,825	XT50	$6\frac{3}{8}$	$3\frac{3}{4}$	75,200	1,085,500	45,100	1.74	10
17,693	16,773	NC46	$6\frac{1}{4}$	$2\frac{3}{4}$	44,900	1,183,900	23,200	0.81	9
17,693	16,773	$4\frac{1}{2}$ OH	6	2	43,400	1,191,100	24,600	0.78	9
17,693	16,773	$4\frac{1}{2}$ FH	$6\frac{1}{4}$	$2\frac{3}{4}$	40,200	1,111,600	20,100	0.72	9
17,693	16,773	$4\frac{1}{2}$ H90	$6\frac{1}{4}$	$2\frac{3}{4}$	51,500	1,221,100	24,600	0.93	9
17,693	16,773	HT46	$6\frac{1}{4}$	$3\frac{1}{4}$	47,600	901,200	28,600	0.86	9
17,693	16,773	NC50	$6\frac{3}{8}$	$3\frac{1}{2}$	44,700	1,109,900	23,400	0.81	9
17,693	16,773	HT50	$6\frac{3}{8}$	$3\frac{1}{2}$	65,700	1,109,900	39,400	1.18	9
17,693	16,773	XT40	$5\frac{1}{4}$	$2\frac{13}{16}$	44,100	751,600	26,500	0.79	10
17,693	16,773	XT46	6	$3\frac{1}{2}$	58,100	910,300	34,900	1.05	10
17,693	16,773	XT50	$6\frac{3}{8}$	$3\frac{3}{4}$	75,200	1,085,500	45,100	1.35	10
17,693	16,773	GPDS46	6	$3\frac{1}{4}$	42,900	901,200	25,700	0.77	9
18,348	17,228	HT46	$6\frac{1}{4}$	$3\frac{1}{4}$	47,600	901,200	28,600	0.83	9
18,348	17,228	HT50	$6\frac{3}{8}$	$3\frac{1}{2}$	65,700	1,109,900	39,400	1.14	9
18,348	17,228	XT40	$5\frac{1}{4}$	$2\frac{13}{16}$	44,100	751,600	26,500	0.77	10
18,348	17,228	XT46	6	$3\frac{1}{2}$	58,100	910,300	34,900	1.01	10
18,348	17,228	XT50	$6\frac{3}{8}$	$3\frac{3}{4}$	75,200	1,085,500	45,100	1.31	10
18,348	17,228	GPDS46	6	$3\frac{3}{16}$	45,500	939,100	27,300	0.79	9
19,658	18,103	HT46	$6\frac{1}{4}$	$3\frac{1}{4}$	47,600	901,200	28,600	0.77	9
19,658	18,103	HT50	$6\frac{3}{8}$	$3\frac{1}{2}$	65,700	1,109,900	39,400	1.07	9
19,658	18,103	XT40	$5\frac{1}{4}$	$2\frac{13}{16}$	44,100	751,600	26,500	0.72	10
19,658	18,103	XT46	$6\frac{1}{4}$	$3\frac{1}{4}$	70,200	1,069,300	42,100	1.14	10
19,658	18,103	XT50	$6\frac{3}{8}$	$3\frac{1}{2}$	81,200	1,256,300	48,700	1.32	10

(Continued)

TABLE 2.6 *(Continued)*

Tool Joint Data									
Internal Pressure	Collapse Pressure	Connection Type	Outside Diameter	Inside Diameter	Torsional Yield Strength	Tensile Yield Strength	Make-up Torque	Torsional Ratio Tool Joint to Pipe	* Pin Tong Space
psi	psi		in.	in.	ft-lb	lb	ft-lb		in.
19,658	18,103	GPDS46	6	$3\frac{1}{8}$	48,000	976,300	28,800	0.78	9
12,542	12,964	NC46	$6\frac{1}{4}$	$3\frac{1}{4}$	34,000	901,200	17,600	0.92	9
12,542	12,964	$4\frac{1}{2}$ OH	6	$3\frac{1}{2}$	34,100	884,800	18,200	0.92	9
12,542	12,964	$4\frac{1}{2}$ H90	6	$3\frac{1}{4}$	39,000	938,400	18,800	1.06	9
12,542	12,964	HT46	$6\frac{1}{4}$	$3\frac{1}{4}$	47,600	901,200	28,600	1.29	9
12,542	12,964	NC50	$6\frac{3}{8}$	$3\frac{5}{8}$	41,200	1,026,000	21,600	1.12	9
12,542	12,964	HT50	$6\frac{1}{4}$	$3\frac{3}{4}$	52,700	939,100	31,600	1.43	9
12,542	12,964	XT46	6	$3\frac{1}{2}$	58,100	910,300	34,900	1.57	10
12,542	12,964	XT50	$6\frac{3}{8}$	$3\frac{1}{2}$	81,200	1,256,300	48,700	2.20	10
15,886	16,421	NC46	$6\frac{1}{4}$	3	39,700	1,048,400	20,500	0.85	9
15,886	16,421	$4\frac{1}{2}$ OH	$6\frac{1}{4}$	$3\frac{1}{4}$	40,700	1,043,800	21,500	0.87	9
15,886	16,421	$4\frac{1}{2}$ H90	6	$3\frac{1}{4}$	39,000	938,400	18,800	0.84	9
15,886	16,421	HT46	$6\frac{1}{4}$	$3\frac{1}{4}$	47,600	901,200	28,600	1.02	9
15,886	16,421	NC50	$6\frac{3}{8}$	$3\frac{5}{8}$	41,200	1,026,000	21,600	0.88	9
15,886	16,421	HT50	$6\frac{1}{4}$	$3\frac{3}{4}$	52,700	939,100	31,600	1.13	9
15,886	16,421	XT46	6	$3\frac{1}{2}$	58,100	910,300	34,900	1.24	10
15,886	16,421	XT50	$6\frac{3}{8}$	$3\frac{1}{2}$	81,200	1,256,300	48,700	1.74	10

17,558	18,149	NC46	$6\frac{1}{4}$	$2\frac{3}{4}$	44,900	1,183,900	23,200	0.87	9
17,558	18,149	$4\frac{1}{2}$ OH	$6\frac{1}{4}$	3	46,600	1,191,100	24,600	0.90	9
17,558	18,149	$4\frac{1}{2}$ H90	$6\frac{1}{4}$	3	45,700	1,085,700	21,800	0.88	9
17,558	18,149	HT46	$6\frac{1}{4}$	$3\frac{1}{4}$	47,600	901,200	28,600	0.92	9
17,558	18,149	NC50	$6\frac{3}{8}$	$3\frac{1}{2}$	44,700	1,109,900	23,400	0.86	9
17,558	18,149	HT50	$6\frac{1}{4}$	$3\frac{3}{4}$	52,700	939,100	31,600	1.02	9
17,558	18,149	XT46	6	$3\frac{1}{2}$	58,100	910,300	34,900	1.12	10
17,558	18,149	XT50	$6\frac{3}{8}$	$3\frac{1}{2}$	81,200	1,256,300	48,700	1.57	10
22,575	23,335	NC46	$6\frac{3}{8}$	$2\frac{1}{2}$	49,900	1,307,600	25,600	0.75	9
22,575	23,335	$4\frac{1}{2}$ OH	$6\frac{3}{8}$	$2\frac{3}{4}$	52,200	1,326,600	27,400	0.79	9
22,575	23,335	$4\frac{1}{2}$ H90	$6\frac{3}{8}$	$2\frac{3}{4}$	51,700	1,221,100	24,600	0.78	9
22,575	23,335	HT46	$6\frac{1}{4}$	3	57,700	1,048,400	34,600	0.87	9
22,575	23,335	NC50	$6\frac{1}{2}$	$3\frac{1}{4}$	51,400	1,269,000	26,800	0.77	9
22,575	23,335	HT50	$6\frac{3}{8}$	$3\frac{1}{2}$	65,700	1,109,900	39,400	0.99	9
22,575	23,335	XT46	6	$3\frac{1}{2}$	58,100	910,300	34,900	0.88	10
22,575	23,335	XT50	$6\frac{3}{8}$	$3\frac{1}{2}$	81,200	1,256,300	48,700	1.22	10
22,575	23,335	GPDS46	6	3	52,900	1,048,400	31,700	0.80	9
23,411	24,199	HT46	$6\frac{1}{4}$	3	57,700	1,048,400	34,600	0.84	9
23,411	24,199	HT50	$6\frac{3}{8}$	$3\frac{1}{2}$	65,700	1,109,900	39,400	0.95	9
23,411	24,199	XT46	6	$3\frac{1}{2}$	58,100	910,300	34,900	0.84	10
23,411	24,199	XT50	$6\frac{3}{8}$	$3\frac{1}{2}$	81,200	1,256,300	48,700	1.18	10
23,411	24,199	GPDS46	6	$2\frac{15}{16}$	55,300	1,083,400	33,200	0.80	9
25,083	25,927	HT46	$6\frac{1}{4}$	3	57,700	1,048,400	34,600	0.78	9

(Continued)

TABLE 2.6 *(Continued)*

Tool Joint Data									
Internal Pressure	Collapse Pressure	Connection Type	Outside Diameter	Inside Diameter	Torsional Yield Strength	Tensile Yield Strength	Make-up Torque	Torsional Ratio Tool Joint to Pipe	* Pin Tong Space
psi	psi		in.	in.	ft-lb	lb	ft-lb		in.
25,083	25,927	HT50	$6\frac{3}{8}$	$3\frac{1}{2}$	65,700	1,109,900	39,400	0.89	9
25,083	25,927	XT46	$6\frac{1}{4}$	$3\frac{1}{4}$	70,200	1,069,300	42,100	0.95	10
25,083	25,927	XT50	$6\frac{3}{8}$	$3\frac{1}{2}$	81,200	1,256,300	48,700	1.10	10
25,083	25,927	GPDS46	6	$2\frac{3}{4}$	60,700	1,183,900	36,400	0.82	9
9,503	9,962	NC50	$6\frac{5}{8}$	$3\frac{3}{4}$	38,100	939,100	19,800	0.92	9
9,503	9,962	HT50	$6\frac{5}{8}$	$3\frac{3}{4}$	53,300	939,100	32,000	1.29	9
9,503	9,962	$5\frac{1}{2}$ FH	7	$3\frac{3}{4}$	62,900	1,448,400	33,400	1.53	10
9,503	9,962	XT46	6	$3\frac{1}{2}$	36,500	910,300	21,900	0.89	10
9,503	9,962	XT50	$6\frac{1}{2}$	4	38,700	902,900	23,200	0.94	10
12,037	12,026	NC50	$6\frac{5}{8}$	$3\frac{1}{2}$	45,100	1,109,900	23,400	0.87	9
12,037	12,026	HT50	$6\frac{5}{8}$	$3\frac{3}{4}$	53,300	939,100	32,000	1.02	9
12,037	12,026	$5\frac{1}{2}$ FH	7	$3\frac{3}{4}$	62,900	1,448,400	33,400	1.21	10
12,037	12,026	XT46	6	$3\frac{1}{2}$	58,100	910,300	34,900	1.12	10
12,037	12,026	XT50	$6\frac{1}{2}$	4	62,500	902,900	37,500	1.20	10
13,304	12,999	NC50	$6\frac{5}{8}$	$3\frac{1}{4}$	51,700	1,269,000	26,800	0.90	9

13,304	12,999	HT50	$6\frac{5}{8}$	$3\frac{3}{4}$	53,300	939,100	32,000	0.93	9
13,304	12,999	$5\frac{1}{2}$ FH	7	$3\frac{3}{4}$	62,900	1,448,400	33,400	1.09	10
13,304	12,999	XT46	6	$3\frac{1}{2}$	58,100	910,300	34,900	1.01	10
13,304	12,999	XT50	$6\frac{1}{2}$	4	62,500	902,900	37,500	1.09	10
13,304	12,999	GPDS50	$6\frac{1}{2}$	$3\frac{3}{4}$	47,500	939,400	28,500	0.82	9
17,105	15,672	NC50	$6\frac{5}{8}$	$2\frac{3}{4}$	63,400	1,551,700	32,900	0.86	9
17,105	15,672	HT50	$6\frac{5}{8}$	$3\frac{1}{2}$	66,200	1,109,900	39,700	0.89	9
17,105	15,672	$5\frac{1}{2}$ FH	$7\frac{1}{4}$	$3\frac{1}{2}$	72,500	1,619,200	37,400	0.98	10
17,105	15,672	XT46	6	$3\frac{1}{2}$	58,100	910,300	34,900	0.78	10
17,105	15,672	XT50	$6\frac{1}{2}$	$3\frac{3}{4}$	77,000	1,085,500	46,200	1.04	10
17,105	15,672	GPDS50	$6\frac{1}{2}$	$3\frac{1}{2}$	60,200	1,110,200	36,100	0.81	9
17,738	16,079	HT50	$6\frac{5}{8}$	$3\frac{1}{2}$	66,200	1,109,900	39,700	0.86	9
17,738	16,079	XT46	6	$3\frac{1}{2}$	58,100	910,300	34,900	0.76	10
17,738	16,079	XT50	$6\frac{1}{2}$	$3\frac{3}{4}$	77,000	1,085,500	46,200	1.00	10
17,738	16,079	GPDS50	$6\frac{1}{2}$	$3\frac{1}{2}$	60,200	1,110,200	36,100	0.78	9
19,005	16,858	HT50	$6\frac{5}{8}$	$3\frac{1}{2}$	66,200	1,109,900	39,700	0.80	9
19,005	16,858	XT46	$6\frac{1}{4}$	$3\frac{1}{4}$	70,200	1,069,300	42,100	0.85	10
19,005	16,858	XT50	$6\frac{1}{2}$	$3\frac{3}{4}$	77,000	1,085,500	46,200	0.94	10
19,005	16,858	GPDS50	$6\frac{1}{2}$	$3\frac{3}{8}$	66,200	1,191,200	39,700	0.80	9
13,125	13,500	NC50	$6\frac{5}{8}$	$3\frac{1}{2}$	45,100	1,109,900	23,400	0.86	9
13,125	13,500	HT50	$6\frac{5}{8}$	$3\frac{3}{4}$	53,300	939,100	32,000	1.02	9
13,125	13,500	$5\frac{1}{2}$ FH	7	$3\frac{1}{2}$	62,900	1,619,200	37,400	1.20	10
13,125	13,500	XT50	$6\frac{5}{8}$	$3\frac{3}{4}$	77,300	1,085,500	46,400	1.48	10

(Continued)

TABLE 2.6 *(Continued)*

Tool Joint Data									
Internal Pressure	Collapse Pressure	Connection Type	Outside Diameter	Inside Diameter	Torsional Yield Strength	Tensile Yield Strength	Make-up Torque	Torsional Ratio Tool Joint to Pipe	* Pin Tong Space
psi	psi		in.	in.	ft-lb	lb	ft-lb		in.
16,625	17,100	NC50	$6\frac{5}{8}$	3	57,800	1,416,200	30,000	0.87	9
16,625	17,100	HT50	$6\frac{5}{8}$	$3\frac{1}{2}$	66,200	1,109,900	39,700	1.00	9
16,625	17,100	$5\frac{1}{2}$ FH	7	$3\frac{1}{2}$	62,900	1,619,200	37,400	0.95	10
16,625	17,100	XT50	$6\frac{5}{8}$	$3\frac{3}{4}$	77,300	1,085,500	46,400	1.17	10
18,375	18,900	NC50	$6\frac{5}{8}$	$2\frac{3}{4}$	63,400	1,551,700	32,900	0.87	9
18,375	18,900	HT50	$6\frac{5}{8}$	$3\frac{1}{4}$	78,000	1,269,000	46,800	1.07	9
18,375	18,900	$5\frac{1}{2}$ FH	$7\frac{1}{4}$	$3\frac{1}{2}$	72,500	1,619,200	37,400	0.99	10
18,375	18,900	XT50	$6\frac{5}{8}$	$3\frac{3}{4}$	77,300	1,085,500	46,400	1.06	10
18,375	18,900	GPDS50	$6\frac{1}{2}$	$3\frac{1}{2}$	60,200	1,110,200	36,100	0.82	9
23,625	24,300	NC50	$6\frac{5}{8}$	$2\frac{3}{4}$	63,400	1,551,700	32,900	0.67	9
23,625	24,300	HT50	$6\frac{5}{8}$	3	88,800	1,416,200	53,300	0.94	9
23,625	24,300	$5\frac{1}{2}$ FH	$7\frac{1}{4}$	$3\frac{1}{4}$	78,700	1,778,300	41,200	0.84	10
23,625	24,300	XT50	$6\frac{5}{8}$	$3\frac{3}{4}$	77,300	1,085,500	46,400	0.82	10
23,625	24,300	GPDS50	$6\frac{1}{2}$	$3\frac{3}{16}$	74,700	1,307,200	44,800	0.79	9
24,500	25,200	HT50	$6\frac{5}{8}$	3	88,800	1,416,200	53,300	0.91	9
24,500	25,200	XT50	$6\frac{5}{8}$	$3\frac{3}{4}$	77,300	1,085,500	46,400	0.79	10
24,500	25,200	GPDS50	$6\frac{1}{2}$	$3\frac{1}{8}$	77,400	1,344,300	46,400	0.79	9

26,250	27,000	HT50	$6\frac{5}{8}$	3	88,800	1,416,200	53,300	0.85	9
26,250	27,000	XT50	$6\frac{5}{8}$	$3\frac{1}{2}$	90,700	1,256,300	54,400	0.87	10
26,250	27,000	GPDS50	$6\frac{5}{8}$	3	82,900	1,416,500	49,700	0.79	9
8,413	8,615	$5\frac{1}{2}$ FH	7	4	57,900	1,265,800	31,200	1.14	10
8,413	8,615	HT55	7	4	77,200	1,265,800	46,300	1.52	10
8,413	8,615	XT54	$6\frac{3}{4}$	$4\frac{1}{4}$	70,400	960,700	42,200	1.39	10
8,413	8,615	XT57	7	$4\frac{3}{8}$	85,600	1,107,100	51,400	1.69	10
10,019	10,912	$5\frac{1}{2}$ FH	7	$3\frac{3}{4}$	65,100	1,448,400	35,700	1.01	10
10,019	10,912	HT55	7	4	77,200	1,265,800	46,300	1.20	10
10,019	10,912	XT54	$6\frac{3}{4}$	$4\frac{1}{4}$	70,400	960,700	42,200	1.10	10
10,019	10,912	XT57	7	$4\frac{3}{8}$	85,600	1,107,100	51,400	1.33	10
10,753	12,061	$5\frac{1}{2}$ FH	$7\frac{1}{4}$	$3\frac{1}{2}$	75,000	1,691,200	40,000	1.06	10
10,753	12,061	HT55	7	4	77,200	1,265,800	46,300	1.09	10
10,753	12,061	XT54	$6\frac{3}{4}$	$4\frac{1}{4}$	70,400	960,700	42,200	0.99	10
10,753	12,061	XT57	7	$4\frac{3}{8}$	85,600	1,107,100	51,400	1.21	10
10,753	12,061	GPDS55	7	4	74,200	1,292,500	44,500	1.05	10
12,679	15,507	$5\frac{1}{2}$ FH	$7\frac{1}{2}$	3	90,200	1,925,500	47,700	0.99	10
12,679	15,507	HT55	7	4	77,200	1,265,800	46,300	0.85	10
12,679	15,507	XT54	$6\frac{3}{4}$	$4\frac{1}{4}$	70,400	960,700	42,200	0.77	10
12,679	15,507	XT57	7	$4\frac{3}{8}$	85,600	1,107,100	51,400	0.94	10
12,679	15,507	GPDS55	7	4	74,200	1,292,500	44,500	0.81	10
12,957	16,081	$5\frac{1}{2}$ FH	$7\frac{1}{2}$	3	90,200	1,925,500	47,700	0.95	10
12,957	16,081	HT55	7	4	77,200	1,265,800	46,300	0.82	10
12,957	16,081	XT54	$6\frac{3}{4}$	$4\frac{1}{4}$	70,400	960,700	42,200	0.74	10

(Continued)

TABLE 2.6 *(Continued)*

Tool Joint Data									
Internal Pressure	Collapse Pressure	Connection Type	Outside Diameter	Inside Diameter	Torsional Yield Strength	Tensile Yield Strength	Make-up Torque	Torsional Ratio Tool Joint to Pipe	* Pin Tong Space
psi	psi		in.	in.	ft-lb	lb	ft-lb		in.
12,957	16,081	XT57	7	$4\frac{3}{8}$	85,600	1,107,100	51,400	0.90	10
12,957	16,081	GPDS55	7	4	74,200	1,292,500	44,500	0.78	10
13,473	17,230	$5\frac{1}{2}$ FH	$7\frac{1}{2}$	3	90,200	1,925,500	47,700	0.89	10
13,473	17,230	HT55	7	4	77,200	1,265,800	46,300	0.76	10
13,473	17,230	XT54	$6\frac{3}{4}$	4	86,600	1,155,100	52,000	0.85	10
13,473	17,230	XT57	7	$4\frac{1}{4}$	94,300	1,208,700	56,600	0.93	10
13,473	17,230	GPDS55	$7\frac{1}{8}$	$3\frac{7}{8}$	82,000	1,385,200	49,200	0.81	10
10,464	9,903	$5\frac{1}{2}$ FH	7	4	57,900	1,265,800	31,200	1.02	10
10,464	9,903	HT55	7	4	77,200	1,265,800	46,300	1.36	10
10,464	9,903	XT54	$6\frac{3}{4}$	$4\frac{1}{4}$	70,400	960,700	42,200	1.24	10
10,464	9,903	XT57	7	$4\frac{3}{8}$	85,600	1,107,100	51,400	1.51	10
12,933	12,544	$5\frac{1}{2}$ FH	$7\frac{1}{4}$	$3\frac{1}{2}$	75,000	1,619,200	40,000	1.05	10
12,933	12,544	HT55	7	4	77,200	1,265,800	46,300	1.08	10
12,933	12,544	XT54	$6\frac{3}{4}$	$4\frac{1}{4}$	70,400	960,700	42,200	0.98	10
12,933	12,544	XT57	7	$4\frac{3}{8}$	85,600	1,107,100	51,400	1.19	10
14,013	13,865	$5\frac{1}{2}$ FH	$7\frac{1}{4}$	$3\frac{1}{2}$	75,000	1,619,200	40,000	0.95	10
14,013	13,865	HT55	7	4	77,200	1,265,800	46,300	0.97	10

14,013	13,865	XT54	$6\frac{3}{4}$	$4\frac{1}{4}$	70,400	960,700	42,200	0.89	10
14,013	13,865	XT57	7	$4\frac{3}{8}$	85,600	1,107,100	51,400	1.08	10
14,013	13,865	GPDS55	7	4	74,200	1,292,500	44,500	0.94	10
17,023	17,826	$5\frac{1}{2}$ FH	$7\frac{1}{2}$	3	90,200	1,925,500	47,700	0.89	10
17,023	17,826	HT55	7	$3\frac{3}{4}$	87,700	1,448,400	52,600	0.86	10
17,023	17,826	XT54	$6\frac{3}{4}$	4	86,600	1,155,100	52,000	0.85	10
17,023	17,826	XT57	7	$4\frac{1}{4}$	94,300	1,208,700	56,600	0.93	10
17,023	17,826	GPDS55	$7\frac{1}{8}$	$3\frac{7}{8}$	82,000	1,385,200	49,200	0.81	10
17,489	18,486	$5\frac{1}{2}$ FH	$7\frac{1}{4}$	$3\frac{1}{2}$	75,000	1,691,200	40,000	0.71	10
17,489	18,486	HT55	7	4	77,200	1,265,800	46,300	0.73	10
17,489	18,486	XT54	$6\frac{3}{4}$	4	86,600	1,155,100	52,000	0.82	10
17,489	18,486	XT57	7	$4\frac{1}{4}$	94,300	1,208,700	56,600	0.89	10
17,489	18,486	GPDS55	$7\frac{1}{8}$	$3\frac{7}{8}$	82,000	1,385,200	49,200	0.78	10
18,386	19,807	$5\frac{1}{2}$ FH	$7\frac{1}{2}$	3	90,200	1,925,500	47,700	0.80	10
18,386	19,807	HT55	7	$3\frac{3}{4}$	87,700	1,448,400	52,600	0.78	10
18,386	19,807	XT54	$6\frac{3}{4}$	4	86,600	1,155,100	52,000	0.77	10
18,386	19,807	XT57	7	$4\frac{1}{4}$	94,300	1,208,700	56,600	0.83	10
18,386	19,807	GPDS55	$7\frac{1}{8}$	$3\frac{3}{4}$	89,300	1,475,100	53,600	0.79	10
7,453	8,065	XT57	7	$4\frac{1}{4}$	94,300	1,208,700	56,600	1.61	10
8,775	10,216	XT57	7	$4\frac{1}{4}$	94,300	1,208,700	56,600	1.27	10
9,362	11,291	XT57	7	$4\frac{1}{4}$	94,300	1,208,700	56,600	1.15	10
10,825	14,517	XT57	7	$4\frac{1}{4}$	94,300	1,208,700	56,600	0.89	10
11,023	15,054	XT57	7	$4\frac{1}{4}$	94,300	1,208,700	56,600	0.86	10

(Continued)

TABLE 2.6 *(Continued)*

Tool Joint Data									
Internal Pressure	Collapse Pressure	Connection Type	Outside Diameter	Inside Diameter	Torsional Yield Strength	Tensile Yield Strength	Make-up Torque	Torsional Ratio Tool Joint to Pipe	* Pin Tong Space
psi	psi		in.	in.	ft-lb	lb	ft-lb		in.
11,376	16,130	XT57	7	$4\frac{1}{4}$	94,300	1,208,700	56,600	0.80	10
9,558	9,271	XT57	7	$4\frac{1}{4}$	94,300	1,208,700	56,600	1.44	10
11,503	11,744	XT57	7	$4\frac{1}{4}$	94,300	1,208,700	56,600	1.14	10
12,414	12,980	XT57	7	$4\frac{1}{4}$	94,300	1,208,700	56,600	1.03	10
14,892	16,688	XT57	7	$4\frac{1}{4}$	94,300	1,208,700	56,600	0.80	10
15,266	17,306	XT57	7	$4\frac{1}{4}$	94,300	1,208,700	56,600	0.77	10
15,976	18,543	XT57	7	4	106,200	1,403,100	63,700	0.81	10
4,788	6,538	$6\frac{5}{8}$ FH	8	5	73,700	1,448,400	38,400	1.04	10
4,788	6,538	HT65	8	5	99,700	1,448,400	59,800	1.41	10
4,788	6,538	XT65	8	5	135,300	1,543,700	81,200	1.92	10
5,321	8,281	$6\frac{5}{8}$ FH	8	5	73,700	1,448,400	38,400	0.82	10
5,321	8,281	HT65	8	5	99,700	1,448,400	59,800	1.12	10
5,321	8,281	XT65	8	5	135,300	1,543,700	81,200	1.51	10
5,500	9,153	$6\frac{5}{8}$ FH	$8\frac{1}{4}$	$4\frac{3}{4}$	86,200	1,678,100	44,600	0.87	10
5,500	9,153	HT65	8	5	99,700	1,448,400	59,800	1.01	10
5,500	9,153	XT65	8	5	135,300	1,543,700	81,200	1.37	10
6,036	11,768	$6\frac{5}{8}$ FH	$8\frac{1}{2}$	$4\frac{1}{4}$	109,200	2,102,300	56,100	0.86	10

6,036	11,768	HT65	8	5	99,700	1,448,400	59,800	0.79	10
6,036	11,768	XT65	8	5	135,300	1,543,700	81,200	1.07	10
6,036	11,768	GPDS65	8	$4\frac{15}{16}$	102,000	1,538,600	61,200	0.80	10
6,121	12,204	$6\frac{5}{8}$ FH	$8\frac{1}{2}$	$4\frac{1}{4}$	109,200	2,102,300	56,100	0.83	10
6,121	12,204	HT65	8	5	99,700	1,448,400	59,800	0.76	10
6,121	12,204	XT65	8	5	135,300	1,543,700	81,200	1.03	10
6,121	12,204	GPDS65	8	$4\frac{7}{8}$	107,500	1,596,400	64,500	0.82	10
6,260	13,075	$6\frac{5}{8}$ FH	$8\frac{1}{2}$	$4\frac{1}{4}$	109,200	2,102,300	56,100	0.77	10
6,260	13,075	HT65	8	5	99,700	1,448,400	59,800	0.71	10
6,260	13,075	XT65	8	5	135,300	1,543,700	81,200	0.96	10
6,260	13,075	GPDS65	$8\frac{1}{4}$	$4\frac{3}{4}$	119,000	1,709,800	71,400	0.84	10
5,894	7,172	$6\frac{5}{8}$ FH	8	5	73,700	1,448,400	38,400	0.97	10
5,894	7,172	HT65	8	5	99,700	1,448,400	59,800	1.31	10
5,894	7,172	XT65	8	5	135,300	1,543,700	81,200	1.77	10
6,755	9,084	$6\frac{5}{8}$ FH	$8\frac{1}{4}$	$4\frac{3}{4}$	86,200	1,678,100	44,600	0.89	10
6,755	9,084	HT65	8	5	99,700	1,448,400	59,800	1.03	10
6,755	9,084	XT65	8	5	135,300	1,543,700	81,200	1.40	10
7,103	10,040	$6\frac{5}{8}$ FH	$8\frac{1}{4}$	$4\frac{3}{4}$	86,200	1,678,100	44,600	0.81	10
7,103	10,040	HT65	8	5	99,700	1,448,400	59,800	0.93	10
7,103	10,040	XT65	8	5	135,300	1,543,700	81,200	1.27	10
7,813	12,909	$6\frac{5}{8}$ FH	$8\frac{1}{2}$	$4\frac{1}{4}$	109,200	2,102,300	56,100	0.80	10
7,813	12,909	HT65	8	5	99,700	1,448,400	59,800	0.73	10
7,813	12,909	XT65	8	5	135,300	1,543,700	81,200	0.99	10

(Continued)

TABLE 2.6 *(Continued)*

Tool Joint Data									
Internal Pressure	Collapse Pressure	Connection Type	Outside Diameter	Inside Diameter	Torsional Yield Strength	Tensile Yield Strength	Make-up Torque	Torsional Ratio Tool Joint to Pipe	* Pin Tong Space
psi	psi		in.	in.	ft-lb	lb	ft-lb		in.
7,813	12,909	GPDS65	8	$4\frac{7}{8}$	107,500	1,596,400	64,500	0.78	10
7,881	13,387	$6\frac{5}{8}$ FH	$8\frac{1}{2}$	$4\frac{1}{4}$	109,200	2,102,300	56,100	0.77	10
7,881	13,387	HT65	8	5	99,700	1,448,400	59,800	0.70	10
7,881	13,387	XT65	8	5	135,300	1,543,700	81,200	0.95	10
7,881	13,387	GPDS65	$8\frac{1}{4}$	$4\frac{3}{4}$	119,000	1,709,800	71,400	0.84	10
7,970	14,343	$6\frac{5}{8}$ FH	$8\frac{1}{2}$	$4\frac{1}{4}$	109,200	2,102,300	56,100	0.72	10
7,970	14,343	HT65	8	5	99,700	1,448,400	59,800	0.65	10
7,970	14,343	XT65	8	5	135,300	1,543,700	81,200	0.89	10
7,970	14,343	GPDS65	$8\frac{1}{4}$	$4\frac{3}{4}$	119,000	1,709,800	71,400	0.78	10

Assembly Data						
* Box Tong Space	Adjusted Weight	Minimum Tool Joint O.D. for Prem. Class	Drift Diameter	Capacity	Displacement	Size O.D.
in.	lb/ft	in.	in.	US gal/ft	US gal/ft	in.
10	5.52	$3\frac{1}{8}$	$1\frac{5}{8}$	0.160	0.085	$2\frac{3}{8}$
10	5.52	3	$1\frac{5}{8}$	0.160	0.085	

10	5.42	$3\frac{1}{16}$	$1\frac{3}{4}$	0.161	0.083	
10	5.30	$2\frac{31}{32}$	$1\frac{7}{8}$	0.162	0.081	
10	5.52	$3\frac{3}{16}$	$1\frac{5}{8}$	0.160	0.085	$2\frac{3}{8}$
10	5.52	$3\frac{1}{16}$	$1\frac{5}{8}$	0.160	0.085	
10	5.42	$3\frac{1}{8}$	$1\frac{3}{4}$	0.161	0.083	
10	5.42	3	$1\frac{3}{4}$	0.161	0.083	
10	5.52	$3\frac{7}{32}$	$1\frac{5}{8}$	0.160	0.085	$2\frac{3}{8}$
10	5.52	$3\frac{3}{32}$	$1\frac{5}{8}$	0.160	0.085	
10	5.42	$3\frac{5}{32}$	$1\frac{3}{4}$	0.161	0.083	
10	5.42	$3\frac{1}{32}$	$1\frac{3}{4}$	0.161	0.083	
10	5.52	$3\frac{9}{32}$	$1\frac{5}{8}$	0.160	0.085	$2\frac{3}{8}$
10	5.72	$3\frac{3}{16}$	$1\frac{3}{8}$	0.158	0.087	
10	5.63	$3\frac{7}{32}$	$1\frac{1}{2}$	0.159	0.086	
10	5.52	$3\frac{1}{8}$	$1\frac{5}{8}$	0.160	0.085	
15	5.70	$2\frac{3}{4}$	$1\frac{3}{8}$	0.157	0.087	$2\frac{3}{8}$
15	5.79	$2\frac{15}{16}$	$1\frac{5}{8}$	0.160	0.089	
12	5.61	NA	$1\frac{5}{8}$	0.160	0.086	
15	5.70	$2\frac{25}{32}$	$1\frac{3}{8}$	0.157	0.087	$2\frac{3}{8}$
15	5.79	$2\frac{31}{32}$	$1\frac{5}{8}$	0.160	0.089	
12	5.61	NA	$1\frac{5}{8}$	0.160	0.086	
10	7.19	$3\frac{3}{16}$	$1\frac{5}{8}$	0.134	0.110	$2\frac{3}{8}$
12	7.27	N/A	$1\frac{5}{8}$	0.134	0.111	
10	7.07	$3\frac{1}{16}$	$1\frac{5}{8}$	0.134	0.108	

(Continued)

TABLE 2.6 (*Continued*)

Assembly Data						
* Box Tong Space	Adjusted Weight	Minimum Tool Joint O.D. for Prem. Class	Drift Diameter	Capacity	Displacement	Size O.D.
in.	lb/ft	in.	in.	US gal/ft	US gal/ft	in.
10	7.02	$3\frac{1}{32}$	$1\frac{11}{16}$	0.134	0.107	
10	7.19	$3\frac{1}{4}$	$1\frac{5}{8}$	0.134	0.110	$2\frac{3}{8}$
12	7.27	NA	$1\frac{5}{8}$	0.134	0.111	
10	7.07	$3\frac{5}{32}$	$1\frac{5}{8}$	0.134	0.108	
10	7.02	$3\frac{3}{32}$	$1\frac{11}{16}$	0.134	0.107	
10	7.19	$3\frac{9}{32}$	$1\frac{5}{8}$	0.134	0.110	$2\frac{3}{8}$
12	7.27	NA	$1\frac{5}{8}$	0.134	0.111	
10	7.07	$3\frac{3}{16}$	$1\frac{5}{8}$	0.134	0.108	
10	7.02	$3\frac{1}{8}$	$1\frac{11}{16}$	0.134	0.107	
10	7.65	$3\frac{13}{32}$	$1\frac{3}{8}$	0.132	0.117	$2\frac{3}{8}$
12	7.27	NA	$1\frac{5}{8}$	0.134	0.111	
10	7.07	NA	$1\frac{5}{8}$	0.134	0.108	
10	7.02	$3\frac{7}{32}$	$1\frac{11}{16}$	0.134	0.107	
10	7.32	$3\frac{11}{32}$	$1\frac{5}{8}$	0.134	0.112	
15	7.35	$2\frac{29}{32}$	$1\frac{3}{8}$	0.131	0.112	$2\frac{3}{8}$
15	7.43	$3\frac{1}{16}$	$1\frac{5}{8}$	0.134	0.114	
12	7.27	NA	$1\frac{5}{8}$	0.134	0.111	

10	7.37	$3\frac{5}{16}$	$1\frac{9}{16}$	0.133	0.113	
15	7.44	$2\frac{29}{32}$	$1\frac{1}{4}$	0.130	0.114	$2\frac{3}{8}$
15	7.43	$3\frac{3}{32}$	$1\frac{5}{8}$	0.134	0.114	
12	7.27	NA	$1\frac{5}{8}$	0.134	0.111	
10	7.42	$3\frac{11}{32}$	$1\frac{1}{2}$	0.133	0.114	
10	7.24	$3\frac{9}{32}$	$1\frac{5}{8}$	0.237	0.111	$2\frac{7}{8}$
12	7.32	NA	$1\frac{5}{8}$	0.236	0.112	
11	7.93	$3\frac{11}{16}$	$2\frac{1}{32}$	0.240	0.121	
10	7.18	3	$1\frac{3}{8}$	0.235	0.110	
11	7.23	$3\frac{1}{2}$	$2\frac{1}{4}$	0.242	0.111	
15	7.48	$2\frac{19}{32}$	$1\frac{5}{8}$	0.235	0.114	
13	7.89	$3\frac{1}{2}$	$2\frac{3}{97}$	0.240	0.121	
15	7.79	$3\frac{7}{16}$	$2\frac{1}{4}$	0.242	0.119	
10	7.55	$3\frac{3}{8}$	$1\frac{3}{8}$	0.235	0.116	$2\frac{7}{8}$
12	7.32	NA	$1\frac{5}{8}$	0.236	0.112	
11	7.93	$3\frac{3}{4}$	$2\frac{1}{32}$	0.240	0.121	
10	7.18	$3\frac{1}{8}$	$1\frac{3}{8}$	0.235	0.110	
11	7.23	$3\frac{9}{16}$	$2\frac{1}{4}$	0.242	0.111	
15	7.48	3	$1\frac{5}{8}$	0.235	0.114	
13	7.89	$3\frac{19}{32}$	$2\frac{1}{32}$	0.240	0.121	
15	7.79	$3\frac{1}{2}$	$2\frac{1}{4}$	0.242	0.119	
10	7.50	$3\frac{13}{32}$	$1\frac{5}{8}$	0.237	0.115	$2\frac{7}{8}$
12	7.32	NA	$1\frac{5}{8}$	0.236	0.112	

(Continued)

TABLE 2.6 (*Continued*)

Assembly Data						
* Box Tong Space	Adjusted Weight	Minimum Tool Joint O.D. for Prem. Class	Drift Diameter	Capacity	Displacement	Size O.D.
in.	lb/ft	in.	in.	US gal/ft	US gal/ft	in.
11	7.93	$3\frac{13}{16}$	$2\frac{1}{32}$	0.240	0.121	
10	7.18	NA	$1\frac{3}{8}$	0.235	0.110	
11	7.61	$3\frac{19}{32}$	$2\frac{1}{32}$	0.240	0.116	
15	7.48	$3\frac{1}{16}$	$1\frac{5}{8}$	0.235	0.114	
13	7.89	$3\frac{5}{8}$	$2\frac{1}{32}$	0.240	0.121	
15	7.79	$3\frac{17}{32}$	$2\frac{1}{4}$	0.242	0.119	
10	7.69	$3\frac{17}{32}$	$1\frac{3}{8}$	0.235	0.118	$2\frac{7}{8}$
12	7.66	$3\frac{5}{16}$	$1\frac{3}{8}$	0.234	0.117	
11	7.97	$3\frac{29}{32}$	2	0.240	0.122	
10	7.18	NA	$1\frac{3}{8}$	0.235	0.110	
11	7.61	$3\frac{23}{32}$	$2\frac{1}{32}$	0.240	0.116	
15	7.48	$3\frac{7}{32}$	$1\frac{5}{8}$	0.235	0.114	
13	7.89	$3\frac{23}{32}$	$2\frac{1}{32}$	0.240	0.121	
15	7.79	$3\frac{21}{32}$	$2\frac{1}{4}$	0.242	0.119	
12	7.66	$3\frac{11}{32}$	$1\frac{3}{8}$	0.234	0.117	$2\frac{7}{8}$
15	7.48	$3\frac{7}{32}$	$1\frac{5}{8}$	0.235	0.114	
13	7.89	$3\frac{3}{4}$	$2\frac{1}{32}$	0.240	0.121	
15	7.79	$3\frac{21}{32}$	$2\frac{1}{4}$	0.242	0.119	

12	7.66	$3\frac{3}{8}$	$1\frac{3}{8}$	0.234	0.117	$2\frac{7}{8}$
15	7.48	$3\frac{9}{32}$	$1\frac{5}{8}$	0.235	0.114	
13	7.89	$3\frac{25}{32}$	$2\frac{1}{32}$	0.240	0.121	
15	7.79	$3\frac{11}{16}$	$2\frac{1}{4}$	0.242	0.119	
11	11.16	$3\frac{13}{16}$	$2\frac{1}{32}$	0.189	0.171	$2\frac{7}{8}$
10	10.80	$3\frac{13}{32}$	$1\frac{3}{8}$	0.183	0.165	
11	10.46	NA	$1\frac{3}{8}$	0.183	0.160	
11	10.84	$3\frac{19}{32}$	$2\frac{1}{32}$	0.189	0.166	
11	11.16	$3\frac{19}{32}$	$2\frac{1}{32}$	0.189	0.171	
13	10.51	$2\frac{31}{32}$	$1\frac{3}{8}$	0.183	0.161	
12	10.89	$3\frac{3}{16}$	$1\frac{3}{8}$	0.183	0.167	
13	11.27	$3\frac{5}{8}$	$2\frac{1}{32}$	0.189	0.172	
15	11.05	$2\frac{31}{32}$	$1\frac{3}{8}$	0.182	0.169	
15	11.25	$3\frac{13}{32}$	$2\frac{1}{32}$	0.189	0.172	
11	11.32	$3\frac{29}{32}$	$1\frac{7}{8}$	0.187	0.173	$2\frac{7}{8}$
10	10.80	NA	$1\frac{3}{8}$	0.183	0.165	
11	10.46	NA	$2\frac{1}{32}$	0.183	0.160	
11	10.84	$3\frac{23}{32}$	$2\frac{1}{32}$	0.189	0.166	
11	11.16	$3\frac{11}{16}$	$1\frac{3}{8}$	0.189	0.171	
13	10.51	$3\frac{1}{8}$	$1\frac{3}{8}$	0.183	0.161	
12	10.89	$3\frac{5}{16}$	$\frac{3}{8}$	0.183	0.167	
13	11.27	$3\frac{23}{32}$	$\frac{3}{8}$	0.189	0.172	
15	11.05	$3\frac{3}{32}$		0.182	0.169	

(Continued)

TABLE 2.6 (Continued)

Assembly Data						
* Box Tong Space	Adjusted Weight	Minimum Tool Joint O.D. for Prem. Class	Drift Diameter	Capacity	Displacement	Size O.D.
in.	lb/ft	in.	in.	US gal/ft	US gal/ft	in.
15	11.25	$3\frac{17}{32}$	$\frac{1}{32}$	0.189	0.172	
11	11.32	$3\frac{15}{16}$	$1\frac{7}{8}$	0.187	0.173	$2\frac{7}{8}$
10	10.80	NA	$1\frac{3}{8}$	0.183	0.165	
11	10.54	NA	$1\frac{1}{4}$	0.183	0.161	
11	10.84	$3\frac{3}{4}$	$2\frac{1}{32}$	0.189	0.166	
11	11.32	$3\frac{23}{32}$	$1\frac{7}{8}$	0.187	0.173	
13	10.60	NA	$1\frac{1}{4}$	0.182	0.162	
12	11.03	$3\frac{3}{8}$	$1\frac{3}{8}$	0.183	0.169	
13	11.27	$3\frac{25}{32}$	$2\frac{1}{32}$	0.189	0.172	
15	11.05	$3\frac{5}{32}$	$1\frac{3}{8}$	0.182	0.169	
15	11.25	$3\frac{9}{16}$	$2\frac{1}{32}$	0.189	0.172	
11	11.98	$4\frac{1}{16}$	$1\frac{1}{2}$	0.184	0.183	$2\frac{7}{8}$
10	10.94	NA	$1\frac{3}{8}$	0.183	0.167	
11	10.46	NA	$1\frac{3}{8}$	0.183	0.160	
11	11.00	NA	$1\frac{7}{8}$	0.187	0.168	
11	11.32	$3\frac{27}{32}$	$1\frac{7}{8}$	0.187	0.173	
13	10.51	NA	$1\frac{3}{8}$	0.183	0.161	
12	11.03	$3\frac{9}{16}$	$1\frac{3}{8}$	0.183	0.169	

13	11.44	$3\frac{27}{32}$	$1\frac{7}{8}$	0.187	0.175	
15	11.15	$3\frac{5}{16}$	$1\frac{1}{4}$	0.181	0.171	$2\frac{7}{8}$
15	11.25	$3\frac{23}{32}$	$2\frac{1}{32}$	0.189	0.172	
11	11.32	$3\frac{15}{16}$	$1\frac{7}{8}$	0.187	0.173	
12	11.12	$3\frac{9}{16}$	$1\frac{1}{4}$	0.182	0.170	$2\frac{7}{8}$
13	11.44	$3\frac{7}{8}$	$1\frac{7}{8}$	0.187	0.175	
15	11.24	$3\frac{5}{16}$	$1\frac{1}{8}$	0.180	0.172	
15	11.25	$3\frac{3}{4}$	$2\frac{1}{32}$	0.189	0.172	
11	11.32	$3\frac{15}{16}$	$1\frac{7}{8}$	0.187	0.173	
12	11.20	$3\frac{9}{16}$	$1\frac{1}{8}$	0.181	0.171	$2\frac{7}{8}$
13	11.44	$3\frac{29}{32}$	$1\frac{7}{8}$	0.187	0.175	
15	11.24	$3\frac{3}{8}$	$1\frac{1}{8}$	0.180	0.172	
15	11.43	$3\frac{23}{32}$	$1\frac{7}{8}$	0.187	0.175	
11	11.44	$3\frac{15}{16}$	$1\frac{3}{4}$	0.186	0.175	
12.5	11.14	$4\frac{13}{32}$	$2\frac{9}{16}$	0.361	0.170	$3\frac{1}{2}$
11	10.58	$3\frac{7}{8}$	2	0.355	0.162	
13	10.70	$3\frac{11}{16}$	2	0.354	0.164	
15.5	11.38	$4\frac{5}{32}$	$2\frac{9}{16}$	0.360	0.174	
12.5	11.14	$4\frac{3}{16}$	$2\frac{9}{16}$	0.361	0.170	
15	10.68	$3\frac{1}{2}$	2	0.353	0.163	
15	10.92	4	$2\frac{11}{16}$	0.362	0.167	
12.5	11.14	$4\frac{15}{32}$	$2\frac{9}{26}$	0.361	0.170	$3\frac{1}{2}$
11	10.70	4	$1\frac{7}{8}$	0.354	0.164	

(Continued)

TABLE 2.6 *(Continued)*

Assembly Data						
* Box Tong Space	Adjusted Weight	Minimum Tool Joint O.D. for Prem. Class	Drift Diameter	Capacity	Displacement	Size O.D.
in.	lb/ft	in.	in.	US gal/ft	US gal/ft	in.
13	10.70	$3\frac{13}{16}$	2	0.354	0.164	
15.5	11.38	$4\frac{1}{4}$	$2\frac{9}{16}$	0.360	0.174	
12.5	11.14	$4\frac{9}{32}$	$2\frac{9}{16}$	0.361	0.170	
15	10.68	$3\frac{5}{8}$	2	0.353	0.163	
15	10.92	$4\frac{3}{32}$	$2\frac{11}{16}$	0.362	0.167	
12.5	11.14	$4\frac{17}{32}$	$2\frac{9}{16}$	0.361	0.170	$3\frac{1}{2}$
11	10.70	$4\frac{1}{16}$	$1\frac{7}{8}$	0.354	0.164	
13	10.83	$3\frac{27}{32}$	$1\frac{7}{8}$	0.353	0.166	
15.5	11.38	$4\frac{9}{32}$	$2\frac{9}{16}$	0.360	0.174	
12.5	11.14	$4\frac{5}{16}$	$1\frac{9}{16}$	0.361	0.170	
15	10.68	$3\frac{11}{16}$	1	0.353	0.163	
15	10.92	$4\frac{5}{32}$	$1\frac{11}{16}$	0.362	0.167	
12.5	11.75	$4\frac{21}{32}$	$2\frac{7}{16}$	0.359	0.180	$3\frac{1}{2}$
11	10.70	NA	$1\frac{7}{8}$	0.354	0.164	
13	10.83	4	$1\frac{7}{8}$	0.353	0.166	
15.5	11.38	$4\frac{7}{16}$	$2\frac{9}{16}$	0.360	0.174	
12.5	11.31	$4\frac{7}{16}$	$2\frac{7}{16}$	0.359	0.173	
15	10.82	$3\frac{13}{16}$	$1\frac{7}{8}$	0.351	0.166	

15	10.92	$4\frac{9}{32}$	$2\frac{11}{16}$	0.362	0.167	
13	10.83	$4\frac{1}{32}$	$1\frac{7}{8}$	0.353	0.166	$3\frac{1}{2}$
15.5	11.38	$4\frac{15}{32}$	$2\frac{9}{16}$	0.360	0.174	
15	10.82	$3\frac{27}{32}$	$1\frac{7}{8}$	0.351	0.166	
15	10.92	$4\frac{5}{16}$	$2\frac{11}{16}$	0.362	0.167	
13	11.26	4	$1\frac{5}{5}$	0.350	0.172	$3\frac{1}{2}$
15.5	11.38	$4\frac{1}{2}$	$2\frac{9}{16}$	0.360	0.174	
15	10.82	$3\frac{29}{32}$	$1\frac{7}{8}$	0.351	0.166	
15	10.92	$4\frac{11}{32}$	$2\frac{11}{16}$	0.362	0.167	
12.5	14.30	$4\frac{1}{2}$	$2\frac{9}{16}$	0.311	0.219	$3\frac{1}{2}$
11	13.88	$4\frac{1}{32}$	$1\frac{7}{8}$	0.303	0.212	
13	13.86	$3\frac{27}{32}$	2	0.304	0.212	
15.5	14.51	$4\frac{1}{4}$	$2\frac{9}{16}$	0.310	0.222	
12.5	14.13	$4\frac{9}{32}$	$2\frac{11}{16}$	0.312	0.216	
15	13.82	$3\frac{21}{32}$	2	0.303	0.211	
15	14.47	$4\frac{1}{32}$	$2\frac{9}{16}7$	0.310	0.221	
12.5	14.90	$4\frac{19}{32}$	$2\frac{7}{16}$	0.309	0.228	$3\frac{1}{2}$
11	13.88	NA	$1\frac{7}{8}$	0.303	0.212	
13	13.99	$3\frac{15}{16}$	$1\frac{7}{8}$	0.302	0.214	
15.5	14.51	$4\frac{3}{8}$	$2\frac{9}{16}$	0.310	0.222	
12.5	14.30	$4\frac{3}{8}$	$2\frac{9}{16}$	0.311	0.219	
15	13.82	$3\frac{13}{16}$	2	0.303	0.211	
15	14.47	$4\frac{5}{32}$	$2\frac{9}{16}$	0.310	0.221	

(Continued)

TABLE 2.6 (*Continued*)

Assembly Data						
* Box Tong Space	Adjusted Weight	Minimum Tool Joint O.D. for Prem. Class	Drift Diameter	Capacity	Displacement	Size O.D.
in.	lb/ft	in.	in.	US gal/ft	US gal/ft	in.
12.5	15.07	$4\frac{21}{32}$	$2\frac{5}{16}$	0.307	0.231	$3\frac{1}{2}$
11	13.88	NA	$1\frac{7}{8}$	0.303	0.212	
13	13.99	4	$1\frac{7}{8}$	0.302	0.214	
15.5	14.51	$4\frac{7}{16}$	$2\frac{9}{16}$	0.310	0.222	
12.5	14.47	$4\frac{7}{16}$	$2\frac{7}{16}$	0.309	0.221	
15	14.16	$3\frac{13}{16}$	$1\frac{7}{8}$	0.301	0.217	
15	14.47	$4\frac{7}{32}$	$2\frac{9}{16}$	0.310	0.221	
12.5	15.45	$4\frac{13}{16}$	2	0.303	0.236	$3\frac{1}{2}$
11	13.88	NA	$1\frac{7}{8}$	0.303	0.212	
13	13.99	NA	$1\frac{7}{8}$	0.302	0.214	
15.5	14.69	$4\frac{9}{16}$	$2\frac{7}{16}$	0.309	0.225	
12.5	14.47	$4\frac{19}{32}$	$2\frac{7}{16}$	0.309	0.221	
15	14.29	4	$1\frac{3}{4}$	0.300	0.219	
15	14.47	$4\frac{13}{32}$	$2\frac{9}{16}$	0.310	0.221	
12.5	14.68	$4\frac{11}{16}$	$2\frac{7}{16}$	0.309	0.225	
13	14.12	NA	$1\frac{3}{4}$	0.301	0.216	$3\frac{1}{2}$
15.5	14.51	$4\frac{5}{8}$	$2\frac{9}{16}$	0.310	0.222	
15	14.42	4	$1\frac{5}{8}$	0.299	0.221	

15	14.25	$4\frac{7}{16}$	$2\frac{9}{16}$	0.311	0.218	
12.5	14.77	$4\frac{11}{16}$	$2\frac{3}{8}$	0.308	0.226	
13	14.42	$4\frac{1}{4}$	$1\frac{5}{8}$	0.300	0.221	$3\frac{1}{2}$
15.5	14.51	$4\frac{11}{16}$	$2\frac{9}{16}$	0.310	0.222	
15	14.42	$4\frac{1}{16}$	$1\frac{5}{8}$	0.299	0.221	
15	14.65	$4\frac{13}{32}$	$2\frac{7}{16}$	0.309	0.224	
12.5	15.07	$4\frac{11}{16}$	$2\frac{5}{16}$	0.307	0.231	
12.5	16.56	$4\frac{17}{32}$	$2\frac{7}{16}$	0.276	0.253	$3\frac{1}{2}$
15.5	16.60	$4\frac{5}{16}$	$2\frac{9}{16}$	0.278	0.254	
15	16.73	$4\frac{1}{32}$	$2\frac{7}{16}$	0.276	0.256	
12.5	17.16	$4\frac{21}{32}$	$2\frac{5}{16}$	0.274	0.262	$3\frac{1}{2}$
15.5	16.60	$4\frac{7}{16}$	$2\frac{9}{16}$	0.278	0.254	
15	16.73	$4\frac{5}{32}$	$2\frac{7}{16}$	0.276	0.256	
12.5	17.54	$4\frac{23}{32}$	2	0.270	0.268	$3\frac{1}{2}$
15.5	16.77	$4\frac{7}{16}$	$2\frac{7}{16}$	0.276	0.256	
12	17.30	$4\frac{15}{16}$	$2\frac{7}{16}$	0.276	0.265	
15	16.73	$4\frac{7}{32}$	$2\frac{7}{16}$	0.276	0.256	
12.5	17.54	$4\frac{29}{32}$	2	0.270	0.268	$3\frac{1}{2}$
12.5	16.72	$4\frac{19}{32}$	$2\frac{5}{16}$	0.274	0.256	
12.5	18.35	$5\frac{3}{32}$	$2\frac{1}{8}$	0.271	0.281	
15	16.73	$4\frac{7}{16}$	$2\frac{7}{16}$	0.276	0.256	
15	17.20	$4\frac{7}{16}$	$2\frac{7}{16}$	0.276	0.263	
12.5	17.24	$4\frac{11}{16}$	$2\frac{1}{4}$	0.273	0.264	

(Continued)

TABLE 2.6 (*Continued*)

Assembly Data						
* Box Tong Space	Adjusted Weight	Minimum Tool Joint O.D. for Prem. Class	Drift Diameter	Capacity	Displacement	Size O.D.
in.	lb/ft	in.	in.	US gal/ft	US gal/ft	in.
12.5	16.72	$4\frac{5}{8}$	$2\frac{5}{16}$	0.274	0.256	$3\frac{1}{2}$
15	16.91	$4\frac{13}{32}$	$2\frac{5}{16}$	0.274	0.259	
15	17.20	$4\frac{15}{32}$	$2\frac{7}{16}$	0.276	0.263	
12.5	17.24	$4\frac{23}{32}$	$2\frac{1}{4}$	0.273	0.264	
12.5	17.16	$4\frac{11}{16}$	$2\frac{5}{16}$	0.274	0.262	$3\frac{1}{2}$
15	16.91	$4\frac{15}{32}$	$2\frac{5}{16}$	0.274	0.259	
15	17.20	$4\frac{17}{32}$	$2\frac{7}{16}$	0.276	0.263	
12.5	17.39	$4\frac{23}{32}$	$2\frac{1}{8}$	0.272	0.266	
12	13.35	$4\frac{3}{4}$	$2\frac{11}{16}$	0.482	0.204	4
12	12.85	$4\frac{3}{8}$	$2\frac{7}{16}$	0.479	0.197	
15.5	13.02	$4\frac{9}{32}$	$2\frac{9}{16}$	0.479	0.199	
15	12.93	$4\frac{1}{16}$	$2\frac{9}{16}$	0.479	0.198	
15	13.11	$4\frac{3}{16}$	$2\frac{3}{4}$	0.482	0.200	
12	13.35	$4\frac{27}{32}$	$2\frac{11}{16}$	0.482	0.204	4
12	12.85	$4\frac{1}{2}$	$2\frac{7}{16}$	0.479	0.197	
15.5	13.02	$4\frac{13}{32}$	$2\frac{7}{16}$	0.479	0.199	
15	12.93	$4\frac{3}{16}$	$2\frac{9}{16}$	0.479	0.198	

15	13.11	$4\frac{5}{16}$	$2\frac{3}{4}$	0.482	0.200	
12	13.35	$4\frac{29}{32}$	$2\frac{11}{16}$	0.482	0.204	4
12	12.85	$4\frac{9}{16}$	$2\frac{7}{16}$	0.479	0.197	
15.5	13.20	$4\frac{13}{32}$	$2\frac{7}{16}$	0.477	0.202	
15	12.93	$4\frac{1}{4}$	$2\frac{9}{16}$	0.479	0.198	
15	13.11	$4\frac{3}{8}$	$2\frac{3}{4}$	0.482	0.200	
12	14.17	$5\frac{1}{16}$	$2\frac{7}{16}$	0.478	0.217	4
12	12.85	$4\frac{23}{32}$	$2\frac{7}{16}$	0.479	0.197	
15.5	13.38	$4\frac{17}{32}$	$2\frac{5}{16}$	0.475	0.202	
15	12.93	$4\frac{7}{16}$	$2\frac{9}{16}$	0.479	0.198	
15	13.11	$4\frac{17}{32}$	$2\frac{3}{4}$	0.482	0.200	
15.5	13.38	$4\frac{9}{16}$	$2\frac{5}{16}$	0.475	0.205	4
15	12.93	$4\frac{15}{32}$	$2\frac{9}{16}$	0.479	0.198	
15	13.11	$4\frac{9}{16}$	$2\frac{3}{4}$	0.482	0.200	
15.5	13.86	$4\frac{5}{8}$	$2\frac{5}{16}$	0.475	0.212	4
15	13.11	$4\frac{15}{32}$	$2\frac{7}{16}$	0.477	0.201	
15	13.11	$4\frac{5}{8}$	$2\frac{3}{4}$	0.482	0.200	
12	15.58	$4\frac{13}{16}$	$2\frac{11}{16}$	0.447	0.238	4
15.5	15.23	$4\frac{3}{8}$	$2\frac{9}{16}$	0.444	0.233	
12	15.25	$4\frac{7}{16}$	$2\frac{5}{16}$	0.442	0.233	
15	15.88	$4\frac{19}{32}$	$2\frac{11}{16}$	0.446	0.243	
12	16.64	$4\frac{9}{32}$	$3\frac{1}{8}$	0.454	0.254	
15	15.20	$5\frac{5}{32}$	$2\frac{9}{16}$	0.444	0.232	

(Continued)

TABLE 2.6 (*Continued*)

Assembly Data						
* Box Tong Space	Adjusted Weight	Minimum Tool Joint O.D. for Prem. Class	Drift Diameter	Capacity	Displacement	Size O.D.
in.	lb/ft	in.	in.	US gal/ft	US gal/ft	in.
15	15.47	$4\frac{1}{4}$	$2\frac{11}{16}$	0.446	0.237	
12	15.58	$4\frac{15}{16}$	$2\frac{11}{16}$	0.447	0.238	4
15.5	15.23	$4\frac{17}{32}$	$2\frac{9}{16}$	0.444	0.233	
12	15.25	$4\frac{19}{32}$	$2\frac{5}{16}$	0.442	0.233	
15	15.88	$4\frac{23}{32}$	$2\frac{11}{16}$	0.446	0.243	
12	16.64	$5\frac{3}{8}$	$3\frac{1}{8}$	0.454	0.254	
15	15.20	$4\frac{5}{16}$	$2\frac{9}{16}$	0.444	0.232	
15	15.47	$4\frac{3}{8}$	$2\frac{11}{16}$	0.446	0.237	
12	15.58	5	$2\frac{11}{16}$	0.447	0.238	4
15.5	15.89	$4\frac{17}{32}$	$2\frac{7}{16}$	0.441	0.243	
12	15.25	$4\frac{21}{32}$	$2\frac{5}{16}$	0.442	0.233	
15	15.88	$4\frac{25}{32}$	$2\frac{11}{16}$	0.446	0.243	
12	16.64	$5\frac{7}{16}$	$3\frac{1}{8}$	0.454	0.254	
15	15.20	$4\frac{3}{8}$	$2\frac{9}{16}$	0.444	0.232	
15	15.47	$4\frac{7}{16}$	$2\frac{11}{16}$	0.446	0.237	
12	16.40	$5\frac{3}{16}$	$2\frac{7}{16}$	0.443	0.251	4
15.5	16.07	$4\frac{11}{16}$	$2\frac{5}{16}$	0.440	0.246	
12	15.25	NA	$2\frac{5}{16}$	0.442	0.233	

15	16.07	$4\frac{29}{32}$	$2\frac{9}{16}$	0.444	0.246	
12	17.04	$5\frac{9}{16}$	$3\frac{7}{8}$	0.450	0.261	
15	15.85	$4\frac{17}{32}$	$2\frac{7}{16}$	0.442	0.242	
15	15.47	$4\frac{21}{32}$	$2\frac{11}{16}$	0.446	0.237	
12	15.76	5	$2\frac{9}{16}$	0.445	0.241	
15.5	16.07	$4\frac{23}{32}$	$2\frac{5}{16}$	0.440	0.246	4
15	16.07	$4\frac{15}{16}$	$2\frac{9}{16}$	0.444	0.246	
15	15.85	$4\frac{9}{16}$	$2\frac{7}{16}$	0.442	0.242	
15	15.47	$4\frac{11}{16}$	$2\frac{11}{16}$	0.446	0.237	
12	15.85	5	$2\frac{1}{2}$	0.444	0.242	
15.5	16.07	$4\frac{25}{32}$	$2\frac{5}{16}$	0.440	0.246	4
15	16.07	5	$2\frac{9}{16}$	0.444	0.246	
15	16.03	$4\frac{19}{32}$	$2\frac{5}{16}$	0.440	0.245	
15	15.47	$4\frac{3}{4}$	$2\frac{11}{16}$	0.446	0.237	
12	16.02	$5\frac{1}{32}$	$2\frac{3}{8}$	0.442	0.245	
12	17.17	$4\frac{7}{8}$	$2\frac{11}{16}$	0.422	0.263	4
15	17.45	$4\frac{5}{8}$	$2\frac{11}{16}$	0.421	0.267	
12	17.63	$4\frac{32}{32}$	$2\frac{11}{16}$	0.421	0.270	
12	17.90	$5\frac{5}{16}$	$3\frac{1}{8}$	0.428	0.274	
15	17.04	$4\frac{9}{32}$	$2\frac{11}{16}$	0.421	0.261	
15	17.54	$4\frac{5}{16}$	$2\frac{11}{16}$	0.421	0.268	
12	17.52	5	$2\frac{7}{16}$	0.418	0.268	4
15	17.45	$4\frac{25}{32}$	$2\frac{11}{16}$	0.421	0.267	

(Continued)

TABLE 2.6 (*Continued*)

Assembly Data						
* Box Tong Space	Adjusted Weight	Minimum Tool Joint O.D. for Prem. Class	Drift Diameter	Capacity	Displacement	Size O.D.
in.	lb/ft	in.	in.	US gal/ft	US gal/ft	in.
12	17.63	$5\frac{3}{32}$	$2\frac{11}{16}$	0.421	0.270	
12	17.90	$5\frac{7}{16}$	$3\frac{1}{8}$	0.428	0.274	
15	17.04	$4\frac{7}{16}$	$2\frac{11}{16}$	0.421	0.261	
15	17.54	$4\frac{15}{32}$	$2\frac{11}{16}$	0.421	0.268	
12	18.14	$5\frac{1}{16}$	$2\frac{5}{16}$	0.416	0.278	4
15	17.45	$4\frac{27}{32}$	$2\frac{11}{16}$	0.421	0.267	
15	17.96	$5\frac{5}{32}$	$2\frac{11}{16}$	0.421	0.275	
12	17.90	$5\frac{15}{32}$	$3\frac{1}{8}7$	0.428	0.274	
15	17.04	$4\frac{17}{32}$	$2\frac{11}{16}$	0.421	0.261	
15	17.54	$4\frac{17}{32}$	$2\frac{11}{16}$	0.421	0.268	
12	18.66	$5\frac{1}{4}$	$1\frac{7}{8}$	0.411	0.285	4
15	18.34	$4\frac{15}{16}$	$2\frac{7}{16}$	0.417	0.280	
15	18.69	$5\frac{5}{16}$	$2\frac{9}{16}$	0.418	0.286	
12	18.30	$5\frac{21}{32}$	$2\frac{7}{8}$	0.424	0.280	
15	17.24	$4\frac{11}{16}$	$2\frac{9}{16}$	0.419	0.264	
15	17.54	$4\frac{3}{4}$	$2\frac{11}{16}$	0.421	0.268	
12	17.52	$5\frac{1}{32}$	$2\frac{7}{16}$	0.418	0.268	
15	18.34	$4\frac{31}{32}$	$2\frac{7}{16}$	0.417	0.280	4

15	17.24	$4\frac{23}{32}$	$2\frac{9}{16}$	0.419	0.264	
15	17.54	$4\frac{25}{32}$	$2\frac{11}{16}$	0.421	0.268	
12	17.83	$5\frac{1}{16}$	$2\frac{3}{8}$	0.417	0.273	
15	18.34	$5\frac{1}{16}$	$2\frac{7}{16}$	0.417	0.280	4
15	17.42	$4\frac{23}{32}$	$2\frac{7}{16}$	0.417	0.266	
15	17.54	$4\frac{27}{32}$	$2\frac{11}{16}$	0.421	0.268	
12	18.14	$5\frac{3}{32}$	$2\frac{5}{16}$	0.416	0.278	
12	19.17	$5\frac{13}{32}$	$3\frac{1}{8}$	0.586	0.293	$4\frac{1}{2}$
12	17.75	$5\frac{15}{32}$	$3\frac{5}{8}$	0.596	0.272	
12	19.05	$5\frac{3}{8}$	$2\frac{7}{8}$	0.583	0.291	
12	18.64	$5\frac{11}{32}$	$3\frac{1}{8}$	0.587	0.285	
15	19.62	$5\frac{13}{32}$	$3\frac{1}{8}$	0.585	0.300	
12	18.80	$5\frac{23}{32}$	$3\frac{5}{8}$	0.596	0.288	
15	18.90	$5\frac{13}{16}$	$3\frac{5}{8}$	0.596	0.289	
15	17.94	$4\frac{7}{8}$	$2\frac{7}{8}$	0.581	0.274	
15	18.68	$5\frac{5}{8}$	$3\frac{3}{8}$	0.590	0.286	
15	19.34	$5\frac{31}{32}$	$3\frac{5}{8}$	0.596	0.296	
12	19.17	$5\frac{17}{32}$	$3\frac{1}{8}$	0.586	0.293	$4\frac{1}{2}$
12	18.20	$5\frac{19}{32}$	$3\frac{3}{8}$	0.591	0.278	
12	19.05	$5\frac{1}{2}$	$2\frac{7}{8}$	0.583	0.291	
12	18.64	$5\frac{15}{32}$	$3\frac{1}{8}$	0.587	0.285	
15	19.62	$5\frac{13}{32}$	$3\frac{1}{8}$	0.585	0.300	
12	18.80	$5\frac{27}{32}$	$3\frac{5}{8}$	0.596	0.288	
15	18.90	$5\frac{13}{16}$	$3\frac{5}{8}$	0.596	0.289	

(Continued)

TABLE 2.6 (*Continued*)

Assembly Data						
* Box Tong Space	Adjusted Weight	Minimum Tool Joint O.D. for Prem. Class	Drift Diameter	Capacity	Displacement	Size O.D.
in.	lb/ft	in.	in.	US gal/ft	US gal/ft	in.
15	17.94	$4\frac{7}{8}$	$2\frac{7}{8}$	0.581	0.274	
15	18.68	$5\frac{5}{8}$	$3\frac{3}{8}$	0.590	0.286	
15	19.34	$5\frac{31}{32}$	$3\frac{5}{8}$	0.596	0.296	
12	19.59	$5\frac{19}{32}$	$2\frac{7}{8}$	0.582	0.300	$4\frac{1}{2}$
12	18.88	$5\frac{21}{32}$	$3\frac{1}{8}$	0.587	0.289	
12	19.97	$5\frac{9}{16}$	$2\frac{5}{8}$	0.578	0.305	
12	18.64	$5\frac{17}{32}$	$3\frac{1}{8}$	0.587	0.285	
15	19.62	$5\frac{13}{32}$	$3\frac{1}{8}$	0.585	0.300	
12	18.80	$5\frac{29}{32}$	$3\frac{5}{8}$	0.596	0.288	
15	18.90	$5\frac{13}{16}$	$3\frac{5}{8}$	0.596	0.289	
15	17.94	$4\frac{7}{8}$	$2\frac{7}{8}$	0.581	0.274	
15	18.68	$5\frac{5}{8}$	$3\frac{3}{8}$	0.590	0.286	
15	19.34	$5\frac{31}{32}$	$3\frac{5}{8}$	0.596	0.296	
12	19.97	$5\frac{25}{32}$	$2\frac{5}{8}$	0.578	0.305	$4\frac{1}{2}$
12	19.28	$5\frac{13}{16}$	$2\frac{7}{8}$	0.583	0.295	
12	19.97	$5\frac{3}{4}$	$2\frac{5}{8}$	0.578	0.305	
12	18.97	$5\frac{11}{16}$	$2\frac{5}{8}$	0.578	0.305	
15	19.62	$5\frac{1}{2}$	$3\frac{1}{8}$	0.585	0.300	

12	19.26	$6\frac{1}{16}$	$3\frac{3}{8}$	0.591	0.295	
15	19.71	$5\frac{13}{16}$	$3\frac{3}{8}$	0.590	0.301	
15	18.24	$4\frac{15}{16}$	$2\frac{11}{16}$	0.578	0.279	
15	18.68	$5\frac{5}{8}$	$3\frac{3}{8}$	0.590	0.286	
15	19.34	$5\frac{31}{32}$	$3\frac{5}{8}$	0.596	0.296	
12	18.64	$5\frac{19}{32}$	$3\frac{1}{8}$	0.587	0.285	
15	19.62	$5\frac{17}{32}$	$3\frac{1}{8}$	0.585	0.300	$4\frac{1}{2}$
15	19.71	$5\frac{13}{16}$	$3\frac{3}{8}$	0.590	0.301	
15	18.24	$4\frac{21}{32}$	$2\frac{11}{16}$	0.578	0.279	
15	18.68	$5\frac{5}{8}$	$3\frac{3}{8}$	0.590	0.286	
15	19.34	$5\frac{31}{32}$	$3\frac{5}{8}$	0.596	0.296	
12	18.75	$5\frac{19}{32}$	$3\frac{1}{16}$	0.586	0.287	
15	19.62	$5\frac{19}{32}$	$3\frac{1}{8}$	0.585	0.300	$4\frac{1}{2}$
15	19.71	$5\frac{13}{16}$	$3\frac{3}{8}$	0.590	0.301	
15	18.24	$5\frac{1}{16}$	$2\frac{11}{16}$	0.578	0.279	
15	19.77	$5\frac{5}{8}$	$3\frac{1}{8}$	0.585	0.302	
15	19.85	$5\frac{31}{32}$	$3\frac{3}{8}$	0.590	0.304	
12	18.85	$5\frac{21}{32}$	3	0.585	0.288	
12	22.51	$5\frac{1}{2}$	$3\frac{1}{8}$	0.533	0.344	$4\frac{1}{2}$
12	21.81	$5\frac{17}{32}$	$3\frac{3}{8}$	0.538	0.334	
12	21.98	$5\frac{7}{16}$	$3\frac{1}{8}$	0.534	0.336	
15	22.93	$5\frac{13}{32}$	$3\frac{1}{8}$	0.533	0.351	
12	22.38	$5\frac{13}{16}$	$3\frac{1}{2}$	0.540	0.342	
15	22.25	$5\frac{13}{16}$	$3\frac{5}{8}$	0.543	0.340	

(Continued)

TABLE 2.6 (*Continued*)

Assembly Data						
* Box Tong Space	Adjusted Weight	Minimum Tool Joint O.D. for Prem. Class	Drift Diameter	Capacity	Displacement	Size O.D.
in.	lb/ft	in.	in.	US gal/ft	US gal/ft	in.
15	21.99	$5\frac{5}{8}$	$3\frac{3}{8}$	0.538	0.336	
15	23.17	$5\frac{31}{32}$	$3\frac{3}{8}$	0.538	0.354	
12	22.92	$5\frac{21}{32}$	$3\frac{7}{8}$	0.529	0.351	$4\frac{1}{2}$
12	22.77	$5\frac{11}{16}$	$3\frac{1}{8}$	0.534	0.348	
12	21.98	$5\frac{9}{16}$	$3\frac{1}{8}$	0.534	0.336	
15	22.93	$5\frac{13}{32}$	$3\frac{1}{8}$	0.533	0.351	
12	22.38	$5\frac{15}{16}$	$3\frac{1}{2}$	0.540	0.342	
15	22.25	$5\frac{13}{16}$	$3\frac{5}{8}$	0.543	0.340	
15	21.99	$5\frac{5}{8}$	$3\frac{3}{8}$	0.538	0.336	
15	23.17	$5\frac{31}{32}$	$3\frac{3}{8}$	0.538	0.354	
12	23.30	$5\frac{23}{32}$	$2\frac{5}{8}$	0.526	0.356	$4\frac{1}{2}$
12	23.17	$5\frac{3}{4}$	$2\frac{7}{8}$	0.530	0.354	
12	22.92	$5\frac{5}{8}$	$2\frac{7}{8}$	0.529	0.351	
15	22.93	$5\frac{7}{16}$	$3\frac{1}{8}$	0.533	0.351	
12	22.61	$6\frac{1}{32}$	$3\frac{3}{8}$	0.538	0.346	
15	22.25	$5\frac{13}{16}$	$3\frac{5}{8}$	0.543	0.340	
15	21.99	$5\frac{5}{8}$	$3\frac{3}{8}$	0.538	0.336	
15	23.17	$5\frac{31}{32}$	$3\frac{3}{8}$	0.538	0.354	

12	23.94	$5\frac{15}{16}$	$2\frac{3}{8}$	0.522	0.366	$4\frac{1}{2}$
12	23.82	$5\frac{31}{32}$	$2\frac{5}{8}$	0.526	0.364	
12	23.58	$5\frac{27}{32}$	$2\frac{5}{8}$	0.525	0.361	
15	23.38	$5\frac{9}{16}$	$2\frac{7}{8}$	0.528	0.358	
12	23.32	$6\frac{7}{32}$	$3\frac{1}{8}$	0.534	0.357	
15	23.03	$5\frac{13}{16}$	$3\frac{3}{8}$	0.538	0.352	
15	21.99	$5\frac{5}{8}$	$3\frac{3}{8}$	0.538	0.336	
15	23.17	$5\frac{31}{32}$	$3\frac{3}{8}$	0.538	0.354	
12	22.39	$5\frac{21}{32}$	$2\frac{7}{8}$	0.530	0.342	
15	23.38	$5\frac{19}{32}$	$2\frac{7}{8}$	0.528	0.358	$4\frac{1}{2}$
15	23.03	$5\frac{27}{32}$	$3\frac{3}{8}$	0.538	0.352	
15	21.99	$5\frac{21}{32}$	$3\frac{3}{8}$	0.538	0.336	
15	23.17	$5\frac{31}{32}$	$3\frac{3}{8}$	0.538	0.354	
12	22.49	$5\frac{21}{32}$	$2\frac{13}{16}$	0.529	0.344	
15	23.38	$5\frac{21}{32}$	$2\frac{7}{8}$	0.528	0.358	$4\frac{1}{2}$
15	23.03	$5\frac{29}{32}$	$3\frac{3}{8}$	0.538	0.352	
15	23.07	$5\frac{5}{8}$	$3\frac{1}{8}$	0.532	0.353	
15	23.17	$5\frac{31}{32}$	$3\frac{3}{8}$	0.538	0.354	
12	22.76	$5\frac{21}{32}$	$2\frac{5}{8}$	0.526	0.348	
12	22.16	$5\frac{7}{8}$	$3\frac{5}{8}$	0.735	0.339	5
15	22.61	$5\frac{13}{16}$	$3\frac{5}{8}$	0.734	0.346	
12	23.24	$6\frac{3}{8}$	$3\frac{5}{8}$	0.734	0.356	
15	21.71	$5\frac{5}{8}$	$3\frac{3}{8}$	0.729	0.332	

(*Continued*)

TABLE 2.6 *(Continued)*

Assembly Data						
* Box Tong Space	Adjusted Weight	Minimum Tool Joint O.D. for Prem. Class	Drift Diameter	Capacity	Displacement	Size O.D.
in.	lb/ft	in.	in.	US gal/ft	US gal/ft	in.
15	21.88	$5\frac{31}{32}$	$3\frac{7}{8}$	0.739	0.335	
12	22.63	$6\frac{1}{32}$	$3\frac{3}{8}$	0.730	0.346	5
15	22.61	$5\frac{13}{16}$	$3\frac{5}{8}$	0.734	0.346	
12	23.24	$6\frac{1}{2}$	$3\frac{5}{8}$	0.734	0.356	
15	21.71	$5\frac{5}{8}$	$3\frac{3}{8}$	0.729	0.332	
15	21.88	$5\frac{31}{32}$	$3\frac{7}{8}$	0.739	0.335	
12	23.08	$6\frac{3}{32}$	$3\frac{1}{8}$	0.726	0.353	5
15	22.61	$5\frac{27}{32}$	$3\frac{5}{8}$	0.734	0.346	
12	23.24	$6\frac{9}{16}$	$3\frac{5}{8}$	0.734	0.356	
15	21.71	$5\frac{5}{8}$	$3\frac{3}{8}$	0.729	0.332	
15	21.88	$5\frac{31}{32}$	$3\frac{7}{8}$	0.739	0.335	
12	21.87	$5\frac{15}{16}$	$3\frac{5}{8}$	0.735	0.335	
12	23.87	$6\frac{5}{16}$	$2\frac{5}{8}$	0.718	0.365	5
15	23.12	$5\frac{15}{16}$	$3\frac{3}{8}$	0.728	0.354	
12	24.41	$6\frac{3}{4}$	$3\frac{3}{8}$	0.729	0.373	
15	21.71	$5\frac{23}{32}$	$3\frac{3}{8}$	0.729	0.332	
15	22.43	$5\frac{31}{32}$	$3\frac{5}{8}$	0.733	0.343	
12	22.35	$6\frac{1}{32}$	$3\frac{3}{8}$	0.730	0.342	

15	23.12	$5\frac{31}{32}$	$3\frac{3}{8}$	0.728	0.354	5
15	21.71	$5\frac{25}{32}$	$3\frac{3}{8}$	0.729	0.332	
15	22.43	$5\frac{31}{32}$	$3\frac{5}{8}$	0.733	0.343	
12	22.35	$6\frac{3}{32}$	$3\frac{3}{8}$	0.730	0.342	
15	23.12	$6\frac{1}{32}$	$3\frac{3}{8}$	0.728	0.354	5
15	22.79	$5\frac{23}{32}$	$3\frac{1}{8}$	0.723	0.349	
15	22.43	$5\frac{31}{32}$	$3\frac{5}{8}$	0.733	0.343	
12	22.57	$6\frac{3}{32}$	$3\frac{1}{4}$	0.728	0.345	
12	28.13	$6\frac{1}{32}$	$3\frac{3}{8}$	0.643	0.430	5
15	28.06	$5\frac{13}{16}$	$3\frac{5}{8}$	0.647	0.429	
12	29.21	$6\frac{1}{2}$	$3\frac{3}{8}$	0.642	0.447	
15	28.20	$5\frac{31}{32}$	$3\frac{5}{8}$	0.647	0.431	
12	28.98	$6\frac{7}{32}$	$2\frac{7}{8}$	0.634	0.443	5
15	28.57	$5\frac{13}{16}$	$3\frac{3}{8}$	0.642	0.437	
12	29.21	$6\frac{21}{32}$	$3\frac{3}{8}$	0.642	0.447	
15	28.20	$5\frac{31}{32}$	$3\frac{5}{8}$	0.647	0.431	
12	29.36	$6\frac{9}{32}$	$2\frac{5}{8}$	0.631	0.449	5
15	29.05	$5\frac{13}{16}$	$3\frac{1}{8}$	0.637	0.444	
12	29.87	$6\frac{23}{32}$	$3\frac{3}{8}$	0.642	0.457	
15	28.20	$5\frac{31}{32}$	$3\frac{5}{8}$	0.647	0.431	
12	27.84	$6\frac{1}{32}$	$3\frac{3}{8}$	0.643	0.426	
12	29.36	$6\frac{17}{32}$	$2\frac{5}{8}$	0.631	0.449	5
15	29.49	$5\frac{31}{32}$	$2\frac{7}{8}$	0.632	0.451	

(Continued)

TABLE 2.6 (*Continued*)

Assembly Data						
* Box Tong Space	Adjusted Weight	Minimum Tool Joint O.D. for Prem. Class	Drift Diameter	Capacity	Displacement	Size O.D.
in.	lb/ft	in.	in.	US gal/ft	US gal/ft	in.
12	30.33	$6\frac{15}{16}$	$3\frac{1}{8}$	0.637	0.464	
15	28.20	$6\frac{1}{16}$	$3\frac{5}{8}$	0.647	0.431	
12	28.39	$6\frac{5}{32}$	$3\frac{1}{16}$	0.638	0.434	
15	29.49	$6\frac{1}{32}$	$2\frac{7}{8}$	0.632	0.451	5
15	28.20	$6\frac{3}{32}$	$3\frac{5}{8}$	0.647	0.431	
12	28.49	$6\frac{3}{16}$	3	0.637	0.436	
15	29.49	$6\frac{3}{32}$	$2\frac{7}{8}$	0.632	0.451	5
15	28.72	$6\frac{1}{16}$	$3\frac{3}{8}$	0.641	0.439	
12	28.98	$6\frac{7}{32}$	$2\frac{7}{8}$	0.634	0.443	
12	24.85	$6\frac{15}{32}$	$3\frac{7}{8}$	0.913	0.380	$5\frac{1}{2}$
15	25.35	$6\frac{13}{32}$	$3\frac{7}{8}$	0.911	0.388	
15	24.08	$6\frac{7}{32}$	$4\frac{1}{8}$	0.917	0.368	
15	24.45	$6\frac{15}{32}$	$4\frac{1}{4}$	0.920	0.374	
12	25.46	$6\frac{5}{8}$	$3\frac{5}{8}$	0.908	0.389	$5\frac{1}{2}$
15	25.44	$6\frac{13}{32}$	$3\frac{7}{8}$	0.911	0.389	
15	24.08	$6\frac{7}{32}$	$4\frac{1}{8}$	0.917	0.368	
15	24.45	$6\frac{15}{32}$	$4\frac{1}{4}$	0.920	0.374	

12	26.61	$6\frac{11}{16}$	$3\frac{3}{8}$	0.902	0.407	$5\frac{1}{2}$
15	25.44	$6\frac{13}{32}$	$3\frac{7}{8}$	0.911	0.389	
15	24.08	$6\frac{7}{32}$	$4\frac{1}{8}$	0.917	0.368	
15	24.45	$6\frac{51}{32}$	$4\frac{1}{4}$	0.920	0.374	
12	24.85	$6\frac{7}{16}$	$3\frac{7}{8}$	0.913	0.380	
12	28.21	$6\frac{29}{32}$	$2\frac{7}{8}$	0.893	0.431	$5\frac{1}{2}$
15	25.44	$6\frac{5}{8}$	$3\frac{7}{8}$	0.911	0.389	
15	24.08	$6\frac{5}{16}$	$4\frac{1}{8}$	0.917	0.368	
15	24.45	$6\frac{15}{32}$	$4\frac{1}{4}$	0.920	0.374	
12	24.85	$6\frac{11}{16}$	$3\frac{7}{8}$	0.913	0.380	
12	28.21	$6\frac{15}{16}$	$2\frac{7}{8}$	0.893	0.431	$5\frac{1}{2}$
15	25.44	$6\frac{21}{32}$	$3\frac{7}{8}$	0.911	0.389	
15	24.08	$6\frac{11}{32}$	$4\frac{1}{8}$	0.917	0.368	
15	24.45	$6\frac{15}{32}$	$4\frac{1}{4}$	0.920	0.374	
12	24.85	$6\frac{23}{32}$	$3\frac{7}{8}$	0.913	0.380	$5\frac{1}{2}$
12	28.21	7	$2\frac{7}{8}$	0.893	0.431	
15	25.44	$6\frac{23}{32}$	$3\frac{7}{8}$	0.911	0.389	
15	24.66	$6\frac{9}{32}$	$3\frac{7}{8}$	0.911	0.377	
15	24.76	$6\frac{15}{32}$	$4\frac{1}{8}$	0.917	0.379	
12	25.44	$6\frac{23}{32}$	$3\frac{3}{4}$	0.910	0.389	$5\frac{1}{2}$
12	27.41	$6\frac{17}{32}$	$3\frac{7}{8}$	0.874	0.419	
15	27.88	$6\frac{13}{32}$	$3\frac{7}{8}$	0.872	0.427	
15	26.51	$6\frac{7}{32}$	$4\frac{1}{8}$	0.879	0.406	

(Continued)

TABLE 2.6 (*Continued*)

Assembly Data						
* Box Tong Space	Adjusted Weight	Minimum Tool Joint O.D. for Prem. Class	Drift Diameter	Capacity	Displacement	Size O.D.
in.	lb/ft	in.	in.	US gal/ft	US gal/ft	in.
15	26.88	$6\frac{15}{32}$	$4\frac{1}{4}$	0.882	0.411	$5\frac{1}{2}$
12	29.07	$6\frac{11}{16}$	$3\frac{3}{8}$	0.863	0.445	
15	27.88	$6\frac{13}{32}$	$3\frac{7}{8}$	0.872	0.427	
15	26.62	$6\frac{7}{32}$	$4\frac{1}{8}$	0.879	0.407	
15	26.99	$6\frac{15}{32}$	$4\frac{1}{4}$	0.882	0.413	$5\frac{1}{2}$
12	29.07	$6\frac{25}{32}$	$3\frac{3}{8}$	0.863	0.445	
15	27.88	$6\frac{15}{32}$	$3\frac{7}{8}$	0.872	0.427	
15	26.62	$6\frac{7}{32}$	$4\frac{1}{8}$	0.879	0.407	
15	26.99	$6\frac{15}{32}$	$4\frac{1}{4}$	0.882	0.413	
12	27.30	$6\frac{17}{32}$	$3\frac{7}{8}$	0.874	0.418	$5\frac{1}{2}$
12	30.66	7	$2\frac{7}{8}$	0.854	0.469	
15	28.44	$6\frac{5}{8}$	$3\frac{5}{8}$	0.867	0.435	
15	27.20	$6\frac{9}{32}$	$3\frac{7}{8}$	0.873	0.416	
15	27.30	$6\frac{15}{32}$	$4\frac{1}{8}$	0.879	0.418	
12	27.88	$6\frac{23}{32}$	$3\frac{3}{4}$	0.871	0.427	$5\frac{1}{2}$
12	29.07	$7\frac{1}{32}$	$3\frac{3}{8}$	0.863	0.445	
15	27.88	$6\frac{25}{32}$	$3\frac{7}{8}$	0.872	0.427	
15	27.20	$6\frac{11}{32}$	$3\frac{7}{8}$	0.873	0.416	

15	27.30	$6\frac{15}{32}$	$4\frac{1}{8}$	0.879	0.418	
12	27.88	$6\frac{25}{32}$	$3\frac{3}{4}$	0.871	0.427	$5\frac{1}{2}$
12	30.66	$7\frac{3}{32}$	$2\frac{7}{8}$	0.854	0.469	
15	28.44	$6\frac{23}{32}$	$3\frac{5}{8}$	0.867	0.435	
15	27.20	$6\frac{7}{16}$	$3\frac{7}{8}$	0.873	0.416	
15	27.30	$6\frac{9}{16}$	$4\frac{1}{8}$	0.879	0.418	
12	28.14	$6\frac{13}{16}$	$3\frac{5}{8}$	0.869	0.430	$5\frac{7}{8}$
15	26.46	$6\frac{15}{32}$	$4\frac{1}{8}$	1.059	0.405	$5\frac{7}{8}$
15	26.46	$6\frac{15}{32}$	$4\frac{1}{8}$	1.059	0.405	$5\frac{7}{8}$
15	26.46	$6\frac{15}{32}$	$4\frac{1}{8}$	1.059	0.405	$5\frac{7}{8}$
15	26.46	$6\frac{15}{32}$	$4\frac{1}{8}$	1.059	0.405	$5\frac{7}{8}$
15	26.46	$6\frac{17}{32}$	$4\frac{1}{8}$	1.059	0.405	$5\frac{7}{8}$
15	26.46	$6\frac{5}{8}$	$4\frac{1}{8}$	1.059	0.405	$5\frac{7}{8}$
15	29.10	$6\frac{15}{32}$	$4\frac{1}{8}$	1.017	0.445	$5\frac{7}{8}$
15	29.10	$6\frac{15}{32}$	$4\frac{1}{8}$	1.017	0.445	$5\frac{7}{8}$
15	29.10	$6\frac{15}{32}$	$4\frac{1}{8}$	1.017	0.445	$5\frac{7}{8}$
15	29.10	$6\frac{5}{8}$	$4\frac{1}{8}$	1.017	0.445	$5\frac{7}{8}$
15	29.10	$6\frac{21}{32}$	$4\frac{1}{8}$	1.017	0.445	$5\frac{7}{8}$
15	29.68	$6\frac{5}{8}$	$3\frac{7}{8}$	1.011	0.454	$6\frac{5}{8}$
13	28.81	$7\frac{7}{16}$	$4\frac{7}{8}$	1.423	0.441	
16	29.40	$7\frac{11}{32}$	$4\frac{7}{8}$	1.420	0.450	
15	29.20	$7\frac{11}{32}$	$4\frac{7}{8}$	1.421	0.447	$6\frac{5}{8}$
13	28.81	$7\frac{5}{8}$	$4\frac{7}{8}$	1.423	0.441	

(Continued)

TABLE 2.6 *(Continued)*

Assembly Data						
* Box Tong Space	Adjusted Weight	Minimum Tool Joint O.D. for Prem. Class	Drift Diameter	Capacity	Displacement	Size O.D.
in.	lb/ft	in.	in.	US gal/ft	US gal/ft	in.
16	29.40	$7\frac{11}{32}$	$4\frac{7}{8}$	1.420	0.450	
15	29.20	$7\frac{11}{32}$	$4\frac{7}{8}$	1.421	0.447	$6\frac{5}{8}$
13	30.25	$7\frac{11}{16}$	$4\frac{5}{8}$	1.415	0.463	
16	29.40	$7\frac{13}{32}$	$4\frac{7}{8}$	1.420	0.450	
15	29.20	$7\frac{11}{32}$	$4\frac{7}{8}$	1.421	0.447	$6\frac{5}{8}$
13	32.32	$7\frac{29}{32}$	$4\frac{1}{8}$	1.402	0.494	
16	29.40	$7\frac{5}{8}$	$4\frac{7}{8}$	1.420	0.450	
15	29.20	$7\frac{11}{32}$	$4\frac{7}{8}$	1.421	0.447	
13	28.98	$7\frac{21}{32}$	$4\frac{13}{16}$	1.421	0.443	$6\frac{5}{8}$
13	32.32	$7\frac{31}{32}$	$4\frac{1}{8}$	1.402	0.494	
16	29.40	$7\frac{11}{16}$	$4\frac{7}{8}$	1.420	0.450	
15	29.20	$7\frac{11}{32}$	$4\frac{7}{8}$	1.421	0.447	
13	29.14	$7\frac{21}{32}$	$4\frac{3}{4}$	1.419	0.446	$6\frac{5}{8}$
13	32.32	$8\frac{1}{32}$	$4\frac{1}{8}$	1.402	0.494	
16	29.40	$7\frac{3}{4}$	$4\frac{7}{8}$	1.420	0.450	
15	29.20	$7\frac{11}{32}$	$4\frac{7}{8}$	1.421	0.447	

13	30.25	$7\frac{11}{16}$	$4\frac{5}{8}$	1.415	0.463	$6\frac{5}{8}$
13	30.64	$7\frac{1}{2}$	$4\frac{7}{8}$	1.394	0.469	
16	31.22	$7\frac{11}{32}$	$4\frac{7}{8}$	1.391	0.478	
15	31.03	$7\frac{11}{32}$	$4\frac{7}{8}$	1.392	0.475	$6\frac{5}{8}$
13	32.08	$7\frac{11}{16}$	$4\frac{5}{8}$	1.386	0.491	
16	31.22	$7\frac{3}{8}$	$4\frac{7}{8}$	1.391	0.478	
15	31.03	$7\frac{11}{32}$	$4\frac{7}{8}$	1.392	0.475	$6\frac{5}{8}$
13	32.08	$7\frac{3}{4}$	$4\frac{5}{8}$	1.386	0.491	
16	31.22	$7\frac{15}{32}$	$4\frac{7}{8}$	1.391	0.478	
15	31.03	$7\frac{11}{32}$	$4\frac{7}{8}$	1.392	0.475	$6\frac{5}{8}$
13	34.14	8	$4\frac{1}{8}$	1.373	0.522	
16	31.22	$7\frac{23}{32}$	$4\frac{7}{8}$	1.391	0.478	
15	31.03	$7\frac{11}{32}$	$4\frac{7}{8}$	1.392	0.475	
13	30.97	$7\frac{23}{32}$	$4\frac{3}{4}$	1.390	0.474	$6\frac{5}{8}$
13	34.14	$8\frac{1}{32}$	$4\frac{1}{8}$	1.373	0.522	
16	31.22	$7\frac{3}{4}$	$4\frac{7}{8}$	1.391	0.478	
15	31.03	$7\frac{11}{32}$	$4\frac{7}{8}$	1.392	0.475	
13	32.08	$7\frac{11}{16}$	$4\frac{5}{8}$	1.386	0.491	$6\frac{5}{8}$
13	34.14	$8\frac{1}{8}$	$4\frac{1}{8}$	1.373	0.522	
16	31.22	$7\frac{27}{32}$	$4\frac{7}{8}$	1.391	0.478	
15	31.03	$7\frac{7}{16}$	$4\frac{7}{8}$	1.392	0.475	
13	32.08	$7\frac{25}{32}$	$4\frac{5}{8}$	1.386	0.491	

TABLE 2.7 Drill Pipe Material Properties [1]

Material	Yield Strength (psi) Min	Yield Strength (psi) Max	Tensile Strength (psi) Min	Elongation (Percent)
Pipe grade				
E	75,000	105,000	100,000	17
X	95,000	125,000	105,000	1605
G	105,000	135,000	115,000	15
S	135,000	165,000	145,000	12.5
Tool joint	120,000		140,000	13

TABLE 2.8 Dimensional Properties of API Drill Pipe Tubes [6]

Drill Pipe Size	Nominal Wt (lb/ft)	Wall Thickness	Plain End Wt (lb/ft)	ID (in.)	Section Area Body of Pipe A (in.^{2})	Polar Section Modulus Z (in.^{3})
$2\frac{3}{8}$	4.85	0.109	4.43	1.995	1.304	1.321
$2\frac{3}{8}$	6.65	0.280	6.26	1.815	1.843	1.733
$2\frac{7}{8}$	6.85	0.217	6.16	2.441	1.812	2.241
$2\frac{7}{8}$	10.40	0.362	9.72	2.151	2.858	3.204
$3\frac{1}{2}$	9.50	0.254	8.81	2.992	2.590	3.923
$3\frac{1}{2}$	13.30	0.368	12.31	2.764	3.621	5.144
$3\frac{1}{2}$	15.50	0.449	14.63	2.602	4.304	5.847
4	11.85	0.262	10.46	3.476	3.077	5.400
4	14.00	0.330	12.93	3.340	3.805	6.458
4	15.70	0.380	14.69	3.240	4.322	7.157
$4\frac{1}{2}$	13.75	0.271	12.24	3.958	3.600	7.184
$4\frac{1}{2}$	16.60	0.337	14.98	3.826	4.407	8.543
$4\frac{1}{2}$	20.00	0.430	18.69	3.640	5.498	10.232
$4\frac{1}{2}$	22.82	0.500	21.36	3.500	6.283	11.345
5	16.25	0.296	14.87	4.408	4.374	9.718
5	19.50	0.362	17.93	4.276	5.275	11.415
5	25.60	0.500	24.03	4.000	7.069	14.491
$5\frac{1}{2}$	19.20	0.304	16.87	4.892	4.962	12.221
$5\frac{1}{2}$	21.90	0.361	19.81	4.778	5.828	14.062
$5\frac{1}{2}$	24.70	0.415	22.54	4.670	6.630	15.688
$5\frac{7}{8}$	23.40	0.361	23.40	5.153	6.254	16.251
$5\frac{7}{8}$	24.17	0.415	24.17	5.045	7.119	18.165
$6\frac{5}{8}$	25.20	0.330	22.19	5.965	6.526	19.572
$6\frac{5}{8}$	27.70	0.362	24.19	5.901	7.123	21.156

TABLE 2.9 Interchangeability Chart for Tool Joints

Common Name: Style	Common Name: Size	Threads per inch	Taper (in./ft)	Thread Form	Same As or Interchangeable with
Internal flush (IF)	$2\frac{3}{8}$	4	2	V-0.065 (V-0.038R)	$2\frac{7}{8}$ slim hole NC26
	$2\frac{7}{8}$	4	2	V-0.065 (V-0.038R)	$3\frac{1}{2}$ slim hole NC31
	$3\frac{1}{2}$	4	2	V-0.065 (V-0.038R)	$4\frac{1}{2}$ slim hole NC38
	4	4	2	V-0.065 (V-0.038R)	$4\frac{1}{2}$ XH NC46
	$4\frac{1}{2}$	4	2	V-0.065 (V-0.038R)	5 XH $5\frac{1}{2}$ double streamline NC50
Full hole (FH)	4	4	2	V-0.065 (V-0.038R)	$4\frac{1}{2}$ double streamline NC40
Extra hole (XH, EH)	$2\frac{7}{8}$	4	2	V-0.065 (V-0.038R)	$3\frac{1}{2}$ double streamline
	$3\frac{1}{2}$	4	2	V-0.065 (V-0.038R)	4 slim hole $4\frac{1}{2}$ external flush
	$4\frac{1}{2}$	4	2	V-0.065 (V-0.038R)	4 internal flush NC46
	5	4	2	V-0.065 (V-0.038R)	$4\frac{1}{2}$ internal flush NC50 $5\frac{1}{2}$ double streamline
Slim hole (SH)	$2\frac{7}{8}$	4	2	V-0.065 (V-0.038R)	$2\frac{3}{8}$ IF NC26
	$3\frac{1}{2}$	4	2	V-0.065 (V-0.038R)	$2\frac{7}{8}$ IF NC31

(Continued)

TABLE 2.9 (*Continued*)

Common Name: Style	Size	Threads per inch	Taper (in./ft)	Thread Form	Same As or Interchangeable With
	4	4	2	V-0.065 (V-0.038R)	$3\frac{1}{2}$ XH
					$4\frac{1}{2}$ IF
	$4\frac{1}{2}$	4	2	V-0.065 (V-0.038R)	$3\frac{1}{2}$ IF
					NC38
Double sreamline (DSL)	$3\frac{1}{2}$	4	2	V-0.065 (V-0.038R)	$2\frac{7}{8}$ XH
	$4\frac{1}{2}$	4	2	V-0.065 (V-0.038R)	4 FH
					NC40
	$5\frac{1}{2}$	4	2	V-0.065 (V-0.038R)	$4\frac{1}{2}$ IF
					5 XH
					NC50
Numbered conn (NC)	26	4	2	V-0.38R	$2\frac{3}{8}$ IF
					$2\frac{7}{8}$ SH
	31	4	2	V-0.38R	$2\frac{7}{8}$ IF
					$3\frac{1}{2}$ SH
	38	4	2	V-0.38R	$3\frac{1}{2}$ IF
					$4\frac{1}{2}$ SH
	40	4	2	V-0.38R	4 FH
					$4\frac{1}{2}$ DSL
	46	4	2	V-0.38R	4 IF
					$4\frac{1}{2}$ XH
	50	4	2	V-0.38R	$4\frac{1}{2}$ IF
					5 XH
					$5\frac{1}{2}$ DSL
External flush (EF)	$4\frac{1}{2}$	4	2	V-0.065 (V-0.038R)	4SH
					$3\frac{1}{2}$ XH

number of tool joint types. Tool joints with the same thread form and pitch diameter at the gage point are usually interchangeable. Drill pipe manufacturers may also offer proprietary tool joints, such as different variations of double shoulder tool joints for increased torsional strength and tool joints with special threads.

Tool joint OD and ID. The OD and ID (in., mm) of the tool joint dictates its strength. Generally, the torsional strength of the box is dictated by the tool joint OD and that of the pin is dictated by the ID. Because the box strength does not depend on its ID, the box ID of most drill pipe assemblies is the maximum available regardless of the pin ID. The tool joint OD affects the fish ability of the length and the equivalent circulation density (ECD). The tool joint ID affects the drilling fluid pressure losses in the string.

Tool joint tong length. The tong length (in, mm) is the length of the cylindrical portion of the tool joint where the tongs grip. API specifies tong lengths for API connections. Drill pipe is often produced with tong lengths greater than the API-specified tong length, usually in 1-inch increments, to allow more thread recuts. Each time a damaged thread is recut, about $\frac{3}{4}$ in. of tong space is lost.

Drill pipe length. API-defined drill pipe is available in three standard lengths: range 1, range 2, and range 3, which are approximately 22, $31\frac{1}{2}$, and 45 ft long, respectively.

Hardbanding. Drill pipe is often produced with hardbanding on the box. The hardbanding reduces the wear rate of the tool joints, reduces casing wear, and reduces the frictional drag of the pipe rotating and sliding in the hole. Sometimes, hardbanding is applied to the pin as well as the box.

2.2.1 Classification of Drill Pipe

Torsional, tensile, collapse, and internal pressure data for new drill pipe are given in Tables 2.10, 2.11, and 2.12. Dimensional data is given in Table 2.13.

Drill pipe is classified based on wear, mechanical damage and corrosion of the tool joint, and tube. The classifications are premium, class 2, and class 3. Descriptions and specifications for the classes are found in Table 2.14 and API RP7G. The pipe is marked with colored bands as shown in Figure 2.12.

The torsion, tension, collapse and internal pressure resistance for new, premium class, and class 2 drill pipe are shown in Tables 2.10, 2.11, and 2.12, respectively.

Calculations for the minimum performance properties of drill pipe are based on formulas given in Appendix A of API RP7G. The numbers in Tables 2.10, 2.11, and 2.12 have been calculated for uniaxial sterss, (e.g.,

TABLE 2.10 New Drill Pipe: Torsional, Tensile, Collapse, and Internal Pressure Data

OD	Nominal Wt (lbs/ft)	Wall Thickness (in.)	Section Area (in.^{2})	Polar Section Modulus (in.^{3})	Torsional Yield Strength (ft-lb)				Tensile Yield Strength (lb)				Collapse Pressure (psi)				Internal Pressure (psi)			
					E-75	X-95	G-105	S-135	E-75	X-95	G-105	S-135	E-75	X-95	G-105	S-135	E-75	X-95	G-105	S-135
$2\frac{3}{8}$	4.85	0.190	1.304	1.321	4,763	6,033	6,668	8,574	97,817	123,902	136,944	176,071	11,040	13,984	15,456	19,035	10,500	13,300	14,700	18,900
$2\frac{3}{8}$	6.65	0.280	1.843	1.733	6,250	7,917	8,751	11,251	138,214	175,072	193,500	248,786	15,599	19,759	21,839	28,079	15,474	19,600	21,663	27,853
$2\frac{7}{8}$	6.85	0.217	1.812	2.241	8,083	10,238	11,316	14,549	135,902	172,143	190,263	244,624	1,503	2,189	2,580	3,941	9,907	12,548	13,869	17,832
$2\frac{7}{8}$	10.4	0.362	2.858	3.204	11,554	14,635	16,176	20,798	214,344	271,503	300,082	385,820	16,509	20,911	23,112	29,716	16,526	20,933	23,137	29,747
$3\frac{1}{2}$	9.5	0.254	2.590	3.923	14,146	17,918	19,805	25,463	194,264	246,068	271,970	349,676	10,001	12,077	13,055	15,748	9,525	12,065	13,335	17,145
$3\frac{1}{2}$	13.3	0.368	3.621	5.144	18,551	23,498	25,972	33,392	271,569	343,988	380,197	488,825	14,113	17,877	19,758	25,404	13,800	17,480	19,320	28,840
$3\frac{1}{2}$	15.5	0.449	4.304	5.847	21,086	26,708	29,520	37,954	322,775	408,848	451,885	580,995	16,774	21,247	23,484	30,194	16,838	21,328	23,573	30,308
4	11.85	0.262	3.077	5.400	19,474	24,668	27,264	35,054	230,755	292,290	323,057	415,360	8,381	9,978	10,708	12,618	8,597	10,889	12,036	15,474
4	14.00	0.330	3.805	6.458	23,288	29,498	32,603	41,918	285,359	361,454	399,502	513,646	11,354	14,382	15,896	20,141	10,828	13,716	15,159	19,491
4	15.70	0.380	4.322	7.157	25,810	32,692	36,134	46,458	324,118	410,550	453,765	583,413	12,896	16,335	18,055	23,213	12,469	15,794	17,456	22,444
$4\frac{1}{2}$	13.75	0.271	3.600	7.184	25,907	32,816	36,270	46,633	270,034	342,043	378,047	486,061	7,173	8,412	8,956	10,283	7,904	10,012	11,066	14,228
$4\frac{1}{2}$	16.60	0.337	4.407	8.543	30,807	39,022	43,130	55,453	330,558	418,707	462,781	595,004	10,392	12,765	13,825	16,773	9,829	12,450	13,761	17,693
$4\frac{1}{2}$	20.00	0.430	5.498	10.232	36,901	46,741	51,661	66,421	412,358	522,320	577,301	742,244	12,964	16,421	18,149	23,335	12,542	15,886	17,558	22,575
$4\frac{1}{2}$	22.82	0.500	6.283	11.345	40,912	51,821	57,276	73,641	471,239	596,903	659,734	848,230	14,815	18,765	20,741	26,667	14,583	18,472	20,417	26,250
5	16.25	0.296	4.374	9.718	35,044	44,389	49,062	63,079	328,073	415,559	459,302	590,531	6,938	8,108	8,616	9,831	7,770	9,842	10,878	13,986
5	19.50	0.362	5.275	11.415	41,167	52,144	57,633	74,100	395,595	501,087	553,833	712,070	9,962	12,026	12,999	15,672	9,503	12,037	13,304	17,105
5	25.60	0.250	3.731	8.441	30,439	38,556	42,614	54,790	279,798	354,411	391,717	503,637	4,831	5,377	5,562	6,090	6,563	8,313	9,188	11,813
$5\frac{1}{2}$	19.20	0.304	4.962	12.221	44,074	55,827	61,703	79,332	372,181	471,429	521,053	669,925	6,039	6,942	7,313	8,093	7,255	9,189	10,156	13,058
$5\frac{1}{2}$	21.90	0.361	5.828	14.062	50,710	64,233	70,994	91,278	437,116	553,681	611,963	786,809	8,413	10,019	10,753	12,679	8,615	10,912	12,061	15,507
$5\frac{1}{2}$	24.70	0.415	6.630	15.688	56,574	71,660	79,204	101,833	497,222	629,814	696,111	894,999	10,464	12,933	14,013	17,023	9,903	12,544	13,865	17,826
$5\frac{7}{8}$	23.40	0.361	6.254	16.251	58,605	74,233	82,047	105,489	469,013	594,083	656,619	844,224	7,453	8,775	9,362	10,825	8,065	10,216	11,291	14,517
$5\frac{7}{8}$	24.17	0.415	7.119	18.165	65,508	82,977	91,711	117,915	533,890	676,261	747,446	961,002	9,558	11,503	12,414	14,892	9,271	11,744	12,980	16,688
$6\frac{5}{8}$	25.20	0.330	6.526	19.572	70,580	89,402	98,812	127,044	489,464	619,988	685,250	881,035	4,788	5,321	5,500	6,036	6,538	8,281	9,153	11,768
$6\frac{5}{8}$	27.70	0.362	7.123	21.156	76,295	96,640	106,813	137,330	534,198	676,651	747,877	961,556	5,894	6,755	7,103	7,813	7,172	9,084	10,040	12,909

TABLE 2.11 Premium Drill Pipe: Torsional, Tensile, Collapse, and Internal Pressure Data

OD	Nominal Wt (lb/ft)	New Wall Thickness (in.)	Premium Wall Thickness (in.)	Premium OD (in.)	ID (in.)	Section Area (in.^{2})	Polar Section Modulus (in.^{3})	Torsional Yield Strength (ft-lb)				Tensile Yield Strength (lb)				Collapse Pressure (psi)				Internal Pressure (psi)			
								E-75	X-95	G-105	S-135	E-75	X-95	G-105	S-135	E-75	X-95	G-105	S-135	E-75	X-95	G-105	S-135
$2\frac{3}{8}$	4.85	0.190	0.152	2.299	1.995	1.025	1.033	3,725	4,719	5,215	6,705	76,875	97,375	107,625	138,375	8,522	10,161	10,912	12,891	9,600	13,400	14,700	17,280
$2\frac{3}{8}$	6.65	0.280	0.224	2.263	1.815	1.434	1.334	4,811	6,093	6,735	8,659	107,550	136,313	150,570	193,590	13,378	16,945	18,729	24,080	14,147	19,806	21,663	25,465
$2\frac{7}{8}$	6.85	0.217	0.174	2.788	2.441	1.425	1.756	6,332	8,020	8,865	11,397	106,875	135,465	149,625	192,375	7,640	9,017	9,633	11,186	9,057	12,680	13,869	16,303
$2\frac{7}{8}$	10.4	0.362	0.290	2.730	2.151	2.220	2.456	8,858	11,220	12,401	15,945	166,500	210,945	233,100	299,700	14,223	18,016	19,912	25,602	15,110	21,153	23,137	27,197
$3\frac{1}{2}$	9.5	0.254	0.203	3.398	2.992	2.039	3.076	11,094	14,052	15,531	19,969	152,925	193,774	214,095	275,265	7,074	8,284	8,813	10,093	8,709	12,192	13,335	15,675
$3\frac{1}{2}$	13.3	0.368	0.294	3.353	2.764	2.828	3.982	14,361	18,191	20,106	25,850	212,100	268,723	296,940	381,780	12,015	15,218	16,820	21,626	12,617	17,664	19,320	22,711
$3\frac{1}{2}$	15.5	0.449	0.359	3.320	2.602	3.341	4.477	16,146	20,452	22,605	29,063	250,575	317,452	350,805	451,035	14,472	18,331	20,260	26,049	15,394	21,552	23,573	27,710
4	11.85	0.262	0.210	3.895	3.476	2.426	4.245	15,310	19,392	21,434	27,557	181,950	230,554	254,730	327,510	5,704	6,508	6,827	7,445	7,860	11,004	12,036	14,148
4	14.00	0.330	0.264	3.868	3.340	2.989	5.046	18,196	23,048	25,474	32,753	224,175	283,963	313,845	403,515	9,012	10,795	11,622	13,836	9,900	13,860	15,159	17,820
4	15.70	0.380	0.304	3.848	3.240	3.384	5.564	20,067	25,418	28,094	36,120	253,800	321,544	355,320	456,840	10,914	13,825	15,190	18,593	11,400	15,960	17,456	20,520
$4\frac{1}{2}$	13.75	0.271	0.217	4.392	3.958	2.843	5.658	20,403	25,844	28,564	36,725	213,225	270,127	298,515	383,805	4,686	5,190	5,352	5,908	7,227	10,117	11,066	13,008
$4\frac{1}{2}$	16.60	0.337	0.270	4.365	3.826	3.468	6.694	24,139	30,576	33,795	43,451	260,100	329,460	364,140	468,180	7,525	8,868	9,467	10,964	8,987	12,581	13,761	16,176
$4\frac{1}{2}$	20.00	0.430	0.344	4.328	3.640	4.305	7.954	28,683	36,332	40,157	51,630	322,875	409,026	452,025	581,175	10,975	13,901	15,350	18,806	11,467	16,053	17,558	20,640
$4\frac{1}{2}$	22.82	0.500	0.400	4.300	3.500	4.900	8.759	31,587	40,010	44,222	56,856	367,500	465,584	514,500	661,500	12,655	16,030	17,718	22,780	13,333	18,667	20,417	24,000
5	16.25	0.296	0.237	4.882	4.408	3.455	7.655	27,607	34,969	38,650	49,693	259,125	328,263	362,775	466,425	4,490	4,935	5,067	5,661	7,104	9,946	10,878	12,787
5	19.50	0.362	0.290	4.855	4.276	4.153	8.953	32,285	40,895	45,200	58,114	311,475	394,612	436,065	560,655	7,041	8,241	8,765	10,029	8,688	12,163	13,304	15,638
5	25.60	0.250	0.200	4.900	4.500	2.953	6.669	24,049	30,462	33,668	43,287	221,475	280,544	310,065	398,655	2,954	3,282	3,387	3,470	6,000	8,400	9,188	10,800
$5\frac{1}{2}$	19.20	0.304	0.243	5.378	4.892	3.923	9.640	34,764	44,035	48,670	62,576	294,225	372,730	411,915	529,605	3,736	4,130	4,336	4,714	6,633	9,286	10,156	11,939
$5\frac{1}{2}$	21.90	0.361	0.289	5.356	4.778	4.597	11.054	39,864	50,494	55,809	71,754	344,775	436,721	482,685	620,595	5,730	6,542	6,865	7,496	7,876	11,027	12,061	14,177
$5\frac{1}{2}$	24.70	0.415	0.332	5.334	4.670	5.217	12.290	44,320	56,139	62,048	79,776	391,275	495,627	547,785	704,295	7,635	9,011	9,626	11,177	9,055	12,676	13,865	16,298
$5\frac{7}{8}$	23.40	0.361	0.289	5.731	5.153	4.937	12.793	46,134	58,437	64,588	83,042	370,275	469,044	518,385	666,495	4,922	5,495	5,694	6,204	7,374	10,323	11,291	13,273
$5\frac{7}{8}$	24.17	0.415	0.332	5.709	5.045	5.608	14.255	51,408	65,116	71,971	92,534	420,600	532,785	588,840	757,080	6,699	7,798	8,269	9,368	8,477	11,867	12,980	15,258
$6\frac{5}{8}$	25.20	0.330	0.264	6.493	5.965	5.166	15.464	55,766	70,637	78,072	100,379	387,450	490,790	542,430	697,410	2,931	3,252	3,353	3,429	5,977	8,368	9,153	10,759
$6\frac{5}{8}$	27.70	0.362	0.290	6.480	5.901	5.632	16.691	60,191	76,242	84,268	108,345	422,400	535,063	591,360	760,320	3,615	4,029	4,222	4,562	6,557	9,180	10,040	11,803

TABLE 2.12 Class 2 Drill Pipe: Torsional, Tensile, Collapse, and Internal Pressure Data

OD	Nominal Wt (lb/ft)	New Wall Thickness (in.)	Class 2 Wall Thickness (in.)	Class 2 OD (in.)	ID (in.)	Section Area (in.^{2})	Polar Section Modulus (in.^{3})	Torsional Yield Strength (ft-lb)				Tensile Yield Strength (lb)				Collapse Pressure (psi)				Internal Pressure (psi)			
								E-75	X-95	G-105	S-135	E-75	X-95	G-105	S-135	E-75	X-95	G-105	S-135	E-75	X-95	G-105	S-135
2 3/8	4.85	0.190	0.133	2.261	1.995	0.889	0.894	3,224	4,083	4,513	5,802	66,686	84,469	93,360	120,035	6,852	7,996	8,491	9,664	8,400	10,640	11,760	15,120
2 3/8	6.65	0.280	0.196	2.207	1.815	1.238	1.145	4,130	5,232	5,782	7,434	92,871	117,636	130,019	167,167	12,138	15,375	16,993	21,849	12,379	15,680	17,331	22,282
2 7/8	6.85	0.217	0.152	2.745	2.441	1.237	1.521	5,484	6,946	7,677	9,871	92,801	117,548	129,922	167,043	6,055	6,963	7,335	8,123	7,925	10,039	11,095	14,265
2 7/8	10.4	0.362	0.253	2.658	2.151	1.914	2.105	7,591	9,615	10,627	13,663	143,557	181,839	200,980	258,403	12,938	16,388	18,113	23,288	13,221	16,746	18,509	23,798
3 1/2	9.5	0.254	0.178	3.348	2.992	1.770	2.666	9,612	12,176	13,457	17,302	132,793	168,204	185,910	239,027	5,544	6,301	6,596	7,137	7,620	9,652	10,668	13,716
3 1/2	13.3	0.368	0.258	3.279	2.764	2.445	3.429	12,365	15,663	17,312	22,258	183,398	232,304	256,757	330,116	10,858	13,753	15,042	18,396	11,040	13,984	15,456	19,872
3 1/2	15.5	0.449	0.314	3.231	2.602	2.879	3.834	13,828	17,515	19,359	24,890	215,967	273,558	302,354	388,741	13,174	16,686	18,443	23,712	13,470	17,062	18,858	24,246
4	11.85	0.262	0.183	3.843	3.476	2.108	3.683	13,282	16,823	18,594	23,907	158,132	200,331	221,385	284,638	4,310	4,702	4,876	5,436	6,878	8,712	9,629	12,380
4	14.00	0.330	0.231	3.802	3.340	2.591	4.364	15,738	19,935	22,034	28,329	194,363	246,193	272,108	349,853	7,295	8,570	9,134	10,520	8,663	10,973	12,123	15,593
4	15.70	0.380	0.266	3.772	3.240	2.929	4.801	17,315	21,932	24,241	31,166	219,738	278,534	307,633	395,528	9,531	11,468	12,374	14,840	9,975	12,635	13,965	17,955
4 1/2	13.75	0.271	0.190	4.337	3.958	2.471	4.912	17,715	22,439	24,801	31,887	185,390	234,827	259,546	333,702	3,397	3,845	4,016	4,287	6,323	8,010	8,853	11,382
4 1/2	16.60	0.337	0.236	4.298	3.826	3.010	5.798	20,908	26,483	29,271	37,634	225,771	285,977	316,080	406,388	5,951	6,828	7,185	7,923	7,863	9,960	11,039	14,154
4 1/2	20.00	0.430	0.301	4.242	3.640	3.726	6.862	24,747	31,346	34,645	44,544	279,501	354,035	391,302	503,103	9,631	11,598	12,520	15,033	10,033	12,709	14,047	18,060
4 1/2	22.82	0.500	0.350	4.200	3.500	4.233	7.532	27,161	34,404	38,026	48,891	317,497	402,163	444,496	571,495	11,458	14,514	16,042	20,510	11,667	14,778	16,333	21,000
5	16.25	0.296	0.207	4.822	4.408	3.004	6.648	23,974	30,368	33,564	43,154	225,316	285,400	315,442	405,568	3,275	3,696	3,850	4,065	6,216	7,874	8,702	11,189
5	19.50	0.362	0.253	4.783	4.276	3.605	7.758	27,976	35,436	39,166	50,356	270,432	342,547	378,605	486,778	5,514	6,262	6,552	7,079	7,602	9,629	10,643	13,684
5	25.60	0.250	0.175	4.850	4.500	2.570	5.799	20,913	26,490	29,279	37,644	192,766	244,170	269,873	346,979	2,248	2,370	2,374	2,374	5,250	6,650	7,350	9,450
5 1/2	19.20	0.304	0.213	5.318	4.892	3.412	8.377	30,208	38,264	42,291	54,375	255,954	324,208	358,335	460,717	2,835	3,128	3,215	3,265	5,804	7,351	8,125	10,447
5 1/2	21.90	0.361	0.253	5.283	4.778	3.993	9.589	34,582	43,804	48,415	62,247	299,533	379,409	419,346	539,160	4,334	4,733	4,899	5,465	6,892	8,730	9,649	12,405
5 1/2	24.70	0.415	0.291	5.251	4.670	4.527	10.644	38,383	48,619	53,737	69,090	339,534	430,076	475,347	611,160	6,050	6,957	7,329	8,115	7,923	10,035	11,092	14,261
5 7/8	23.40	0.361	0.253	5.658	5.153	4.291	11.105	40,049	50,729	56,069	72,088	321,861	407,691	450,605	579,350	3,608	4,023	4,215	4,553	6,452	8,172	9,033	11,613
5 7/8	24.17	0.415	0.291	5.626	5.045	4.869	12.356	44,559	56,441	62,382	80,206	365,201	462,588	511,282	657,362	5,206	5,863	6,105	6,561	7,417	9,395	10,384	13,351
6 5/8	25.20	0.330	0.231	6.427	5.965	4.496	13.448	48,497	61,430	67,896	87,295	337,236	427,166	472,131	607,026	2,227	2,343	2,346	2,346	5,230	6,625	7,322	9,414
6 5/8	27.70	0.362	0.253	6.408	5.901	4.899	14.505	52,308	66,257	73,231	94,154	367,454	465,442	514,436	661,418	2,765	3,037	3,113	3,148	5,737	7,267	8,032	10,327

TABLE 2.13 Dimensional Data and Mechanical Properties Heavy-Weight Drill Pipe [Smith International]

	Tube							Tool Joint						Approximate Weight Including Tube and Tool Joints (lb)	
	Tube Dimensions					Mechanical Properties Tube Section					Mechanical Properties				
Size (in.)	ID (in.)	Wall Thick-ness	Area (in.^{2})	Center Upset (in.^{2})	Elevator Upset (ln).	Tensile Yield (lb)	Torsional Yield (ft-lb)	OD (in.)	ID (in.)	Connection Size	Tensile Yield (lb)	Torsional Yield (ft-lb)	Makeup Torque (ft-lb)	Wt/ft	Wt/jt
$3\frac{1}{2}$	$2\frac{1}{4}$	0.625	5.645	4	$3\frac{5}{8}$	310,475	18,460	$4\frac{3}{4}$	$2\frac{3}{8}$	NC38	675,045	17,575	10,000	23.4	721
4	$2\frac{9}{16}$	0.719	7.41	$4\frac{1}{2}$	$4\frac{1}{8}$	407,550	27,635	$5\frac{1}{4}$	$2\frac{11}{16}$	NC40	711,475	23,525	13,300	29.9	920
$4\frac{1}{2}$	$2\frac{3}{4}$	0.875	9.965	5	$4\frac{5}{8}$	548,075	40,715	$6\frac{1}{4}$	$2\frac{7}{8}$	NC60	1,024,500	38,800	21,800	41.1	1,265
5	3	1	12.566	$\frac{5}{12}$	$5\frac{1}{8}$	691,185	56,495	$6\frac{3}{8}$	$3\frac{1}{16}$	NC50	1,266,000	51,375	29,200	50.1	1,543
$5\frac{1}{2}$	$3\frac{3}{8}$	1.063	14.812	6	$5\frac{11}{16}$	814,660	74,140	7	$3\frac{1}{2}$	5-$\frac{1}{2}$ FH	1,349,365	53,080	32,800	57.8	1,770
$6\frac{5}{8}$	$4\frac{1}{2}$	1.063	18.567	$7\frac{1}{8}$	$6\frac{3}{4}$	1,021,185	118,845	8	$4\frac{5}{8}$	6-$\frac{5}{8}$ FH	1,490,495	73,215	45,800	71.3	2,193
								Range 3							
$4\frac{1}{2}$	$2\frac{3}{4}$	0.875	9.965	5	$4\frac{5}{8}$	548,075	40,715	$6\frac{1}{4}$	$2\frac{7}{8}$	NC 46	1,024,500	38,800	21,800	39.9	1,750
5	3	1	12.566	$5\frac{1}{2}$	$5\frac{1}{8}$	691,185	56,495	$6\frac{3}{8}$	$3\frac{1}{16}$	NC 50	1,266,000	51,375	29,200	48.5	2,130

	Capacity – Volume of Fluid Necessary to Fill the ID				Displacement – Open Ended (Metal Displacement Only)			
Size (in.)	Gal per Joint*	BBL per Joint*	Gal per 100 ft	BBL per 100 ft	Gal per Joint*	BBL per Joint*	Gal per 100 ft	BBL per 100 ft
$3\frac{1}{2}$	6.29	0.150	21.0	0.500	10.62	0.253	35.4	0.843
4	8.13	0.194	27.1	0.645	12.62	0.300	42.1	1.002
$4\frac{1}{2}$	9.37	0.223	31.2	0.743	18.82	0.448	62.7	1.493
5	11.14	0.265	37.1	0.883	22.62	0.539	75.4	1.796
$5\frac{1}{2}$	14.08	0.335	46.9	1.117	26.48	0.630	88.3	2.102
$6\frac{5}{8}$	24.79	0.590	82.6	1.967	32.77	0.780	109.2	2.600

TABLE 2.14 Classification of Used Drill Pipe [4]

Classification Condition	Premium Class, Two White Bands	Class 2, One Yellow Band
Exterior Conditions		
OD wear	Remaining wall not less than 80%	Remaining wall not less than 70%
Dents and mashes	OD not less than 97%	OD not less than 96%
Crushing and necking	OD not less than 97%	OD not less than 96%
Slip area: cuts and gouges	Depth not more than 10% of average adjacent wall, and remaining wall not less than 80%	Depth not more than 20% of average adjacent wall, and remaining wall not less than 80% for transverse (70% for longitudinal)
Stretching	OD not less than 97%	OD not less than 96%
String shot	OD not more than 103%	OD not more than 104%
External corrosion	Remaining wall not less than 80%	Remaining wall not less than 70%
Longitudinal cuts and gouges	Remaining wall not less than 80%	Remaining wall not less than 70%
Cracks	None	None
Internal Conditions		
Corrosion pitting	Remaining wall not less than 80%	Remaining wall not less than 70%
Erosion and internal wall wear	Remaining wall not less than 80%	Remaining wall not less than 70%
Cracks	None	None

torsion only, tension only). The tensile strength is decreased when the drill string is subjected to both axial tension and torque; the collapse pressure rating is also decreased when the drill pipe is simultaneously affected by collapse pressure and tensile loads.

2.2.2 Load Capacity of Drill Pipe

In normal drilling operations and in operations such as DST or washover, drill pipe is subjected to combined effects of stresses.

To evaluate the load capacity of drill pipe (e.g., allowable tensile load while simultaneously applying torque), the maximum distortion energy theory is usually applied. This theory is in good agreement with experiments on ductile materials such as steel. According to this theory, the equivalent stress may be calculated from the following formula [10]:

$$2\sigma_e^2 = (\sigma_z - \sigma_t)^2 + (\sigma_t - \sigma_r)^2 + (\sigma_r - \sigma_z)^2 + 6\tau^2 \quad (2.7)$$

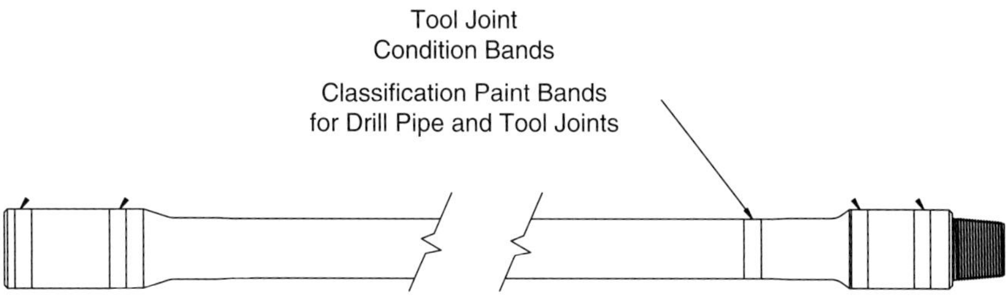

Tool Joint and Drill Pipe Classification	Number and Color of Bands
Premium Class	Two White
Class 2	One Yellow

Tool Joint Condition	Color of Bands
Scrap or Shop Repairable	Red
Field Repairable	Green

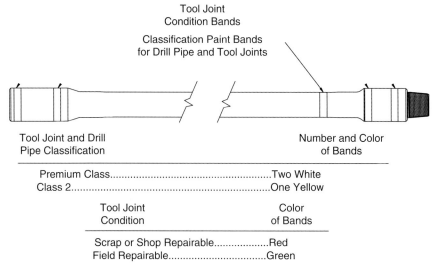

FIGURE 2.12 Color classification of drill pipe.

where σ_e = equivalent stress in psi
σ_z = axial stress in psi ($\sigma_z > 0$ for tension, $\sigma_z < 0$ for compression)
σ_t = tangential stress in psi ($\sigma_t > 0$ for tension, $\sigma_t < 0$ for compression)
σ_r = radial stress in psi (usually neglected for the drill pipe strength analysis)
τ = shear stress in psi

Yielding of pipe does not occur provided that the equivalent stress is less than the yield strength of the drill pipe. For practical calculations, the equivalent stress is taken to be equal to the minimum yield strength of the pipe as specified by API. The stresses being considered in Equation 2.7 are effective stresses that exist beyond any isotropic stresses caused by hydrostatic pressure of the drilling fluid.

Consider a case in which the drill pipe is exposed to an axial load (P) and a torque (T). The axial stress (σ_z) and the shear stress (τ) are given by the following formulas:

$$\sigma_z = \frac{P}{A} \tag{2.8}$$

$$\tau = \frac{T}{Z} \tag{2.9}$$

where P = axial load (lb)
A = cross-sectional area of drill pipe (in.^{2})

T = torque (in.-lb)

Z = polar section modulus of drill pipe (in.^{3})

$Z = 2J/D_{dp}$

$J = (\pi/32)(D_{dp}^4 - d_{dp}^4)$

D_{dp} = outside diameter of drill pipe (in.)

d_{dp} = inside diameter of drill pipe (in.)

Substituting Equation 2.8 and Equation 2.9 into Equation 2.7 and putting $\sigma_z = Y_m, \sigma_t = 0$, (tangential stress equals zero in this case), the following formulas are obtained:

$$P = A\left[Y_m^2 - 3\left(\frac{T}{Z}\right)^2\right]^{\frac{1}{2}} \tag{2.10}$$

$$P = \left[P_t^2 - 3\left(\frac{A \cdot T}{Z}\right)^2\right]^{\frac{1}{2}} \tag{2.11}$$

where $P_t = Y_m, A$ = tensile load capacity of drill pipe in uniaxial tensile stress (lb).

Equation 2.11 permits calculation of the tensile load capacity when the pipe is subjected to rotary torque (T).

Example 2.5

Determine the tensile load capacity of a $4\frac{1}{2}$ in., 16.60 lb/ft grade X-95 premium drill pipe subjected to a rotary torque of 12,000 ft-lb if the required safety factor is 2.0.

Solution

From Table 2.11, the following data are obtained:

$$A = 3.468\,\text{in.}^2$$

$$P_t = 329{,}460\,\text{lb}$$

$$Z = 6.694\,\text{in.}^4$$

Using Equation 2.11,

$$P = \left[(329,460)^2 - 3\left(\frac{3.468 \times 144{,}000}{6.694}\right)^2\right]^{\frac{1}{2}} = 303{,}063$$

Because of the safety factor of 1.2, the tensile capacity of the drill pipe is 303,063/1.2 = 252,552 lb.

Example 2.6

Calculate the maximum value of a rotary torque that may be applied to the drill pipe as specified in Example 2.5 if the actual working tension load

$P = 275{,}000$ lb. (e.g., pulling and trying to rotate a differentially stuck drill string).

Solution

From Equation 2.11, the magnitude of rotary torque is

$$T = \frac{Z}{A}\left(\frac{P_T^2 - P^2}{3}\right)^{\frac{1}{2}}$$

so

$$T = \frac{6.694}{3.468}\left[\frac{329{,}460^2 - 275{,}000^2}{3}\right]^{\frac{1}{2}}$$
$$= 202{,}194 \text{ in./lb or } 16{,}850 \text{ ft/lb}$$

Caution: No safety factor is included in this example calculation. Additional checking must be done if the obtained value of the torque is not greater than the recommended makeup torque for tool joints.

During normal rotary drilling processes, because of frictional pressure losses and the pressure drop across the bit nozzles, the pressure inside the drill string is greater than that in the annulus outside drill string. The greatest difference between these pressures is at the surface.

If the drill string is thought to be a thin-walled cylinder with closed ends, the drill pipe pressure produces the axial stress and tangential stress given by the following formulas (for stress calculations, the pressure in the annulus may be ignored):

$$\sigma_a = \frac{P_{dp} D_{dp}}{4t} \tag{2.12}$$

$$\sigma_t = \frac{P_{dp} D_{dp}}{2t} \tag{2.13}$$

where σ_a = axial stress (psi)
σ_t = tangential stress (psi)
P_{dp} = internal drill pipe pressure (psi)
D_{dp} = outsdie diameter of drill pipe (in.)
t = wall thickness of drill pipe (in.)

Substituting Equations 2.12, 2.13, and 2.8 into Equation 2.7 and solving for the tensile load capacity of drill pipe yields

$$P = \left[P_t^2 - \frac{3}{16}\left(\frac{P_{dp} D_{dp} A}{t}\right)^2 - 3\left(\frac{AT}{Z}\right)^2\right]^{\frac{1}{2}} \tag{2.14}$$

Example 2.7

Find the tensile load capacity of a 5-in., 19.50 lb/ft S-135 premium class drill pipe with an internal pressure of 3000 psi and an applied torque of 15,000 ft-lb.

Solution

From Table 2.11, 5-in. premium class pipe has a wall thickness of 0.290 in. and an OD of 4.855 in. The cross-sectional area is 4.153 in.^{2} The polar section modulus is 8.953 in.^{3} The tensile capacity is 560,655 lb. Using Equation 2.14,

$$P = \left\{ 560{,}655^2 - \frac{3}{16}\left[\frac{(3000)(4.855)(4.153)}{0.290}\right]^2 - 3\left[\frac{(4.513)(15{,}000)(12)}{8.953}\right]^2 \right\}^{\frac{1}{2}}$$

$$= 534{,}099 \text{ lb}$$

The reduction in tensile capacity of the drill pipe is 560,655 − 534,099 = 26,556 lb. For practical purpose, depending on drilling conditions, a reasonable safety factor should be applied.

During DST operations, the drill pipe may be affected by a combined effect of collapse pressure and tensile load. For such a case,

$$\frac{\sigma_t}{Y_m} = \frac{P_{cc}}{P_c} \tag{2.15}$$

or

$$\sigma_t = Y_m \frac{P_{cc}}{P_t} \tag{2.16}$$

where P_c = minimum collapse pressure resistance. (psi) (see Table 2.11)

P_{cc} = corrected collapse pressure resistance for effect of tension (psi)

Y_m = minimum yield strength of pipe (psi)

Substituting Equation 2.16 in Equation 2.8 and solving for P_{cc} yields (note: $\sigma_r = 0, \tau = 0, \sigma_c = Y_m$)

$$P_{cc} = P_c\left\{\left[-\left(\frac{P}{2AY_m}\right)^2\right]^{\frac{1}{2}} - \frac{P}{2AYm}\right\} \tag{2.17}$$

or

$$P_{cc} = P_c\left\{\left[1 - 0.75\left(\frac{\sigma_z}{Y_m}\right)^2\right]^{\frac{1}{2}} - 0.5\frac{\sigma_z}{Y_m}\right\} \tag{2.18}$$

Equation 2.18 indicates that increased tensile load results in decreased collapse pressure resistance. The decrement of collapse pressure resistance during normal DST operations is relatively small; nevertheless, under certain conditions, it may be considerable.

Example 2.8

Determine if the drill pipe is strong enough to satisfy the safety factor on collapse of 1.1 for the DST conditions as below:

- Drill pipe: $4\frac{1}{2}$ in., 16.60 lb/ft, G-105, premium class
- Drilling fluid with a density of 12 lb/gal and drill pipe empty inside
- Packer set at a depth of 8,500 ft
- Tension load of 45,000 lb applied the drill pipe

Solution

From Table 2.11, the collapse pressure resistance in uniaxial state of stress (P_c) is 9,467 psi. The cross-sectional area is 3.468 in^2. The axial tensile stress at the packer level is

$$\sigma_Z = \frac{45,000}{3.468} = 12,946 \cdot \text{psi}$$

The corrected collapse pressure resistance from Equation 2.18 is

$$P_c = 9,467 \left\{ \left[1 - 0.75 \left(\frac{12,946}{105,000} \right)^2 \right] - 0.5 \frac{12,946}{105,000} \right\}$$
$$= 8,775 \, \text{psi}$$

Hydrostatic pressure of the drilling fluid behind the drill string at the packer level is

$$P_h = (0.52)(12)(8,500) = 5,304 \text{ psi}$$
$$\text{Safety factor} = 8,775/5,304 = 1.65$$

The required safety factor is 1.1; therefore, the pipe may be run empty.

2.2.3 Tool Joints

Individual lengths of drill pipe are connected to each other with threaded rotary shouldered connections called *tool joints* (Figure 2.13). Four types of tool joints (or thread forms) are used commonly used.

API tool joints. Tool joints with specifications defined by the API include all NC connections, $5\frac{1}{2}$ FH, $6\frac{5}{8}$ FH, and API regular. connections. Detailed information on the dimensional properties of these tool joints can be found in API Spec 7 and Section B of the IADC Drilling Manual.

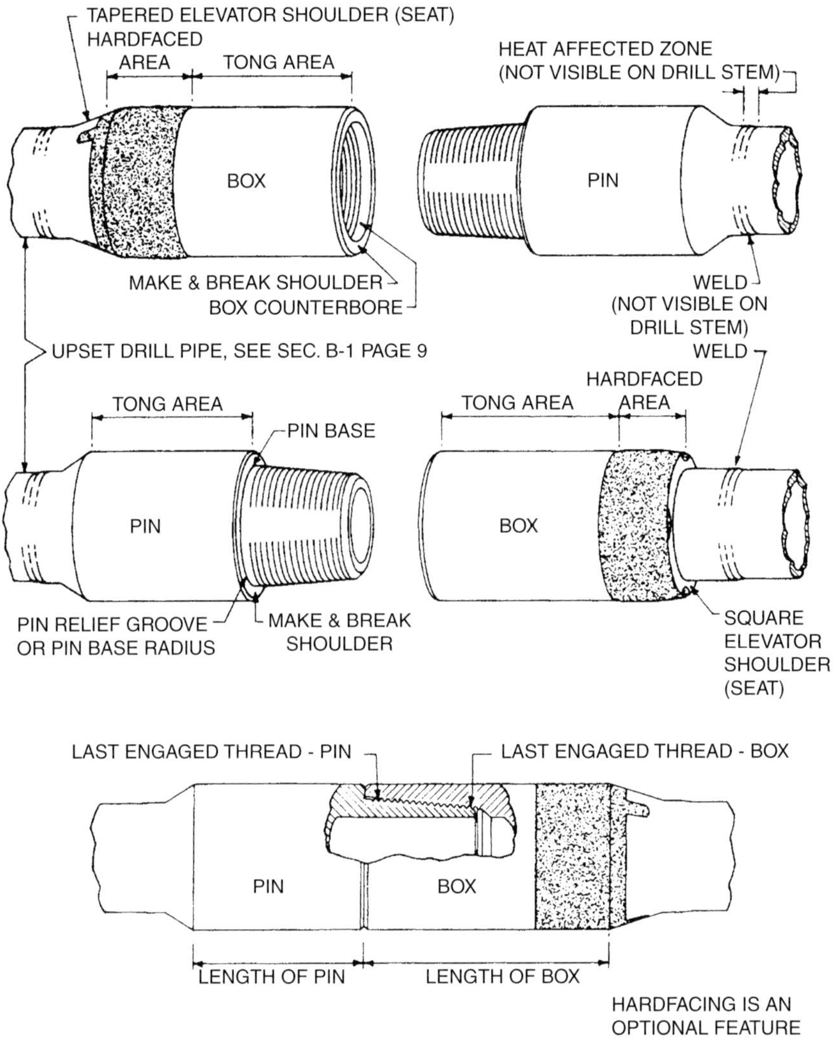

FIGURE 2.13 Tool joint nomenclature.

Single shoulder non-API tool joints. For tool joints similar to (and sometimes interchangeable with) API tool joints, detailed information on the dimensional properties can be found in Section B of the IADC Drilling Manual. Examples of non-API tool joints are extra hole (EH or XH), open hole (OH), and many others. Table 2.9 shows the interchangability of single-shoulder tool joints with API tool joints.

Double shoulder tool joints. Double-shoulder tool joints have a shoulder on the nose of the pin that contacts an internal shoulder in the box.

These additional shoulders limit the amount of axial deformation in the pin and provide more torsional strength.

Other proprietary threads. There are other tool joints that do not fall into the previous categories.

Table 2.6 contains the dimensions and performance characteristics of many possible tool joint, pipe size, pipe weight, and pipe grade combinations.

2.2.4 Makeup Torque

Makeup toque provides a preload in the connection that prevents downhole makeup, shoulder separation, wobble, or relative movement between the pin and box when subjected to the varied service loads the pipe sees and when rotating in doglegs. The preload provides the bearing stress in the shoulders to provide the seal in the tool joint. Tapered threads are used on tool joints to make it easier to put the pin and the box when adding lengths of pipe to the string, the taper does not seal.

Tool joints are generally made up to about 60% of their yield torque. Makeup torque values are shown in Table 2.6 and in API RP7G. The makeup torque for tool joints is calculated to induce a tensile stress at the last engaged thread of the pin or counterbore section of the box of 72,000 psi. A tolerance of ±10% is often used to speed up adding additional lengths of pipe to the drill string. As a deterrent to tool joint fatigue failures, Grant Prideco, who produced Table 2.6, includes this footnote in the drill pipe table of their product catalogue: "The makeup torque of the tool joint is based on the lower of 60% of the tool joint torsional yield strength or the 'T3' value calculated per the equation paragraph A.8.3 of API RP7G. Minimum makeup torques of 50% of the tool joint torsional strength may also be used."

The T3 value is the torsional load required to produce additional makeup of the tool joint when the shoulders are separated by an external tensile load that produces yield stress in the tool joint pin. This unlikely load case would occur when the pipe is stuck.

An explanation of these formulas and dimensional values for tool joints needed to solve them can be found in API RP7G.

2.2.5 Heavy-Weight Drill Pipe

Heavy-weight drill pipe with wall thicknesses of approximately 1 in. is frequently used for drilling vertical and directional holes (see Figure 2.14). It is often placed directly above the drill collars in vertical holes because it has been found to reduce the rate of fatigue failures in drill pipe just above the collars.

The best performance of individual members of the drill string is achieved when the bending stress ratio (BSR) of adjoining lengths of pipe

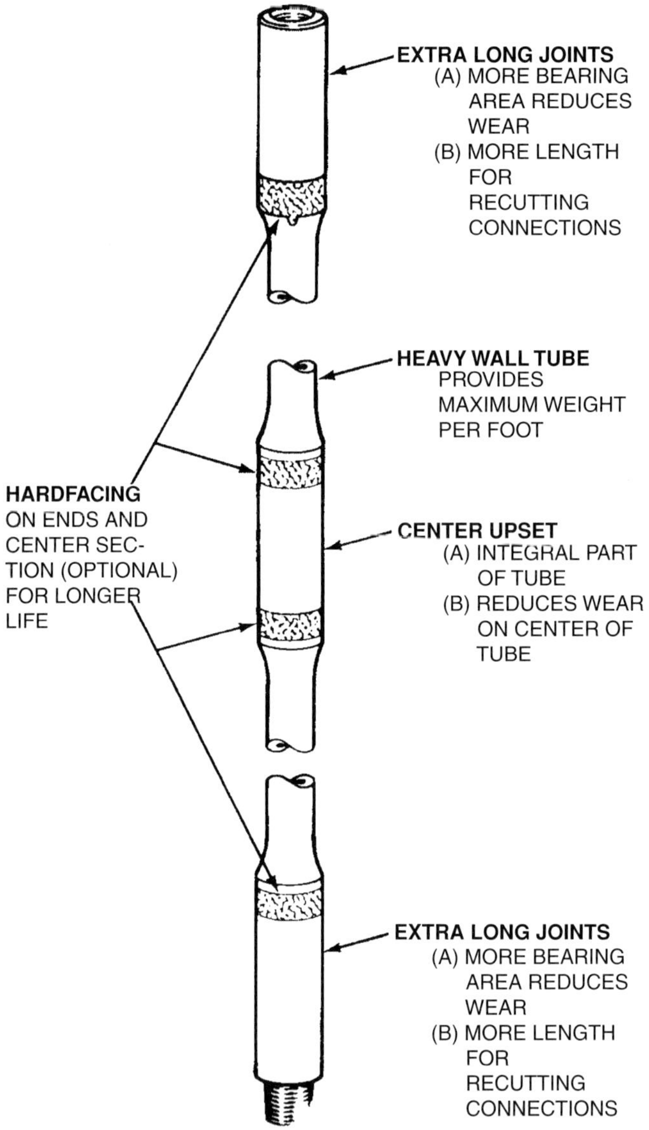

FIGURE 2.14 Drilco's Hevi-Wate® drill pipe [9].

is less than 5.5 [2]. BSR in this case is defined as a ratio of the bending section moduli of two connecting members (e.g., between the top most drill collar and the pipe just above it). For severe drilling conditions such as hole enlargement, corrosive environment, or hard formations, reduction of the BSR to 3.5 helps to reduce the frequency of drill pipe failure. To maintain the BSR of less than 5.5 between adjacent members of the drill string, the drill collar string must sometimes be made up of collars with different sizes. Heavy-weight pipe can be used at the transition from drill collars to drill pipe to bring the BSR into the desired range.

Sometimes, heavy-weight pipe is used instead of drill collars to provide weight to the bit. Pipe can be used because drill collars create excessive torque and drag or because there may be a deferential sticking problem. In directional holes, heavy-weight pipe is often used to provide weight to the bit. In many directional holes in which the deviation from vertical is too much for the pipe to slide, heavy-weight pipe is placed above the build zone.

The dimensional and mechanical properties of heavy-weight drill pipe manufactured by the Drilco division of Smith International are given in Table 2.13.

Example 2.9

Calculate the required length of $4\frac{1}{2}$ in. heavy-weight drill pipe for the following conditions:

- Hole size: $9\frac{1}{2}$ in.
- Hole angle: 40deg
- Desired weight on bit: 40,000 lb
- Drill collars: $7 \times 2\frac{13}{6}$ in.
- Length of drill collars: 330 ft.
- Drilling fluid specific gravity: 1.2
- Desired safety factor for neutral point: 1.15

Solution

Check to see if the BSR of drill collar and heavy-weight drill pipe is less that 5.5. The bending section modulus of heavy-weight drill pipe is

$$\frac{\pi}{16}\left[\frac{(4.5)^4-(2.75)^4}{4.5}\right]=15.397 \text{ in.}^3$$

$$\text{BSR}=\frac{65.592}{15.397}=4.26<5.5$$

The unit weight of drill collar in drilling fluid is

$$110\left(1-\frac{1.2}{7.85}\right)=93.18 \text{ lb/ft.}$$

The unit weight of heavy-weight drill pipe in drilling fluid is $(41)(0.847)=34.72$ lb/ft. Part of the weight on bit may be created by using drill collars $=(93.18)(330)(\cos 40)=23{,}555$ lb. The required length of heavy-weight drill pipe is

$$\frac{(40{,}000-23{,}555)(1.5)}{(34.72)(\cos\cdot 40)}=711 \text{ ft}$$

Assuming an average length of one joint of heavy-weight drill pipe to be 30 ft, 24 joints are required.

2.2.6 Fatigue Damage to Drill Pipe

Most drill pipe and tool joint failures occur as a result of fatigue. Fatigue damage is caused by the cyclic bending loads induced in the drill pipe during service. Cyclic stress results in a crack, which is the first stage of a fatigue failure. The crack grows to the point where the remaining cross-section is not great enough to support the service loads, and the drill pipe separates, this is the second stage. A washout often precedes a catastrophic failure. If the driller is able to detect a drop in standpipe pressure or a change in the sound of the pumps, the pipe can be pulled out of the hole intact.

Fatigue cracks most often occur on the outside surface because that is where bending stresses are highest and because material defects such as corrosion pitting are on the surface. Fatigue cracks can also begin on the inside surface of stress raisers such as pitting exist.

The resistance of a material to fatigue, or its endurance limit, is proportional to its ultimate strength. The endurance limit of a material is the cyclic stress below which a fatigue failure will not occur or the pipe will run under those conditions "forever." Forever in terms of drill pipe is usually between 2 and 3 million cycles in the laboratory. Endurance limits of particular materials are determined by conducting cyclic bending tests on a small specimen. These tests show that the endurance limit of the small, carefully machined samples is about 50% of the material's ultimate strength. The endurance limit of a full-size joint of drill pipe tested in the laboratory free from a corrosive environment is between 20,000 and 30,000 psi.

In general terms, the endurance limit is an indicator of how long, if ever, it will take a crack to form under cyclic loading. The stronger the pipe, the greater the endurance limit. The impact strength of the material is an indicator of how long it will take the crack to propagate to the point of separation. Generally, the stronger the pipe, the quicker the crack will propagate.

Based on work done by A. Lubinsky, J. E. Hansford, and R. W. Nicholson, API RP7G contains the formulas for the maximum permissible hole curvature to avoid fatigue damage to drill pipe.

$$c = \frac{432{,}000}{\pi} \frac{\sigma_b}{ED_{dp}} \frac{\tanh(KL)}{KL} \tag{2.19}$$

$$K = \left(\frac{T}{EI}\right)^{\frac{1}{2}} \tag{2.20}$$

For grade E drill pipe

$$\sigma_b = 19{,}500 - (0.149)\sigma_t \tag{2.21}$$

For grade S drill pipe

$$\sigma_b = 20{,}000\left(1 - \frac{\sigma_t}{145{,}000}\right) \tag{2.22}$$

$$\sigma_t = \frac{T}{A} \tag{2.23}$$

where

- c = maximum permissible dogleg severity in (degrees/100 ft)
- σ_b = maximum permissible bending stress (psi)
- σ_t = tensile stress from weight of suspended drill string (psi)
- E = Young's modulus, $E = 30 \times 10^6$ (psi)
- D_{dp} = outside diameter of drill pipe (in.)
- L = half the distance between tool joints (in.), $L = 180$ in. for range 2 drill pipe.
- T = weight of drill pipe suspended below the dogleg (lb)
- I = drill pipe moment of inertia with respect to its diameter (in.^{4})
- A = cross-sectional area of drill pipe (in.^{2})

By intelligent application of these formulas, several practical questions can be answered at the borehole design state and while drilling.

Example 2.10

Calculate the maximum permissible hole curvature for the following drill string:

- 5-in., 19.50 lb/ft S-135 range 2 premium class pipe with $6\frac{5}{8}$ OD $\times$ $2\frac{3}{4}$ ID tool joints
- Drill collars: $7 \times 2\frac{1}{4}$ in., unit weight of 117 lb/ft
- Length of drill collars: 550 ft
- Drilling fluid density: 12 lb/gal
- Anticipated length of hole below the dogleg: 8,000 ft
- Assume hole is vertical below dogleg

Solution

From Table 2.11, $D_{dp} = 4.855$ in., $d_{dp} = 4.276$ in., and $A = 4.153$ in.^{2}, and from Table 2.6, the unit weight of drill pipe adjusted for tool joint (W_{dp}) is 23.87 lb/ft. Even though this example is for premium class pipe, the weight for new pipe is used because it cannot be known for certain that the wall thickness of all the pipe is uniformly worn to 80% of its new thickness.

The weight of the drill collar string is

$$(550)(117)\left(1-\frac{12}{65.4}\right)=52{,}543\ \text{lb}$$

The weight of the drill pipe is

$$(8{,}000-550)(23.87)\left(1-\frac{12}{65.4}\right)=145{,}202\ \text{lb}$$

Weight suspended below the dogleg (T = 197,745 lb) is

$$\text{Tensile stress } \sigma_t=\frac{197{,}745}{4.153}=47{,}615\ \text{psi}$$

Maximum permissible bending stress is

$$\sigma_b=20{,}000\left(1-\frac{47{,}615}{145{,}000}\right)=13{,}432\ \text{lb}$$

Drill pipe moment of inertia is

$$I=\frac{\pi}{64}\left[(4.855)^4-(4.276)^4\right]=10.862\ \text{in.}^4$$

$$K=\sqrt{\frac{197{,}745}{(30)(10)^6(10.862)}}=2.463\times10^{-2}\text{in.}^{-1}$$

Maximum permissible hole curvature is

$$c=\frac{432{,}000}{\pi}\ \frac{13{,}432}{(30)(10)^6(5)}\ \frac{\tanh[(2.406\times10^{-2})(180)]}{(2.406\times10^{-2})(180)}$$
$$=2.842^\circ/100\ \text{ft}$$

The results of these calculations indicate that if a length of drill pipe were rotated in a dogleg under these conditions, a fatigue crack eventually would develop.

2.3 DRILL STRING INSPECTION PROCEDURE

Drill pipe should be periodically inspected to determine the following:

- Reduction of pipe body wall thickness that reduces pipe strength
- Reduction of tool joint OD that reduces tool joint strength
- Cracks in pipe body or tool joint threads that will lead to washouts or a parted string
- Damage to tool joint threads and makeup shoulders that could cause galling or washouts

- Extent of corrosion damage that could reduce pipe strength or lead to washouts
- Excessive slip damage that could lead to washouts
- Condition of internal plastic coating. Look for cracking, specifically under the pin threads (indicates stretched pin) and near the internal upset area where corrosion could lead to washouts.
- Condition of hardbanding
- Changes in dimensions caused by excessive service loads such as stretched pipe, stretched pin threads, and swollen boxes.

The frequency of inspection for the above varies. Some are visual inspections that can be done when the pipe is tripped out of the hole or while tripping back in. Generally, pipe is inspected after each job. More detailed information on types of inspections, procedures and frequencies can be found in API RP7G and ISO 10407-2.

Table 2.14, taken from API RP7G, shows the allowable wear and damage limits for Premium and Class 2 pipe.

2.3.1 Drill String Design

The drill string design should determine the optimum combination of drill pipe sizes, weights, and grades for the lowest cost or incorporate the performance characteristics to successfully accomplish the expected goals. The drill string is subjected to many service loads that may exist as static loads, cyclic loads, or dynamic loads. These loads include tensile loads, torsional loads, bending loads, internal pressure, and compressive loads. In certain circumstances, the drill string experiences external pressure. In addition to the varied service loads the pipe will see, the designer must consider hole drag and the risk of becoming stuck.

The designer should simultaneously consider the following conditions:

1. The working load at any part of the string must be less than or equal to the load capacity of the drill string member under consideration divided by the safety factor.
2. Ratio of section moduli of individual string members should be less than 5.5.
3. Pressure losses through the pipe going downhole and around the pipe going back to surface should be calculated so that the combination of pipe size and tool joint size are right for the diameter of the hole.

Selecting the size, weight, and grade of drill pipe and the tool joint OD and ID may be an iterative process, but the pipe size usually can be determined early on based on past experience and pipe availability. After calculating the strength requirements of the pipe and the lengths of each section of tapered strings, it may be necessary to chose a different pipe size from the one initially picked.

In a vertical hole, the maximum length of the pipe above the heavy-weight pipe or the drill collars is determined by the following equation:

$$(L_{dc}W_{dc} + L_{hw}W_{hw} + L_{dp1}W_{dp1})K_b = \frac{P_1}{SF} \tag{2.24}$$

where

L_{dc} = length of drill collar string (ft)
W_{dc} = unit weight of drill collar in air (lb/ft)
L_{hw} = length of heavy-weight drill pipe if used in the string (ft)
W_{hw} = unit weight of heavy-weight drill pipe (lb/ft)
L_{dp1} = length of drill pipe under consideration above the heavy-weight drill pipe (ft)
W_{dp1} = unit weight of drill pipe in section 1 (lb/ft)
K_b = buoyancy factor
P_1 = tensile load capacity of drill pipe in section 1 (lb)
SF = safety factor

Solving Equation 2.24 for L_{dp1} yields

$$L_{dp1} = \frac{P_1}{SF \cdot K_b \cdot W_{dp1}} - \frac{L_{dc}W_{dc}}{W_{dp1}} - \frac{L_{hw}W_{hw}}{W_{dp1}} \tag{2.25}$$

If the sum of $L_{dc} + L_{hw} + L_{dp1}$ is less than the planned borehole depth, stronger pipe must selected; that is, P_1 must be increased, or a stronger section of pipe must be placed above section 1.

The maximum length of the upper part in a tapered string in a vertical hole may be calculated from Equation 2.26:

$$L_{dp2} = \frac{P_2}{SF \cdot K_b \cdot W_{dp2}} - \frac{L_{dc}W_{dc}}{W_{dp2}} - \frac{L_{hw}W_{hw}}{W_{dp2}} - \frac{L_{dp1}W_{dp1}}{W_{dp2}} \tag{2.26}$$

where

P_2 = tensile load capacity of next (upper) section of drill pipe (lb)
L_{dp2} = length of drill pipe section 2 (ft)
W_{dp2} = unit weight of drill pipe section 2 (lb/ft)

Normally, not more that two sections are designed, but if necessary, three or more sections can be used. To calculate the tensile load capacity of drill pipe, it is suggested to apply Equation 2.11 and use the recommended makeup torque of the weakest tool joint for the rotary torque. The rotary torque should not exceed the makeup torque of any tool joint in the string.

The magnitude of the safety factor is very important and usually ranges from 1.4 to 2.8, depending on downhole conditions, drill pipe condition, and acceptable degree of risk. It is recommended that a value of safety factor be selected to produce a margin of overpull of at least about 70,000 lb.

$$MOP = P_n - K_b WT_{string} \tag{2.27}$$

where MOP = margin of overpull (lb)
P_n = tensile load capacity of top section of pipe (lb)
WT_{string} = combined air weight of entire drill string

Horizontal or deviated holes. Drill string design is more difficult on horizontal and deviated holes. Calculating frictional drag, the effects of compressive loading, and other factors by hand can be cumbersome and inaccurate. For these type holes, it is better to use torque and drag software for drill string design.

Slip loading. An additional check, especially in deep drilling, should be done to avoid drill pipe crushing in the slip area. The maximum load that can be suspended in the slips can be found from Equation 2.28:

$$W_{max} = \frac{P_1}{SF\left[1 + \frac{D_{dp}}{2}\frac{K}{L_s} + \left(\frac{D_{dp}}{2}\frac{K}{L_s}\right)^2\right]^{1/2}} \tag{2.28}$$

$$K = \frac{(1 - f \cdot \tan\alpha)}{f + \tan\alpha} \tag{2.29}$$

where

W_{max} = maximum allowable drill string load that can be suspended in the slips (lb)
P_t = tensile load capacity of drill pipe (lb)
D_{dp} = outside diameter of drill pipe (in)
L_s = length of slips (in)
K = lateral load factor of slip
F = coefficient of friction between slips and slip bowl, often taken as 0.08
α = slip taper (9° 27′45″)
SF = safety factor to account for dynamic loads when slips are set on moving drill pipe (SF = 1.1)

Example 2.11

Design a drill string for the conditions specified:

- Hole depth: 10,000 ft
- Hole size: $9\frac{7}{8}$ in.
- Mud weight: 12 lb/gal
- Maximum weight on bit: 60,000 lb

- Neutral point design factor: 1.15
- No crooked hole tendency
- Safety factor for tension: 1.4
- Required margin of overpull: 100,000 lb
- From offset wells, it is known that six joints of heavy-weight drill pipe are desirable
- Assume a vertical hole

Solution

Select the drill collar size from Table 2.3: $7\frac{3}{4} \times 2\frac{13}{16}$ in., 139 lb/ft. Such drill collars can be caught with overshot or washed over with washpipe.

$$\text{Length of drill collars} = \frac{(60,000)(1.15)}{(139)(0.816)} = 608 \text{ ft}$$

$$\text{Buoyancy factor } K_b = \left(1 - \frac{12}{65.4}\right) = 0.816$$

Select 21 joints of $7\frac{3}{4} \times 2\frac{13}{16}$ in. drill collars. This is a drill collar length of 630 ft. The calculated section modulus of the drill collars is 89.6 in.^{3}

Example 2.12

To maintain a BSR of less than 5.5, select 5-in. heavy-weight drill pipe with a section modulus of 21.4 in^3 and a unit weight of 49.3 lb/ft (see Table 2.13).

$$\text{Length of heavy-weight drill pipe } L_{hw} = (6)(30) = 180 \text{ ft}$$

Select 5 in., 19.50 lb/ft grade G-105 premium class pipe with $6\frac{5}{8} \times 3\frac{1}{2}$ tool joints (see Table 2.6).

The unit weight of drill pipe with tool joints is 22.63 lb/ft (weight calculation based on new pipe with no wear). The section modulus of this pipe can be calculated to be 4.5 in.^{3} From Table 2.11, the minimum tensile load capacity of the selected pipe is 436,065 lb. From Table 2.6, the recommended makeup torque for this tool joint is 23,400 ft-lb.

The tensile load capacity of the drill pipe, corrected for the effect of the maximum allowable torque, according to Equation 2.11 is

$$P = \left\{(436,065)^2 - 3\left[\frac{(4.153)(23,400)(12)}{8.953}\right]^2\right\}^{\frac{1}{2}}$$
$$= 373,168 \text{ lb}$$

Determine the maximum allowable length of the selected drill pipe from Equation 2.25:

$$L_{dp} = \frac{373,168}{(1.4)(0.816)(22.63)} - \frac{(630)(139)}{22.63} - \frac{(180)(49.3)}{22.63}$$
$$= 10,172 \text{ ft}$$

Required length of drill pipe $L_{dp} = 10{,}000 - (630 + 180) = 9{,}190\,\text{ft}$.

Because the required length of drill pipe (9,190 ft) is less than the maximum allowable length (10,172 ft), it is apparent that the selected drill pipe satisfies the tensile load requirements.

Margin of overpull can be calculated in this way:

$$\begin{aligned} MOP &= (0.90)(436{,}065) - [(630)(139) \\ &\quad + (180)(49.3) + (9{,}190)(22.63)](0.816) \\ &= 144{,}057\ \text{lb (greater than required 100,000 lb)} \end{aligned}$$

In this example, the cost of the drill string is not considered. From a practical standpoint, the calculations outlined should be performed for various drill pipe unit weights and steel grades, and the design that produces the lowest cost should be selected.

The maximum load that can be suspended in the slips, from Equation 2.28 (assume $K = 2.36$, $L_s = 12$) is

$$W_{max} = \frac{436{,}065}{1.1\left[1 + \frac{5}{2}\frac{2{,}36}{12} + \left(\frac{5}{2}\frac{2.36}{12}\right)^2\right]^{\frac{1}{2}}} = 301{,}099\ \text{lb}$$

Total weight of string = 248,401 lb

The drill pipe will not be crushed in the slips. The drill string design satisfies the specified criteria.

References

[1] API Specification 7, 40th Edition, "Specification for Rotary Drill Stem Elements," November 2001.

[2] *Drilco Drilling Assembly Handbook*, Drilco Division of Smith International, Inc., 1977.

[3] Wilson, G. E., and W. R. Garrett, "How To Drill A Usable Hole," Part 3, *World Oil*, October 1976.

[4] API Recommended Practice 7G, 16th Edition, "Recommended Practice for Drill Stem Design and Operating Limits." December 1998.

[5] Grant Prideco Product Catalog, December 2002.

[6] API Specification 5D, 4th Edition, "Specification for Drill Pipe." January 2000.

[7] Dawson R, and P. R. Paslay, "Drilling Pipe Buckling in Included Holes," SPE Paper 11167 (presented at the 57th Annual Fall Technical Conference and Exhibition of the SPE of AIME held in New Orleans, September 26–29, 1983).

[8] Lubinski, A., "Maximum Permissible Dog-legs in Rotary Boreholes," *Journal of Petroleum Technology*, February 1961.

[9] Rowe, M. E., "Heavy Wall Drip Pipe, a Key Member of the Drill Stem," Publ. No.45, Drilco, Division of Smith International, Inc., Houston, 19XX.

[10] Timoshenko, S., and D. H. Young, *Elements of Strength of Materials*, Fifth Edition, D. Van Nostrand Co., New York, 1982.

CHAPTER

3

Air and Gas Drilling

OUTLINE

Air and gas drilling refers to the use of compressed air (or other gases) as the circulating drilling fluid for rotary drilling operations. The majority of these drilling operations use compressed air as the circulating drilling fluid. In some oil and natural gas recovery drilling operations it is necessary to drill with a gas that will not support downhole combustion. This is particularly the case when drilling in or near the hydrocarbon producing formations. This objective has been realized by using natural gas or inert atmospheric air (air stripped of most oxygen) as drilling gases. Natural gas from pipelines has been used as a drilling gas since the 1930's. The use of inert air as a drilling gas is another technological development [1]. Drilling location deployable equipment units are available to the oil and natural gas resource recovery industry. These units are also called "nitrogen generators." The units strip most of the oxygen content from the compressed air output from standard positive displacement primary compressors (rotary or reciprocation piston compressors). These inert air generating field units have only been in use for a few years.

The basic planning steps for the drilling of a deep well with compressed gases are as follows [2]:

1. Determine the geometry of the borehole section or sections to be drilled with air or other gases (i.e., openhole diameters, the casing or liner inside diameters, and depths).
2. Determine the geometry of the associated drill strings for the sections to be drilled with air or other gases (i.e., drill bit size and type, the drill collar size, drill pipe size, and maximum depth).
3. Determine the type of rock formations to be drilled and estimate the anticipated drilling rate of penetration. Also, estimate the quantity and depth location of any formation water that might be encountered.
4. Determine the elevation of the drilling site above sea level, the temperature of the air during the drilling operation, and the approximate geothermal temperature gradient.
5. Establish the objective of the air (or other gas) drilling operation:
   - To eliminate loss of circulation problems,
   - To reduce formation damage,
   - To allow formation fluids to be produced as the formation is drilled.
6. Determine whether direct or reverse circulation techniques will be used to drill the various sections of the well.
7. Determine the required approximate minimum volumetric flow rate of air (or other gas) to carry the rock cuttings from the well when drilling at the maximum depth.
8. Select the contractor compressor(s) that will provide the drilling operation with a volumetric flow rate of air that is greater than the required minimum volumetric flow rate (use a factor of safety of at least 1.2).
9. Using the compressor(s) air volumetric flow rate to be injected into the well, determine the bottomhole and surface injection pressures as a function of drilling depth (over the interval to be drilled). Also, determine the maximum power required by the compressor(s) and the available maximum derated power from the prime mover(s).
10. Determine the approximate volume of fuel required by the compressor(s) to drill the well.
11. In the event formation water is encountered, determine the approximate volumetric flow rate of "mist" injection water needed to allow

formation water or formation oil to be carried from the well during the drilling operation.

12. Determine the approximate volumetric flow rate of formation water or formation oil that can be carried from the well during the drilling operation (assuming the injection air will be saturated with water vapor at bottomhole conditions).

The circulation system for an air (or other gas) drilling operation is a typical compressible fluid flow calculation problem. In these problems, the pressure and temperature is usually known at the exit to the system. In this case, the exit is at the end of the blooey line. At this exit, the pressure and temperature are the local atmospheric pressure and temperature at the drilling location. The calculation procedures for air drilling circulation problems are to start at these known exit conditions and work upstream through the system. In these calculations, the volumetric flow rate must be assumed or known. If compressors are used to provide compressed air, then the volumetric flow rate is the sum of the outputs of all the primary compressors. For direct circulation drilling operations, the equations given in the "*Bottomhole Pressure*" subsection below are applied from the top of the well annulus to each constant cross-section section in sequence starting from the top well. The pressure found at the bottom of each constant cross-section section is used as the initial pressure for the next deeper constant cross-section section until the bottomhole pressure at the bottom of the annulus is determined. In this calculation, all major and minor flow losses should be considered. The minimum volumetric flow rate air (or gas) can be determined using the equations in the "*Bottomhole Pressure*" subsection and the kinetic energy equation given in the "*Minimum Volumetric Flow Rate*" subsection [1, 2]. The actual volumetric flow rate to the well must be greater than the minimum by a factor of at least 1.2. Working upstream, the pressure above the drill bit orifices (or nozzles) inside the drill bit is found using the equations in "*Drill Bit Orifices and Nozzles*" subsection. Care must be taken to determine whether the flow through these bit openings is sonic or subsonic. The equations given in the "*Injection Pressure*" subsection are applied from the bottom of the inside of the drill string to each constant cross-section section in sequence starting from the pressure above the drill bit. The pressure found at the top of each constant cross-section section is used as the initial pressure for the pressure at the top of next constant cross-section section until the injection pressure at the top of the inside of the drill string is found.

The methodology outline above has been successful in predicting bottomhole and injection pressures with an accuracy of about 5%. In order to attain this accuracy, it is necessary to consider all major and minor flow losses in the circulating system, and to account for any water or other incompressible fluids being carried from the well by the air or gas drilling fluid.

3.1 BOTTOMHOLE PRESSURE

The perfect gas law is used as the basis of the air and gas drilling equations presented. The perfect gas law can be written as

$$\frac{P}{\gamma} = \frac{RT}{S_g} \tag{3.1}$$

where

P is pressure (lb/ft^2 absolute, or N/m^2 absolute),
γ is specific weight (lb/ft^3, or N/m^3),
T is absolute temperature (°R, or K),
R is the universal gas constant (53.36 lb-ft/lb-°R, or 29.28 N-m/N-K),
S_g is the specific gravity of the gas.

The equation for the pressure in the air (or gas) flow at the entrance end to the blooey line (just after the Tee from the annulus) can be approximated by

$$P_b = \left[\left(f_b\frac{L_b}{D_b} + K_t + \sum K_v\right)\left(\frac{\dot{w}_g^2 R T_r}{g A_b^2 S_g}\right) + P_{at}^2\right]^{0.5} \tag{3.2}$$

where

P_b is the pressure at the entrance to the blooey line (lb/ft^2 absolute, or N/m^2 absolute).
P_{at} is the atmospheric pressure at the exit to the blooey line (lb/ft^2 absolute, or N/m^2 absolute).
f_b is the Darcy-Weisbach friction factor for the blooey line.
L_b is the length of the blooey line (ft, or m).
D_b is the inside diameter of the blooey line (ft, or m).
K_t is the minor loss factor for the Tee turn at the top of the annulus.
K_v is the minor loss factor for the valves in the blooey line.
$\dot{w}_g$ is the weight rate of flow of gas (lb/sec, or N/sec).
A_b is the cross-sectional area of the inside to the blooey line (ft^2, or m^2).
T_r is the average temperature of the gas flow in the blooey line (°R, or K)

The approximate value of the K_t can be determined from Figures 3.1 and 3.2.

The friction factor is determined from the empirical von Karman relationship. This relationship is

$$f_b = \left[\frac{1}{2\log_{10}\left(\frac{D_b}{e}\right) + 1.14}\right]^2 \tag{3.3}$$

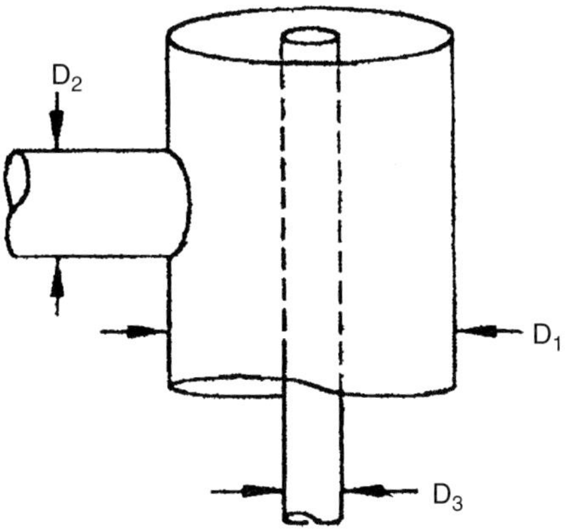

FIGURE 3.1 Dimensions of the blind Tee at top of annulus.

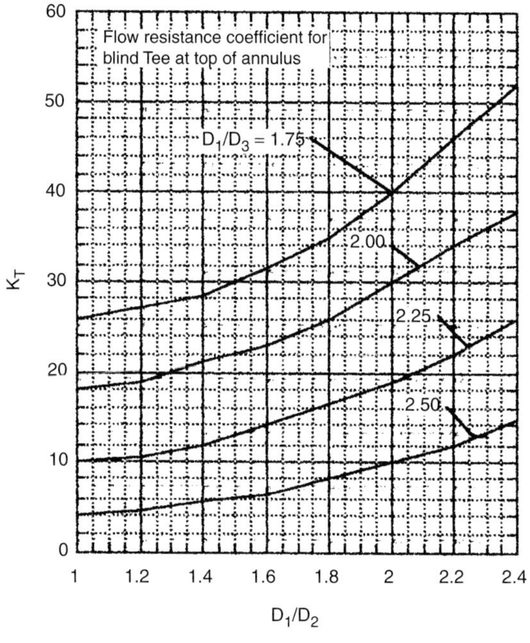

FIGURE 3.2 Flow resistance coefficient for the blind Tee at top of annulus.

where e is the approximate absolute roughness of the blooey line inside surface (ft, or m).

For direct (conventional) circulation for air and gas drilling, the bottomhole pressure in the annulus is given by

$$P_{bh} = \left[\left(P_{at}^2 + b_a T_{av}^2\right) e^{\frac{2a_a H}{T_{av}}} - b_a T_{av}^2\right]^{0.5} \tag{3.4}$$

where

P_{bh} is the bottomhole pressure in the annulus (lb/ft^2 absolute, or N/m^2 absolute). (This pressure can also be the pressure at the bottom of any calculation interval with a uniform cross-sectional flow area.)
P_{at} is the pressure at the top of the annulus (lb/ft^2 absolute, or N/m^2 absolute). (This is the atmospheric pressure when the calculation interval starts at the top of the well. However, it is the pressure at the bottom of the calculation interval of the interval above if numerous intervals with various cross-sectional flow areas are being considered.)
T_{av} is the average temperature of the calculation interval (°R, or K).
H is the length (or height) of the calculation interval (ft, or m).

The constants a_a and b_a are

$$a_a = \left(\frac{S_g}{R}\right)\left[1 + \left(\frac{\dot{W}_s}{\dot{W}_g}\right)\right] \tag{3.5}$$

and

$$b_a = \frac{f}{2g(D_h - D_p)}\left(\frac{R}{S_g}\right)\frac{\dot{W}_g^2}{\left(\frac{\pi}{4}\right)^2\left(D_h^2 - D_p^2\right)^2} \tag{3.6}$$

where

f is the Darcy-Weisbach (or Moody) friction factor for the annulus (this can be a weighted average if the annulus has dissimilar surfaces).
g is the acceleration of gravity (i.e., 32.2 ft/sec^2, or 9.81 m/sec^2)
$\dot{w}_s$ is the weight rate of flow of the solids (lb/sec, or N/sec).
$\dot{w}_g$ is the weight rate of flow of gas (lb/sec, or N/sec).
D_h is the inside diameter of the annulus borehole (ft, or m).
D_p is the outside diameter of the pipe in the annulus (ft, or m).

The values of $\dot{w}_s$ and $\dot{w}_g$ are determined from

$$\dot{W}_s = \frac{\pi}{4} D_h^2 \gamma_w S_s \kappa \tag{3.7}$$

and

$$\dot{W}_g = \gamma_g Q_g \tag{3.8}$$

where

γ_w is the specific weight of the water (62.4 lb/ft^3, or 9810 N/ft^3).
γ_g is the specific weight of the gas (lb/ft^3, or N/ft^3).
S_s is the specific gravity of the solids cuttings.

Q_g is the volumetric flow rate of gas at reference pressure and temperature conditions (ft^3/sec, or m^3/sec).
κ is the drilling rate of penetration (ft/sec, or m/sec).

The compressed air or other gas flowing through the annulus is assumed to be wholly turbulent, therefore, the friction factor is determined from the empirical von Karman relationship for annulus flow. This relationship for the annulus is [3]

$$f = \left[\frac{1}{2 \log_{10}\left(\frac{D_h - D_p}{e_{av}} \right) + 1.14} \right]^2 \tag{3.9}$$

where e_{av} is the average absolute roughness of the annulus surfaces (ft, or m).

The average annulus open hole absolute roughness can be determined from a surface area weight average relationship between the inside openhole surface area and its roughness and the outside surface area of the drill string and its roughness. The weight average relationship is [2]

$$e_{av} = \frac{e_{oh} D_h^2 + e_p D_p^2}{D_h^2 + D_p^2} \tag{3.10}$$

The openhole borehole inside surface absolute roughness can be approximated from Table 3.1. The outside absolute roughness of the outside of the steel pipe is usually taken as 0.00015 ft [3].

TABLE 3.1 Openhole Wall Approximate Absolute Roughness for Rock Formation Types [1]

Rock Formation Types	**Surface Roughness (ft)**
Competent, low fracture	
Igneous (e.g., granite, basalt)	0.001 to 0.02
Sedimentary (e.g., limestone, sandstone)	
Metamorphic (e.g., gneiss)	
Competent, medium fracture	
Igneous (e.g., granite, basalt)	
Sedimentary (e.g., limestone, sandstone)	0.02 to 0.03
Metamorphic (e.g., gneiss)	
Competent, high fracture	
Igneous (e.g., breccia)	
Sedimentary (e.g., sandstone, shale)	0.03 to 0.04
Metamorphic (e.g., schist)	

3.2 MINIMUM VOLUMETRIC FLOW RATE

The minimum volumetric flow rate can be approximated by determining the volumetric flow rate that will give a minimum kinetic energy per unit volume in the annulus equal to 3.0 lb-ft/ft^3 (or 141.9 N-m/m^3). This is usually at the section of the annulus that has the largest cross-sectional area. The kinetic energy per unit volume is

$$KE = \frac{1}{2}\frac{\gamma_g}{g}V^2 \tag{3.11}$$

where

KE is the kinetic energy per unit volume (lb-ft/ft^3, or N-m/m^3).
V is the velocity of the gas in the annulus (ft/sec, or m/sec).

3.3 DRILL BIT ORIFICES OR NOZZLES

Gas flow through an orifice or nozzle constriction can be either sonic or subsonic. These flow conditions depend upon the critical pressure ratio. The critical pressure ratio is the relationship between the downstream pressure and the upstream pressure. This ratio is a function of the properties of the gas and is given by [3]

$$\left(\frac{P_2}{P_1}\right)_c = \left(\frac{2}{k+1}\right)^{\frac{k}{k-1}} \tag{3.12}$$

where

P_1 is the upstream pressure (lb/ft^2 absolute, or N/m^2 absolute).
P_2 is the downstream pressure (lb/ft^2 absolute, or N/m^2 absolute).
k is the ratio of specific heats of the gas.

The flow conditions through the constriction are sonic if

$$\left(\frac{P_2}{P_1}\right)_c \leq \left(\frac{2}{k+1}\right)^{\frac{k}{k-1}} \tag{3.13}$$

The flow conditions through the constriction are subsonic if

$$\left(\frac{P_2}{P_1}\right)_c \geq \left(\frac{2}{k+1}\right)^{\frac{k}{k-1}} \tag{3.14}$$

If the gas flow through the orifice or nozzle throats is sonic, the pressure above the drill bit inside the drill string can be determined from

$$P_{ai} = \frac{\dot{w}_g T_{bh}^{0.5}}{A_n \left[\left(\frac{gkS_g}{R} \right) \left(\frac{2}{k+1} \right)^{\left(\frac{k+1}{k-1} \right)} \right]^{0.5}} \tag{3.15}$$

where

P_{ai} is the pressure above the drill bit inside the drill string (lb/ft^2 absolute, or N/m^2 absolute).
A_n is the total flow area of the orifices or nozzles (ft^2, or m^2).
T_{bh} is the temperature of the gas at the bottom of the borehole (°R, or K).

If the gas flow through the orifice or nozzle throats is subsonic, the pressure above the drill bit inside the drill string can be determined from

$$P_{ai} = P_{bh} \left[\frac{\left(\frac{\dot{w}_g}{A_n} \right)^2}{2g \left(\frac{k}{k-1} \right) P_{bh} \gamma_{bh}} + 1 \right]^{\frac{k}{k-1}} \tag{3.16}$$

where γ_{bh} is the specific weight of the gas at bottomhole conditions (lb/ft^3, or N/m^3).

3.4 INJECTION PRESSURE

The pressure above the bit at the bottom of the inside of the drill string is used as the initial pressure for a sequence of calculations that determine the pressure at the top of each section of drill string having constant cross-section. The pressure at the top of each section is used as the initial pressure for the section above. This procedure is used until the injection pressure is determined. The injection pressure is obtained from [1]

$$P_{in} = \left[\frac{P_{ai}^2 + b_i T_{av}^2 \left(e^{\frac{2a_i H}{T_{av}}} - 1 \right)}{e^{\frac{2a_i H}{T_{av}}}} \right]^{0.5} \tag{3.17}$$

where

P_{in} is the injection pressure into the inside of the drill string (lb/ft^2 absolute, or N/m^2 absolute). (This pressure can also be the pressure at the top of any calculation interval with a uniform cross-sectional flow area.)

The constants a_i and b_i are

$$a_i = \frac{S_g}{R} \tag{3.18}$$

and

$$b_i = \frac{f}{2gD_i}\left(\frac{R}{S_g}\right)^2 \frac{w_g^2}{\left(\frac{\pi}{4}\right)^2 D_i^4} \tag{3.19}$$

where

D_i is the inside diameter of the constant cross-section section (ft, or m).

3.5 WATER INJECTION

Water is injected into the volumetric flow rate of air (or other gases) flowing from the compressors to the top of the inside of the drill string for three important reasons:

- Saturate the air or other gas with water vapor at bottomhole annulus pressure and temperature conditions
- Eliminate the stickiness of the small rock cuttings flour generated by the advance of the drill bit
- Assist in suppressing the combustion of the mixture of produced hydrocarbons and oxygen rich air

Water injection (with additives) is accomplished with a liquid pump that injects water into the compressed gas flow line to the well. The liquid pump draws its water from a liquid suction tank. The typical additives are corrosion inhibitors, polymer, and a foaming agent. The foaming agent causes the gas/liquid mixture to develop into an unstable foam as the mixture passes through the bit orifices. Table 3.2 gives a typical additives mixture used to create unstable foam drilling. "Mist drilling" was the old term used to describe the practice of injecting water (with additives) into the compressed gas to the well. Later it was found that a foaming agent improved

TABLE 3.2 Typical Approximate Additives Volumes per 20 bbl of Water for Unstable Foam Drilling (actual commercial product volumes may vary)

Additives	Volume per 20 bbl of Water
Foamer	4.2 to 8.4 gal
Polymer	1 to 2 quarts
Corrosion Inhibitor	0.5 gal

the hole cleaning effectiveness of water injection. The water injection practice is now known as "unstable foam drilling" [4].

3.6 SATURATION OF GAS

Water is injected into the air or other circulation gases at the surface in order to saturate the gas with water vapor at bottomhole conditions. The reason this is done is to assure that the circulation gas as it flows out of the drill bit orifices into the annulus will be able to carry formation water coming into the annulus as whole droplets. If the gas entering the annulus is not saturated, the gas will use its internal energy to convert the formation water to vapor. This will reduce the capability of the gas to expand as it enters the annulus. If the gas cannot expand properly, it will not create adequate gas flow velocity in the annulus to carry cuttings (and formation water) to the surface. Water saturated gas will carry formation water to the surface as droplets.

The empirical formula for determining the saturation of various gases including air can be found in a variety of chemistry handbooks and other literature. The empirical formula for the saturation pressure of air, p_{sat}, can be written as [5]

$$P_{sat} = 10^{\left[6.39416 - \left(\frac{1750.286}{217.23 + 0.555 t_{bh}}\right)\right]} \tag{3.20}$$

where P_{sat} is saturation pressure of the air at annulus bottomhole conditions (psia) and t_{bh} is the temperature of the air at annulus bottomhole conditions (°F). The approximate volumetric flow rate of injected water to an air drilling operation is determined by the relationship between the above saturation pressure and the bottomhole pressure in the annulus, and the weight rate of flow of gas being injected into the top of the inside of drill string. Thus, the flow rate of injected water, q_{iw}, is determined from [6]

$$q_{iw} = \left(\frac{P_{sat}}{P_{bh} - P_{sat}}\right)\left(\frac{3600}{8.33}\right)\dot{w}_g \tag{3.21}$$

where q_{iw} is the volumetric flow rate of injected water (gal/hr) and P_{bh} is the pressure at annulus bottomhole conditions (psia).

3.7 ELIMINATE STICKINESS

The next higher volumetric flow rate of injected water is that amount needed to eliminate borehole stickiness. As the drill bit advances in some types of rock formations, the cutting action of the bit creates rock cuttings and a small amount of "rock flour". Rock flour are very small rock particles that act mechanically very much like the flour one cooks with in the kitchen. If a borehole is originally dry, the circulation gas will efficiently carry the rock cutting particles and the rock flour up the annulus to the surface as the drill bit is advanced (this is called "dust drilling"). If a water bearing formation is penetrated, formation water will begin to flow into the annulus. When the water combines with rock flour, the flour particles begin to stick to each other and the borehole wall. This is very much like placing cooking flour in a bowl and putting a small amount of water in with it and mixing it. The cooking flour will become sticky and nearly impossible to work with a spoon. In the open borehole, the slightly wetted rock flour sticks to the nonmoving inside surface of the borehole. Because gas flow eddy currents form just above the top of the drill collars, "mud rings" develop from this sticky rock flour at this location on the borehole wall. These mud rings can build up and create a constriction to the annulus gas flow. This constriction will in turn cause the injection pressure to increase slightly (by 5 to 10 psi) in a matter of a minute or so. This rather sharp increase in pressure should alert the driller that mud rings are forming due to an influx of formation water (or perhaps even crude oil). If mud rings are allowed to continue to form they will begin to resist the rotation of the drill string (since cuttings are not being efficiently remove from the well). This in turn will increase the applied torque at the top of the drill string and increase the danger of a drill string torque failure. Also, the existence of mud rings creates a confined chamber of high pressure gas. If hydrocarbon rock formations are being penetrated, the potential for combustion in the chamber increases. The solution to this operational problem is to begin to inject water into the circulation gas. Additional injected water (above that given by Equations 3.20 and 3.21) is needed to reduce the stickiness (in much the same way the cooking flour stickiness is reduced by adding more water). The amount of water added to eliminate the mud ring must be determined empirically and will be somewhat unique for each drilling operation.

The procedure for eliminating mud rings is as follows:

1. Begin injecting sufficient water to saturate the gas flow with water vapor.
2. Curtail drilling ahead but continue gas circulation.
3. Bring the rotations of the drill string up to about 100 rpm and lift the drill string up to the top of the drilling mast and lower it several times. This will allow the drill collars to smash into the mud rung structures and break them off the borehole wall.
4. Return to drilling ahead.
5. If the mud rings begin to form again (the injection pressure increases again), increase the water injection flow rate and repeat the above sequence.

6. Continue the above five steps until the volumetric flow rate of injected water reduces the stickiness of the rock flour so that the mud rings no longer form on the open borehole wall (this will be indicted by a return to a nearly constant gas injection pressure).

The typical injected water volumetric flow rate to eliminate rock flour stickiness in a $7\frac{7}{8}$ in. borehole would be of the order of approximately 2 to 10 bbl/hr.

3.8 SUPPRESSION OF HYDROCARBON COMBUSTION

The next higher volumetric flow rate of injected water is that needed to suppress the downhole combustion of hydrocarbons (e.g., downhole fire or explosion). This combustion is created when the advance of the drill bit gives a mixture of drilling air with produced oil, natural gas, and/or coal dust. The ignition spark is easily caused by the action of the steel drill bit cutters on the rock face at the bottom or the sides of the borehole. Figure 3.3 gives the ignition (ignition zone) parameters of pressure versus the percent mixture for natural gas mixed with standard atmosphere air (at ASME Standard Conditions) [6, 7]. Increasing the water injection volumetric flow rate (with additives, particularly the foam agent) to the borehole wets down

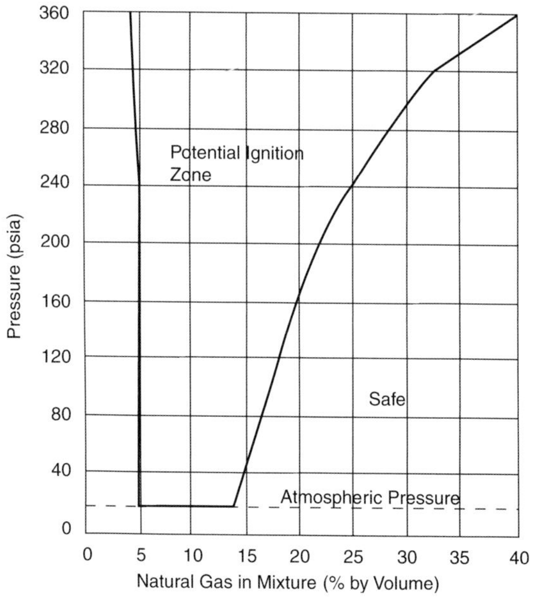

FIGURE 3.3 Ignition mixture by volume of natural gas and atmospheric air.

rock and steel surfaces reducing the risk of ignition sparks, and creates a bottomhole foam that deprives the combustion process of unlimited source of oxygen. It is thought that the water injection can be successful in suppressing hydrocarbon combustion in vertical wells where the hydrocarbon producing rock formation section has a thickness of approximately 200 ft or less. Horizontal drilling in hydrocarbon bearing rock formations is considered quite risky because of the long time exposure of the spark producing drill bit.

In a drilling operation where air is being used as the drilling fluid, as the drill bit is advanced and hydrocarbon bearing rock formations are about to be penetrated, the following steps should be taken to reduce the risk of downhole hydrocarbon combustion:

1. Drilling should be immediately stopped.
2. Air injection should be shut off and the gas flare monitored. If the flare is sustained by hydrocarbons from the well, the operator should note the wetness of the cuttings at the sample catcher (indicting water or distillate in the gas).
3. If the gas flare will not sustain or burn when the air is turned off, air can be turned back on, and water injection operations initiated, followed by careful bit advancement.
4. If the gas flare is sustained and/or the cuttings are wet, precautions must be taken to eliminate any mud rings before drilling forward. If water injection has not commenced, it should be initiated. With the air on and water injection operations underway, the drill string should be raised and lowered while rotating the drill string at approximately 100 rpm to smash any existing mud rings in the hole above the hydrocarbon bearing formations. If the injection pressure has not come down after this attempt to clear the mud rings, repeat the above sequence until the operator is satisfied that any mud rings have been successfully cleared. Once satisfied no existing mud rings are present in the borehole, carefully advance the drill bit through the hydrocarbon bearing formations.

Note that an alternative to the above is to switch the drilling gas to natural gas or liquid nitrogen.

There is a new technology that allows underbalanced drilling operations to be carried out using inert atmosphere. The technology is the development of large industrial membrane filters that strip most of the oxygen from the compressed air output of the primary compressors [8]. The compressed inert air is either injected directly into the top of the drill string or injected into the top of the drill string after passing through the booster compressor. Details of this technology and its engineering applications are presented in Section 3.12, "*Compressor and Inert Air Generator Units,*" given below.

3.9 AERATED DRILLING (GASIFIED FLUID DRILLING)

Aerated drilling (or gasified fluid drilling) refers to the use of compressed air (or other gases) injected into an incompressible drilling fluid that flows in the annulus. This is accomplished by two basic methods. (1) Drill pipe injection method requires the injection of compressed gas into the incompressible drilling fluid flow as this fluid is injected into the inside of the drill string at the top of the well. This allows the mixture of the gas and incompressible drilling fluid to flow throughout the circulating system. A variation on the drill pipe injection method is the use of a drill string jet sub above the drill collar/drill pipe interface to allow the injected gas to gasify the annulus above the jet sub position in the drill string. This later technique minimizes the aerated flow friction losses inside of drill collars and through drill bit nozzles. (2) Annulus injection method requires the use of parasite tubing installed on the backside of the casing. This allows the upper section of the annulus to be aerated directly from the bottom of the casing.

The basic advantage of aerated drilling is the control of bottomhole pressure in the annulus (via surface adjustments). Through calculations and trial and error adjustments in the field, the inflow of formation fluids into the well can be controlled with the pressure exerted on the bottom of the annulus by the mixture of gas and incompressible drilling fluid. The gas injected into the incompressible drilling fluid is usually air.

Another important application for aerated drilling has been in lost circulation situations. When incompressible drilling fluids are being lost in the annulus to thief formations, aerated drilling techniques have been used successfully to minimize or eliminate the problem. The bubbles created by gasifying the incompressible drilling fluid tend to fill in the fracture or pore openings in the borehole wall. This flow obstruction effect often controls the loss of incompressible drilling fluid to the thief zones.

The basic planning steps for the drilling of a deep well with aerated drilling fluids are as follows [1]:

1. Determine the geometry of the borehole section or sections to be drilled with aerated drilling fluids (i.e., openhole diameters, the casing or liner inside diameters, and depths).
2. Determine the geometry of the associated drill strings for the sections to be drilled with aerated drilling fluids (i.e., drill bit size and type, the drill collar size, drill pipe size, and maximum depth).
3. Determine the type of rock formations to be drilled and estimate the anticipated drilling rate of penetration. Also, estimate the quantity and depth location of any formation water that might be encountered.

4. Determine the elevation of the drilling site above sea level, the temperature of the air during the drilling operation, and the approximate geothermal temperature gradient.
5. Establish the objective of the aerated drilling operation:
    - To allow formation fluids to be produced as the formation is drilled.
    - To control loss of circulation problems,
    - To reduce formation damage.
6. Determine whether direct or reverse circulation techniques will be used to drill the various sections of the well.
7. If underbalanced drilling is the objective, determine the bottomhole pressure limit that must be maintained in order to allow minimal production of formation fluids into the well bore.
8. For either of the above objectives, determine the required approximate volumetric flow rate of incompressible fluid to be used in the aerated fluid drilling operation. This is usually the minimum volumetric flow rate required to clean the rock cuttings from the bottom of the well and transport the cuttings to the surface. In most aerated drilling operations, the incompressible fluid volumetric flow rate is held constant as drilling progresses through the openhole interval (as the gas injection flow rate is increased).
9. Determine the approximate volumetric flow rate of air (or other gas) to be injected with the flow of incompressible fluid into the top of the drill string (or into the annulus) as a function of drilling depth.
10. Using the incompressible fluid and air volumetric flow rates to be injected into the well, determine the bottomhole pressure and the associated surface injection pressure as a function of drilling depth (over the openhole interval to be drilled).
11. Select the contractor compressors that will provide the drilling operation with the appropriate air or gas volumetric flow rate needed to properly aerate the drilling fluid. Also, determine the maximum power required by the compressor(s) and the available maximum derated power from the prime movers.
12. Determine the approximate volume of fuel required by the compressor(s) to drill the well.

The circulation system for an aerated drilling operation can be modeled by as a multiphase flow calculation problem. In these problems, the pressure and temperature of the gas is usually known at the exit to the system. In this case, the exit is at the end of the return flow line to the mud tank. At this exit, the pressure and temperature are the local atmospheric pressure and temperature at the drilling location. The calculation procedures for aerated drilling circulation problems are to start at these known exit conditions and work backwards (or upstream) through the system. In these calculations, the volumetric flow rates of both the incompressible fluid and the injected gas must be assumed or known. If compressors are used to provide compressed air, then the volumetric flow rate is the sum of the outputs of all the primary compressors. Unlike the air and gas drilling calculations described in the subsection above, most aerated drilling fluid operations are designed on a basis of an incompressible drilling mud (or other liquid phase) circulation rate that can carry the anticipated rock bit generated cuttings to the surface [1]. For direct circulation, the geometry of the well and the physical properties of the drilling mud are used to determine the minimum volumetric flow rate for the incompressible drilling mud in the "*Minimum Volumetric Flow Rate*" subsection below. Once this drilling mud minimum flow rate is known, the gas phase volumetric flow rate to be injected into the well is selected (trial and error) to give the appropriate bottomhole pressure or other drilling conditions (see No. 5 earlier). The equations given in the "*Bottomhole Pressure*" subsection below are used to determine the bottomhole pressure. These equations are applied from the top of the well annulus to each constant cross-section section in sequence starting from the top well. This is a trial and error calculation of determining the pressure at the bottom of the constant cross-section section of the well annulus. The pressure found at the bottom of each constant cross-section section is used as the initial pressure for the next deeper constant cross-section section until the bottomhole pressure at the bottom of the annulus is determined. In this calculation, all major and minor flow losses should be considered. Working upstream, the pressure above the drill bit orifices (or nozzles) inside the drill bit is found using the equations in the "*Drill Bit Orifices and Nozzle*" subsection. The equations given in the *Injection Pressure* subsection are applied from the bottom of the inside of the drill string to each constant cross-section section in sequence starting from the pressure above the drill bit. The pressure found at the top of each constant cross-section section is used as the initial pressure for the pressure at the top of next constant cross-section section until the injection pressure at the top of the inside of the drill string is found. These are trial and error calculations.

Aerated drilling (or multiphase flow) calculation models are complex and cumbersome to apply and obtain predictions for field operations. The methodology outline above has been used to predict bottomhole and injection pressures with an accuracy of about 15% to 20%.

3.9.1 Minimum Volumetric Flow Rate

Aerated drilling fluid operations are designed using the minimum volumetric flow rate of incompressible drilling mud (or other liquid) that can carry the anticipated bit cuttings to the surface. This design ensures that the borehole will be cleaned even if the injection of air is interrupted. Therefore, the minimum velocity required to carry cuttings form the borehole must be determined for the largest annulus space in the borehole profile. The average incompressible fluid velocity in the annulus is

$$V_f = V_c + V_t \tag{3.22}$$

where

V_f is the minimum fluid velocity (ft/sec, or m/sec).
V_c is the solids critical concentration velocity (ft/sec, or m/sec).
V_t is the terminal velocity of solids (particles) in the drilling fluid (ft/sec, or m/sec).

The solids critical concentration velocity is

$$V_c = \frac{\kappa}{3600\ C} \tag{3.23}$$

where

κ is the drilling rate of penetration (ft/sec, or m/sec)
C is the concentration factor (usually assumed to be 0.04)

The drilling cuttings particle average diameter can be estimated using the following expression

$$D_c = \frac{\kappa}{(60)\,N} \tag{3.24}$$

where

D_c is the average diameter of drill bit cuttings (ft, or m),
N is the average drill bit rotary speed (rpm).

The fluid flow regions are classified as Laminar, transitional, or turbulent. These flow regions can be approximately defined using the non-dimensional Reynolds Number. The Reynolds Number is

$$N_R = \frac{DV}{\nu} \tag{3.25}$$

where

D is the diameter of the flow channel (ft, or m).
V is the velocity of the flow (ft/sec, or m/sec).
ν is the kinematic viscosity of the flowing fluid (ft^2/sec, or m^2/sec).

The generally used empirically derived terminal velocity expressions in English Units are given below. For the laminar region, the expression is

$$V_{t1} = 0.0333 D_c^2 \left(\frac{\gamma_s - \gamma_f}{\mu_e} \right) \qquad 0 < N_R < 2100 \tag{3.26}$$

For the transition region, the expression is

$$V_{t2} = 0.492 D_c \left(\frac{(\gamma_s - \gamma_f)^{\frac{2}{3}}}{(\gamma_f \mu_e)^{\frac{1}{3}}} \right) \qquad 2100 < N_R < 4000 \qquad (3.27)$$

For the turbulent region, the expression is

$$V_{t3} = 5.35 \left[D_c \left(\frac{\gamma_s - \gamma_f}{\gamma_f} \right) \right]^{\frac{1}{2}} \qquad N_R > 4000 \qquad (3.28)$$

where

V_{t1}, V_{t2}, V_{t3} are terminal velocities (ft/sec).
γ_s is the specific weight of the solids (lb/ft^3).
γ_f is the specific weight of the fluid (lb/ft^3).
μ_e is the effective absolute viscosity (lb-sec/ft^2).

The terminal velocity expressions in SI units are given below. For the laminar region the expression is

$$V_{t1} = 0.0333\, D_c^2 \left(\frac{\gamma_s - \gamma_f}{\mu_e} \right) \qquad 0 < N_R < 2100 \qquad (3.29)$$

For the transition region, the expression is

$$V_{t2} = 0.331 D_c \left[\frac{(\gamma_s - \gamma_f)^{\frac{2}{3}}}{(\gamma_f \mu_e)^{\frac{1}{3}}} \right] \qquad 2100 < N_R < 4000 \qquad (3.30)$$

For the turbulent region, the expression is

$$V_{t3} = 2.95 \left[D_c \left(\frac{\gamma_s - \gamma_f}{\gamma_f} \right) \right]^{\frac{1}{2}} \qquad N_R > 4000 \qquad (3.31)$$

where

V_{t1}, V_{t2}, V_{t3} are terminal velocities (m/sec).
γ_s is the specific weight of the solids (N/m^3).
γ_f is the specific weight of the fluid (N/m^3).
μ_e is the effective absolute viscosity (N-sec/m^2).

Equations 3.22 to 3.31 allow for the determination of the approximate minimum annulus velocity of drilling mud (or other incompressible fluid) that can carry the drill bit cuttings from the borehole. This minimum annulus velocity determination is independent of any gas injection for aerated drilling. Normal aerated drilling operations practice is to have a drilling mud or incompressible fluid volumetric flow rate that will clean the bottom of the well without aeration with injected gas. This is usually the beginning design point for an aerated drilling operation.

3.9.2 Bottomhole Pressure

As in air and gas drilling calculations discussed above, the losses in the entire aerated drilling circulation system must be considered. Thus, aerated drilling calculations are initiated with the known atmospheric pressure at the exit to the surface return flow line. It is assumed that the drilling mud and air mixture will exit the well annulus and enter the surface return flow line to the mud tank (or separator). Knowing the exit pressure, Equation 3.30 can be used to determine the upstream return line entrance pressure. This is a trial-and-error calculation procedure:

$$\int_{P_{ex}}^{P_{en}} \frac{dP}{B_s(P)} = \int_0^{L_{sr}} dl \tag{3.32}$$

where

$$B_s(P) = \left[\frac{\dot{w}_t}{\left(\frac{P_g}{P}\right)\left(\frac{T_r}{T_g}\right)Q_g + Q_m}\right] \times \left\{\frac{f}{2gD_{sr}}\left[\frac{\left(\frac{P_g}{P}\right)\left(\frac{T_r}{T_g}\right)Q_g + Q_m}{\frac{\pi}{4}D_{sr}^2}\right]^2\right\}$$

and

$$\dot{w}_t = \dot{w}_g + \dot{w}_m + \dot{w}_s$$

$$\dot{w}_g = \gamma_g Q_g$$

$$\dot{w}_m = \gamma_m Q_m$$

$$\dot{w}_s = \frac{\pi}{4} D_h^2 \gamma_w S_s \kappa$$

where

P_{ex} is the pressure in the aerated fluid as it exits the end of the surface return line (lb/ft^2 or N/m^2 absolute). This pressure is usually assumed to be the atmospheric pressure at the exit.

P_{en} is the pressure in the aerated fluid as it enters the surface return line at the top of the annulus (lb/ft^2, or N/m^2 absolute). Trial and error methods must be used to find the value of this pressure that allows both sides of Equation 3.30 to be satisfied.

P_g is the pressure in the gas in the aerated fluid at reference surface conditions (lb/ft^2, or N/m^2 absolute). This is usually assumed to be the atmospheric pressure at the location.

T_g is the temperature of the gas in the aerated fluid at reference surface conditions (lb/ft^2, or N/m^2 absolute). This is usually assumed to be the atmospheric pressure at the location.

T_{av} is the average temperature of the aerated fluid flow in the surface return line (°R, or K),
Q_g is the volumetric flow rate of gas (ft^3/sec, or m^3/sec),
Q_m is the volumetric flow rate of drilling mud (ft^3/sec, or m^3/sec),
D_{sr} is the inside diameter of the surface return flow line (ft, or m),
L_{sr} is the length of the surface return flow line (ft, or m).

The empirical expression for the Darcy-Weisbach friction factor for flow in the laminar region is [Section 3.5]

$$f = \frac{64}{N_R} \tag{3.33}$$

Equation 3.31 is valid for $0 \leq N_R \leq 2100$.

The Colebrook empirical expression is used to determine the friction factor for flow in the transition region. This expression is [3]

$$\frac{1}{\sqrt{f}} = -2\log_{10}\left[\frac{\left(\frac{e}{D_{sr}}\right)}{3.7} + \frac{2.51}{N_R\sqrt{f}}\right] \tag{3.34}$$

where *e* is the absolute roughness of the inside surface of the surface return flow line (ft, or m). Note that Equation 3.34 must be solved by trial and error methods. Equation 3.34 is valid for $N_R \geq 4000$.

The von Karman empirical expression is used to determine the friction factor for flow in the wholly turbulent region. This expression is [3]

$$f = \left[\frac{1}{2\log_{10}\left(\frac{D_{sr}}{e}\right) + 1.14}\right]^2 \tag{3.35}$$

Equation 3.35 is valid for $N_R \geq 4000$. In practice, the greater of the f values obtained from Equation 3.34 and Equation 3.35 must be used in Equation 3.30.

For direct (conventional) circulation, the bottomhole pressure in the annulus is given by

$$\int_{P_e}^{P_{bh}} \frac{dP}{B_a(P)} = \int_0^H dh \tag{3.36}$$

where

$$B_a(P) = \left[\frac{w_t}{\left(\frac{P_g}{P}\right)\left(\frac{T_{av}}{T_g}\right)Q_g + Q_m} \right] \times \left\{ 1 + \frac{f}{2g\,(D_h - D_p)} \left[\frac{\left(\frac{P_g}{P}\right)\left(\frac{T_{av}}{T_g}\right)Q_g + Q_m}{\frac{\pi}{4}\left(D_h^2 - D_p^2\right)} \right]^2 \right\}$$

and

P_{bh} is the bottomhole pressure in the annulus (lb/ft^2 absolute, or N/m^2 absolute). (This pressure can also be the pressure at the bottom of any calculation interval with a uniform cross-sectional flow area.)

P_e is the exit pressure at the top of the annulus (lb/ft^2 absolute, or N/m^2 absolute). (This pressure can also be the pressure at the bottom of any previous calculation interval with a uniform cross-sectional flow area. When several intervals of different cross-sections are being considered, this pressure is the bottomhole pressure calculated in the previous calculation above the interval under consideration. It can also be the entrance pressure to the surface return line.

T_{av} is the average temperature of the calculation interval (°R, or K).

H is the length (or height) of the calculation interval (ft, or m).

The empirical expression for the Darcy-Weisbach friction factor for flow in the laminar region is

$$f = \frac{64}{N_R} \tag{3.37}$$

Equation 3.35 is valid for $0 \le N_R \le 2100$.

The empirical expression for the friction factor for the flow in the transition region is

$$\frac{1}{\sqrt{f}} = -2\,\log_{10}\left[\frac{\left(\frac{e_{av}}{D_h - D_p}\right)}{3.7} + \frac{2.51}{N_R\sqrt{f}} \right] \tag{3.38}$$

where e_{av} is the average absolute roughness of the annulus surfaces (ft, or m). See Equation 3.10 for details. Equation 3.38 is valid for $N_R \ge 4000$.

The friction factor is determined from the empirical von Karman relationship. This relationship is

$$f = \left[\frac{1}{2 \log_{10}\left(\frac{D_h - D_p}{e_{av}} \right) + 1.14} \right]^2 \tag{3.39}$$

See Equation 3.10 for details. Equation 3.39 is valid for $N_R \geq 4000$. In practice, the greater of the f values obtained from Equation 3.38 and Equation 3.39 must be used in Equation 3.36.

The solution of Equation 3.36 is by trial and error. Over each section of the annulus having constant geometry, the pressure at the bottom of the section (the upper limit in Equation 3.36) is selected that allows the left side of Equation 3.36 to equal the right side of that equation. This trial and error process is repeated for each geometric section down the annulus until the pressure at the bottom of the annulus is obtained. If the bottom of the annulus pressure (bottomhole pressure) is to be a given known value (for a given depth), then the entire trial and error process must be repeated with adjusted Q_g or Q_m to allow the correct bottomhole pressure to be obtained.

To obtain an initial approximate mixture of Q_g and Q_m for a given bottomhole pressure value, a non-friction solution can be used. Setting $f = 0$ in Equation 3.36 gives an expression that can be integrated to yield a closed form solution. Solving this closed form solution for Q_g, gives

$$Q_g = \frac{(\dot{w}_m + \dot{w}_s)H - (P_{bh} - P_e)Q_m}{\left[P_g \left(\frac{T_{av}}{T_g} \right) \ln\left(\frac{P_{bh}}{P_e} \right) - \gamma_g H \right]} \tag{3.40}$$

3.9.3 Drill Bit Orifices and Nozzles

The mixture of incompressible fluid and the compressed gas passing through the drill bit orifices or nozzles can be assumed to act as a single phase incompressible fluid. However, this assumption is valid only when the friction losses in the flow through the bit orifices are also assumed to be higher. Thus, borrowing from mud drilling technology, the pressure change through the drill bit is

$$\Delta P_b = \frac{(\dot{w}_g + \dot{w}_m)^2}{2g\gamma_{mixbh} C^2 \left(\frac{\pi}{4} \right)^2 D_e^4} \tag{3.41}$$

where

ΔP_b is the pressure drop through the drill bit (lb/ft^2 absolute, or N/m^2 absolute).

γ_{mixbh} is the specific weight of the fluid mixture under bottomhole conditions (lb/ft^3, or N/m^3).

C is the loss coefficient for the aerated fluid flow through the drill bit orifices or nozzles (the values for C for aerated fluid flow should be taken as 0.70 to 0.85).

D_e is the effective orifice diameter (ft).

For drill bits with n equal diameter orifices (or nozzles), D_e is

$$D_e = \sqrt{nD_n^2} \tag{3.42}$$

The pressure change obtained from Equation 3.38 is added to the bottomhole pressure P_{bh} obtained from Equation 3.34. Thus, the pressure above the drill bit inside the drill string, P_{ai}, is

$$P_{ai} = P_{bh} + \Delta P_b \tag{3.43}$$

3.9.4 Injection Pressure

Knowing the pressure at the bottom of the inside of the drill string, Equation 3.42 can be used to determine the upstream injection pressure at the top of the inside of the drill string (or at the top of the section of constant cross-section). This is a trial and error calculation procedure. This equation is

$$\int_{P_{in}}^{P_{ai}} \frac{dP}{B_i(P)} = \int_0^H dh \tag{3.44}$$

where

$$B_i(P) = \left[\frac{\dot{w}_g + \dot{w}_m}{\left(\frac{P_g}{P}\right)\left(\frac{T_r}{T_g}\right)Q_g + Q_m}\right] \times \left\{1 - \frac{f}{2gD_i}\left[\frac{\left(\frac{P_g}{P}\right)\left(\frac{T_r}{T_g}\right)Q_g + Q_m}{\frac{\pi}{4}D_i^2}\right]^2\right\}$$

where

P_{in} is the pressure in the aerated fluid as it is injected into the top of the inside of the drill string (lb/ft^2, or N/m^2 absolute).

The empirical expression for the Darcy-Weisbach friction factor for flow in the laminar region is [3]

$$f = \frac{64}{N_R} \tag{3.45}$$

Equation 3.43 is valid for $0 \leq N_R \leq 2100$.

The Colebrook empirical expression is used to determine the friction factor for flow in the transition region. This expression is [3]

$$\frac{1}{\sqrt{f}} = -2\log_{10}\left[\frac{\left(\frac{e}{D_i}\right)}{3.7} + \frac{2.51}{N_R\sqrt{f}}\right] \tag{3.46}$$

where e is the absolute roughness of the inside surface of the surface return flow line (ft, or m). Note that Equation 3.46 must be solved by trial and error methods. Equation 3.44 is valid for $N_R \geq 4000$.

The von Karman empirical expression is used to determine the friction factor for flow in the wholly turbulent region. This expression is [3]

$$f = \left[\frac{1}{2\log_{10}\left(\frac{D_i}{e}\right) + 1.14}\right]^2 \tag{3.47}$$

Equation 3.47 is valid for $N_R \geq 4000$. In practice, the greater of the f values obtained from Equation 3.46 and Equation 3.47 must be used in Equation 3.44.

3.10 STABLE FOAM DRILLING

From a calculation point of view, stable foam drilling is a special case of the aerated fluid drilling predictive theory given above. In stable foam drilling, a mixture of gas (usually air or nitrogen) and an incompressible fluid (water and a foam agent) is further specified by the foam quality at the top annulus and foam quality at the bottom of the annulus (see Table 3.2). This foam quality must be maintained through the annulus in order for a stable foam to exist in this return flow space. Foam quality, Γ, is defined as

$$\Gamma = \frac{Q_g}{Q_s + Q_L} \tag{3.48}$$

where

Q_s is the volumetric flow rate of gas (ft^3/sec).
Q_L is the volumetric flow rate of the incompressible fluid (ft^3/sec).

The control of the foam quality at the top allows the foam quality at the bottom of the annulus to be calculated. Operationally this control is accomplished by placing a valve on the return flow line from the annulus (back pressure valve). Upstream of the valve is a pressure gauge and by maintaining a specified back pressure at this position the foam quality at the top of the annulus can be determined. Knowing the foam quality at this position (and the other flow characteristics of the circulating system), the foam quality at any position in the annulus (particularly at the bottom of the annulus) can be determined. The foam quality at the bottom of the annulus (once developed) must be maintained at approximately 0.60 or greater [9–12]. If the foam quality at the bottom of the annulus drops much below this level, the foam will collapse and the flow will become aerated fluid drilling. To maintain the bottomhole foam quality in the annulus at a magnitude of approximately 0.60 or greater, the foam quality immediately upstream of the back pressure valve must usually be maintained at a magnitude in the range of 0.90 to 0.98.

Stable foam drilling operations can use a variety of incompressible fluids and compressed gases to develop a stable foam. The majority of the operations use fresh water and a commercial foam agent specifically for drilling use (a surfactant) with injected compressed air. Commercial surfactants for drilling can be obtained from a variety of drilling service companies. Inert atmosphere generators have been used to provide the injected gas for stable foam drilling operations. Using inert atmosphere gas in the stable foam will reduce corrosion of the drill string and the borehole casing and reduce the risk of downhole combustion when drilling through hydrocarbon bearing rock formations.

The basic planning steps for a deep well are as follows:

1. Determine the geometry of the borehole section or sections to be drilled with the stable foam drilling fluids (i.e., open hole diameters, the casing inside diameters, and maximum depths).
2. Determine the geometry of the associated drill string for the sections to be drilled with stable foam drilling fluids (i.e., drill bit size and type, the drill collar size, drill pipe size and description, and maximum depth).
3. Determine the type of rock formations to be drilled in each section and estimate the anticipated drilling rate of penetration.
4. Determine the elevation of the drilling site above sea level, the temperature of the air during the drilling operation, and the approximate geothermal temperature gradient.
5. Establish the objective of the stable foam drilling fluids operation:
    - To drill through loss of circulation formations,
    - To counter formation water entering the annulus (by injecting additional surfactant to foam the formation water in the annulus),

- To maintain low bottom hole pressure to either preclude fracturing of the rock formations, or to allow underbalanced drilling operations.

6. If underbalanced drilling is the objective, it should be understood that stable foam drilling operations cannot maintain near constant bottomhole annulus pressures.
7. For either of the above objectives, determine the required approximate volumetric flow rate of the mixture of incompressible fluid (with surfactant) and the compressed air (or other gas) to be used to create the stable foam drilling fluid. This required mixture volumetric flow rate is governed by the foam quality at the top of annulus (i.e., return flow line back pressure) and the rock cuttings carrying capacity of the flowing mixture in the critical annulus cross-sectional area (usually the largest cross-sectional area of the annulus). The rock cuttings carrying capability of the stable foam can be estimated using a minimum kinetic energy per unit volume value in the critical annulus cross-sectional.
8. Using the incompressible fluid and air volumetric flow rates to be injected into the well, determine the bottomhole pressure and the surface injection pressure as a function of drilling depth (over the openhole interval to be drilled).
9. Select the contractor compressor(s) that will provide the drilling operation with the appropriate air or gas volumetric flow rate needed to create the stable foam drilling fluid. Also, determine the maximum power required by the compressor(s) and the available maximum derated power from the prime mover(s).
10. Determine the approximate volume of fuel required by the compressor(s) to drill the well.

Stable foam drilling predictive calculations are carried out in the annulus space only. It is assumed that in deep drilling operations the flow of the mixed gas and incompressible fluid (and foam agent additive) inside the drill string and through the nozzle openings acts as an aerated fluid (governed by Equations 3.41 to 3.47). However, for flow in the return line and the annulus the flow is stable foam. For modeling stable foam drilling, the basic aerated fluid return line and annulus flow equations (i.e., Equations 3.32 to 3.39) are modified using the additional limitations imposed by the requirements that the foam qualities be specified at the top and bottom of the annulus. Equations 3.32 to 3.39 (or Equation 3.40) are restricted by Equation 3.48 and the specified foam qualities at the top and bottom of the annulus. These restrictions result in the inability to be able to specify the bottomhole annulus pressure for a foam drilling operations. In essence, the bottomhole annulus pressure cannot be controlled. Stable foam flow calculations for the return line and annulus spaces are carried out with the same trial and error methodology that was described above in the aerated drilling subsection.

Given below are the advantages and disadvantages of the stable foam drilling technique. The advantages are as follows:

- The technique does not generally require any additional downhole equipment.
- Nearly the entire annulus is filled with the stable foam drilling fluid, thus, low bottomhole pressures can be achieved.
- Since the bubble structures of stable foam drilling fluids have a high fluid yield point, these structures can support rock cuttings in suspension when drilling operations are discontinued to make connections. Stable foams can have seven to eight times the rock cutting carrying capacity of water.
- Rock cuttings retrieved from the foam at the surface are easy to analyze for rock properties information.

The disadvantages are as follows:

- Bottomhole annulus pressure cannot be specified and maintained. The specified foam qualities at the top and bottom of the annulus and the geometry of the will result in a unique bottomhole annulus pressure for each drilling depth.
- Stable foam fluids injection cannot be continued when circulation is discontinued during connections and tripping. Therefore, it can be difficult to maintain underbalanced conditions during connections and trips.
- Since the injected gas is trapped under pressure inside the drill string by the various string floats, time must be allowed for the pressure bleed-down when making connections and trips. The bleed-down makes it difficult to maintain underbalanced conditions.
- The flow down the inside of the drill string is two phase flow and, therefore, high pipe friction losses are present. The high friction losses result in high pump and compressor pressures during injection.
- The gas phase in the stable foam attenuates the pulses of conventional (measure-while-drilling) MWD systems. Therefore, conventional mud pulse telemetry MWD cannot be used.

3.10.1 Foam Models

In order to adequately model the flow of stable foam flow in the annulus it is necessary to have absolute viscosity that will describe the foam two-phase flow absolute viscosity. Most models are based on the assumption that the two-phase flow absolute viscosity is a strong function of foam quality. Foam is effectively Newtonian at qualities from 0.55 to as high as 0.74 (this foam change over value is dependent upon the type of foam used). In this lower range of foam quality the effective absolute viscosity of the foam can be approximated by

$$\mu_f = \mu_L (1 + 3.6\Gamma) \tag{3.49}$$

where

μ_f is the absolute viscosity of the foam (lb-sec/ft^2).
μ_L is the absolute viscosity of the liquid (incompressible fluid) (lb-sec/ft^2).

For foam qualities greater than the above discussed value (up to 0.98), the effective absolute viscosity of the foam can be approximated by

$$\mu_f = \frac{\mu_L}{1 - \Gamma^n} \tag{3.50}$$

where n is an exponent that depends on the type of foam being used and can vary from 0.33 to 0.49. The higher values of n are more associated with stiff foam (i.e., foam using drilling mud as the incompressible fluid base).

Drilling foam can be modeled to act as a Bingham plastic fluid. However, power law rheology has been also been used [12]. It cannot be overstated that the modeling of foam drilling using Equations 3.31 to 3.50 above (or any other similarly constituted set of equations) will very dependent upon the type incompressible fluid (liquid) used as the bases for the foam and the type of foaming agent added to create the foam.

3.10.2 Bottomhole Pressure

Equations 3.32 to 3.39 together with the limitations of Equations 3.48 to 3.50 do not allow the control of the bottomhole pressure during stable foam drilling operations. Thus, stable foam drilling operations are not amenable to situations where precisely controlled bottomhole pressures are necessary for the success of the drilling operations.

3.10.3 Minimum Volumetric Flow Rate

Most stable foam vertical drilling operations are drilled over a depth interval with variable incompressible fluid volumetric flow rates and variable compressible gas volumetric flow rates. These variable volumetric flow rates are necessary in order to keep the annulus surface exit foam quality and the annulus bottomhole quality at predetermined values. The requirement of variable volumetric flow rates versus depth make the control of stable foam drilling operations complicated. In addition to keeping the foam qualities at the top and bottom of the annulus at predetermined values, the control of the flow rates must also assure that the foam flow in the annulus have sufficient rock cuttings carrying capacity to clean borehole as the drill bit is advanced. Stable foam has rock cuttings carrying capabilities during circulation and when circulation is stopped. Because of the variety of foaming agents and base incompressible fluids that can be used for stable foam operations, the question of a minimum volumetric flow rate is usually determined empirically at the rig floor by the driller during actual operations. Since stable foam circulating fluids create a structure that can support rock cuttings, the flowing velocity in the annulus need not be high relative to air and gas drilling, and aerated fluid drilling. Assuming that the bottomhole annulus foam quality to be greater than approximately 0.60 (to prevent foam collapse), it is likely that successful stable foam drilling operations have the minimum bottomhole kinetic energy per unit volume of the order of 1.0 to 2.0 ft-lb/ft^3.

3.10.4 Drill Bit Orifices and Nozzles

In deep wells the base incompressible fluid (with foam agent additive) and injected gas flow flows down the inside of the drill string as a aerated fluid. Such an aerated mixture can be assumed to pass through the nozzles in much the same manner as an incompressible fluid. Thus, Equations 3.41 to 3.43 can be used to determine the pressure drop across the drill bit. For stable foam drilling operation, nozzles are usually used in the drill bit to increase the fluid shear as the fluid passes through the bit. This fluid shear aids in the creation of the foam at the bottom of the annulus.

3.10.5 Injection Pressure

The flow condition in the inside of the drill string is two phase (gas and fluid) flow. This aerated fluid flow is modeled by Equations 3.44 to 3.47. Because of the need to maintain a continuous stable foam in the annulus space during the drilling operation, it will be necessary to continuously increase the volumetric flow rate of both incompressible fluid and gas to the well as the drill bit is advanced. The control of the drilling operation will rest on the reliability of the data from the injection pressure gauge used together with the data from the back pressure gauge on the return line.

3.11 COMPLETIONS OPERATIONS

Underbalance drilling operations require that the well completion operations also be underbalanced. In general, underbalanced completion techniques do not have to be used until the drill bit advance approaches the production rock formations (production zones). Most underbalanced wells are completed with either, openhole, well screen, noncemented slotted casing, or slotted tubing strings across the production zones. The placement of well screen, slotted casing, or slotted tubing in wells usually must be accomplished with snubbing and stripping techniques.

- Snubbing is the inserting of tubulars or other downhole tools into a well that is under pressure. In order to maintain a required minimum bottomhole static pressure in the well as tubulars are inserted, the well must be vented (usually through the choke line to the burn pit).

- Stripping is the inserting of tubulars or other downhole tools into a flowing well. Keeping the well flowing as the tubulars are inserted assures that the well is underbalanced.

Chapter 6 "Well Pressure Control" has additional information concerning snubbing and stripping.

Inserting tubulars into open holes through exposed freshly drilled production zones present some unique problems.

3.11.1 Sloughing Shales

Since the drilling circulation fluid is not heavy, there is a constant threat of caving and sloughing of the openhole borehole wall. Air and gas drilling operations will have drilling penetration rates that can be twice that of mud drilling operations. This faster drilling penetration rate is an important feature since openhole integrity is very dependent upon the length of time the hole remains open and unsupported by cement and casing.

When drilling with air and gas the shale sequences of rock formations are usually the most susceptible to caving or sloughing. This is due mainly to bedding layered texture of shale and the generally weak bonding between these layers. When these shales are penetrated with a drill bit, the openhole wall surfaces of the exposed shale formations tend to break off and the large fragments and fall into the annulus space between the openhole wall and the drill collar and drill pipe outside surfaces. This sloughing of shale formations can be temporarily controlled by injecting additional additives into water being injected into the circulation air or gas (in addition to those given in Table 3.2). Table 3.3 gives the formula for these additional additives. This formula has been successfully used in the San Juan Basin, New Mexico, USA.

3.11.2 Casing and Cementing

When drilling with air or gas as the circulation fluid, the borehole will be basically dry prior to casing and cementing operations. Therefore, there will be no water or drilling mud in the well to float the casing into the

TABLE 3.3 Typical approximate additive weights or volumes per 20 bbl of water for controlling sloughing shales (actual commercial product may vary)

Additives	Weights of Volumes per 20 bbl of Water
Foamer	8.5 gallons
Bentonite	40 lb
CMC	2 lb
Corn starch	5 lb
Soda ash	1 quart

well (making use of buoyancy). This presents some special completions problems for air and gas drilling operations.

When an openhole section of a gas drilled well is to be ceased, the casing with a casing shoe on the bottom of the string is lowered into the dry well. A pre-lush of about 20 bbl of CMC (carboxymethyl-cellulose) treated water must be pumped ahead of the cement. A diaphragm (bottom) plug is usually run between the CMC treated water pre-flush and the cement and a "bumper" plug run at the top of the cement. Fresh water is pumped directly behind the cement and is used to help balance the cement in the annulus between the openhole wall and the outside of the inserted casing. The CMC treated water pre-flush seals the surface of the dry borehole walls prior to the cement entering the open annulus space. This pre-flush limits the rapid hydrating of the cement as it flows from the inside of the casing to the annulus space. Once the casing and cementing operations are properly carried out and the cement successfully sets up in the annulus, it is necessary to remove the water from inside the casing in order to return to air and gas drilling operations (drill out the cement at the casing shoe and continue drilling ahead). There are several safe operational procedures that can be used to remove water from the inside of the casing after a successful cementing operation.

Aerated Fluid Procedure. The aerated fluid procedure is as follows:

1. Run the drill string made up with the appropriate bottomhole assembly and drill bit to a depth a few tens of feet above the last cement plug.
2. Start the mud pump running as slowly as possible, to pump water at a rate of 1.5 to 2.0 bbl/min. This reduces fluid friction resistance to the moving fluids in the circulation system.
3. Bring one compressor and booster on line to aerate the water being pumped to the top of the drill string. The air rate to the well should be about 100 to 150 acfm per barrel of water. If the air volumetric flow rate is too high, the standpipe pressure will exceed the pressure rating of the compressor and the compressor will shut down. Therefore, the compressor must be slowed down until air is mixed with the water going into the drill string.
4. As the fluid column in the annulus (between the inside of the casing and the outside of the drill string) is aerated, the standpipe pressure will drop. Additional compressors can be added (i.e., increasing air volumetric flow rate) to further lighten the fluid column and unload the water form the casing.
5. After the hole has been unloaded, the water injection pumps should be kept in operation to clean the borehole.
6. At this point, begin air or mist drilling. Drill out the cement plug at the bottom of the casing and drill an additional 20 to 100 ft to allow any sloughing walls of the borehole to clean up.

7. Once the hole has been stabilized, stop drilling and blow the hole with air and injected water to eliminate rock cuttings. Continue this drilling and cleaning procedure for 15 to 30 minutes or until the air flow (with injected water) returning to the surface is clean (i.e., shows a fine spray and white color).
8. With the drill bit directly on bottom, continue flowing air with no injected water into the drill string. Air should flow to the well at normal drilling volumetric flow rates until the water and surfactant remaining in the well are swept to the surface.
9. Continuously blow the hole with air for about 30 minutes to an hour.
10. Begin normal air drilling. After 5 to 10 ft have been drilled, the hole should go to dry dust drilling (although it is sometimes necessary to drill as much as 60 to 90 ft before dry dust appears at the surface). If the hole does not dust after these steps have been carried out, inject another surfactant slug into the air flow to the well. If dry dusting cannot be achieved, unstable foam drilling may be required to complete the air drilling operation.

Gas Lift Procedure. The air lift procedure is as follows:

1. Calculate the lifting capability of the primary and booster compressor on the drilling location. Run the drill string made up with the appropriate bottomhole assembly and drill bit to a depth a few hundred feet above this calculated compressor pressure limit.
2. Start the compressors and force compressed air to the bottom of the drill string and begin aerating the water column in the annulus and flow this aerated water column to the surface (removing this portion of the column from the well).
3. Once this column of water has been removed, shut down the compressors and lower the drill string a similar distance as defined by the lifting capability limit determined in No.1 above. Start up the compressors and remove his next column of water from the well.
4. Continue lowering the drill string in increments and air lifting the entire water column from the well.
5. With the drill bit directly on bottom, continue flowing air into the drill string. Air should flow to the well at normal drilling volumetric flow rates until the water and surfactant remaining in the well are swept to the surface.
6. Continuously blow the hole with air for about 30 minutes to 1 hour.
7. Begin normal air drilling. After 5 to 10 ft have been drilled, the hole should go to dry dust drilling (although it is sometimes necessary to drill as much as 60 to 90 ft before dry dust appears at the surface). If the hole does not dust after these steps have been carried out, inject another surfactant slug into the air flow to the well. If dry dusting cannot be

achieved, unstable foam drilling may be required to complete the air drilling operation.

3.11.3 Drilling with Casing

There are some new technologies entering the underbalanced drilling and completions operations. These involve using casing (that will be left in the well) as the "drill pipe." Casing while drilling has been used for many decades in shallow water and geotechnical drilling and completions operations. Oil and gas service companies have adopted the drilling with casing concept and developed new technologies to allow this type of drilling to be safely used for deep pressured wells. Figure 3.4 shows a downhole deployment valve that can be used at the bottom of the drill pipe/casing string. This flapper type valve allows the back pressure of sensitive production rock formations to be isolated as tools are run through the inside of the drill pipe/casing string. Figure 3.5 shows a Wireline Retrievable Float Valve that can be used at the top of the drill pipe/casing string to isolate the gas pressure in the drill string (between the top and bottom valves).

Drilling with casing can utilize a variety of methods to rotate the drill bit (and underreamers). Usually the drill pipe/casing string can be rotated with a rotary table, a top drive swivel, or a hydraulic rotary head. Also a downhole motor can be used to rotate the drill bit. The drilling assembly can be on a separate drill string inside the casing, or attached directly to the bottom of the drill pipe/casing string.

3.12 COMPRESSOR AND INERT AIR GENERATOR UNITS

There are a variety of compressors available commercially. However, the most useful for air and gas drilling operations are the reciprocating piston compressor and the rotary compressor. These compressors are used as primary compressors or as booster compressors. The primary compressor intakes atmospheric air and compresses usually to a pressure of about 200 psig to 350 psig via two or three internal compression stages. The primary compressor can be either a reciprocating piston or a rotary type compressor. If higher pressures are required for the drilling operation, the compressed air from the primary compressor can be passed through a booster compressor for compression up to as high as 2000 psig. Only the reciprocating piston compressor can be used as a booster.

3.12.1 Compressor Units

Compressors are rated by their intake volumetric flow rate (of atmospheric air at a specified standard condition (e.g., API, ASME, EU). This

FIGURE 3.4 Downhole deployment valve (Courtesy of Weatherford International Limited).

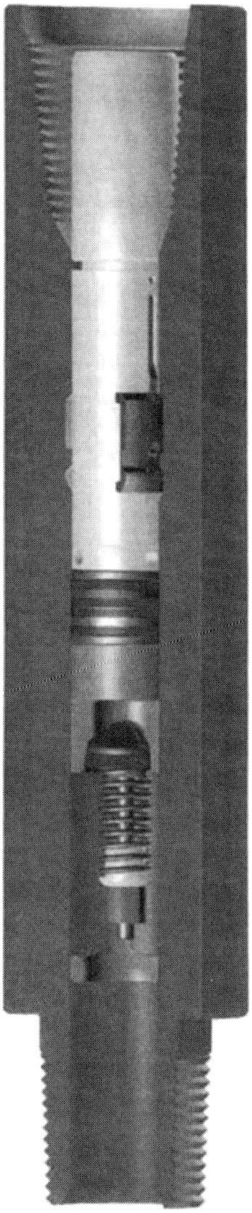

FIGURE 3.5 Wireline retrievable float valve (Courtesy of Weatherford International Limited).

volumetric flow rate is usually specified by the manufacturer in units of standard ft^3/m (usually written as scfm) or in standard $m^3/minute$ (usually written as scm/m). Be sure to use the manufacturer's specified standard conditions to make accurate engineering calculations.

Reciprocating Piston. The advantages of the reciprocating piston compressor are

1. Dependable volumetric flow rates at output line back pressures near compressor maximum pressure capability.
2. Compressor will match its output flow rate pressure with the flow line back pressure.
3. More prime mover fuel efficient.

The disadvantages are

1. High capital costs
2. Requires high maintenance
3. Bulky to transport

Rotary. The advantages of the rotary compressor are

1. Low capital costs
2. Low maintenance
3. Easy to transport and small site footprint on location

The disadvantages are

1. Volumetric flow rates will decrease as flow line back pressure near compressor fixed pressure output
2. Because of fixed output pressure, fuel consumption is the same for all flow line back pressures

3.12.2 Allowable Oxygen Content

Compressed air is combustible when mixed with hydrocarbons in the downhole environment. Downhole combustion suppression is extremely important in air and gas drilling applications. Downhole combustion can cause drilling rig and production personnel injury and death, and damage to equipment associated with drilling production operations. Combustion hazards are prevalent in wells that are not drilling mud filled. When drilling with compressed air, the oxygen in the air mixes with methane (or other hydrocarbon gases) to create combustible mixtures. Three basic methods of combustion suppression are used in modern drilling operations. These are using natural gas from a nearby pipeline as the drilling fluid, using inert atmospheric air provided by inert air generator units, and using liquid nitrogen injection (used only for drilling of short distances).

Using natural gas as the drilling fluid essentially eliminates all oxygen from the borehole and, therefore, eliminates the possibility of all fires and explosions. However, using natural gas as a drilling fluid creates new danger problems on the rig floor and in the area around the drilling location. The drilling location fire and explosion problems and the expense of using a marketable gas usually make natural gas uneconomical.

Liquid nitrogen can be mixed with compressor air to reduce the oxygen content of the mixture below the level to support combustion in the event the drilling operation using the mixture encounters hydrocarbons. The expense of this market gas makes liquid nitrogen uneconomical for prolong drilling.

The allowable oxygen content for a mixture of oxygen, nitrogen, and methane is a function of the maximum pressure of the mixture in the borehole during the drilling operation. The minimum oxygen percent of a mixture with nitrogen and methane is given by [8, 9]

$$O_{2\,min} = 13.98 - 1.68 \log_{10} P \tag{3.51}$$

where

O_{2min} is the minimum percent of oxygen content in the mixture (%).
P is the absolute pressure of the mixture (psia).

The minimum percent of oxygen content to support combustion for pressures of less than 3000 psia is approximately 8% or greater. Therefore, to prevent combustion in any drilling or completion operation the oxygen content must be kept below 8%.

3.12.3 Inert Air Generator Units

Inert air generator units (or nitrogen generator units) use membrane filtration technology to remove oxygen from atmospheric air. This results in an inert atmospheric gas low in oxygen content [9]. The oxygen percentage in this mixture must be lower than 8% for the mixture to be inert. Because other gas components in atmospheric air are removed in the membrane filtration process, the efficiency of these inert air generator units is a function of the percentage of oxygen content required in the final output of inert atmospheric air exiting the units. This efficiency must be considered in making engineering predictive calculations. The efficiency of the inert air generator unit refers to the ratio of the volumetric flow rate of inert atmospheric air exiting the unit to the volumetric flow rate of atmospheric air entering the unit. The atmospheric air entering the primary compressor is passed to the inert air generator after exiting the compressor. Figure 3.6 the typical inert atmospheric air drilling location schematic [9]. Field test data demonstrates a linear relationship between the percentage of oxygen remaining in the output of the inert air generator unit and the percent efficiency of the unit (Figure 3.7) [9]. Equation 3.52 gives the relationship of

FIGURE 3.6 Typical compressor and inert air generator unit layout at drilling location. (Courtesy of Weatherford International Limited.)

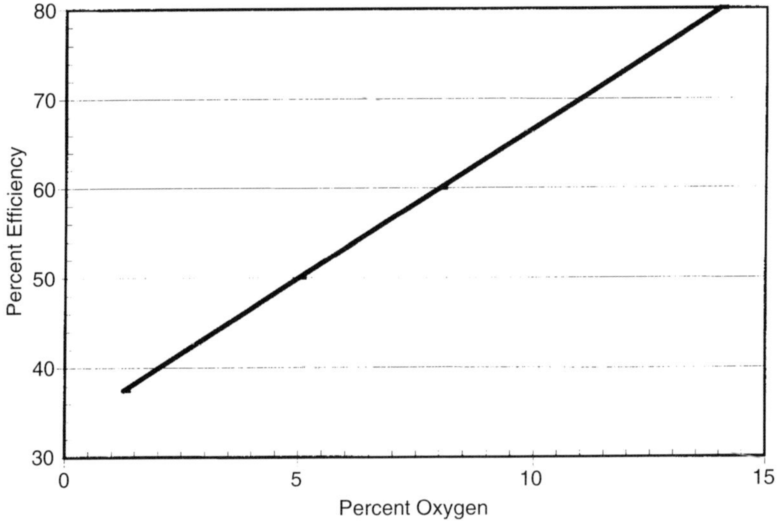

FIGURE 3.7 Inert air generator unit efficiency versus percentage of oxygen remaining in unit output. (Courtesy of SPE.)

percent efficiency of the unit to the oxygen percentage remaining in the inert air output from the unit. Equation 3.52 is

$$\% \text{ Efficiency} = (\% \text{ Oxygen})(3.33) + (33.33) \tag{3.52}$$

where

% Efficiency is the percent of volumetric flow rate of inert atmospheric air exiting the unit relative to the volumetric flow rate of atmospheric entering the unit.

% Oxygen is the percent of oxygen remaining in the inert atmospheric air exiting the unit.

FIGURE 3.8 Sea Wolf combined primary compressor and inert air generation unit (Courtesy of Weatherford International Limited).

Figure 3.8 shows the unique Sea Wolf combined primary compressor (on the bottom) and inert air generator (on the top) unit. Such combined units allow for small footprints at land drilling locations and for offshore platforms.

3.12.4 Liquid Nitrogen

Liquid nitrogen is used to suppress downhole combustion. The two main reasons for using liquid nitrogen in a drilling operation are that only a short interval needs to be drilled using a non-combustible drilling fluid and that atmospheric air generation units are not available.

Liquid nitrogen is injected into the compressed atmospheric air from the compressors in order to reduce the overall oxygen content in the resulting flow stream of gas to the well. The objective is to reduce the oxygen content in the resulting flow of gases to the well to a level where combustion cannot be supported. Thus, specifying the desired oxygen content (percent) in the resulting flow of gases to the well, the approximate volumetric flow rate of pure nitrogen to be injected into the compressed air to a well can be obtained from

$$q_n = q_t - \frac{q_t\,(\%O_2)}{0.21} \tag{3.53}$$

where

q_n is the volumetric flow rate of injected nitrogen gas to the well (scfm).
q_t is the total volumetric flow rate of gas needed to adequately clean the bottom of the hole for the drilling operation (scfm).
O_2 is the desired oxygen content in the resulting mixture (%)

The compressor produced volumetric flow rate of air required to drill the well is determined by subtracting the nitrogen flow rate from the total flow rate needed to safely clean the borehole. Note that this calculation is for standard atmospheric conditions (scfm, usually API mechanical equipment standard conditions). Therefore, the actual atmospheric conditions at the drilling location must be considered and the volumetric flow rates determined in acfm. Since primary compressor come in specific flow rate segments, care must be taken not to select compressor units that will increase the oxygen content beyond the level desired.

Pure nitrogen comes to the drilling location in a liquid form (at very low temperatures). Therefore, the above volumetric flow rate of pure nitrogen in scfm must be converted to gallons of liquid nitrogen. This conversion is

$$\text{One gallon liquid nitrogen} = 93.11 \text{ scf nitrogen gas} \tag{3.54}$$

3.13 HIGHLY DEVIATED WELL DRILLING AND COMPLETIONS

Many technologies have been developed through the past decades directed at improving directional drilling and the completions of directional wells using conventional incompressible drilling fluids (e.g., water-based drilling muds and oil-based drilling muds). Air and gas drilling technology has been a small niche area of the drilling industry and, therefore, up until the late 1980's little attention was given to the development of directional drilling technologies for air and gas drilling operations. As the North American oil and gas fields deplete, underbalanced drilling and completion operations have become more important in extending the useful life of these fields. Although there have been some development activities to develop air and gas directional drilling technologies, these have not been entirely successful or accepted commercially.

3.13.1 Drilling Operations

The mud pulse communication systems between the surface and bottomhole assembly are the major technologies used for MWD and LWD. No similar technologies have been developed for drilling highly deviated wells with air and gas drilling fluids, aerated drilling fluids, or stable foam drilling fluid. There are two technologies that were developed as possible alternatives to the mud pulse technology for mud drilling operations that have had limited success in air and gas, aerated fluid, and stable foam directional drilling operations. These are electromagnetic earth transmission MWD and wireline steering and logging tools. One of the most effective methods of making hole using air and gas as the drilling fluid is the downhole air hammer. At present there are no methods for utilizing this drilling tool in deviated wells. Progressive cavity motors have been adapted for use with air and gas, aerated, or stable foam drilling fluids. However, the lack of a reliable MWD technology is still the primary limitation to directional drilling using air and gas drilling technology.

The risk of downhole fires and explosion exists for both vertical and horizontal drilling operations using air and gas drilling technology. However, the risk is higher when drilling long horizontal boreholes using. This is due to the fact that during a typical horizontal drilling operation, the horizontal interval drilled in the hydrocarbon bearing rock formations is several times longer than in typical vertical interval drilled in a vertical drilling operation (assuming similar hydrocarbon bearing rock formations). Further, the drilling rate of penetration for a horizontal drilling operation will be about half that of vertical drilling (assuming the similar rock type). Thus, this increased exposure time in a horizontal borehole in hydrocarbon bearing rock formation is far greater and, therefore, the risk is far greater. The development of membrane filter technology that basically eliminates the risk of downhole fires and explosions has increased interest in developing more reliable MWD technologies for air and gas drilling operations.

3.13.2 Completions Operations

Wells that have been directionally drilled with air and gas drilling technology are completed either open hole completions, or with slotted tubing liners (not cemented). The development of expandable casing and liners holds the prospect of greatly improving completions for highly deviated boreholes drilled with air and gas technology (see Chapters 8 and 10).

3.14 DOWNHOLE MOTORS

3.14.1 Background

In 1873, an American, C. G. Cross, was issued the first patent related to a downhole turbine motor for rotating the drill bit at the bottom of a drillstring with hydraulic power [13]. This drilling concept was conceived nearly 30 years before rotary drilling was introduced in oil well drilling. Thus the concept of using a downhole motor to rotate or otherwise drive a drill bit at the bottom of a fluid conveying conduit in a deep borehole is not new.

The first practical applications of the downhole motor concept came in 1924 when engineers in the United States and the Soviet Union began to design, fabricate and field test both single-stage and multistage downhole turbine motors [14]. Efforts continued in the United States, the Soviet Union and elsewhere in Europe to develop an industrially reliable downhole turbine motor that would operate on drilling mud. But during the decade to follow, all efforts proved unsuccessful.

In 1934 in the Soviet Union a renewed effort was initiated to develop a multistage downhole turbine motor [14–17]. This new effort was successful. This development effort marked the beginning of industrial use of the downhole turbine motor. The Soviet Union continued the development of the downhole turbine motor and utilized the technology to drill the majority of its oil and gas wells. By the 1950s the Soviet Union was drilling nearly 80% of their wells with the downhole turbine motors using surface pumped drilling mud or freshwater as the activating hydraulic power.

In the late 1950s, with the growing need in the United States and elsewhere in the world for directional drilling capabilities, the drilling industry in the United States and elsewhere began to reconsider the downhole turbine motor technology. There are presently three service companies that offer downhole turbine motors for drilling of oil and gas wells. These motors are now used extensively throughout the world for directional drilling operations and for some straight-hole drilling operations.

The downhole turbine motors that are hydraulically operated have some fundamental limitations. One of these is high rotary speed of the motor and drill bit. The high rotary speeds limit the use of downhole turbine motors when drilling with roller rock bits. The high speed of these direct drive motors shortens the life of the roller rock bit.

In the 1980s in the United States an effort was initiated to develop a downhole turbine motor that was activated by compressed air. This motor was provided with a gear reducer transmission. This downhole pneumatic turbine has been successfully field tested [18, 19].

The development of positive displacement downhole motors began in the late 1950s. The initial development was the result of a United States patent filed by W. Clark in 1957. This downhole motor was based on the original work of a French engineer, René Monineau, and is classified as a helimotor. The motor is actuated by drilling mud pumped from the surface. There are two other types of positive displacement motors that have been used, or are at present in use today: the vane motor and the reciprocating motor. However, by far the most widely used positive displacement motor is the helimotor [14].

The initial work in the United States led to the highly successful single-lobe helimotor. From the late 1950s until the late 1980s there have been a number of other versions of the helimotor developed and fielded. In general, most of the development work in helimotors has centered around multilobe motors. The higher the lobe system, the lower the speed of these direct drive motors and the higher the operating torque.

There have been some efforts over the past three decades to develop positive development vane motors and reciprocating motors for operation with drilling mud as the actuating fluid. These efforts have not been successful.

In the early 1960s efforts were made in the United States to operate vane motors and reciprocating motors with compressed air. The vane motors experienced some limited test success but were not competitive in the market of that day [19]. Out of these development efforts evolved the reciprocating (compressed) air hammers that have been quite successful and are operated extensively in the mining industry and have some limited application in the oil and gas industry [20]. The air hammer is not a motor in the true sense of rotating equipment. The reciprocating action of the air hammer provides a percussion effect on the drill bit, the rotation of the bit to new rock face location is carried out by the conventional rotation of the drill string.

In this section the design and the operational characteristics and procedures of the most frequently used downhole motors will be discussed. These are the downhole turbine motor and the downhole positive displacement motor.

3.14.2 Turbine Motors

Figure 3.9 shows the typical rotor and stator configuration for a single stage of a multistage downhole turbine motor section. The activating drilling mud or freshwater is pumped at high velocity through the motor

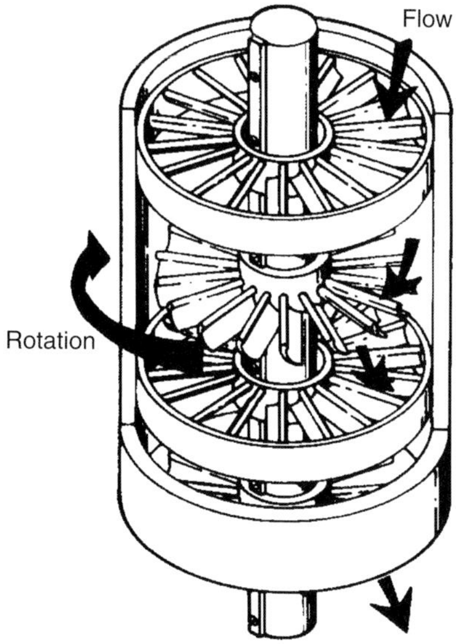

FIGURE 3.9 Basic turbine motor design principle. (courtesy of Smith International, Inc).

section, which, because of the vane angle of each rotor and stator (which is a stage), causes the rotor to rotate the shaft of the motor. The kinetic energy of the flowing drilling mud is converted through these rotor and stator stages into mechanical rotational energy.

3.14.2.1 *Design*

The rotational energy provided by the flowing fluid is used to rotate and provide torque to the drill bit. Figure 3.10 shows the typical complete downhole turbine motor actuated with an incompressible drilling fluid.

In general, the downhole turbine motor is composed of two sections: (1) the turbine motor section and (2) the thrust-bearing and radial support bearing. These sections are shown in Figure 3.10. Sometimes a special section is used at the top of the motor to provide a filter to clean up the drilling mud flow before it enters the motor, or to provide a by-pass valve.

The turbine motor section has multistages of rotors and stators, from as few as 25 to as many as 300. For a basic motor geometry with a given flowrate, an increase in the number of stages in the motor will result in an increase in torque capability and an increase in the peak horsepower. This performance improvement, however, is accompanied by an increase in the differential pressure through the motor section (Table 3.4). The turbine motor section usually has bearing groups at the upper and lower ends of the rotating shaft (on which are attached the rotors). The bearing groups have only radial load capabilities.

The lower end of the rotating shaft of the turbine motor section is attached to the upper end of the main shaft. The drilling fluid after passing through the turbine motor section is channeled into the center of the shaft through large openings in the main shaft. The drill bit is attached to the lower end of the main shaft. The weight on the bit is transferred to the downhole turbine motor housing via the thrust-bearing section. This bearing section provides for rotation while transferring the weight on the bit to the downhole turbine motor housing.

In the thrust-bearing section is a radial support bearing section that provides a radial load-carrying group of bearings that ensures that the main shaft rotates about center even when a side force on the bit is present during directional drilling operations.

There are of course variations on the downhole turbine motor design, but the basic sections discussed above will be common to all designs.

The main advantages of the downhole turbine motor are:

1. Hard to extremely hard competent rock formations can be drilled with turbine motors using diamond or the new polycrystalline diamond bits.
2. Rather high rates of penetration can be achieved since bit rotation speeds are high.
3. Will allow circulation of the borehole regardless of motor horsepower or torque being produced by the motor. Circulation can even take place when the motor is installed.

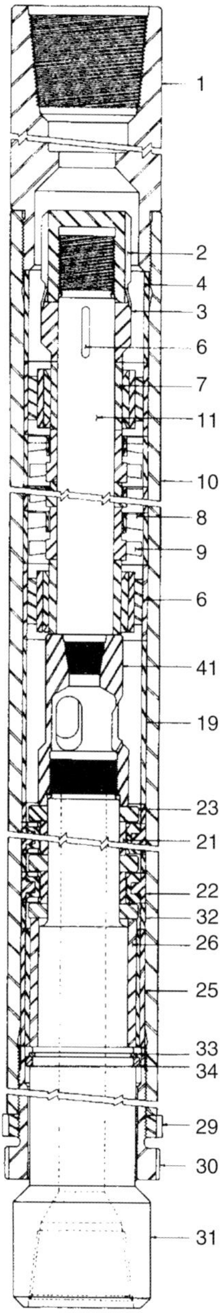

Item No.	Description
1	Top Sub
2	Shaft Cap
3	Lockwasher-Turbine Section
4	Stator Spacer
5	Shaft Key-Turbine Section
6	Intermediates Bearing Body
7	Intermediates Bearing Sleeve
8	Stator - Assembly
9	Rotor -
10	Turbine Housing
11	Turbine Shaft
19	Spacer-Bearing Section
21	Thrust Bearing Sleeve
22	Thrust Bearing Body
23	Thrust Disc
25	Lower Bearing Body
26	Lower Bearing Sleeve
29	Lower Sub Lock Ring
30	Lower Sub
31	Bearing Shaft
32	Lower Bearing Spacer
33	Retaining Ring
34	Catch Ring
40	Float Retainer Ring
41	Shaft Coupling
45	Shaft Cap Lock Screw
46	Eastco Float
	Turbodrill Complete
	[1]Optimal, order by topf sub too
	Repair Accessories
	Rubber Lubricant
	Assembly Compound
	Joint Compound
	Retaining Ring Pliers-External
	Retaining Ring Pliers-Internal
	1/4" Lock Screw Wrench
	Eastco Float Repair Kit

FIGURE 3.10 Downhole turbine motor design (courtesy of Baker Hughes Co.)

The main disadvantages of the downhole turbine motor are:

TABLE 3.4 Turbine Motor, $6\frac{3}{4}$-in. Outside Diameter, Circulation Rate 400 gpm, Mud Weight 10 lb/gal

Number of Stages	Torque* (ft-lb)	Optimum Bit speed (rpm)	Differential Pressure (psi)	Horse-power*	Thrust Load (1000 lb)
212	1412	807	1324	217	21
318	2118	807	1924	326	30

*At optimum speed. Courtesy of Eastman-Christensen.

1. Motor speeds and, therefore, bit speeds are high, which limits the use of roller rock bits.
2. The required flowrate through the downhole turbine motor and the resulting pressure drop through the motor require large surface pump systems, significantly larger pump systems than are normally available for most land and for some offshore drilling operations.
3. Unless a measure while drilling instrument is used, there is no way to ascertain whether the turbine motor is operating efficiently since rotation speed and/or torque cannot be measured using normal surface data (i.e., standpipe pressure, weight on bit).
4. Because of the necessity to use many stages in the turbine motor to obtain the needed power to drill, the downhole turbine motor is often quite long. Thus the ability to use these motors for high-angle course corrections can be limited.
5. Downhole turbine motors are sensitive to fouling agents in the mud; therefore, when running a turbine motor steps must be taken to provide particle-free drilling mud.
6. Downhole turbine motors can only be operated with drilling mud.

3.14.2.2 *Operations*

Figure 3.11 gives the typical performance characteristics of a turbine motor. The example in this figure is a $6\frac{3}{4}$-in. outside diameter turbine motor having 212 stages and activated by a 10-lb/gal mud flowrate of 400 gal/min.

For this example, the stall torque of the motor is 2,824 ft-lb. The runaway speed is 1,614 rpm and coincides with zero torque. The motor produces its maximum horsepower of 217 at a speed of 807 rpm. The torque at the peak horsepower is 1,412 ft-lb, or one-half of the stall torque.

A turbine device has the unique characteristic that it will allow circulation independent of what torque or horsepower the motor is producing. In the example where the turbine motor has a 10-lb/gal mud circulating at 400 gal/min, the pressure drop through the motor is about 1,324 psi. This pressure drop is approximately constant through the entire speed range of the motor.

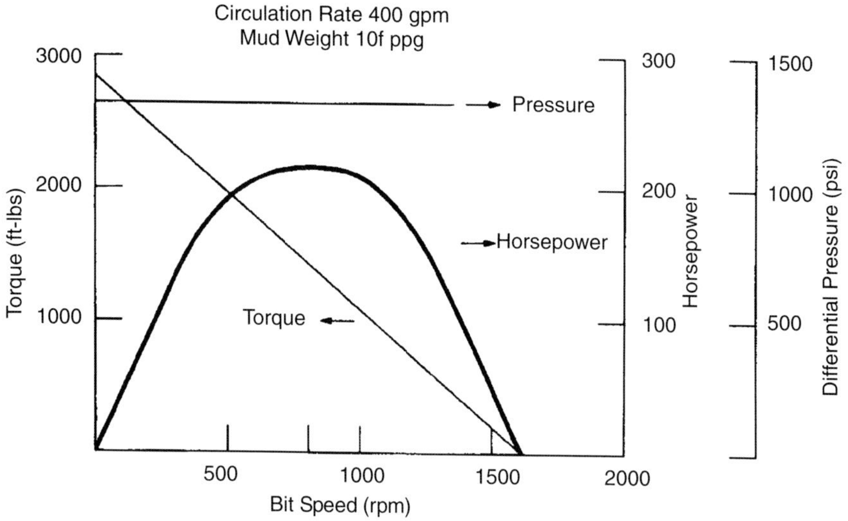

FIGURE 3.11 Turbine motor, $6\frac{3}{4}$-in. outside diameter, two motor sections, 212 stages, 400 gal/min, 10-lb/gal mud weight (courtesy of Baker Hughes Co.).

If the turbine motor is lifted off the bottom of the borehole and circulation continues, the motor will speed up to the runaway speed of 1,614 rpm. In this situation the motor produces no drilling torque or horsepower.

As the turbine motor is lowered and weight is placed on the motor and thus the bit, the motor begins to slow its speed and produce torque and horsepower. When sufficient weight has been placed on the turbine motor, the example motor will produce its maximum possible horsepower of 217. This will be at a speed of 807 rpm. The torque produced by the motor at this speed will be 1,412 ft-lb.

If more weight is added to the turbine motor and the bit, the motor speed and horsepower output will continue to decrease. The torque, however, will continue to increase.

When sufficient weight has been placed on the turbine motor and bit, the motor will cease to rotate and the motor is described as being stalled. At this condition, the turbine motor produces its maximum possible torque. Even when the motor is stalled, the drilling mud is still circulating and the pressure drop is approximately 1,324 psi.

The stall torque M_s (ft-lb) for any turbine motor can be determined from [21].

$$M_s = 1.38386 \times 10^{-5} \frac{\eta_h \eta_m n_s \bar{\gamma}_m q^2 \tan\beta}{h} \tag{3.55}$$

where η_h = hydraulic efficiency
η_m = mechanical efficiency
n_s = number of stages
$\bar{\gamma}_m$ = specific weight of mud in lb/gal
q = circulation flowrate in gal/min
β = exit blade angle in degrees
h = radial width of the blades in in.

Figure 3.10 is the side view of a single-turbine stage and describes the geometry of the motor and stator.

The runaway speed N_r (rpm) for any turbine motor can be determined from

$$N_r = 5.85 \frac{\eta_v q \tan\beta}{r_m^2 h} \tag{3.56}$$

where η_v = volumetric efficiency
r_m = mean blade radius in in.

The turbine motor instantaneous torque M (ft-lb) for any speed N (rpm) is

$$M = M_s \left(1 - \frac{N}{N_r}\right) \tag{3.57}$$

The turbine motor horsepower HP (hp) for any speed is

$$HP = \frac{2\pi M_s N}{33{,}000}\left(1 - \frac{N}{N_r}\right) \tag{3.58}$$

The maximum turbine motor horsepower is at the optimum speed, N_o, which is one-half of the runaway speed. This is

$$N_0 = \frac{N_r}{2} \tag{3.59}$$

Thus, the maximum horsepower HP_{max} is

$$HP_{max} = \frac{\pi M_s N_r}{2\,(33{,}000)} \tag{3.60}$$

The torque at the optimum speed M_0 is one-half the stall torque. Thus

$$M_0 = \frac{M_s}{2} \tag{3.61}$$

The pressure drop Δp (psi)/through a given turbine motor design is usually obtained empirically. Once this value is known for a circulation flowrate and mud weight, the pressure drop for other circulation flowrates and mud weights can be estimated.

If the above performance parameters for a turbine motor design are known for a given circulation flowrate and mud weight (denoted as 1), the performance parameters for the new circulation flowrate and mud weight (denoted as 2) can be found by the following relationships:

Torque

$$M_2 = \left(\frac{q_2}{q_1}\right)^2 M_1 \quad (3.62)$$

$$M_2 = \left(\frac{\bar{\gamma}_2}{\bar{\gamma}_1}\right) M_1 \quad (3.63)$$

Speed

$$N_2 = \left(\frac{q_2}{q_1}\right) N_1 \quad (3.64)$$

Power

$$HP_2 = \left(\frac{q_2}{q_1}\right)^3 HP_1 \quad (3.65)$$

$$HP_2 = \left(\frac{\bar{\gamma}_2}{\bar{\gamma}_1}\right) HP_1 \quad (3.66)$$

Pressure drop

$$\Delta p_2 = \left(\frac{q_2}{q_1}\right)^2 \Delta p_1 \quad (3.67)$$

$$\Delta p_2 = \left(\frac{\bar{\gamma}_2}{\bar{\gamma}_1}\right) \Delta p_1 \quad (3.68)$$

Table 3.5 gives the performance characteristics for various circulation flowrates for the 212 state, $6\frac{3}{4}$-in. outside diameter turbine motor described briefly in Table 3.6 and shown graphically in Figure 3.11.

Table 3.6 gives the performance characteristics for various circulation flowrates for the 318-stage, $6\frac{3}{4}$-in. outside diameter turbine motor described

TABLE 3.5 Turbine Motor, $6\frac{3}{4}$-in. Outside Diameter, Two Motor Sections, 212 Stages, Mud Weight 10 lb/gal

Circulation Rate (gpm)	Torque* (ft-lb)	Optimum Bit Speed (rpm)	Differential Pressure (psi)	Maximum Horsepower*	Thrust Load (1000 lb)
200	353	403	331	27	5
250	552	504	517	53	8
300	794	605	745	92	12
350	1081	706	1014	145	16
400	1421	807	1324	217	21
450	1787	908	1676	309	26
500	2206	1009	2069	424	32

*At optimum speed.

TABLE 3.6 Turbine Motor, $6\frac{3}{4}$-in. Outside Diameter, Three Motor Sections, 318 Stages, Mud Weight 10 lb/gal

Circulation Rate (gpm)	Torque* (ft-lb)	Optimum Bit Speed (rpm)	Differential Pressure (psi)	Maximum Horsepower*	Thrust Load (1000 lb)
200	529	403	485	40	8
250	827	504	758	79	12
300	1191	605	1092	137	17
350	1622	706	1486	218	23
400	2118	807	1941	326	30
450	2681	908	2457	464	38

*At optimum power.

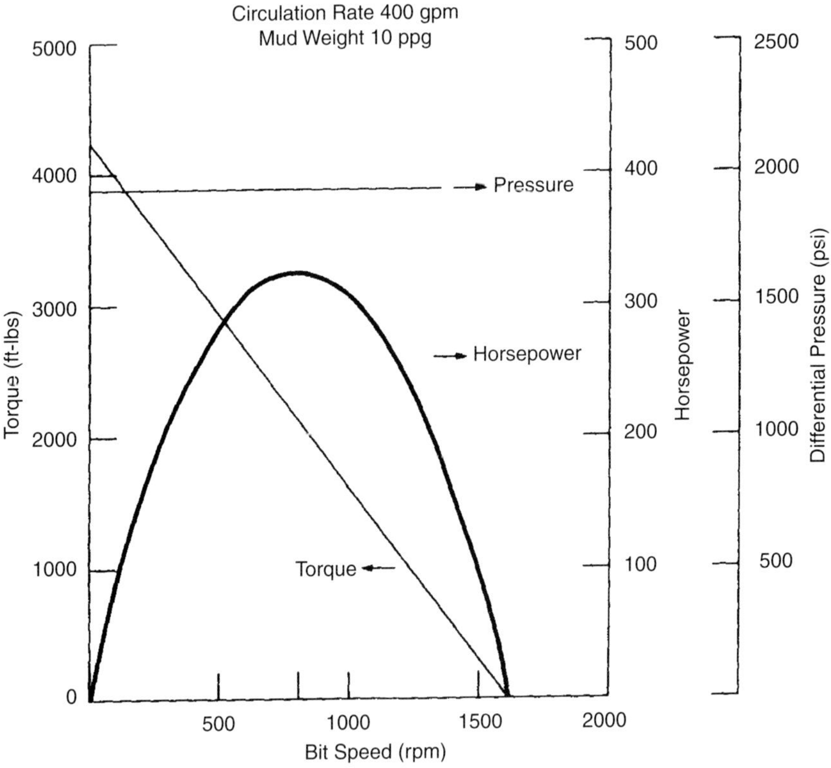

FIGURE 3.12 Turbine motor, $6\frac{3}{4}$-in. outside diameter, three motor sections, 318 stages, 480 gal/min, 10-lb/gal mud weight (courtesy of Baker Hughes Co.).

briefly in Table 3.4. Figure 3.12 shows the performance of the 318-stage turbine motor at a circulation flowrate of 400 gal/min and mud weight of 10 lb/gal.

The turbine motor whose performance characteristics are given in Table 3.5 is made up of two motor sections with 106 stages in each section.

The turbine motor whose performance characteristics are given in Table 3.6 is made up of three motor sections.

The major reason most turbine motors are designed with various add-on motor sections is to allow flexibility when applying turbine motors to operational situations.

For straight hole drilling the turbine motor with the highest possible torque and the lowest possible speed is of most use. Thus the turbine motor is selected such that the motor produces the maximum amount of power for the lowest possible circulation flowrate (i.e., lowest speed). The high power increase rate of penetration and the lower speed increase bit life particularly if roller rock bits are used.

For deviation control drilling the turbine motor with a lower torque and the shortest overall length is needed.

Example 3.1

Using the basic performance data given in Table 3.5 for the $6\frac{3}{4}$-in. outside diameter turbine motor with 212 stages, determine the stall torque, maximum horsepower and pressure drop for this motor if only one motor section with 106 stages were to be used for a deviation control operation. Assume the same circulation flow rate of 400 gal/min, but a mud weight of 14 lb/gal is to be used.

Stall Torque. From Table 3.5 the stall torque for the turbine motor with 212 stages will be twice the torque value at optimum speed. Thus the stall torque for 10 lb/gal mud weight flow is

$$\begin{aligned} M_S &= 2(1421) \\ &= 2{,}842 \text{ ft-lb} \end{aligned} \tag{3.69}$$

From Equation 3.55, it is seen that stall torque is proportional to the number of stages used. Thus the stall torque for a turbine motor with 106 stages will be (for the circulation flowrate of 400 gal/min and mud weight of 10 lb/gal)

$$\begin{aligned} M_S &= 2{,}842\left(\frac{106}{212}\right) \\ &= 1{,}421 \text{ ft-lb} \end{aligned} \tag{3.70}$$

and from Equation 3.63 for the 14-lb/gal mud weight

$$\begin{aligned} M_S &= 1{,}421\left(\frac{14}{10}\right) \\ &= 1{,}989 \text{ ft-lb} \end{aligned} \tag{3.71}$$

Maximum Horsepower. From Table 3.5 the maximum horsepower for the turbine motor with 212 stages is 217. From Equation 3.60, it can be seen that the maximum power is proportional to the stall torque and the runaway speed. Since the circulation flow rate is the same, the runaway speed is the same for this case. Thus, the maximum horsepower will be proportional to the stall torque. The maximum power will be (for the circulation flowrate of 400 gal/min and mud weight of 10 lb/gal)

$$HP_{max} = 217\left(\frac{1421}{2842}\right)$$
$$= 108.5 \tag{3.72}$$

and from Equation 3.66 for the 14-lb/gal mud weight

$$HP_{max} = 108.5\left(\frac{14}{10}\right)$$
$$= 152 \tag{3.73}$$

Pressure Drop. Table 3.5 shows that the 212-stage turbine motor has a pressure drop of 1,324 psi for the circulation flowrate of 400 gal/min and a mud weight of 10 lb/gal. The pressure drop for the 106 stage turbine motor should be roughly proportional to the length of the motor section (assuming the motor sections are nearly the same in design). Thus the pressure drop in the 106-stage turbine motor should be proportional to the number of stage. Therefore, the pressure drop should be

$$\Delta p = 1{,}324\left(\frac{106}{212}\right)$$
$$= 662 \text{ psi} \tag{3.74}$$

and from Equation 3.68 for the 14-lb/gal mud weight

$$\Delta p = 662\left(\frac{14}{10}\right)$$
$$= 927 \text{ psi} \tag{3.75}$$

The last column in the Tables 3.5 and 3.6 show the thrust load associated with each circulation flowrate (i.e., pressure drop). This thrust load is the result of the pressure drop across the turbine motor rotor and stator blades. The magnitude of this pressure drop depends on the individual internal design details of the turbine motor (i.e., blade angle, number of stages, axial height of blades and the radial width of the blades) and the operating conditions. The additional pressure drop results in thrust, T (lb), which is

$$T = \pi r_m^2 \Delta p \tag{3.76}$$

Example 3.2

A $6\frac{3}{4}$-in outside diameter turbine motor (whose performance data are given in Tables 3.5 and 3.6) is to be used for a deviation control direction drilling operation. The motor will use a new $8\frac{1}{2}$-in. diameter diamond bit for the drilling operation. The directional run is to take place at a depth of 17,552 ft (measured depth). The rock formation to be drilled is classified as extremely hard, and it is anticipated that 10 ft/hr will be the maximum possible drilling rate. The mud weight is to be 16.2 lb/gal. The drilling rig has a National Supply Company, triplex mud pump Model 10-P-130 available. The details of this pump are given in Table 3.7. Because this is a deviation control run, the shorter two motor section turbine motor will be used.

Determine the appropriate circulation flowrate to be used for the diamond bit, turbine motor combination and the appropriate liner size to be used in the triplex pump. Also, prepare the turbine motor performance graph for the chosen circulation flowrate. Determine the total flow area for the diamond bit.

Bit Pressure Loss. To obtain the optimum circulation flowrate for the diamond bit, turbine motor combination, it will be necessary to consider the bit and the turbine motor performance at various circulation flowrates: 200, 300, 400 and 500 gal/min.

Since the rock formation to be drilled is classified as extremely hard, 1.5 hydraulic horsepower per square inch of bit area will be used as bit cleaning and cooling requirement [22]. The projected bottomhole area of the bit $A_b(\text{in.}^2)$ is

$$A_b = \frac{\pi}{4}(8.5)^2$$
$$= 56.7 \text{ in.}^2 \quad (3.77)$$

For a circulation flowrate of 200 gal/min, the hydraulic horsepower for the bit HP_b (hp) is

$$HP_b = 1.5(56.7)$$
$$= 85.05 \quad (3.78)$$

TABLE 3.7 Triplex Mud Pump, Model 10-P-13, National Supply Company, Example 3.2

Input Horsepower, 1300; Maximum Strokes per Minute, 140; Length of Stroke, 10 Inches					
	Liner Size (inches)				
	$5\frac{1}{4}$	$5\frac{1}{2}$	$5\frac{3}{4}$	6	$6\frac{1}{4}$
Output per Stroke (gals)	2.81	3.08	3.37	3.67	3.98
Maximum Pressure (psi)	5095	4645	4250	3900	3595

The pressure drop across the bit Δp_b (psi) to produce this hydraulic horsepower at a circulation flowrate of 200 gal/min is

$$\Delta p_b = \frac{85.05(1{,}714)}{200}$$
$$= 729 \text{ psi} \tag{3.79}$$

Similarly, the pressure drop across the bit to produce the above hydraulic horsepower at a circulation flowrate of 300 gal/min is

$$\Delta p_b = \frac{85.05(1{,}714)}{300}$$
$$= 486 \text{ psi} \tag{3.80}$$

The pressure drop across the bit at a circulation flowrate of 400 lb/gal is

$$\Delta p_b = 364 \text{ psi} \tag{3.81}$$

The pressure drop across the bit at a circulation flowrate of 500 gal/min

$$\Delta p_b = 292 \text{ psi} \tag{3.82}$$

Total Pressure Loss. Using Table 3.5 and Equations 3.57 and 3.68, the pressure loss across the turbine motor can be determined for the various circulation flowrates and the mud weight of 16.2 lb/gal. These data together with the above bit pressure loss data are presented in Table 3.8. Also presented in Table 3.8 are the component pressure losses of the system for the various circulation flowrates considered. The total pressure loss tabulated in the lower row represents the surface standpipe pressure when operating at the various circulation flowrates.

Pump Limitations. Table 3.7 shows there are five possible liner sizes that can be used on the Model 10-P-130 mud pump. Each liner size must be considered to obtain the optimum circulation flowrate and appropriate liner size. The maximum pressure available for each liner

TABLE 3.8 Drill String Component Pressure Losses at Various Circulation Flowrates for Example 3.2

	Pressure (psi)			
Components	200 gpm	300 gpm	400 gpm	500 gpm
Surface Equipment	4	11	19	31
Drill Pipe Bore	460	878	1401	2021
Drill Collar Bore	60	117	118	272
Turbine Motor	536	1207	2145	3352
Drill Bit	729	486	364	292
Drill Collar Annulus	48	91	144	207
Drill Pipe Annulus	133	248	391	561
Total Pressure Loss	1970	3038	4652	6736

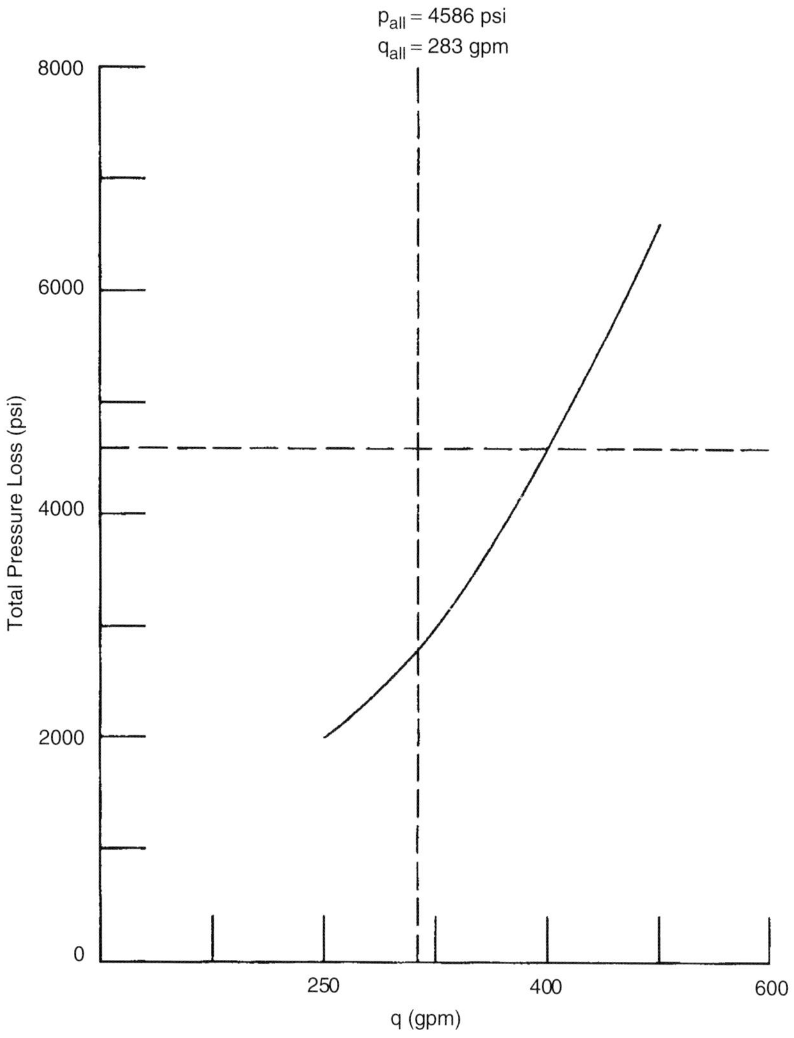

FIGURE 3.13 A $5\frac{3}{4}$-in. liner, total pressure loss versus flowrate, Example 3.2 (courtesy of Baker Hughes Co.).

size will be reduced by a safety factor of 0.90.[1] The maximum volumetric flowrate available for each liner size will also be reduced by a volumetric efficiency factor of 0.80 and an additional safety factor of 0.90.[2] Thus, from Table 3.7, the allowable maximum pressure and allowable maximum volume, metric flowrates will be those shown in Figures 3.13 through 3.17,

1. This safety factor is not necessary for new, well-maintained equipment.
2. The volumetric efficiency factor is about 0.95 for precharged pumps.

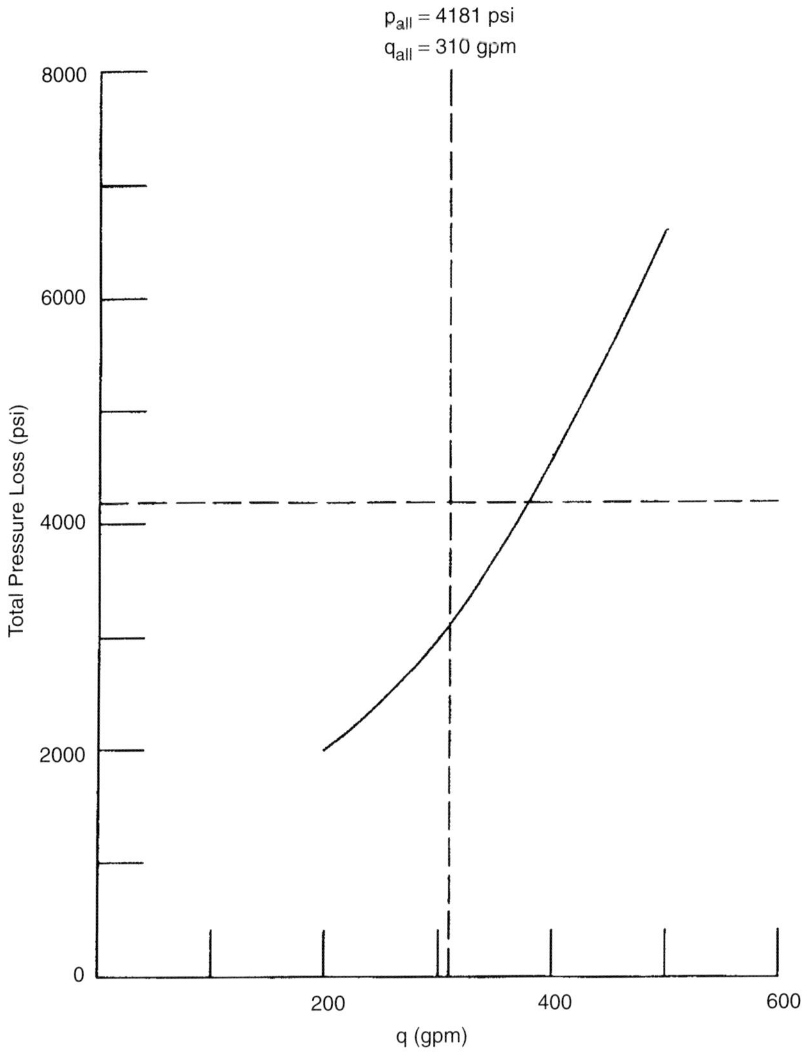

FIGURE 3.14 A $5\frac{1}{2}$-in. liner, total pressure loss versus flowrate, Example 3.2 (courtesy of Baker Hughes Co.).

which are the liner sizes $5\frac{1}{4}$, $5\frac{1}{2}$, $5\frac{3}{4}$, 6 and $6\frac{1}{4}$ in., respectively. Plotted on each of these figures are the total pressure losses for the various circulation flowrates considered. The horizontal straight line on each figure is the allowable maximum pressure for the particular liner size. The vertical straight line is the allowable maximum volumetric flowrate for the particular liner size. Only circulation flowrates that are in the lower left quadrant of the figures are practical. The highest circulation flowrate (which produces

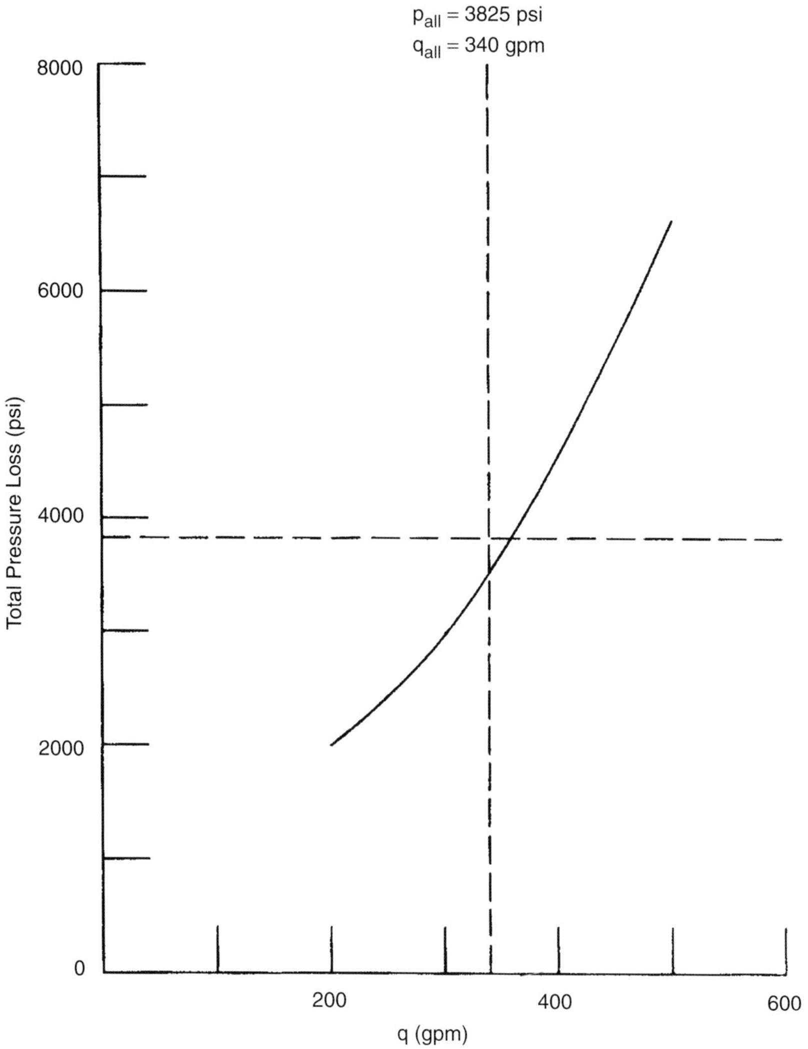

FIGURE 3.15 A $5\frac{3}{4}$-in. liner, total pressure loss versus flowrate, Example 3.2 (courtesy of Baker Hughes Co.).

the highest turbine motor horsepower) is found in Figure 3.15, the $5\frac{3}{4}$ in. liner. This optimal circulation flowrate is 340 gal/min.

Turbine Motor Performance. Using the turbine motor performance data in Table 3.5 and the scaling relationships in Equations 3.62 through 3.68, the performance graph for the turbine motor operating with a circulation flowrate of 340 gal/min and mud weight of 16.2 lb/gal can be prepared. This is given in Figure 3.18.

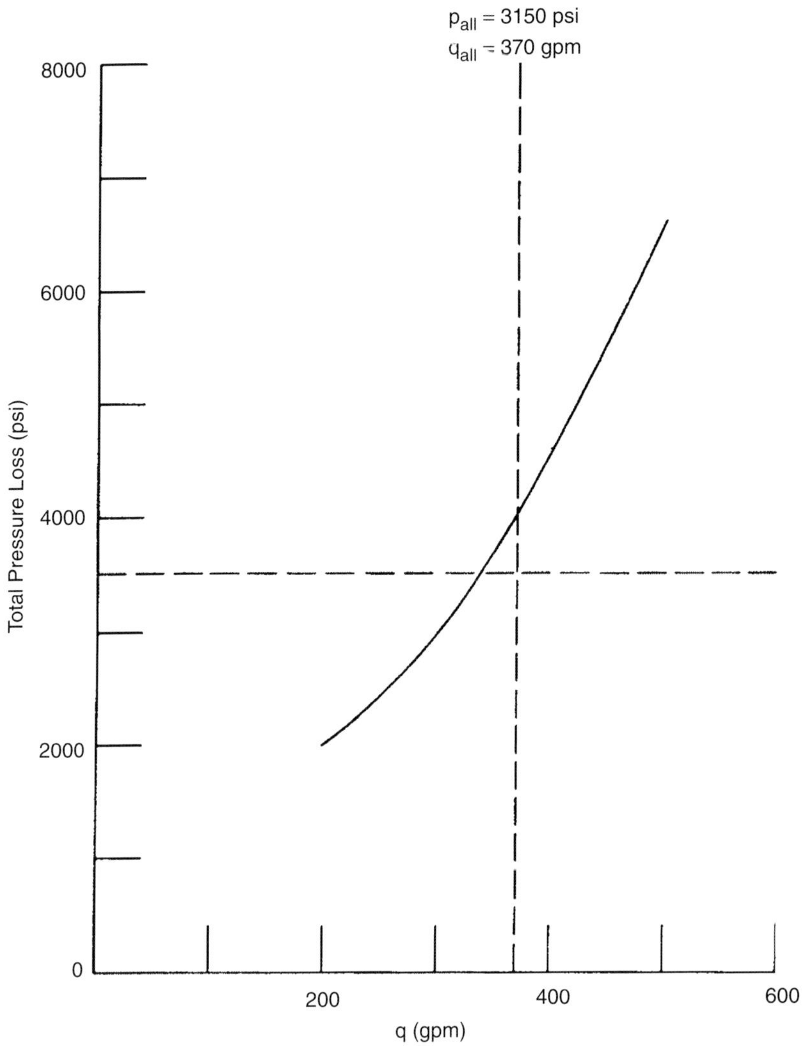

FIGURE 3.16 A 6-in. liner, total pressure loss versus flowrate, Example 3.2 (courtesy of Baker Hughes Co.).

Total Flow Area for Bit. Knowing the optimal circulation flowrate, the actual pressure loss across the bit can be found as before in the above. This is

$$\Delta p_b = \frac{85.05(1{,}714)}{340}$$

$$= 429 \text{ psi} \tag{3.83}$$

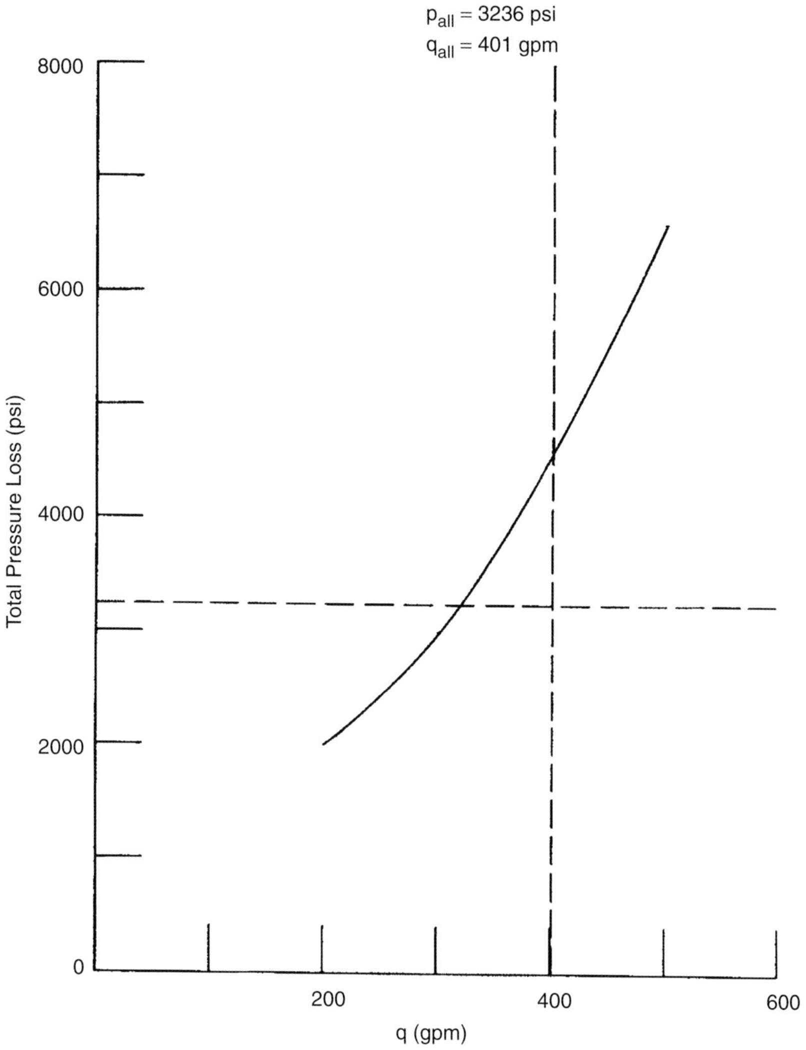

FIGURE 3.17 A $6\frac{1}{4}$-in. liner, total pressure loss versus flowrate, Example 3.2 (courtesy of Baker Hughes Co.).

With the flowrate and pressure loss across the bit, the total flow area of the diamond bit A_{tf} (in.^{2}) can be found using [23]

$$\Delta p_b = \frac{q^2 \overline{\gamma}_m}{8795\left(A_{tf} e^{-0.832 \frac{ROP}{N_b}}\right)^2} \tag{3.84}$$

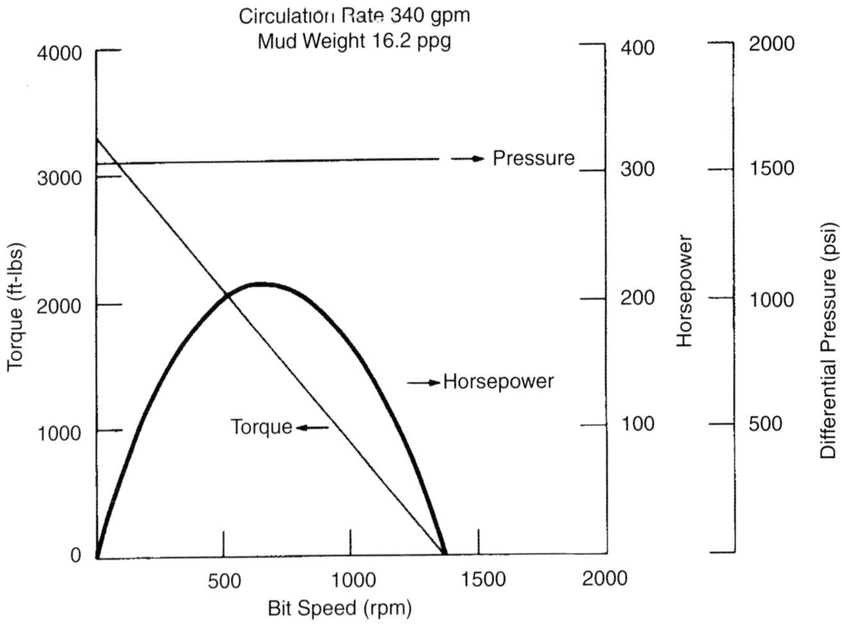

FIGURE 3.18 Turbine motor, $6\frac{3}{4}$-in. outside diameter, two motor sections, 212 stages, 340 gal/min, 16.2-lb/gal mud weight, Example 3.2 (courtesy of Baker Hughes Co.).

where ROP = rate of penetration in ft/hr
N_b = bit speed in rpm

The bit speed will be the optimum speed of the turbine motor, 685 rpm. The total flow area A_{tf} for the diamond bit is

$$A_{tf} = \left[\frac{(340)^2\, 16.2}{8795\,(429)}\right]^{1/2} \frac{1}{0.9879}$$
$$= 0.713 \text{ in.}^2 \tag{3.85}$$

3.14.3 Positive Displacement Motor

Figure 3.19 shows the typical rigid rotor and flexible elastomer stator configuration for a single chamber of a multichambered downhole positive displacement motor section. All the positive displacement motors presently in commercial use are of Moineau type, which uses a stator made of an elastomer. The rotor is made of rigid material such as steel and is fabcricated in a helical shape. The activating drilling mud, freshwater, aerated mud, foam or misted air is pumped at rather high velocity through the motor section, which, because of the eccentricity of the rotor and stator configuration, and the flexibility of the stator, allows the hydraulic pressure of the flowing fluid to impart a torque to the rotor. As the rotor rotates the

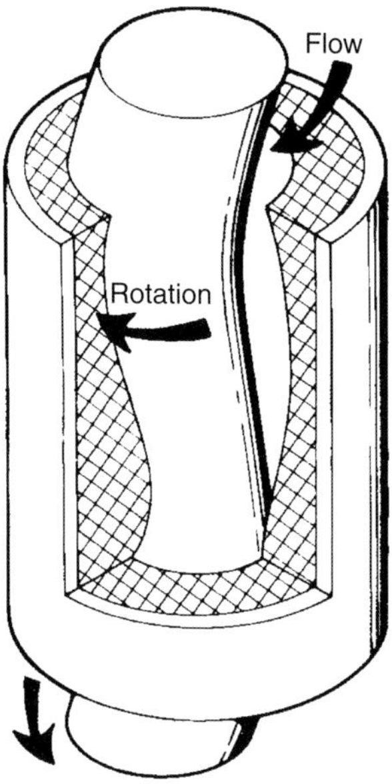

FIGURE 3.19 Basic positive displacement motor design principle (courtesy of Smith International, Inc.).

fluid is passes from chamber to chamber (a chamber is a lengthwise repeat of the motor). These chambers are separate entities and as one opens up to accept fluid from the preceding, the preceding closes up. This is the concept of the positive displacement motor.

3.14.3.1 Design

The rotational energy of the positive displacement motor is provided by the flowing fluid, which rotates and imparts torque to the drill bit. Figure 3.20 shows the typical complete downhole positive displacement motor.

In general, the downhole positive displacement motor constructed on the Moineau principle is composed of four sections: (1) the dump valve section, (2) the multistage motor section, (3) the connecting rod section and (4) the thrust and radial-bearing section. These sections are shown in Figure 3.20. Usually the positive displacement motor has multichambers, however, the number of chambers in a positive displacement motor is much less than the number of stages in a turbine motor. A typical positive displacement motor has from two to seven chambers.

The dump valve is a very important feature of the positive displacement motor. The positive displacement motor does not permit fluid to flow through the motor unless the motor is rotating. Therefore, a dump valve at the top of the motor allows drilling fluid to be circulated to the annulus even if the motor is not rotating. Most dump valve designs allow the fluid to circulate to the annulus when the pressure is below a certain threshold,

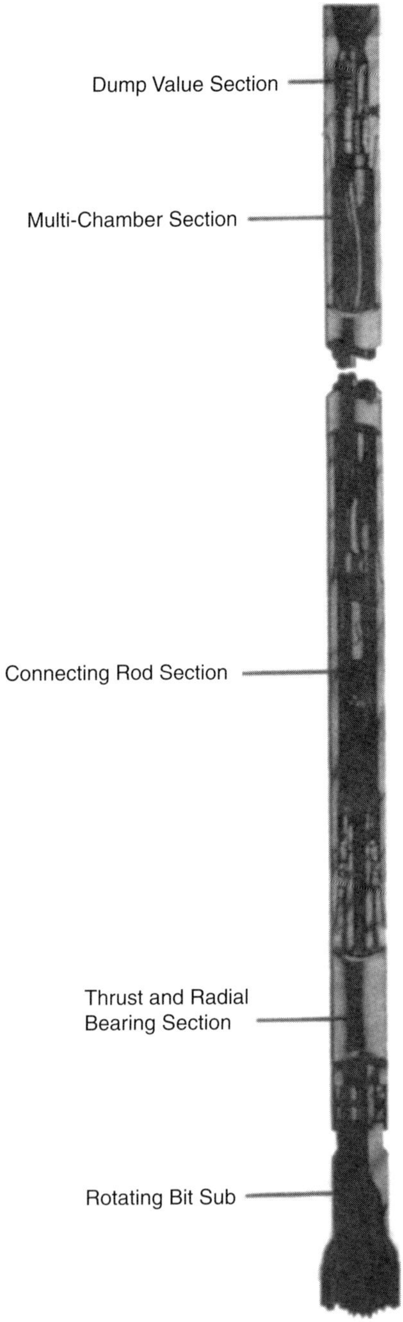

FIGURE 3.20 Downhole positive displacement motor design (courtesy of Smith International, Inc.).

say below 50 psi or so. Only when the surface pump is operated does the valve close to force all fluid through the motor.

The multichambered motor section is composed of only two continuous parts, the rotor and the stator. Although they are continuous parts, they usually constitute several chambers. In general, the longer the motor section, the more chambers. The stator is an elastomer tube formed to be the inside surface of a rigid cylinder. This elastomer tube stator is of a special material and shape. The material resists abrasion and damage from drilling muds containing cuttings and hydrocarbons. The inside surface of the stator is of an oblong, helical shape. The rotor is a rigid steel rod shaped as a helix. The rotor, when assembled into the stator and its outside rigid housing, provides continuous seal at contact points between the outside surface of the rotor and the inside surface of stator (see Figure 3.19). The rotor or driving shaft is made up of n_r lodes. The stator is made up of n_s lodes, which is equal to one lobe more than the rotor. Typical cross-sections of positive displacement motor lobe profiles are shown in Figure 3.21. As drilling fluid is pumped through the cavities in each chamber that lies open between the stator and rotor, the pressure of the flowing fluid causes the rotor to rotate within the stator. There are several chambers in a positive displacement motor because the chambers leak fluid. If the first chamber did not leak when operating, there would be no need for additional chambers.

In general, the larger lobe profile number ratios of a positive displacement motor, the higher the torque output and the lower the speed (assuming all other design limitations remain the same).

The rotors are eccentric in their rotation at the bottom of the motor section. Thus, the connecting rod section provides a flexible coupling between the rotor and the main drive shaft located in the thrust and radial bearing section. The main drive shaft has the drill bit connected to its bottom end.

The thrust and radial-bearing section contains the thrust bearings that transfer the weight-on-bit to the outside wall of the positive displacement motor. The radial support bearings, usually located above the thrust bearings, ensure that the main drive shaft rotates about a fixed center. As in

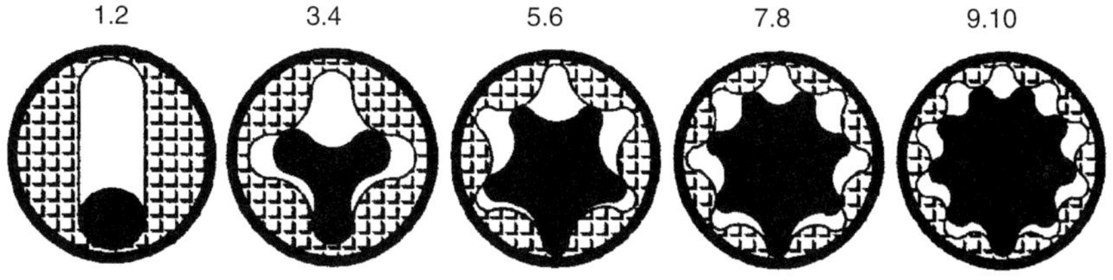

FIGURE 3.21 Typical positive displacement motor lobe profiles (courtesy of Smith International, Inc.).

most turbine motor designs, the bearings are cooled by the drilling fluid. There are some positive displacement motor designs that are now using grease-packed, sealed bearing assemblies. There is usually a smaller upper thrust bearing that allows rotation of the motor while pulling out of the hole. This upper thrust bearing is usually at the upper end of thrust and radial bearing section.

There are, of course, variations on the downhole positive displacement motor design, but the basic sections discussed above will be common to all designs.

The main advantages of the downhole positive displacements motor are:

1. Soft, medium and hard rock formations can be drilled with a positive displacement motor using nearly and type of rock bit. The positive displacement motor is especially adaptable to drilling with roller rock bits.
2. Rather moderate flow rates and pressures are required to operate the positive displacement motor. Thus, most surface pump systems can be used to operate these downhole motors.
3. Rotary speed of the positive displacement motor is directly proportional to flowrate. Torque is directly proportional to pressure. Thus, normal surface instruments can be used to monitor the operation of the motor downhole.
4. High torques and low speeds are obtainable with certain positive displacement motor designs, particularly, the higher lobe profiles (see Figure 3.21).
5. Positive displacement motors can be operated with aerated muds, foam and air mist.

The main disadvantages of the downhole positive displacement motors are

1. When the rotor shaft of the positive displacement motor is not rotating, the surface pump pressure will rise sharply and little fluid will pass through the motor.
2. The elastomer of the stator can be damaged by high temperatures and some hydrocarbons.

3.14.3.2 *Operations*

Figure 3.22 gives the typical performance characteristics of a positive displacement motor. The example in this figure is a $6\frac{3}{4}$ in. outside diameter positive displacement motor having five chambers activated by a 400-lb/gal flowrate of drilling mud.

For this example, a pressure of about 100 psi is required to start the rotor shaft against the internal friction of the rotor moving in the elastomer stator (and the bearings). With constant flowrate, the positive displacement

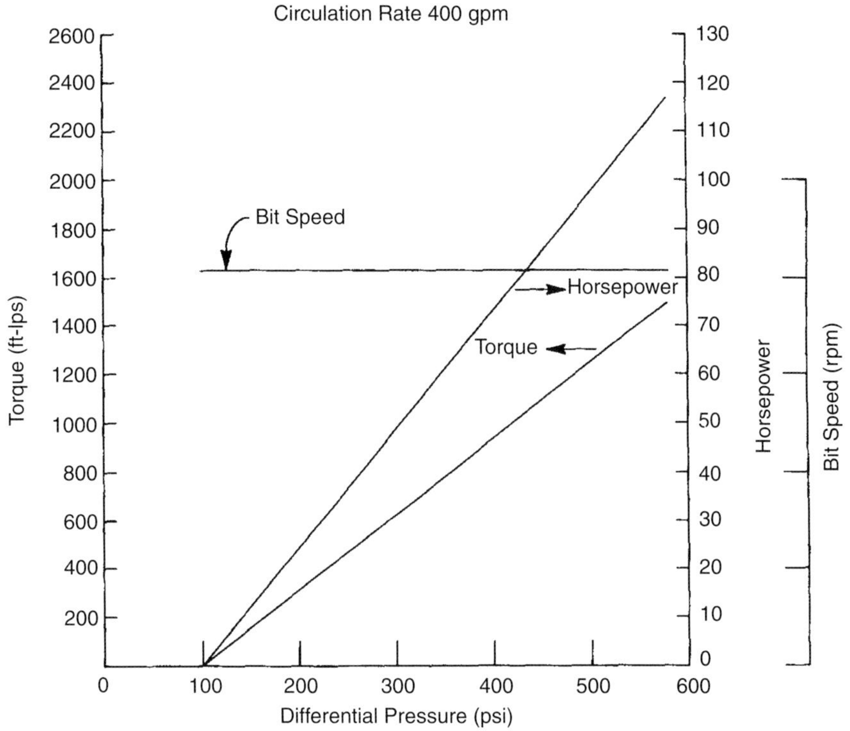

FIGURE 3.22 Positive displacement motor, $6\frac{3}{4}$-in. outside diameter, 1:2 lobe profile, 400 gal/min, differential pressure limit 580 psi (courtesy of Baker Hughes Co.).

motor will run at or near constant speed. Thus, this 1:2 lobe profile example motor has an motor speed of 408 rpm. The torque and the horsepower of the positive displacement motor are both linear with the pressure drop across the motor. Therefore, as more weight is placed on the drill bit (via the motor), the greater is the resisting torque of the rock. The mud pumps can compensate for this increased torque by increasing the pressure on the constant flowrate through the motor. In this example the limit in pressure drop across the motor is about 580 psi. Beyond this limit there will be either extensive leakage or damage to the motor, or both.

If the positive displacement motor is lifted off the bottom of the borehole and circulation continues, the motor will simply continue to rotate at 408 rpm. The differential pressure, however, will drop to the value necessary to overcome internal friction and rotate, about 100 psi. In this situation the motor produces no drilling torque or horsepower.

As the positive displacement motor is lowered and weight is placed on the motor and thus the bit, the motor speed continues but the differential pressure increases, resulting in an increase in torque and horsepower. As more weight is added to the positive displacement motor and bit, the torque and horsepower will continue to increase with increasing differentiated pressure (i.e., standpipe pressure). The amount of torque and power can be determined by the pressure change at the standpipe at the surface between the unloaded condition and the loaded condition. If too much weight is placed on the motor, the differential pressure limit for the motor will be reached and there will be leakage or a mechanical failure in the motor.

The rotor of the Moineau-type positive displacement motor has a helical design. The axial wave number of the rotor is one less than the axial wave number for the stator for a given chamber. This allows the formation of a series of fluid cavities as the rotor rotates. The number of stator wave lengths n_s and the number of rotor wave length n_r per chamber are related by [14, 21].

$$n_s = n_r + 1 \tag{3.86}$$

The rotor is designed much like a screw thread. The rotor pitch is equivalent to the wavelength of the rotor. The rotor lead is the axial distance that a wave advances during one full revolution of the rotor. The rotor pitch and the stator pitch are equal. The rotor lead and stator lead are proportional to their respective number of waves. Thus, the relationship between rotor pitch t_r (in.) and stator pitch, t_s (in.) is [21]

$$t_r = t_s \tag{3.87}$$

The rotor lead L_r (in.) is

$$L_r = n_r t_r \tag{3.88}$$

The stator lead L_s (in.) is

$$L_s = n_s t_s \tag{3.89}$$

The specific displacement per revolution of the rotor is equal to the cross-sectional area of the fluid multiplied by the distance the fluid advances. The specific displacement s (in.^{3}) is

$$s = n_r n_s t_r A \tag{3.90}$$

where A is the fluid cross-sectional area (in.^{2}). The fluid cross-sectional area is approximately

$$A \approx 2ne_r^2\left(n_r^2 - 1\right) \tag{3.91}$$

where e_r is the rotor rotation eccentricity (in.). The special case of a 1:2 lobe profile motor has a fluid cross-sectional area of

$$A \approx 2e_r d_r \tag{3.92}$$

where d_r is the reference diameter of the motor (in.). The reference diameter is

$$d_r = 2e_r n_s \tag{3.93}$$

For the 1:2 lobe profile motor, the reference diameter is approximately equal to the diameter of the rotor shaft.

The instantaneous torque of the positive displacement motor M(ft-lb) is

$$M = 0.0133\ s\ \Delta p \eta \tag{3.94}$$

where Δp = differential pressure loss through the motor in psi
η = total efficiency of the motor. The 1:2 lobe profile motors have efficiencies around 0.80. The higher lobe profile motors have efficiencies that are lower (i.e., of the order of 0.70 or less)

The instantaneous speed of the positive displacement motor N (rpm) is

$$N = \frac{231.016q}{s} \tag{3.95}$$

where q is the circulation flowrate (gal/min).

The positive displacement motor horsepower HP (hp) for any speed is

$$HP = \frac{q\Delta p}{1{,}714}\eta \tag{3.96}$$

The number of positive displacement motor chambers n_c is

$$n_c = \frac{L}{t_s} - (n_s - 1) \tag{3.97}$$

where L is the length of the actual motor section (in.).

The maximum torque M_{max} will be at the maximum differential pressure Δp_{max}, which is

$$M_{max} = 0.133 s \Delta p_{max} \eta \tag{3.98}$$

The maximum horse power HP_{max} will also be at the maximum differential pressure Δp_{max}, which is

$$HP_{max} = \frac{q\Delta p_{max}}{1{,}714}\eta \tag{3.99}$$

It should be noted that the positive displacement motor performance parameters are independent of the drilling mud weight. Thus, these performance parameters will vary with motor design values and the circulation flowrate.

If the above performance parameters for a positive displacement motor design are known for a given circulation flowrate (denoted as 1), the performance parameters for the new circulation flowrate (denoted as 2) can be found by the following relationships:
Torque

$$M_2 = M_1 \tag{3.100}$$

Speed

$$N_2 = \left(\frac{q_2}{q_1}\right) N_1 \tag{3.101}$$

Power

$$HP_2 = \left(\frac{q_2}{q_1}\right) HP_1 \tag{3.102}$$

Table 3.9 gives the performance characteristics for various circulation flowrates for the 1:2 lobe profile $6\frac{3}{4}$-in. outside diameter positive displacement motor. Figure 3.22 shows the performance of the 1:2 lobe profile positive displacement motor at a circulation flowrate of 400 gal/min.

Table 3.10 gives the performance characteristics for various circulation flowrates for the 5:6 lobe profile, $6\frac{3}{4}$-in. outside diameter positive displacement motor. Figure 3.23 shows the performance of the 5:6 lobe profile positive displacement motor at a circulation flow rate of 400 gal/min.

The positive displacement motor whose performance characteristics are given in Table 3.9 is a 1:2 lobe profile motor. This lobe profile design is usually used for deviation control operations. The 1:2 lobe profile design yields a downhole motor with high rotary speeds and low torque. Such a combination is very desirable for the directional driller. The low torque minimizes the compensation that must be made in course planning which must be made for the reaction torque in the lower part of the drill string.

TABLE 3.9 Positive Displacement Motor, $6\frac{3}{4}$-in. Outside Diameter, 1:2 Lobe Profile, Five Motor Chambers

Circulation Rate (gpm)	Speed (rpm)	Maximum Differential Pressure (psi)	Maximum Torque (ft-lb)	Maximum Horse-power
200	205	580	1500	59
250	255	580	1500	73
300	306	580	1500	87
350	357	580	1500	102
400	408	580	1500	116
450	460	580	1500	131
500	510	580	1500	145

TABLE 3.10 Positive Displacement Motor, $6\frac{3}{4}$-in. Outside Diameter, 5:6 Lobe Profile, Five Motor Chambers

Circulation Rate (gpm)	Speed (rpm)	Maximum Differential Pressure (psi)	Maximum Torque (ft-lb)	Maximum Horse-power
200	97	580	2540	47
250	122	580	2540	59
300	146	580	2540	71
350	170	580	2540	82
400	195	580	2540	94

Courtesy of Eastman-Christensen.

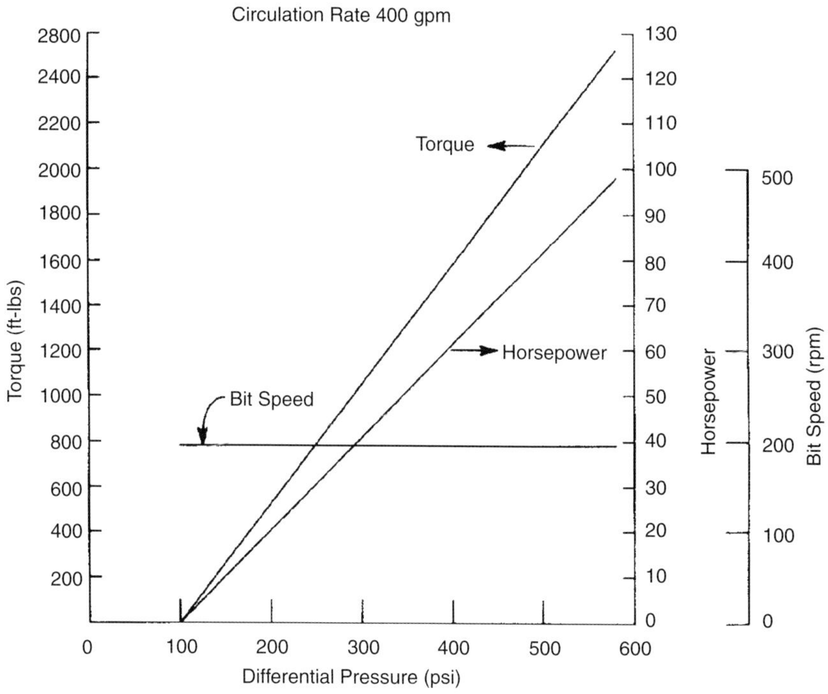

FIGURE 3.23 Positive displacement motor, $6\frac{3}{4}$-in. outside diameter, 5:6 lobe profile, 400 gal/min, differential pressure limit 580 psi (courtesy of Baker Hughes Co.).

This reactive torque when severe can create difficulties in deviation control planning. The tradeoff is, however, that higher speed reduces the bit life, especially roller rock bit life.

The positive displacement motor whose performance characteristics are given in Table 3.10 is a 5:6 lobe profile motor. This lobe profile design is usually used for straight hole drilling with roller rock bits, or for deviation control operations where high torque polycrystalline diamond compact bit or diamond bits are used for deviation control operations.

Example 3.3

A $6\frac{3}{4}$-in. outside diameter positive displacement motor of a 1:2 lobe profile design (where performance data are given in Table 3.7) has rotor eccentricity of 0.60 in., a reference diameter (rotor shaft diameter) of 2.48 in. and a rotor pitch of 38.0 in. If the pressure drop across the motor is determined to be 500 psi at a circulation flowrate of 350 gal/min with 12.0 lb/gal, find the torque, rotational speed and the horsepower of the motor.

Torque. Equation 3.91 gives the fluid cross-sectional area of the motor, which is

$$\begin{aligned} A &= 2(0.6)(2.48) \\ &= 2.98 \text{ in.}^2 \end{aligned} \tag{3.103}$$

Equation 3.91 gives the specific displacement of the motor, which is

$$\begin{aligned} s &= (1)(2)(38.0)(2.98) \\ &= 226.5 \text{ in.}^3 \end{aligned} \tag{3.104}$$

The torque is obtained from Equation 3.94, assuming an efficiency of 0.80 for the 1:2 lobe profile motor. This is

$$\begin{aligned} M &= 0.0133(226.5)(500)(0.80) \\ &= 1205 \text{ ft-lb} \end{aligned} \tag{3.105}$$

Speed. The rotation speed is obtained from Equation 3.95. This is

$$\begin{aligned} N &= \frac{231.016\,(350)}{226.5} \\ &= 357 \text{ rpm} \end{aligned} \tag{3.106}$$

Horsepower. The horsepower the motor produces is obtained from Equation 3.96. This is

$$\begin{aligned} HP &= \frac{350\,(500)}{1714}(0.80) \\ &= 82 \end{aligned} \tag{3.107}$$

Planning for a positive displacement motor run and actually drilling with such a motor is easier than with a turbine motor. This is mainly due to the fact that when a positive displacement motor is being operated, the operator can know the operating torque and rotation speed via surface data. The standpipe pressure will yield the pressure drop through the motor, thus the torque. The circulation flowrate will yield the rotational speed.

Example 3.4

A $6\frac{3}{4}$-in. outside diameter positive displacement motor (whose performance data are given in Tables 3.9 and 3.10) is to be used for a deviation control direction drill operation. The motor will use an $8\frac{1}{2}$-in. diameter roller rock bit for the drilling operation. The directional run is to take place at a depth of 10,600 ft (measured depth). The rock formation to be drilled is classified as medium, and it is anticipated that 30 ft/hr will be the maximum possible drilling rate. The mud weight is to be 11.6 lb/gal. The drilling rig has a National Supply Company duplex mud pump Model E-700 available. The details of this pump are given in Table 3.11. Because this is a deviation control run, the 1:2 lobe profile positive displacement motor will be used since it has the lowest torque for a given circulation flowrate (see Table 3.9). Determine the appropriate circulation flowrate to be used for the roller rock bit, positive displacement motor combination and the appropriate liner size to be used in the duplex pump. Also, prepare the positive displacement motor performance graph for the chosen circulation flowrate. Determine the bit nozzle sizes.

Bit Pressure Loss. It is necessary to choose the bit pressure loss such that the thrust load created in combination with the weight on bit will yield an on-bottom load on the motor thrust bearings, which is less than the maximum allowable load for the bearings. Since this is a deviation control run and, therefore, the motor will be drilling only a relatively short time and distance, the motor thrust bearings will be operated at their maximum rated load for on-bottom operation. Figure 3.24 shows that maximum allowable motor thrust bearing load is about 6,000 lb. To have the maximum weight on bit, the maximum recommended bit pressure loss of 500 psi will be used. This will give maximum weight on bit of about 12,000 lb. The higher bit pressure loss will, of course, give the higher cutting face cleaning via jetting force (relative to the lower recommended bit pressure losses).

Total Pressure Loss. Since bit life is not an issue in a short deviation control motor run operation, it is desirable to operate the positive

TABLE 3.11 Duplex Mud Pump, Model E-700, National Supply Company, Example 3.4

Input Horsepower, 825; Maximum Strokes per Minute, 65; length of Stroke, 16 Inches						
	$5\frac{3}{4}$	6	$6\frac{1}{4}$	$6\frac{1}{2}$	$6\frac{3}{4}$	7
Output per Stroke (gals)	6.14	6.77	7.44	8.13	8.85	9.60
Maximum Pressure (psi)	3000	2450	2085	1780	1535	1260

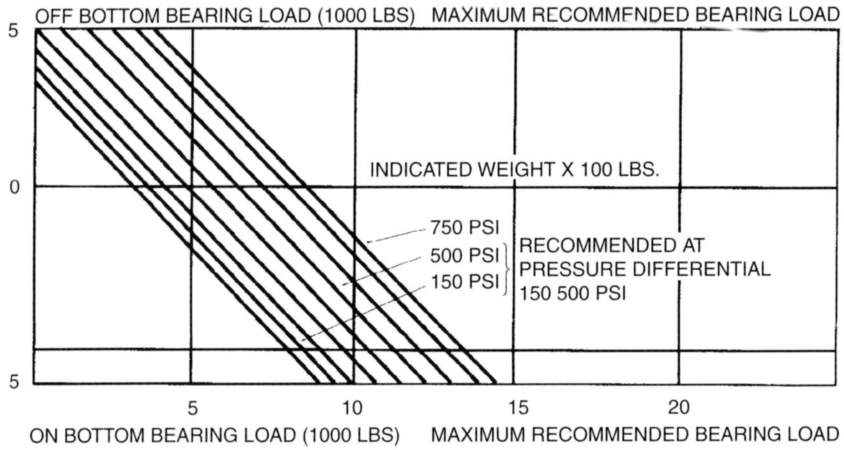

FIGURE 3.24 Hydraulic thrust and indicated weight balance for positive displacement motor (courtesy of Smith International, Inc.).

TABLE 3.12 Drillstring Component Pressure Losses at Various Circulation Flowrates for Example 3.4

	Pressure (psi)			
Components	**200 gpm**	**300 gpm**	**400 gpm**	**500 gpm**
Surface Equipment	5	11	19	30
Drill Pipe Bore	142	318	566	884
Drill Collar Bore	18	40	71	111
PDM	580	580	580	580
Drill Bit	500	500	500	500
Drill Collar Annulus	11	25	45	70
Drill Pipe Annulus	32	72	128	200
Total Pressure Loss	1288	1546	1909	2375

displacement motor at as high a power level as possible during the run. The motor has a maximum pressure loss with which it can operate. This is 580 psi (see Table 3.9). It will be assumed that the motor will be operated at the 580 psi pressure loss in order to maximum the torque output of the motor. To obtain the highest horsepower for the motor, the highest circulation flowrate possible while operating within the constraints of the surface mud pump should be obtained. To obtain this highest possible, or optimal, circulation flowrate, the total pressure losses for the circulation system must be obtained for various circulation flowrates. These total pressure losses tabulated in the lower row of Table 3.12 represent the surface standpipe pressure when operating at the various circulation flowrates.

Pump Limitations. Table 3.11 shows there are six possible liner sizes that can be used on the Model E-700 mud pump. Each liner size must be considered to obtain the optimum circulation flowrate and a appropriate liner size. The maximum pressure available for each liner size will be reduced by a safety factor of 0.90. The maximum volumetric flowrate available for each liner size will also be reduced by a volumetric efficiency factor of 0.80 and an additional safety factor of 0.90. Thus, from Table 3.11, the allowable maximum pressures and allowable maximum volumetric flowrates will be those shown in Figures 3.25 through 3.30 which are the liner sizes $5\frac{3}{4}$, 6, $6\frac{1}{2}$ and 7 in., respectively. Plotted on each of these figures are the total pressure losses for the various circulation flowrates considered. The horizontal straight line on each figure is the allowable maximum pressure for the particular liner size. The vertical straight line is the allowable maximum volumetric flowrate for the particular liner size. Only circulation flowrates that are in the lower left quadrant of the figures are practical. The highest circulation flowrate (which produces the highest positive displacement motor horsepower) is found in Figure 3.27, the $6\frac{1}{4}$-in. liner. The optimal circulation flow rate is 348 gal/min.

Positive Displacement Motor Performance. Using the positive displacement motor performance data in Table 3.9 and the scaling relationships in Equations 3.100 through 3.102, the performance graph for

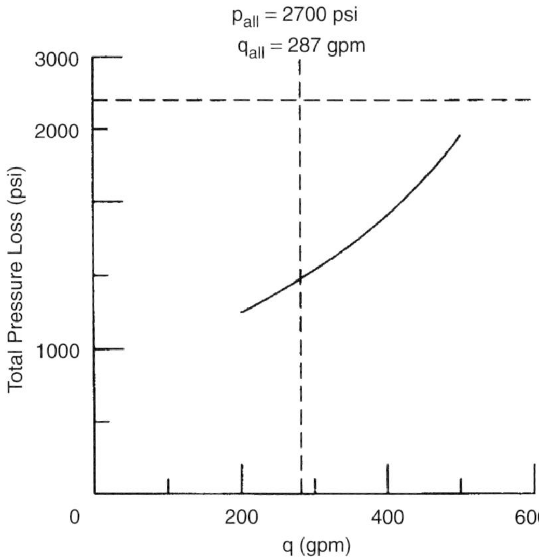

FIGURE 3.25 A $5\frac{3}{4}$-in. liner, total pressure loss vs. flowrate, Example 3.4 (courtesy of Baker Hughes Co.).

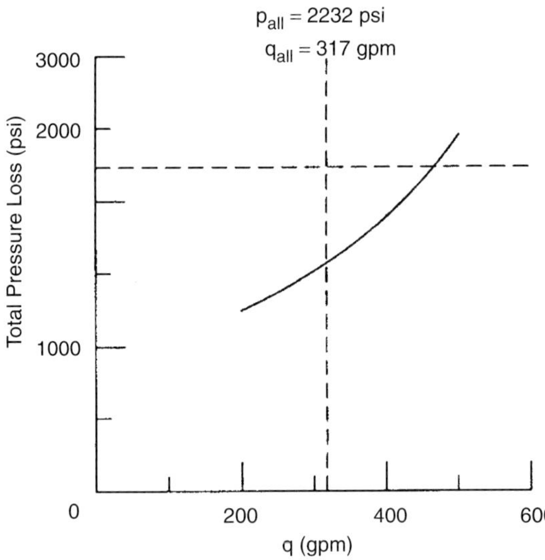

FIGURE 3.26 A 6-in. liner, total pressure loss vs. flowrate, Example 3.4 (courtesy of Baker Hughes Co.).

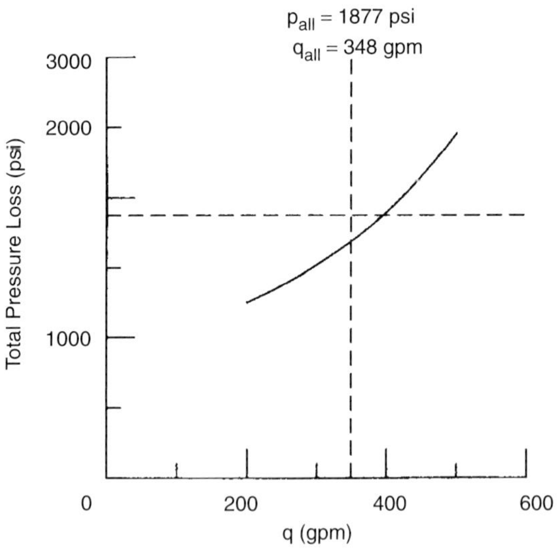

FIGURE 3.27 A $6\frac{1}{4}$-in. liner, total pressure loss vs. flowrate, Example 3.4 (courtesy of Baker Hughes Co.).

the positive displacement motor operating with a circulation flowrate of 348 gal/min can be prepared. This is given in Figure 3.31.

Bit Nozzle Sizes. The pressure loss through the bit must be 500 psi with a circulation flowrate of 348 gal/min with 11.6-lb/gal mud weight. The

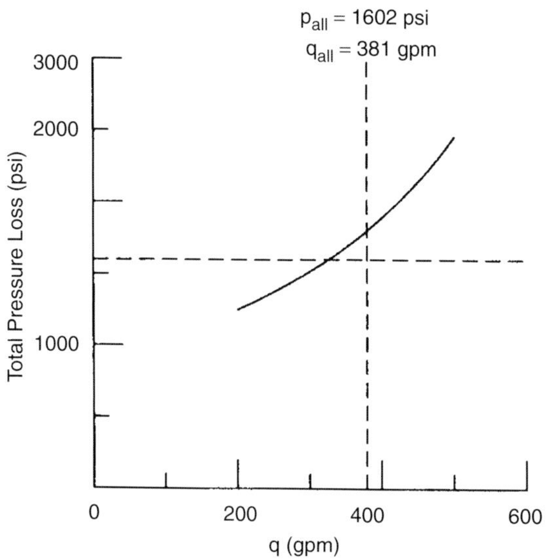

FIGURE 3.28 A $6\frac{1}{2}$-in. liner, total pressure loss vs. flowrate, Example 3.4 (courtesy of Baker Hughes Co.).

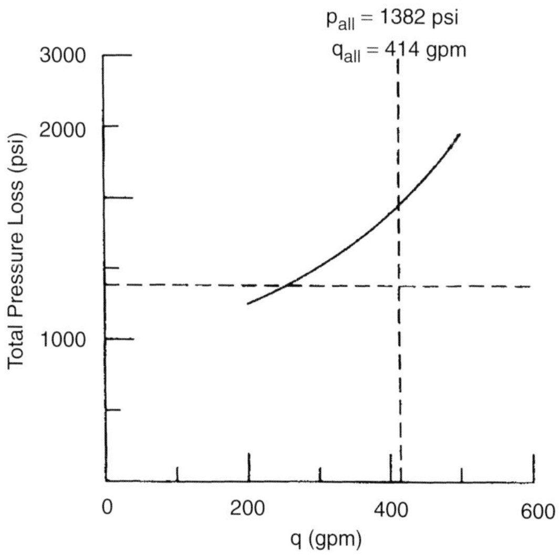

FIGURE 3.29 A $6\frac{3}{4}$-in. liner, total pressure loss vs. flowrate, Example 3.4 (courtesy of Baker Hughes Co.).

pressure loss through a roller rock bit with three nozzles is

$$\Delta p_b = \frac{q^2 \overline{\gamma}_m}{7430 C^2 d^4} \tag{3.108}$$

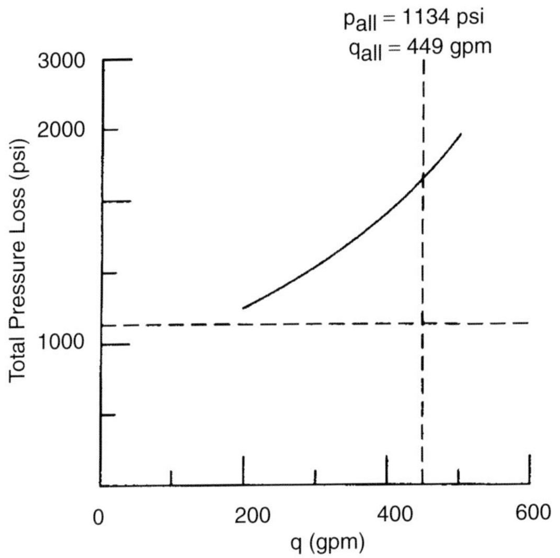

FIGURE 3.30 A 7-in. liner, total pressure loss vs. flowrate, Example 3.4 (courtesy of Baker Hughes Co.).

where C = nozzle coefficient (usually taken to be 0.95)
d_e = hydraulic equivalent diameter in in.

Therefore, Equation 3.108 is

$$500 = \frac{(348)^2 (11.6)}{7{,}430 (0.98)^2 de^4} \tag{3.109}$$

which yields

$$d_e = 0.8045 \text{ in.} \tag{3.110}$$

The hydraulic equivalent diameter is related to the actual nozzle diameters by

$$d_e = \left[a d_1^2 + b d_2^2 + c d_3^2 \right]^{1/2} \tag{3.111}$$

where
a = number of nozzles with diameter d_1
b = number of nozzles with diameter d_2
c = number of nozzles with diameter d_3
d_1, d_2 and d_3 = three separate nozzle diameters in in.

Nozzle diameters are usually in 32nds of an inch. Thus, if the bit has three nozzles with $\frac{15}{32}$ of an inch diameter, then

$$\begin{aligned} d_e &= [(3)(0.4688)^2]^{1/2} \\ &= 0.8120 \text{ in.} \end{aligned} \tag{3.112}$$

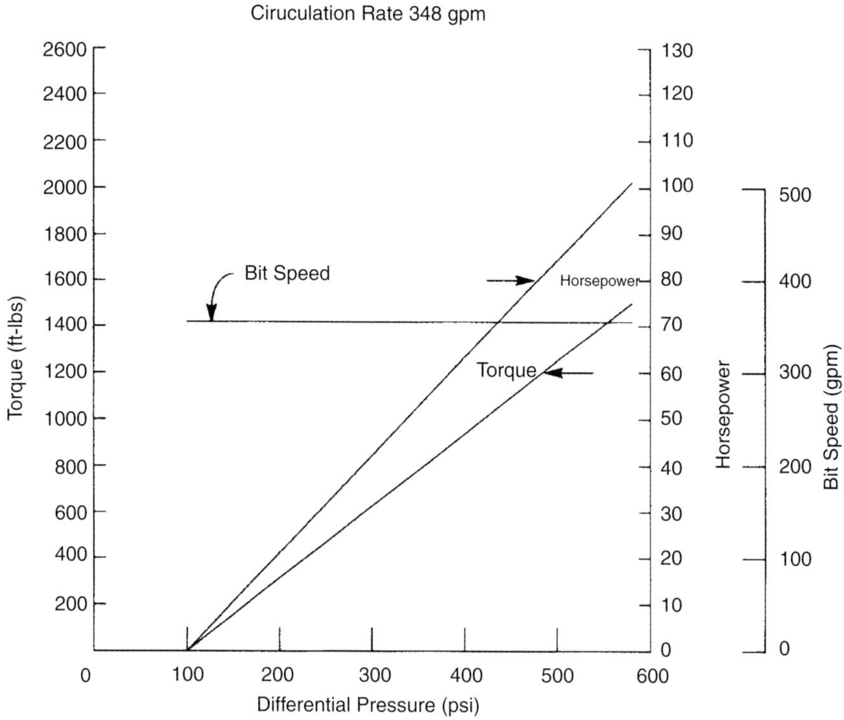

FIGURE 3.31 Positive displacement motor, $6\frac{3}{4}$-in. outside diameter, 1:2 lobe profile, 348 gal/min, differential pressure limit 580 psi, Example 3.4 (courtesy of Baker Hughes Co.).

The above hydraulic equivalent diameter is close enough to the one obtained with Equation 3.108. Therefore, the bit should have three $\frac{5}{32}$-in. diameter nozzles.

3.14.4 Down the Hole Air Hammers

3.14.4.1 Design

There are two basic designs for the downhole air hammer. One design uses a flow path of the compressed air through a control rod (or feed tube) down the center hammer piston (or through passages in the piston) and then through the hammer bit. The other design uses a flow path through a housing annulus passage (around the piston) and then through the hammer bit. Figure 3.32 shows a schematic of a typical control rod flow design downhole air hammer. The hammer action of the piston on the top of the drill bit shank provides an impact force that is transmitted down the shank to the bit studs, which, in turn, crush the rock at the rock face. In shallow boreholes where there is little annulus back pressure, the piston impacts the

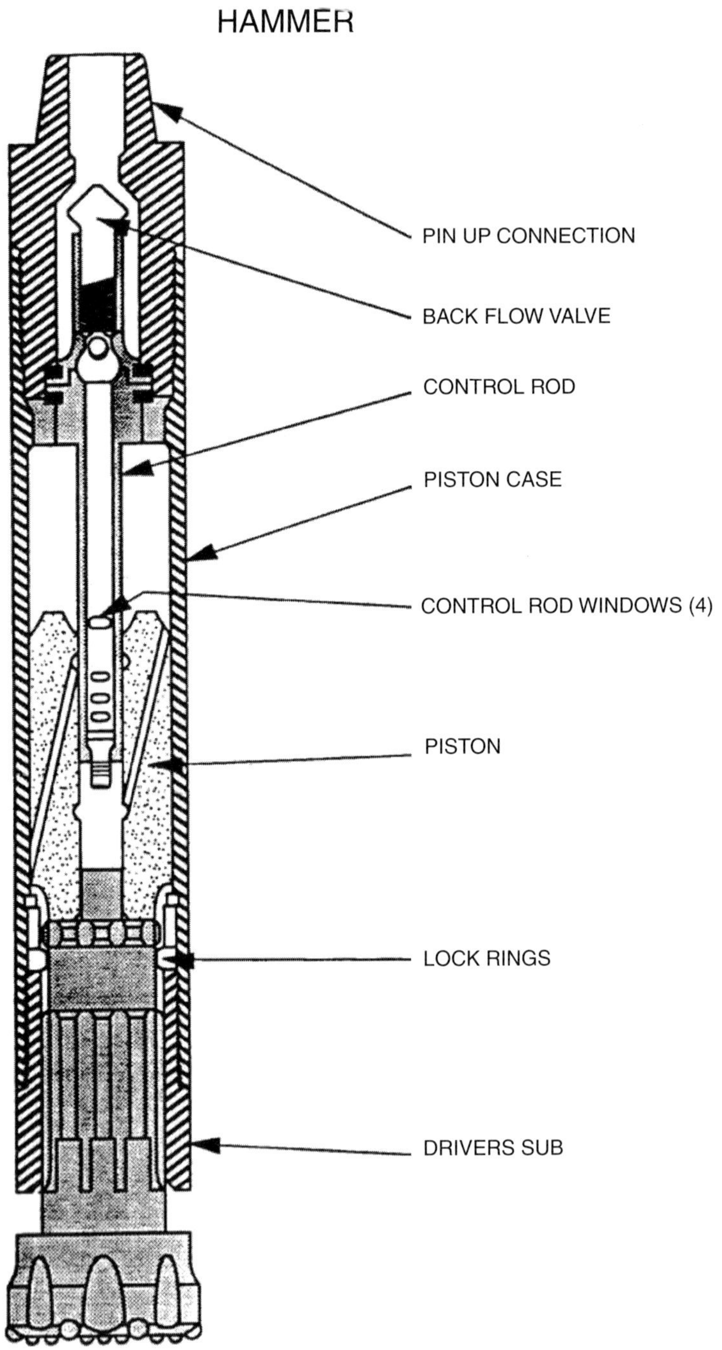

FIGURE 3.32 Schematic cutaway of a typical air hammer (courtesy of Diamond Air Drilling Services, Inc.).

top of the bit shank at a rate of from about 600 to 1,700 strikes per minute (depending on volumetric flow rate of gas). However, in deep boreholes where the annulus back pressure is usually high, impacts can be as low as 100 to 300 strikes per minute. This unique air drilling device requires that the drill string with the air hammer be rotated so that the drill bit studs impact over the entire rock face. In this manner an entire layer of the rock face can be destroyed and the drill bit advanced.

Figure 3.32 shows the air hammer suspended from a drill string lifted off the bottom (the shoulder of the bit is not in contact with the shoulder of the driver sub). In this position, compressed air flows through the pin connection at the top of the hammer to the bit without actuating the piston action (i.e., to blow the borehole clean). When the hammer is placed on the bottom of the borehole and weight placed on the hammer, the bit shank will be pushed up inside the hammer housing until the bit shoulder is in contact with the shoulder of the driver sub. This action aligns one of the piston ports (of one of the flow passage through the piston) with one of the control rod windows. This allows the compressed air to flow to the space below the piston which in turn forces the piston upward in the hammer housing. During this upward stroke of the piston, no air passes through the bit shank to the rock face. In essence, rock cuttings transport is suspended during this upward stroke of the piston.

When the piston reaches the top of its stroke, another one of the piston ports aligns with one of the control rod windows and supplies compressed air to the open space above the piston. This air flow forces the piston downward until it impacts the top of the bit shank. At the same instant the air flows to the space above the piston, the foot valve at the bottom of the control rod opens and air inside the drill string is exhaust through the control rod, bit shank, and the bit orifices to the rock face. This compressed air exhaust entrains the rock cuttings created by the drill bit for transport up the annulus to the surface. This impact force on the bit allows the rotary action of the drill bit to be very effective in destroying rock at the rock face. This in turn allows the air hammer to drill with low WOB. Typically for a $6\frac{3}{4}$ inch outside diameter air hammer drilling with a $7\frac{7}{8}$ inch air hammer bit, the WOB can be as low as 1,500 lb. Downhole air hammers must have an oil type lubricant injected into the injected air during the drilling operation. This lubricant is needed to lubricate the piston surfaces as it moves in the hammer housing. Air hammers are used exclusively for vertical drilling operations.

This piston cycle is repeated as the drill string (and thus the drill bit) is rotated and in this manner continuous layers of rock on the rock face are crushed and the cuttings removed. The air flow through the air hammer is not continuous. When the piston is being lifted in the hammer, air is not being exhausted through the drill bit. For example, at a piston impact rate of 600 strikes per minute the air is shut down for about 0.050 seconds per cycle. This is so short a time that the air flow rate through the annulus can be assumed as a continuous flow.

Downhole air hammers are available in housing outside diameters from 3 inches to 16 inches. The 3 inch housing outside diameter hammer can drill a borehole as small as $3\frac{5}{8}$ inches. The 16 inch housing outside diameter hammer can drill a boreholes from $17\frac{1}{2}$ inches to 33 inches. For shallow drilling operations, conventional air hammer bits are adequate. For deep drilling operations (usually oil and gas recovery wells), higher quality oil field air hammer drill bits are required.

There are a variety of manufacturers of downhole air hammers. These manufacturers use several different designs for their respective products. The air hammer utilizes very little power in moving the piston inside the hammer housing. For example, a typical $6\frac{3}{4}$ inch outside diameter air hammer with a 77 lb piston operating at about 600 strikes per minute, will use less than 2 horsepower driving the piston. This is a very small amount of power relative to the total needed for the actual rotary drilling operation. Thus, it is clear that the vast majority of the power to the drill string is provided by the rotary table. Therefore, any pressure loss (i.e., energy loss) due to the piston lifting effort can usually be ignored. The major pressure loss in the flow through an air hammer is due to the flow energy losses from the constrictions in the flow path when the air is allowed to exit the hammer (on the down stroke of the piston). All air hammer designs have internal flow constrictions. These flow constrictions can be used to model the flow losses through the hammer. In most designs these constrictions can be approximately represented by a set of internal orifice diameters in the flow passages to the drill bit. These internal orifices are usually the ports (and associated channels) through the piston and the orifice at the open foot valve.

3.14.4.2 *Operations*

Hammer Selection. With such a vast array of hammers on the market it is quite a task to pick the best hammer for your application. Some manufacturers may have up to six different hammers to drill the same size hole. Why so many? Read on!

Size of Hammer. Generally, try to choose a hammer size nearest to the intended hole size. The bigger the hammer diameter, the bigger the piston diameter, the bigger the performance. Figure 3.33 shows how rapidly performance drops when using an oversize bit on a small hammer, Figure 3.34 shows how performance can be better maintained by going to a bigger hammer, Not only this, a bigger hammer allows a bigger bit shank, better energy transfer resulting in greater strength and lower stresses—more reliability and lower cost per foot. The only reasons hammer selection should

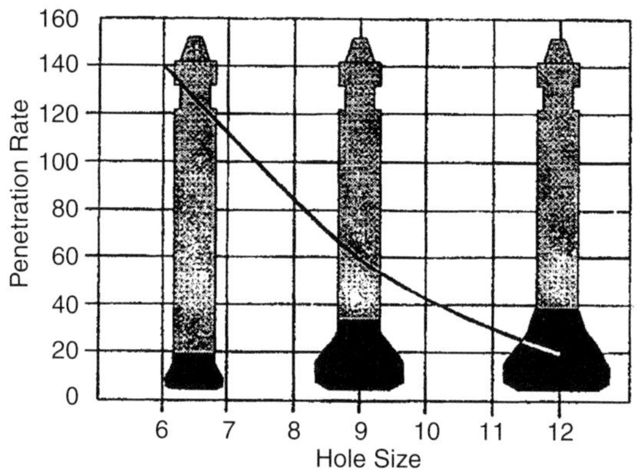

FIGURE 3.33 Poor Hammer/Bit Ratio. (Courtesy of DrilMaster.)

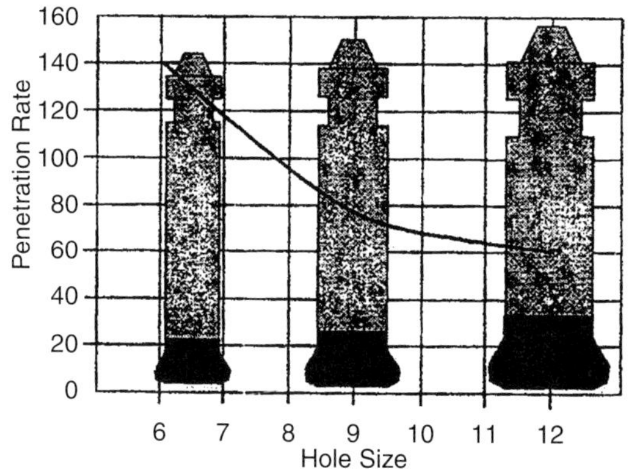

FIGURE 3.34 Correct Hammer/Bit Ratio. (Courtesy of DrilMaster.)

not be based entirely on hole size is because of flexibility and capital cost of other hammers. Special hammer sizes such as 5.5 or 11 inch hammers are often used in such situations to maintain performance without sacrificing flexibility.

Hammer Air Consumption. The amount of air available from the air compressor will affect the choice of hammer. Generally, it is best to select a hammer nearest the maximum operating conditions of the compressor. The higher the hammer operating pressure, the better the performance, However, just like your car engine, the manufacturer's ingenuity and technology will play a part in the efficiency and performance of the hammer under different conditions.

Some manufacturers offer low volume (LV) or Mizer hammers. These are usually variations of standard hammers and may contain different pistons,

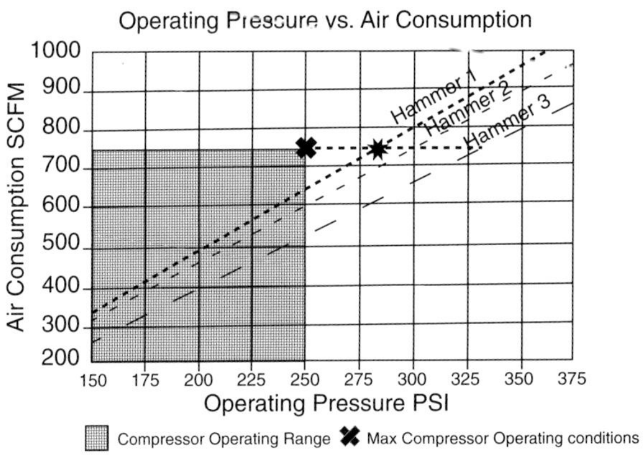

FIGURE 3.35 Case 1: 750 SCFM X 250 PSI Compressor. (Courtesy of DrilMaster.)

piston cases and/or rigid valves. Another approach by manufacturers is to change the internal port timing and volumes by substituting one part to adjust air consumption characteristics. Apart from the advantage of fewer stocked parts, this gives much greater flexibility and greater opportunity to optimize drilling for different rigs and drilling conditions. The following graphs show performance curves for three typical hammers plotted from manufacturer's data. For example, take three typical cases and show how to estimate operating pressure for each and make a hammer selection.

Case 1: 750SCFM X 250 PSI Compressor. Figure 3.35 shows that the compressor produces more air than any of the hammers can use. Hammer 1 is the nearest to the maximum operating conditions of the compressor and will give the best performance. Hammer 2 or 3 will work of course, and could rely on the compressor unloader to control the excess air of the compressor (not a desirable situation with a piston compressor). There are two other reasons why Hammer 2 or 3 should be considered: (1) the lower air consumption will result in lower air velocity and less erosion on the bit and the hammer, and (2) as a means of reducing power requirements and fuel.

Expected performance index:

Hammer 1	190%
Hammer 2	104%
Hammer 3	100%

Case 2: 450 SCFM X 250 PSI. Form a performance aspect, Hammer 3 can be expected to give a far better performance and hold around 220 psi when drilling, The user, for operational reasons, prefers to operate at around 180–200 psi with this specification we would select hammer 2 or possible 3 (Figure 3.36).

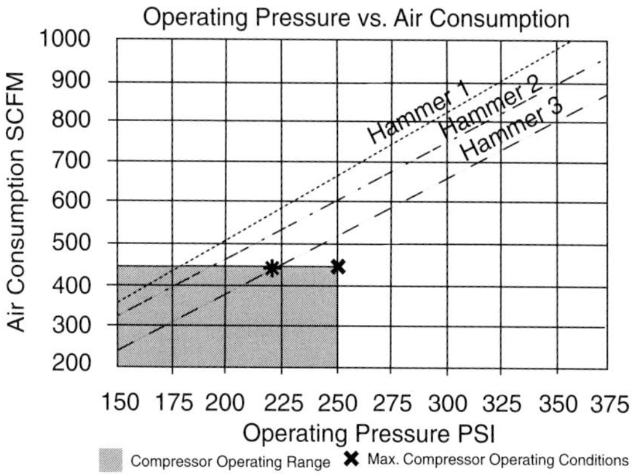

FIGURE 3.36 Case 2: 450 SCFM X 250 PSI Compressor. (Courtesy of DrilMaster.)

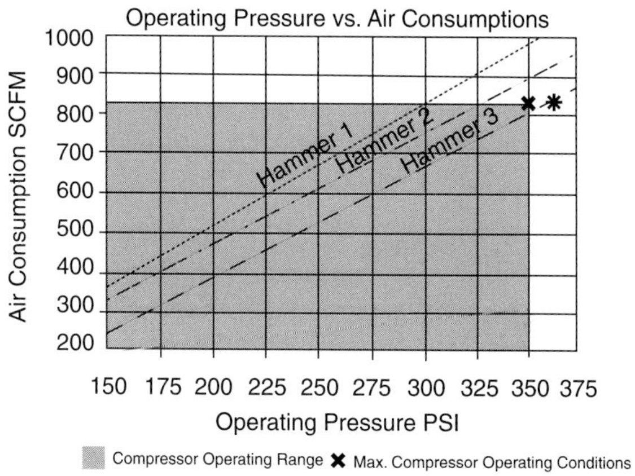

FIGURE 3.37 Case 3: 825 SCFM X 350 PSI Compressor. (Courtesy of DrilMaster.)

Expected performance index:

Hammer 1	190%
Hammer 2	142%
Hammer 3	100%

Case 3: 825 SCFM X 350 PSI. According to Figure 3.37, Hammer 3 would not use all the air the compressor can produce causing the compressor to partially unload. Hammer 2 on the other hand needs more air than the compressor can make so it will only operate at 320 psi. On paper, both of these hammers could be expected to have the same performance. However, experience dictates that compressors rarely deliver the amount of air to the hammer that is claimed by the stickers on the side, Leaks at drill joints, altitude, blocked intake filters, oil scavenge system, even the weather will

affect the air delivery. Knowing this, and the fact that the amount of air might be as little as 660 SCFM, Hammer 3 would be the right choice.

Expected performance index:

Hammer 1	100%
Hammer 2	118%
Hammer 3	136%

Heavy Duty Hammers. These hammers are slightly bigger on the outside diameter to permit more wear allowance. Some manufacturers refer to them as "half inch bigger hammers," for example 6.5 inch hammers. In fact, these may not be true 6.5 inch hammers, but the heavy duty hammers with the same internal parts.

Deep Hole Hammers. These hammers are specifically designed to maintain performance where high back pressures are encountered such at high volumes of ground water or very deep holes. Back pressure greatly reduces the impact velocity of the piston eventually causing the piston to cycle but not drill or "wash out". These hammers have a special piston design incorporating a very large bottom volume to reduce the compression ration making them less sensitive to back pressure. DTH hammers are not as efficient as normal drilling because of the extra piston mass and air consumption. A standard hammer can be made to work in high back pressure applications by boosting the air pressure thereby increasing air consumption and improving flushing.

Hammer Characteristics. Every hammer model has its own characteristics, dependent mostly on design and operating pressure. Two values used to describe these characteristics are blow energy and blow frequency (Blows/Minute). How does this affect performance? In basic terms, percussive hammers with a light-blow energy and a high blow frequency drill well in soft, broken formations where a hard-hitting, high-blow energy hammer will bury itself. In hard rock conditions, without a hard-hitting, high-blow energy hammer, power is insufficient to break rock. Great care must be taken when comparing two hammers on the basis of blow energy and blow frequency. Values measure in test beds or calculated may bear no resemblance to counters gained from drilling in actual rock conditions. To make matters more confusing, the bit face configuration and condition of the carbides play a big part in how efficiently energy is transferred to the rock for maximum penetration rates. Use the figures to select a hammer based on air consumption. This will play the biggest part in performance. Hammer and bit characteristics will effect performance, but trial and error is really the only sure way to find out if the performance can be improved.

3.14.5 Special Applications

As it becomes necessary to infill drill the maturing oil and gas reservoirs in the continental United States and elsewhere in the world, the need to minimize or eliminate formation damage will become an important engineering goal. To accomplish this goal, air and gas drilling techniques will have to be utilized. It is very likely that the future drilling in the maturing oil and gas reservoirs will be characterized by extensive use of high-angle directional drilling coupled with air and gas drilling techniques.

The downhole turbine motor designed to be activated by the flow of incompressible drilling mud cannot operate on air, gas, unstable foam or stable foam drilling fluids. These downhole turbine motors can only be operated on drilling mud or aerated mud.

A special turbine motor has been developed to operate on air, gas and unstable foam [17]. This is the downhole pneumatic turbine motor. This motor has been tested in the San Juan Basin in New Mexico and the Geysers area in Northern California. Figure 3.38 shows the basic design of this drilling device. The downhole pneumatic turbine motor is equipped with a gear reduction transmission. The compressed air or gas that actuates the single stage turbine motor causes the rotor of the turbine to rotate at very high speeds (i.e., ~20,000 rpm). A drill bit cannot be operated at such speeds; thus it is necessary to reduce the speed with a series of planetary gears. The prototype downhole pneumatic turbine motor has a gear reduction transmission with an overall gear ratio of 168 to 1. The particular version of this motor concept that is undergoing field testing is a 9-in. outside diameter motor capable of drilling with a $10\frac{5}{8}$-in. diameter bit or larger. The downhole pneumatic turbine motor will deliver about 40 hp for drilling with a compressed air flowrate of 3,600 scfm. The motor requires very little additional pressure at the surface to operate (relative to normal air drilling with the same volumetric rate).

The positive displacement motor of the Moineau-type design can be operated with unstable foam (or mist) as the drilling fluid. Some liquid must be placed in the air or gas flow to lubricate the elastomer stator as the metal rotor rotates against the elastomer. Positive displacement motors have been operated quite successfully in many air and gas drilling situations. The various manufacturers of these motors can give specific information concerning the performance characteristics of their respective motors operated with air and gas drilling techniques. The critical operating characteristic of these motors, when operated with unstable foam, is that these motors must be loaded with weight on bit when circulation is initiated. If the positive displacement motor is allowed to be started without weight on bit, the rotor will speed up quickly to a very high speed, thus burning out the bearings and severely damaging the elastomer stator.

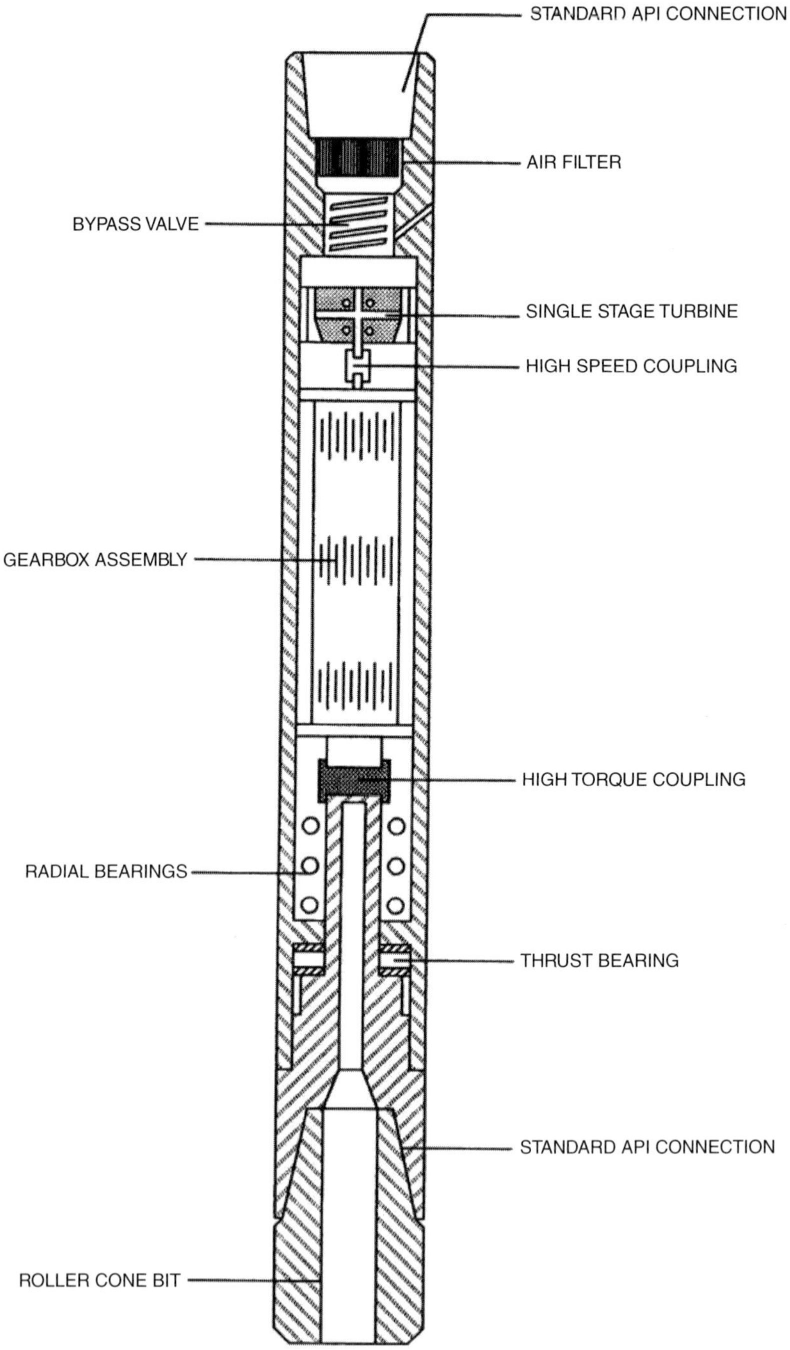

FIGURE 3.38 Downhole pneumatic turbine motor design (courtesy of Pneumatic Turbine Partnership).

References

[1] Lyons, W. C., Guo, B., and Seidel, F. A., *Air and Gas Drilling Manual*, McGraw-Hill, 2000.

[2] Guo, B., and Ghalambor, A., *Gas Volume Requirements for Underbalanced Drilling Deviated Holes*, PennWell, 2002.

[3] Daugherty, R. L., Franzini, J. B., and Finnemore, E. J., *Fluid Mechanics with Engineering Applications*, 8th Edition, McGraw-Hill, 1985.

[4] *API Recommended Practice for Testing Foam Agents for Mist Drilling*, API RP-46, 1st Edition, November 1966.

[5] *Handbook of Chemistry*, McGraw-Hill, 1956.

[6] U. S. Bureau of Mines Report Investigations N. 3798.

[7] Coward, H. P., and Jones, G. W., "Limits of Flammability of Gases and Vapors," Bureau of Mines Bulletin 503, Washington D.C., 1952.

[8] Allan, P. D., "Nitrogen Drilling System for Drilling Applications," SPE Paper 28320, presented at the SPE 69th Annual Technical Conference and Exhibition, New Orleans, Louisiana, September 25–28, 1994.

[9] *Underbalanced Drilling Manual*, Gas Research Institute, GRI Reference No. GRI 97/0236, 1997.

[10] Beyer, A. H., Millhone, R. S., and Foote, R. W., "Flow Behavior of Foam as a Well Circulating Fluid," SPE Paper 3986, Presented at the SPE 47th Annual Fall Meeting, San Antonio, Texas, October 8–11, 1972.

[11] Mitchell, B. J., "Test Data Fill Theory Gap on Using Foam as a Drilling Fluid," *Oil and Gas Journal*, September 1971.

[12] Kuru, E., Miska, S., Pickell, M., Takach, N., and Volk, M., "New Directions in Foam and Aerated Mud Research and Development," SPE Paper 53963, Presented at the 1999 SPE Latin American and Caribbean Petroleum Engineering Conference, Caracas, Venezuela, April 21–23, 1999.

[13] Cross, C. G., "Turbodrill," U.S. Patent 142,992 (September 1873).

[14] Tiraspolsky, W., *Hydraulic Downhole Drilling Motors*, Gulf Publishing Company, Editions Technip, Houston, 1985.

[15] Ioannesion, R. A., *Osnory-Teorii i Tekhnik: Turbinnogo Bureniya (Foundations of Theory and Technology of Turbodrilling)*, Gostopetekhizdat, Moscow, 1954 (in Russian).

[16] Ioannesion, R. A., and Y. V. Vadetsky, "Turbine drilling equipment development in the USSR," *Oil and Gas Journal*, September 28, 1981.

[17] Jurgens, R., and C. Marx, "Neve Bohrmotoren für die erdölindustrie (New drilling motors for the petroleum industry)," *Erdol-Erdges Zeitschrift*, April 1979 (in German).

[18] Lyons, W. C., et al., "Field testing of a downhole pneumatic turbine motor," Geothermal Energy Symposium, ASME/GRC, January 10–13, 1988.

[19] Magner, N. J., "Air motor drill," *The Petroleum Engineer*, October 1960.

[20] Downs, H. F., "Application and evaluation of air-hammer drilling in the Permian Basin," *API Drilling and Production Practices*, 1960.

[21] Bourgoyne, A. T., et al., *Applied Drilling Engineering*, SPE Textbook Series, Vol. 2, First Printing, SPE Richardson, Texas, 1986.

[22] *Drilling Manual*, 11th Edition, IADC, 1992.

[23] Winters, W., and T. M. Warren, "Determining the true weight-on-bit for diamond bits," SPE Paper 11950, presented at the 1983 SPE Annual Technical Conference and Exhibition, San Francisco, October 5–8, 1983.

CHAPTER

4

Directional Drilling

OUTLINE

4.1 GLOSSARY OF TERMS USED IN DIRECTIONAL DRILLING

The glossary of terms used in directional drilling [1] has been developed by the API Subcommittee on Controlled Deviation Drilling under the jurisdiction of the American Petroleum Institute Production Department's Executive Committee on Drilling and Production Practice. The most frequently used terms listed below.

Angle of inclination (angle of drift). The angle, in degrees, taken at one at several points of variation from the vertical as revealed by a deviation survey, sometimes called the inclination or angle of deviation.

Angle of twist. The azimuth change through which the drillstring must be turned to offset the twist caused by the reactive torque of the downhole motor.

Anisotrospic formation theory. Stratified or antisotropic formations are assumed to posses different drillabilities parallel and normal to the bedding planes with the result that the bit does not drill in the direction of the resultant force.

Azimuth. Direction of a course measured in a clockwise direction from 0° to 360°; also called bearing.

Back-torque. Torque on a drill string causing a twisting of the string.

Bent sub. Sub used on top of a downhole motor to give a nonstraight bottom assembly. One of the connecting threads in machined at an angle to the axis of the body of the sub.

Big-eyed bit. Drill bit with one large-sized jet nozzle, used for jet deflection.

Bit stabilization. Refers to stabilization of the downhole assembly near the bit; a stabilized bit is forced to rotate around its own axis.

Borehole direction. Refers to the azimuth in which the borehole is heading.

Borehole directional survey. Refers to the measurements of the inclinations, azimuths and specified depths of the stations through a section of borehole.

Bottom-hole assembly (BHA). Assembly composed of the drill bit, stabilizers, reamers, drill collars, subs, etc., used at the bottom of the drillstring.

Bottomhole location. Position of the bottom of the hole with respect to some known surface location.

Bottomhole orientation sub (BHO). A sub in which a free-floating ball rolls to the low side and opens a port indicating an orientation position.

Build-and-hold wellbore. A wellbore configuration where the inclination is increased to some terminal angle of inclination and maintained at that angle to the specified target.

Buildup. That portion of the hole in which the angle of inclination is increased.

Buildup rate. Rate of change (°/100 ft) of the inclination angle in the section of the hole where the inclination from the vertical is increasing.

Clearance. Space between the outer diameter of the tool in question and the side of the drilled hole; the difference in the diameter of the hole and the tool.

Clinograph. An instrument to measure and record inclination.

Closed traverse. Term used to indicate the closeness of two surveys; one survey going in the hole and the second survey coming out of the hole.

Corrective jetting runs. Action taken with a directional jet bit to change the direction or inclination of the borehole.

Course. The axis of the borehole over an interval length.

Course bearing. The azimuth of the course.

Crooked-hole. Wellbore that has been inadvertently deviated from a straight hole.

Crooked-hole area. An area where subsurface formations are so composed or arranged that it is difficult to drill a straight hole.

Cumulative fatigue damage. The total fatigue damage caused by repeated cyclic stresses.

Deflection tools. Drilling tools and equipment used to change the inclination and direction of the drilled wellbore.

Departure. Horizontal displacement of one station from another.

Deviation angle. See "Angle of inclination."

Deviation control techniques

Fulcrum technique. Utilizes a bending moment principle to create a force on that the bit to counteract reaction forces that are tending to push the bit in a given direction.

Mechanical technique. Utilizes bottomhole equipment which is not normally a part of the conventional drillstring to aid deviation control. This equipment acts to force the bit to turn the hole in direction and inclination.

Packed-hole technique. Utilizes the hole wall to minimize bending of the bottomhole assembly.

Pendulum techniques. The basic principle involved is gravity or the "plumb-bob effect."

Directional drilling contractor. A service company that supplies the special deflecting tools, BHA, survey instruments and a technical representative to perform the directional drilling aspects of the operation.

Direction of inclination. Direction of the course.

Dogleg. Total curvature in the wellbore consisting of a change of inclination and/or direction between two points.

Dogleg severity. A measure of the amount of change in the inclination and/or direction of a borehole; usually expressed in degrees per 100 ft of course length.

Drag. The extra force needed to move the drillstring resulting from the drillstring being in contact with the wall of the hole.

Drainholes. Several high-angle holes drilled laterally form a single wellbore into the producing zone.

Drift angle. The angle between the axis of the wellbore and the vertical (see "Inclination").

Drop off. The portion of the hole in which the inclination is reduced.

Drop-off rate. Rate of change (°/100 ft) of the inclination angle in the section of the wellbore that is decreasing toward vertical.

Goniometer. An instrument for measuring angles, as in surveying.

Gyroscopic survey. A directional survey conducted using a gyroscope for directional control, usually used where magnetic directional control cannot be obtained.

Hole curvature. Refers to changes in inclination and direction of the borehole.

Hydraulic orienting sub. Used in directional holes with inclination greater than 6° to find the low side of the hole. A ball falls to the low side of the sub and restrict an orifice, causing an increase in the circulating pressure. The position of the tool is know with relation to the low side of the hole.

Hydraulically operated bent sub. A deflection sub that is activated by hydraulic pressure of the drilling fluid.

Inclination angle. The angle of the wellbore from the vertical.

Inclinometer. An instrument that measures an angle of deviation from the vertical.

Jet bit deflection. A method of changing the inclination angle and direction of the wellbore by using the washing action of a jet nozzle at one side of the bit.

Keyseat. A condition wherein the borehole is abraded and extended sideways, and with a diameter smaller than the drill collars and bit; usually caused by the tool joints on the drill pipe.

Kickoff point (kickoff depth). The position in the well bore where the inclination of the hole is first purposely increased (KOP).

Lead angle. A method of setting the direction of the wellbore in anticipation of the bit walking.

Magnetic declination. Angular difference, east or west, at any geographical location, between true north or grid north and magnetic north.

Magnetic survey. A directional survey in which the direction is determined by a magnetic compass aligning with the earth's magnetic field.

Measured depth. Actual length of the wellbore from its surface location to any specified station.

Mechanical orienting tool. A device to orient deflecting tools without the use of subsurface surveying instruments.

Methods of orientation

Direct method. Magnets embedded in the nonmagnetic drill collar are used to indicate the position of the tool face with respect to magnetic north. A picture of a needle compass pointing to the magnets is superimposed on the picture of a compass pointing to magnetic north. By knowing the position of the magnets in the tool, the tool can be positioned with respect to north.

Indirect method. A method of orienting deflecting tools in which two survey runs are needed, one showing the direction of the hole and the other showing the position of the tool.

Surface readout. A device on the rig floor to indicate the subsurface position of the tool.

Stoking. Method of orienting a tool using two pipe clamps, a telescope with a hair line, and an aligning bar to determine the orientation at each section of pipe run in the hole.

Monel (K monel). A nonmagnetic alloy used in making portions of downhole tools in the bottomhole assembly (BHA), where the magnetic survey tools are placed for obtaining magnetic direction information. Monel refers to a family of nickel-copper alloys.

Mud motor. Usually a positive displacement or turbine-type motor, positioned above the bit to provide (power) torque and rotation to the bit without rotating the drillstirng.

Mule shoe. A shaped form used on the bottom of orienting tools to position the tool. The shape resembles a mule shoe or the end of a pipe that has been cut both diagonally and concave. The shaped end forms a wedge to rotate the tool when lowered into a mating seat for the mule shoe.

Multishot survey. A directional survey in which multiple data points are recorded with one trip into the wellbore. Data are usually recorded on rolls of film.

Near-bit stabilizer. A stabilizer placed in the bottomhole assembly just above the bit.

Ouija board (registered trademark of Eastern Whipstock). An instrument composed of two protractors and a straight scale that is used to determine the positioning for a deflecting tool in a inclined wellbore.

Permissible dogleg. A dogleg through which equipment and/or tubulars can be operated without failure.

Pendulum effect. Refers to the pull of gravity on a body; tendency of a pendulum to return to vertical position.

Pendulum hookup. A bit and drill collar with a stabilizer to attain the maximum effect of the pendulum.

Rat hole. A hole that is drilled ahead of the main wellbore and which is of a smaller diameter than the bit in the main borehole.

Reamer. A tool employed to smooth the wall of a wellbore, enlarge the hole, stabilize the bit and straighten the wellbore where kinks and abrupt doglegs are encountered.

Rebel tool (registered trademark of Eastman Whipstock). A tool designed to prevent and correct lateral drift (walk) of the bit tool. It consists of two paddles on a common shaft that are designed to push the bit in the desired direction.

Roll off. A correction in the facing of the deflection tool, usually determined by experience, and which must be taken into consideration to give the proper facing to the tool.

Setting off course. A method of setting the direction of the wellbore in anticipation of the bit walking.

Side track. An operation performed to redirect the wellbore by starting a new hole; at a position above the bottom of the original hole.

Slant hole. A nonvertical hole; usually refers to a wellbore purposely inclined in a specific direction; also used to define a wellbore that is nonvertical at the surface.

Slant rig. Drilling rig specifically designed to drill a wellbore that is nonvertical at the surface. The mast is slanted and special pipe-handling equipment is needed.

Spiraled wellbore. A wellbore that has attained a changing configuration such as a helical form.

Spud bit. In directional drilling, a special bit used to change the direction and inclination of the wellbore.

Stabilizer. A tool placed in the drilling assembly to

Change or maintain the inclination angle in a wellbore by controlling the location of the contact point between the hole and drill collars.

Center the drill collars near the bit to improve drilling performance.

Prevent wear and differential sticking of the drill collars.

Surveying frequency. Refers to the number of feet between survey records.

Target area. A defined area, at a prescribed vertical depth, that is planned to be intersected by the wellbore.

Tool azimuth angle. The angle between north and the projection of the tool reference axis onto a horizontal plane.

Tool high-side angle. The angle between the tool reference axis and a line perpendicular to the hole axis and lying the vertical plane.

Total curvature. Implies three-dimensional curvature.

True north. The direction from any geographical location on the earth's surface to the north geometric pole.

True vertical depth (TVD). The actual vertical depth of an inclined wellbore.

Turbodrill. A downhole motor that utilizes a turbine for power to rotate the bit.

Turn. A change in bearing of the hole; usually spoken of as the right or left turn with the orientation that of an observer who views the well course from the surface site.

Walk (of hole). The tendency of a wellbore to deviate in the horizontal plane.

Wellbore survey calculation method. Refers to the mathematical method and assumptions used in reconstructing the path of the wellbore and in generating the space curve path of the wellbore from inclination and direction angle measurements taken along the wellbore. These measurements are obtained from gyroscopic or magnetic instruments of either the single-shot or multishot type.

Whipstock. A long wedge and channel-shaped piece of steel with a collar at its top through which the subs and drillstring may pass. The face of the whipstock sets an angle to deflect the bit.

Woodpecker drill collar (indented drill collar). Round drill collar with a series of indentations on one side to form an eccentrically weighted collar.

4.2 DOGLEG SEVERITY (HOLE CURVATURE) CALCULATIONS

There are several analytical methods available for calculating dogleg severity:

- Tangential
- Radius of curvature

- Average angle
- Trapezoidal (average tangential)
- Minimum curvature

The tangential and radius of curvature methods are outlined here.

4.2.1 Tangential Method

The overall angle change is calculated from

$$\beta = 2\arcsin\sqrt{\sin^2\left(\frac{\Delta v}{2}\right) + \sin^2\left(\frac{\Delta h}{2}\right)\sin^2\left(\frac{V_0 + V_1}{2}\right)} \tag{4.1}$$

where β = overall angle change (total curvature)
Δh = change of horizontal angle (in horizontal plane)
Δv = change in vertical angle (in vertical plane)
V_0, V_1 = hole inclination angle in two successive surveying stations.

The hole curvature is

$$C = 100\frac{\beta}{L} \tag{4.2}$$

where L = course length between the surveying stations.

Example 4.1

Two surveying measurements were taken 30 ft apart. The readings are as below.
Station 1:

Hole inclination 3°30′
Hole direction, N11°E

Station 2:

Hole inclination 4°30′
Hole direction, N23°E

Find the dogleg severity.

Solution

Change in horizontal angle is

$$\Delta h = 23° - 11° = 12°$$

Change in vertical angle is

$$\Delta v = 4.5° - 3.5° = 1°$$

The overall angle change is

$$\beta = 2\arcsin\sqrt{\sin^2\left(\frac{1.0}{2}\right)+\sin^2\left(\frac{12}{2}\right)\sin^2\left(\frac{3.5+4.5}{2}\right)}$$
$$= 1.3^\circ$$

The hole curvature is

$$c = 100\frac{1.3}{30} = 4.33^\circ/100\,\text{ft} = 4^\circ 20'/100\,\text{ft}$$

4.2.2 Radius of Curvature Method

The dogleg severity is calculated from

$$c = 100\sqrt{a^2\sin^4\phi + b^2} \tag{4.3}$$

where a = rate of change in direction angle in degrees/ft
b = rate of change in inclination angle in degrees/ft
ϕ = inclination angle in degrees.

The sequence of computations involved is explained in the following example.

Example 4.2

From two successive directional survey stations is obtained:

	Station 1	**Station 2**
Hole inclination angle	30° (ϕ_1)	40° (ϕ_2)
Hole direction angle	N11°E (θ_1)	N18°E (θ_2)

The distance between the stations is 60 ft. Determine the dogleg severity.

Solution

Rate of change in inclination angle is

$$b = \frac{40-30}{60} = 0.1667^\circ/\text{ft}$$

Radius of curvature in vertical plane is

$$R_v = \frac{180}{\pi b} = \frac{180}{\pi \times 0.1667} = 363\,\text{ft}$$

Horizontal departure (arc length of projection of wellbore in horizontal plane) is

$$H_d = R_v(\cos\phi_1 - \cos\phi_2) = 363(\cos 30 - \cos 40)$$
$$= 36.29\,\text{ft}$$

Rate of change in hole direction is

$$a = \frac{\theta_2 - \theta_1}{H_d} = \frac{18 - 11}{36.29} = 0.1929^\circ/\text{ft}$$

Hole curvature at the first station is

$$c_1 = 100\sqrt{a^2 \sin^4 \phi_1 + b^2} = 100\sqrt{0.1929^2 \times \sin^4 30 + 0.1667^2}$$
$$= 17.35^\circ/100\,\text{ft}$$

Hole curvature at the first station is

$$c = 100\sqrt{a^2 \sin^4 \phi_2 + b^2} = 100\sqrt{0.1929^2 \times \sin^4 40 + 0.1667^2}$$
$$= 18.48^\circ/100\,\text{ft}$$

The average value is

$$c = \frac{17.35 + 18.48}{2} = 17.92^\circ/100\,\text{ft}$$

4.2.3 Deflection Tool Orientation

Application of a deflecting tool (e.g., downhole motor with a bent sub) requires determining the orientation of the tool so that the hole takes the desired course. There are three effects to consider when setting a deflection tool.

1. The existing borehole inclination angle
2. The existing borehole direction angle
3. The bent sub angle of the deflection tool itself

These three effects in combination will result in a new dogleg of a wellbore.

The deflection tool orientation parameters can be obtained using the vectorial method of D. Ragland, the Ouija Board, or the three-dimensional mathematical deflecting model.

4.2.4 Vectorial Method of D. Ragland

This method is explained by solving two example problems.

Example 4.3

Determine the deflection tool-face orientation, tool deflection angle and tool facing change from original course line angle if the data are as follows:

Existing hole inclination angle: 10°
Existing hole direction: N300W
Desired change of azimuth: 25° (to the left)
Desired hole inclination: 8°

Solution

The solution to this problem is shown in Figure 4.1. The following steps are involved in preparing a Ragland diagram.

1. Lay a quadrant N-S-W-E.
2. Select a scale for the angles.
3. With a protractor layoff an angle 30° from N to W.
4. Using the selected angle scale find p. B (10° from p. A).
5. With a protractor layoff an angle 25° to the left of line AB.
6. Find p. C using the angle scale (8° from p. A).

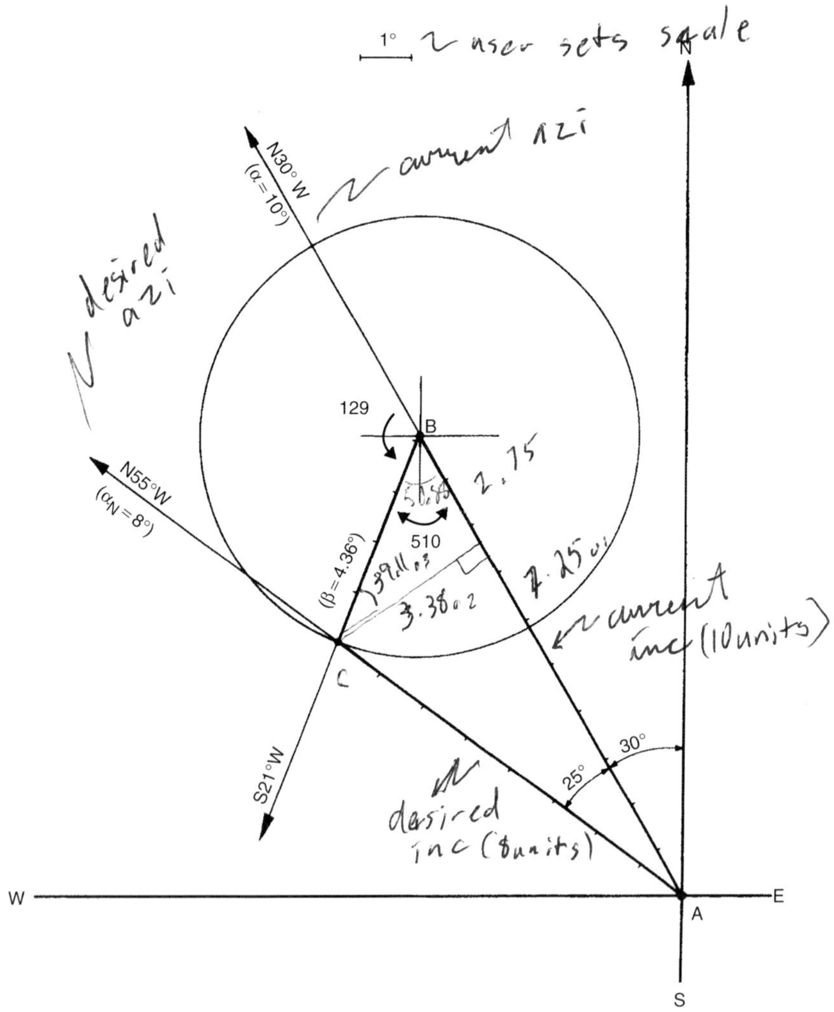

FIGURE 4.1 Graphic solution of Example 4.3.

7. Describe a circle about p. B with a radius of CB = 4.36° (read from the diagram). This is the desired overall angle change.
8. Read of required tool-face orientation: S21°W.
9. Read off required tool facing change from original course line: 129° to the left.

Example 4.4

The original hole direction and inclination were measured to be S50°W and 6°, respectively. It is desired to obtain a new hole direction of S65°W with an inclination angle of 7° after drilling 90 ft. For this purpose, a whipstock was oriented correctly.

After drilling 90 ft, a checkup measurement was performed that revealed that the hole new direction is S62°W and the new inclination is 6°30′. Determine the expected hole curvature delivered by the whipstock, the magnitude of roll-off for this system and the real hole curvature delivered by the whipstock. How should the whipstock be oriented to drill a hole with a direction of S37°W in 90 ft? What should be the inclination of the hole?

Solution

The solution to this problem is shown in Figure 4.2. Steps 1 and 2 are the same as in Example 4.3.

1. Layoff an angle 50° from S to W.
2. Draw a line 00′.
3. Draw a line OA that will represent the desired new hole (S65°W = 15° to the right of the current direction). Point A is found at a distance of 7° from point O. Radius OA represents the expected overall angle changed of 2.0°.
4. Read off the original whipstock orientation N63°W (67° to the right of the current hole direction).
5. Draw line OB (direction, S62°W, inclination, 6.5°).
6. Describe a circle about p. O′ with a radius of OB. Radius OB = 1.40° represents the real overall angle change.
7. The angle AO′B is the roll-off angle. Read off the roll-off angle = 8° to the right.
8. Draw line OC (S37°W). The whipstock orientation should be S68°E, and the final angle is 5.5°. If the roll-off effect is not considered, the whipstock direction is S76°E.

4.2.5 Three-Dimensional Deflecting Model

A mathematical model has been presented [2] that enables one to analyze and plan deflection tool runs. The model is a set of equations that relate the original hole inclination angle (α), new hole inclination angle (α_N), overall

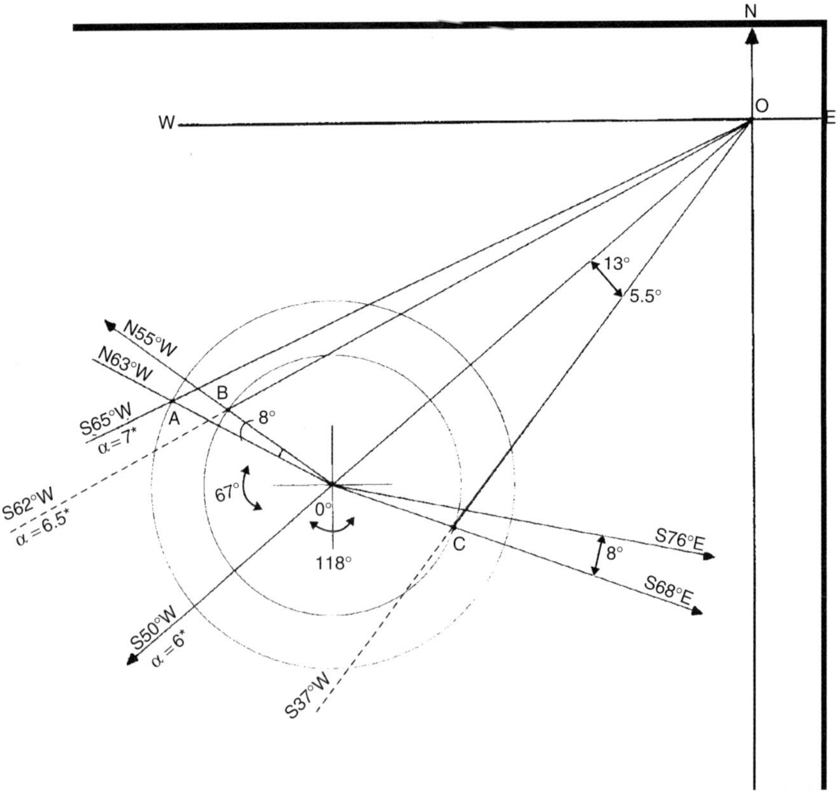

FIGURE 4.2 Graphical solution of Example 4.4. With a downhole motor, there will be a left-hand reaction torque; therefore, the tool should be turned clockwise from the ideal position.

angle change (β), change of direction ($\Delta\varepsilon$), and tool-face rotation from original course direction (γ). These equations are given below:

$$\beta = \arccos(\cos\alpha\cos\alpha_N + \sin\alpha\sin\alpha_N\cos\Delta\varepsilon) \tag{4.4}$$

$$\alpha_N = \arccos(\cos\alpha\cos\beta - \sin\alpha\sin\beta\cos\gamma) \tag{4.5}$$

$$\Delta\varepsilon = \arctan\left(\frac{\tan\beta\sin\gamma}{\sin\alpha + \cos\alpha\tan\beta\cos\gamma}\right) \tag{4.6}$$

The above equations can be rearranged to the form suitable for a solution of a particular problem. A hand-held calculator may be used to perform required calculations [3].

Practical usefulness of this model is presented below.

Example 4.5

Original hole direction and inclination is N20°E and 10°, respectively. It is desired to deviate the borehole so that the new hole inclination is 13° and direction N30°E in 90 ft. What is the tool orientation and deflecting angle (dogleg) necessary to achieve this turn and build?

Solution

From Equation 4.4, the overall angle change is

$$\beta = \arccos[\cos 10 \times \cos 13 + \sin 10 \times \sin 13 \cos(30-20)]$$
$$= 3.59°$$

The dogleg severity (hole curvature) is

$$C = 100\frac{\beta}{L} = 100\frac{3.59}{90} = 3.99°/100\,\text{ft}$$

To obtain the tool orientation we solve Equation 4.5 for γ:

$$\cos\gamma = \arccos\left(\frac{\cos\alpha\cos\beta - \cos\alpha_N}{\sin\alpha\sin\beta}\right)$$
$$= \arccos\left(\frac{\cos 10 \times \cos 3.59 - \cos 13}{\sin 10 \times \sin 3.59}\right) = 38.54°$$

Consequently, the tool-face orientation is 38.54° to the right of the high side of the hole.

Example 4.6

Surveying shows that the hole drift angle is 22° and direction S36°W. It is desired to turn the hole of 6° to the right and build angle during next 90 ft drilling. For this purpose a deflection tool that can deliver a hole curvature of 3.333°/100 ft will be used. What is the expected new hole inclination angle? What is the required tool direction?

Solution

The expected overall angle change in the interval is

$$\beta = 3.333°/100\text{ ft} \times 90\text{ ft} = 3°$$

From Equation 4.4, we have

$$\cos\beta = \cos\alpha\cos\alpha_N + \sin\alpha\sin\alpha_N\cos\Delta\varepsilon$$

Considering the coefficients $a = \cos\alpha$ and $b = \sin\alpha\ \cos\Delta\varepsilon$, we obtain

$$\cos\beta = a\cos\alpha_N + b\sin\alpha_N$$

Now consider dividing both sides by $\sqrt{a^2+b^2}$. We obtain

$$\frac{\cos\beta}{\sqrt{a^2+b^2}} = \frac{a}{\sqrt{a^2+b^2}}\cos\alpha_N + \frac{b}{\sqrt{a^2+b^2}}\sin\alpha_N$$

Making $\cos\phi = \frac{a}{\sqrt{a^2+b^2}}$ and $\sin\phi = \frac{b}{\sqrt{a^2+b^2}}$, we have

$$\frac{\cos\beta}{\sqrt{a^2+b^2}} = \cos\phi\cos\alpha_N + \sin\phi\sin\alpha_N = \cos(\phi - \alpha_N)$$

where $\tan\phi = \frac{b}{a} = \frac{\sin\alpha\cos\Delta\varepsilon}{\cos\alpha}$.
Solving for α_N, we have

$$\alpha_N = \phi - \arccos\left(\frac{\cos\beta}{\sqrt{a^2+b^2}}\right)$$
$$= \arctan\left(\frac{\sin\alpha\cos\Delta\varepsilon}{\cos\alpha}\right) - \arccos\left(\frac{\cos\beta}{\sqrt{a^2+b^2}}\right)$$

Using the expressions for a and b, we obtain

$$\alpha_N = \arctan\left(\frac{\sin\alpha\cos\Delta\varepsilon}{\cos\alpha}\right) - \arccos\left(\frac{\cos\beta}{\sqrt{\cos^2\alpha + \sin^2\alpha\cos^2\Delta\varepsilon}}\right)$$

Substituting the data of the problem, we obtain for the new angle

$$\alpha_N = \arctan\left(\frac{\sin 22 \times \cos 6}{\cos 22}\right) - \arccos\left(\frac{\cos 3}{\sqrt{\cos^2 22 + \sin^2 22 \times \cos^2 6}}\right)$$
$$= 23.88°$$

Using Equation 4.5 and solving for γ, we obtain for the tool-face orientation

$$\gamma = \arccos\left(\frac{\cos\alpha\cos\beta - \cos\alpha_N}{\sin\alpha\sin\beta}\right)$$
$$= \arccos\left(\frac{\cos 22 \times \cos 3 - \cos 23.88}{\sin 22 \times \sin 3}\right) = 54.02°$$

Therefore, the tool-face should be oriented at 54.02° to the right of the high side because it is desired to turn the hole to S42°W.

References

[1] *Sii Datadrill Directional and Drilling Manual*, Second Revision, December 1983.

[2] Millheim, K.K., Gubler, F.H., and Zaremba, H.B., "Evaluating and planning directional wells utilizing post analysis techniques and a three dimensional bottom hole assembly program," SPE Paper 8339, presented at the 54th Annual Fall Technical Conference and Exhibition of the Society of Petroleum Engineers of AIME held in Las Vegas, September 23–26, 1979.

[3] Nicholson, J.T., "Calculator program developed for directional drilling," *Oil and Gas Journal*, September 28, 1981.

CHAPTER

5

Selection of Drilling Practices

OUTLINE

The objective of optimizing drilling practices is to safely deliver a product capable of the highest production capacity in a cost-effective manner. Throughout all phases of well planning and construction, the following ideas must be considered: health, safety, and environment (HSE), production capability, and drilling optimization. These criteria are requisite to delivering a successful drilling outcome. Several choices are normally available to achieve the required outcomes for each criterion, but fewer options, perhaps only one, may be viable to meet the needs of all three.

In today's environment, HSE requirements play an equal role in the planning and implementation of drilling plans. Significant planning is required to protect operating personnel, retain licenses to operate, control costs, and coexist with the environment. Accidents are costly, more so than the up-front costs to mitigate the risk to people, property, and environment.

Production capacity, both rate and volume, are affected by decisions made during planning and construction phases. Proper selection of various cost control options can minimize the negative influences that some of these measures can have on production. Analyzing potential gains in capacity associated with incremental risk and investment is part of the optimization process.

Optimization requires understanding, prioritization, and implementation of the practices most appropriate for the specific well being drilled to accomplish the stated objectives. You must solve potential problems that remove limits or lessen their impact on the outcome. The right answer for one circumstance (e.g., well design, geographical area, hole size) is possibly the wrong answer in another circumstance. Problem solving must be approached with the overall objective in mind while understanding the causes and effects and the relative magnitudes of these effects. No matter how well the planning process is conducted, adjustment in real time is expected and will be required. The parameters of drilling operations parameters must be optimized. Noticing, understanding, and adapting to observed trends will yield significant value to the outcome. Real-time ability to respond to problems during drilling will minimize their impact and not allow these problems to cause other problems, which are often substantially more costly to solve. The days of characterizing drilling practices as simply operational parameters such as weight on bit (WOB), revolutions per minute (rpm), and flow rate are long gone. Drilling optimization is a multivariant analysis that requires diligent effort, but the challenge is rewarding personally and financially.

5.1 HEALTH, SAFETY AND ENVIRONMENT

An effective HSE program is characterized by no injury to people, no loss of property, and no harm to the environment. Great HSE performance is an indication of great leadership. It is much more than statistics, although measurements are necessary to facilitate performance improvement. HSE must be considered a core responsibility for all business participants, operators, contractors, and service companies and be accepted by all individuals on a personal basis. It is imperative that all parties are committed from the top management down throughout their organizations.

The main reasons companies in the current era support strong HSE programs include humanitarian reasons (i.e., not hurting people), legal or regulatory requirements, the company's public image, employee morale, and economic reasons (i.e., loss of business or the cost of poor HSE performance).

The modern era of HSE management for oil field operations began with the Piper Alpha disaster in the North Sea, where 167 men lost their lives in one incident. The subsequent investigation, conducted and published by Lord Cullen, contained the framework of HSE management embodied in much subsequent legislation and regulation [1]. This was the birth of safety cases for oil field facilities, including offshore MODU drilling rigs.

One of the key recommendations was that companies should have *safety management systems* (SMSs) that control a company's operations from top to bottom. The systems were recommended to include the elements of ISO 9001 (a standard for management of quality in organizations) and include elements such as management responsibility and commitment, design control, documentation and procedures, process control, control of nonconformance, corrective actions, internal auditing, and training. All the elements of a safety management system are not directly applicable to drilling optimization and are therefore not addressed here. Examples include emergency response plans and oil spill response procedures that companies should have in place as part of their normal procedures.

Generally, each drilling department should have a set of operating guidelines that control the drilling work processes. The operating guidelines are a subset of the SMSs, and the documents may vary from multivolume sets at the major oil companies to much smaller documents for smaller companies. The drilling engineer is responsible for ensuring that all his work complies with these guidance instructions. Variations from these procedures usually will require a higher level of management approval.

The three components of HSE-related drilling issues that are normally considered during drilling operations are discussed separately in the following sections.

5.1.1 Health

Health issues related to drilling operations can include industrial hygiene issues related to onsite conditions and may include exposures to drilling fluid components such as oil-base mud fumes and skin contact, highly toxic completion brines, or oils, gas, and toxic materials such as hydrogen sulfide originating from the well. Naturally occurring radioactive materials (NORM) can also be encountered during workover operations, and metals such as mercury are encountered periodically in gas streams and may be found in production separators. An optimized drilling program includes careful consideration of the health impact on employees and workers at any well site.

5.1.2 Safety

The general safety culture adapted in the current era was derived from the Dont manufacturing culture and adapted to oil field operations when the company purchased Conoco. Dont was originally in the dynamite manufacturing business, and serious accidents in the past strongly motivated management to adopt a "best practice" approach to safety management. The primary concept is that *all accidents are preventable*. Accidents do not just happen, and with work, resources, and management commitment, accidents can be minimized or eliminated. In most all oil companies and service companies, management personnel leave no doubt that they are committed to providing a safe working environment. A poor safety record leads to the suffering of injured employees, the financial impact of lawsuits, and the loss of business and shareholder support. It is not unusual for service companies to be removed from approved bidding lists if the safety performance is not up to the company's requirements. The second concept generally accepted is that *safety (or HSE) must be considered equally with production and profits* in the decision-making process. "Safety first" is not plausible. If we wanted to only be safe, we would never leave the comfort of our homes in the first place. Safety must be considered as an equal to other factors in the decision-making process to ensure that all jobs meet the company's safety goals.

Safety management is an engineering profession unto itself, with many disciplines and areas of coverage. For the drilling engineer or drilling foreman, safety at the well site is accomplished through sound engineering, formal safety reviews (as appropriate), contracting of reputable firms with strong safety cultures, and appropriate training of all personnel.

Operational risks for drilling operations can vary from simple, shallow onshore jobs to extremely complex offshore operations in remote hostile areas that can be extremely expensive and carry significant risk. Consequently, a "fit for purpose" approach must be considered for the safety management of each operation. A simple onshore job may rely completely on the drilling contract and include relatively simple tools such as a toolbox safety discussion and daily safety meeting. At the other end of the scale, in large offshore production platforms with simultaneous operation of wells and drilling in a remote or hostile area, a large amount of safety engineering may be required. Tools may include a full safety case preparation and hazard and operability (HAZOP) studies [2]. A HAZOP study is an examination procedure. Its purpose is to identify all possible deviations from the way in which a design is expected to work and to identify all the hazards associated with these deviations. When deviations arise that result in hazards, actions are generated that require design engineers to review and suggest solutions to remove the hazard or reduce its risk to an acceptable level. These solutions are reviewed and accepted by the HAZOP team before implementation. HAZOP techniques have been adopted by many countries and are required to provide assurance of safe operations.

It should be understood by all operation personnel that 96% of all accidents are related to unsafe behaviors and that only 4% of accidents are caused by unsafe conditions. Most drilling contractors have adopted policies that focus on encouraging safe behaviors.

5.1.3 Environment

Dramatic changes in environmental impact management have occurred continually from the earliest days of the oil field. Common practice in the early days was to produce oil into open pits for storage. Saltwater was routinely dumped into the nearest creek, killing everything in it. Today's best practices include serious management and minimization of all waste streams. Development of projects today may require full environmental impact assessments (EIA) before approval. The first legal requirement in the United States for an EIA was imposed on the Trans-Alaska oil pipeline, which was delayed for years and experienced cost overruns from an initial estimate of $900 million to a final installed cost of $9 billion. Today, most of the East and West coastlines of the United States and Florida are off limits for drilling because of environmental concerns. The Exxon Valdez incident will not leave the collective public memory any time soon and is an example of the negative impact to the environment from oil and gas operations.

Most companies have management systems in place for environmental management of their operations. Drilling personnel are responsible for planning and conducting operations to ensure optimization and compliance.

5.2 PRODUCTION CAPACITY

Production capacity requires investment. Performance measures of optimization may include one or several of the following: days to pay out, cash flow, profit, finding cost (cost per barrel oil equivalent), and return on investment (ROI). The idea of *diminishing returns* plays a key role in the decision-making process. Delivering rate and reservoir volume carry associated cost and risks. Fundamental to drilling practices optimization is understanding how specific cause-and-effect relationships influence this objective.

Key issues affecting production capability, which therefore must receive a high priority when planning the drilling phase, include permeability (K) and porosity (Φ) of the zone of interest (ZOI). Formation damage resulting from poor planning may be permanent or very costly to correct. The right equipment, resources, knowledge, and skills are prerequisites for optimization.

Several key issues may establish constraints on the drilling program, narrowing choices for the drilling plan. For example, mud types and properties and exposure time to the ZOI may be critical. A special "drill-in" fluid (i.e., drilling mud that can minimize formation damage) may be required. The time and cost to change the drilling mud is deemed high value for achieving the final production capacity versus cost objective. Another example is the determination of the optimum hole and tubular sizes. Hole and tubular sizes must be optimized based on expected initial production (e.g., rate, fluid type, pressure), availability, potential future remediation, and consideration of how production may change throughout the life of the well. The geology and pore pressures determine the number of casing strings required to arrive at the ZOI, and the expected production rate will determine the size of production string that the drilling department must be delivered to the company. In some cases, exploration wells will be designed as expendable and will never be produced. All of these factors influence the spectrum of possible casing plans and potential costs, and they will require evaluation.

Using a "cradle to grave" approach to well design considers the drilling phases and looks at all aspects of the well's life, from initial production through permanent abandonment [3]. Close association is required between the multiple departments (e.g., geology, asset management, drilling, completion, production) of the oil company and the service provider representatives. Communicating and sharing their respective areas of expertise and requirements with the others is paramount to optimization of the outcome.

New technologies can significantly enhance production capacity when properly applied and should not be forgotten. Underbalanced drilling (UBD), expandable sand screens (ESS), expandable drilling liners (EDL), coiled-tubing drilling (CTD), hole enlarging devices (e.g., bi-center drill bits, hole openers), casing drilling, and fiberoptic "intelligent" completions are a few examples.

5.3 WELL PLANNING AND IMPLEMENTATION

5.3.1 Optimum Well Planning

Drilling optimization requires detailed engineering in all aspects of well planning, drilling implementation, and post-run evaluation (Figure 5.1) [4, 5]. Effective well planning optimizes the boundaries, constraints, learning, nonproductive time, and limits and uses new technologies as well as tried and true methods. Use of *decision support packages*, which document the reasoning behind the decision-making, is key to shared learning and continuous improvement processes. It is critical to anticipate potential difficulties, to understand their consequences, and to be prepared with contingency plans. Post-run evaluation is required to capture learning. Many of the processes used are the same as used during the well planning phase, but are conducted using new data from the recent drilling events.

Depending on the phase of planning and whether you are the operator or a service provider, some constraints will be out of your control to alter

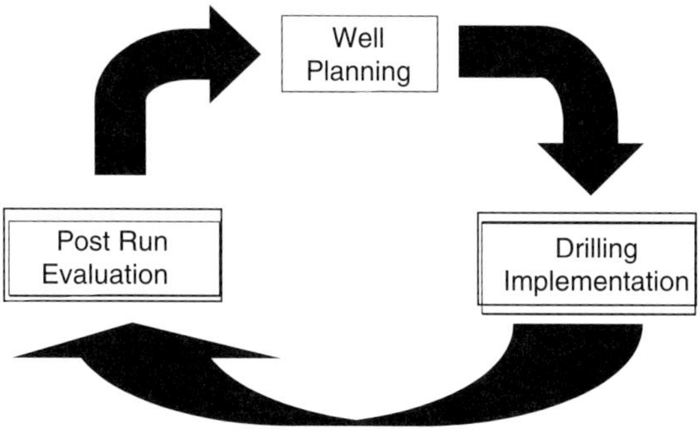

FIGURE 5.1 Drilling Optimization Cycle.

or influence (e.g., casing point selection, casing sizes, mud weights, mud types, directional plan, drilling approach such as BHA types or new technology use). There is significant value in being able to identify alternate possibilities for improvement over current methods, but well planning must consider future availability of products and services for possible well interventions. When presented properly to the groups affected by the change, it is possible to learn why it is not feasible or to alter the plan to cause improvement. Engineers must understand and identify the correct applications for technologies to reduce costs and increase effectiveness. A correct application understands the tradeoffs of risk versus reward and costs versus benefits.

Boundaries Boundaries are related to the "rules of the game" established by the company or companies involved. Boundaries are criteria established by management as "required outcomes or processes" and may relate to behaviors, costs, time, safety, and production targets.

Constraints Constraints during drilling may be preplanned trip points for logs, cores, casing, and BHA or bit changes. Equipment, information, human resource knowledge, skills and availability, mud changeover, and dropping balls for downhole tools are examples of constraints on the plan and its implementation.

The Learning Curve Optimization's progress can be tracked using learning curves that chart the performance measures deemed most effective for the situation and then applying this knowledge to subsequent wells. Learning curves provide a graphic approach to displaying the outcomes. Incremental learning produces an exponential curve slope. Step changes may be caused by radically new approaches or unexpected trouble. With understanding and planning, the step change will more likely be in a

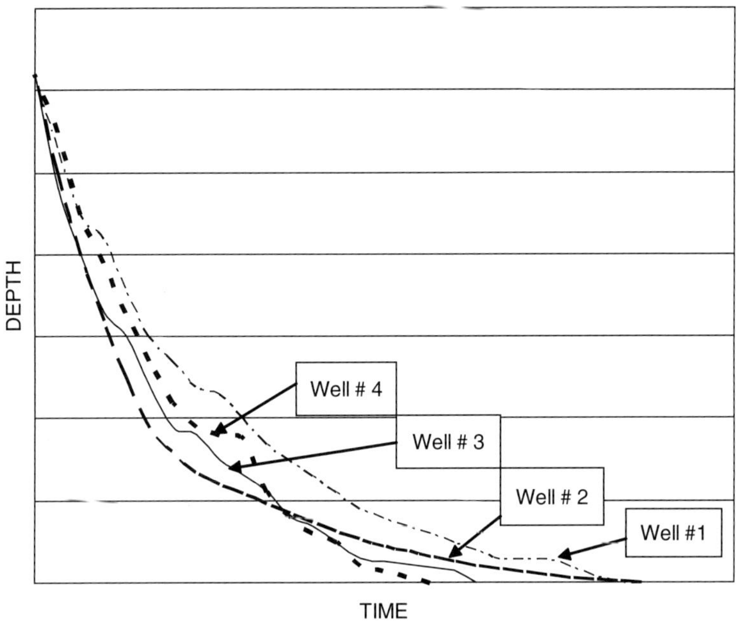

FIGURE 5.2 Learning Curves.

positive direction, imparting huge savings for this and future wells. The curve slope defines the optimization rate.

The learning curve can be used to demonstrate the overall big picture or a small component that affects the overall outcome. In either case, the curve measures the rate of change of the parameter you choose, typically the "performance measures" established by you and your team. Each performance measure is typically plotted against time, perhaps the chronological order of wells drilled as shown in Figure 5.2.

Cost Estimating One of the most common and critical requests of drilling engineers is to provide accurate cost estimates, or *authority for expenditures* (AFEs). The key is to use a systematic and repeatable approach that takes into account all aspects of the client's objectives. These objectives must be clearly defined throughout the organization before beginning the optimization and estimating process. Accurate estimating is essential to maximizing a company's resources. Overestimating a project's cost can tie up capital that could be used elsewhere, and underestimating can create budget shortfalls affecting overall economics.

Integrated Software Packages With the complexity of today's wells, it is advantageous to use integrated software packages to help design all aspects of the well. Examples of these programs include

- Casing design
- Torque and drag
- Directional planning
- Hydraulics
- Cementing
- Well control

Decision Support Packages Decision support packages document the reasoning behind the decisions that are made, allowing other people to understand the basis for the decisions. When future well requirements change, a decision trail is available that easily identifies when new choices may be needed and beneficial.

Performance Measures Common drilling optimization performance measures are cost per foot of hole drilled, cost per foot per casing interval, trouble time, trouble cost, and AFEs versus actual costs [6].

Systems Approach Drilling requires the use of many separate pieces of equipment, but they must function as one system. The borehole should be included in the system thinking. The benefit is time reduction, safety improvement, and production increases as the result of less nonproductive time and faster drilling. For example, when an expected average rate of penetration (ROP) and a maximum instantaneous ROP have been identified, it is possible to ensure that the tools and borehole will be able to support that as a plan. Bit capabilities must be matched to the rpm, life, and formation. Downhole motors must provide the desired rpm and power at the flow rate being programmed. Pumps must be able to provide the flow rate and pressure as planned [7].

Nonproductive Time Preventing trouble events is paramount to achieving cost control and is arguably the most important key to drilling a cost-effective, safe well. Troubles are "flat line" time, a terminology emanating from the days versus depth curve when zero depth is being accomplished for a period of days, creating a horizontal line on the graph. Primary problems invariably cause more serious associated problems. For example, surge pressures can cause lost circulation, which is the most common cause of blowouts. Excessive mud weight can cause differential sticking, stuck pipe, loss of hole, and sidetracking. Wellbore instability can cause catastrophic loss of entire hole sections. Key seating and pipe washouts can cause stuck pipe and a fishing job.

When a trouble event leads to a fishing job, "fishing economics" should be performed. This can help eliminate emotional decisions that lead to overspending. Several factors should be taken into account when determining whether to continue fishing or whether to start in the first place. The most important of these are replacement or lost-in-hole cost of tools and equipment, historical success rates (if known), and spread rate cost of daily operations. These can be used to determine a risk-weighted value of fishing versus the option to sidetrack.

Operational inefficiencies are situations for which better planning and implementation could have saved time and money. Sayings such as "makin' hole" and "turnin' to the right" are heard regularly in the drilling business. These phases relate the concept of *maximizing progress*. Inefficiencies which hinder progress include

- Poor communications
- No contingency plans and "waiting on orders" (WOO)
- Trips
- Tool failure
- Improper WOB and rpm (magnitude and consistency)
- Mud properties that may unnecessarily reduce ROP (spurt loss, water loss and drilled solids)
- Surface pump capacities, pressure and rate (suboptimum liner selection and too small pumps, pipe, drill collars)
- Poor matching of BHA components (hydraulics, life, rpm, and data acquisition rates)
- Survey time

Limits Each well to be drilled must have a plan. The plan is a baseline expectation for performance (e.g., rotating hours, number of trips, tangibles cost). The baseline can be taken from the learning curves of the best experience that characterizes the well to be drilled. The baseline may be a widely varying estimate for an exploration well or a highly refined measure in a developed field. Optimization requires identifying and improving on the limits that play the largest role in reducing progress for the well being planned. Common limits include

1. **Hole Size.** Hole size in the range of $7\frac{7}{8}-8\frac{1}{2}$ in. is commonly agreed to be the most efficient and cost-effective hole size to drill, considering numerous criteria, including hole cleaning, rock volume drilled, downhole tool life, bit life, cuttings handling, and drill string handling. Actual hole sizes drilled are typically determined by the size of production tubing required, the required number of casing points, contingency strings, and standard casing decision trees. Company standardization programs for casing, tubing, and bits may limit available choices.
2. **Bit Life.** Measures of bit life vary depending on bit type and application. Roller cones in soft to medium-soft rock often use KREVs (i.e., total revolutions, stated in thousands of revolutions). This measure fails to consider the effect of WOB on bearing wear, but soft formations typically use medium to high rpm and low WOB; therefore, this measure has become most common. Roller cones in medium to hard rock often use

a multiplication of WOB and total revolutions, referred to as the WR or WN number, depending on bit vender. Roller cone bits smaller than $7\frac{7}{8}$ in. suffer significant reduction in bearing life, tooth life, tooth size, and ROP. PDC bits, impregnated bits, natural diamond bits, and TSP bits typically measure in terms if bit hours and KREVs. Life of all bits is severely reduced by vibration. Erosion can wear bit teeth or the bit face that holds the cutters, effectively reducing bit life.

3. **Hole Cleaning.** Annular velocity (AV) rules of thumb have been used to suggest hole-cleaning capacity, but each of several factors, including mud properties, rock properties, hole angle, and drill string rotation, must be considered. Directional drilling with steerable systems require "sliding" (not rotating) the drill string during the orienting stage; hole cleaning can suffer drastically at hole angles greater than 50. Hole cleaning in large-diameter holes, even if vertical, is difficult merely because of the fast drilling formations and commonly low AV.
4. **Rock Properties.** It is fundamental to understand formation type, hardness, and characteristics as they relate to drilling and production. From a drilling perspective, breaking down and transporting rock (i.e., hole cleaning) is required. Drilling mechanics must be matched to the rock mechanics. Bit companies can be supplied with electric logs and associated data so that drill bit types and operating parameters can be recommended that will match the rock mechanics. Facilitating maximum production capacity is given a higher priority through the production zones. This means drilling gage holes, minimizing formation damage (e.g., clean mud, less exposure time), and facilitating effective cement jobs.
5. **Weight on Bit.** WOB must be sufficient to overcome the rock strength, but excessive WOB reduces life through increased bit cutting structure and bearing wear rate (for roller cone bits). WOB can be expressed in terms of weight per inch of bit diameter. The actual range used depends on the "family" of bit selected and, to some extent, the rpm used. *Families* are defined as natural diamond, PDC, TSP (thermally stable polycrystalline), impregnated, mill tooth, and insert.
6. **Revolutions per Minute (rpm).** Certain ranges of rpm have proved to be prudent for bits, tools, drill strings, and the borehole. Faster rpm normally increases ROP, but life of the product or downhole assembly may be severely reduced if rpm is arbitrarily increased too high. A too-low rpm can yield slower than effective ROP and may provide insufficient hole cleaning and hole pack off, especially in high-angle wells.
7. **Equivalent Circulating Density (ECD).** ECDs become critical when drilling in a soft formation environment where the fracture gradient is not much larger than the pore pressure. Controlling ROP, reducing pumping flow rate, drill pipe OD, and connection OD may all be considered or needed to safely drill the interval.

8. **Hydraulic System.** The rig equipment (e.g., pumps, liners, engines or motors, drill string, BHA) may be a given. In this case, optimizing the drilling plan based on its available capabilities will be required. However, if you can demonstrate or predict an improved outcome that would justify any incremental costs, then you will have accomplished additional optimization. The pumps cannot provide their rated horsepower if the engines providing power to the pumps possess inadequate mechanical horsepower. Engines must be down rated for efficiency. Changing pump liners is a simple cost-effective way to optimize the hydraulic system. Optimization involves several products and services and the personnel representatives. This increases the difficulty to achieve an optimized parameter selection that is best as a system.

New Technologies Positive step changes reflected in the learning curve are often the result of effective implementation of new technologies:

1. **Underbalanced Drilling.** UBD is implemented predominantly to maximize the production capacity variable of the well's optimization by minimizing formation damage during the drilling process. Operationally, the pressure of the borehole fluid column is reduced to less than the pressure in the ZOI. ROP is also substantially increased. Often, coiled tubing is used to reduce the tripping and connection time and mitigate safety issues of "snubbing" joints of pipe.
2. **Surface Stack Blowout Preventer (BOP).** The use of a surface stack BOP configurations in floating drilling is performed by suspending the BOP stack above the waterline and using high-pressure risers (typically $13\frac{3}{8}$ in. casing) as a conduit to the sea floor. This method, generally used in benign and moderate environments, has saved considerable time and money in water depths to 6,000 ft [8].
3. **Expandable Drilling Liners.** EDLs can be used for several situations. The casing plan may start with a smaller diameter than usual, while finishing in the production zone as a large, or larger, final casing diameter. Future advances may allow setting numerous casing strings in succession, all of the exact same internal diameter. The potential as a step change technology for optimizing drilling costs and mitigating risks is phenomenal.
4. **Rig Instrumentation.** The efficient and effective application of weight to the bit and the control of downhole vibration play a key role in drilling efficiency. Excessive WOB applied can cause axial vibration, causing destructive torsional vibrations. Casing handling systems and top drives are effective tools.
5. **Real-Time Drilling Parameter Optimization.** Downhole and surface vibration detection equipment allows for immediate mitigation. Knowing actual downhole WOB can provide the necessary information to perform improved drill-off tests [9].

6. **Bit Selection Processes.** Most bit venders are able to use the electric log data (Sonic, Gamma Ray, Resistivity as a minimum) and associated offset information to improve the selection of bit cutting structures. Formation type, hardness, and characteristics are evaluated and matched to the application needs as an optimization process [10, 11].

5.4 DRILLING IMPLEMENTATION

Most of the well drilling cost is time dependent rather than product cost dependent. Time is often the biggest influence. Rigs, boats, many tools, and personnel costs are charged as a function of time. Drilling mud is also discussed in terms of cost per day due to daily maintenance costs. The sum of the daily time-based costs is referred to as the operation's *spread rate*. In floater operations, the rig rate is the big influence. As spread rate increases, it becomes easier to economically justify higher-priced products that will save time. These may include more expensive bits, downhole turbines, rotary steerable systems, or a standard steerable system versus steering tools. The potential rewards of new technologies can be great.

5.4.1 Rate of Penetration

It is all about ROP, but how should we define ROP? Drilling optimization relies on minimizing total time. It is understood that we must be safe and must not damage production capacity. Our discussion focuses on drilling processes from spud to TD [12]. First consider each casing interval and then identify any subinterval where a substantially different drilling process would be beneficial or required (e.g., due to formation drillability or pressure changes, hole angle, mud type) and any planned events that would cause a trip. Where trips are required, an automatic opportunity is presented to change the drilling assembly and drilling approach or to replace the current tools with new ones if tool life may not be sufficient to reach the next planned trip point. Alternatively, perhaps a planned trip may be challenged and found to be unneeded.

Discussing drilling "time" implies that the time is used to accomplish an outcome (i.e., creating hole)—hence the concept and measure of ROP. There is more than one measure and definition of ROP. This can create misunderstandings, but there are needs for these various measures. ROP, as historically defined, includes the hours after a bit reaches bottomhole divided by the distance drilled until a decision is made to trip out of the hole. By this definition, ROP includes the time spent actually drilling rock plus any back reaming, taking surveys with the steerable system, and making connections. It does not include time tripping in and out of the hole or circulating before tripping out. The potential for any optimization to occur using this criterion is severely limited.

Overall ROP (sometimes referred to as *effective ROP*) reflects the time to drill an entire interval, start to finish, divided by the distance drilled. This measure includes all time from start to finish, the rotating time, and the nonproductive time, and it is of great importance to the drilling engineer and operator management. The interval may start at drill out and conclude with starting the trip out of the hole to run casing, or it may be a subinterval as defined in the planning stage.

Instantaneous ROP is the ROP being achieved at any point in time. This can be correlated to the bit features, operating parameters, mud type and properties, and formation to optimize choices for each of these parameters in the interval in real time and in future wells. Instantaneous ROP can be studied over the length of each joint (or stand if using a top drive) to assess changes in ROP, usually as a function of hole cleaning. This concept can be taken one step further to compare the ROP of respective joints as drilling progresses to assess bit wear or look for formation changes.

5.4.2 Special Well Types

The boundaries, constraints, limits, risks, drilling practices, equipment selection, and costs change significantly, depending on the type of well to be drilled. Because the relative importance and effect of these variables also change, the optimization approach and prioritization become more specific to the particular well type. Common groupings are

- Extended reach drilling (ERD)
- Horizontal
- Deepwater

Extended Reach Drilling An ERD well is typically defined as one in which the horizontal departure is at least twice its TVD. Be aware that this is two-dimensional thinking and does not account for complex well paths or a degree of difficulty based on the equipment. However, Figure 5.3 provides examples of TVD versus horizontal displacement for several groups of drilled assets. ERD wells are expensive and typically are pushing the envelope [13–16]. The "envelope" has been pushed over time, as shown in Figure 5.4. Requirements and drilling objectives become very critical. Drill strings are designed differently [17, 18], rig capacity may be pushed to the limit, hole cleaning processes and capabilities will be tested, dogleg severity and differential sticking take on heightened sensitivity, and casing wear is a distinct possibility. Rotation of the drill string is still a key component to accomplish hole cleaning. Rotary steerable systems have provided tremendous value and are allowing ERD wells to be drilled longer than ever before.

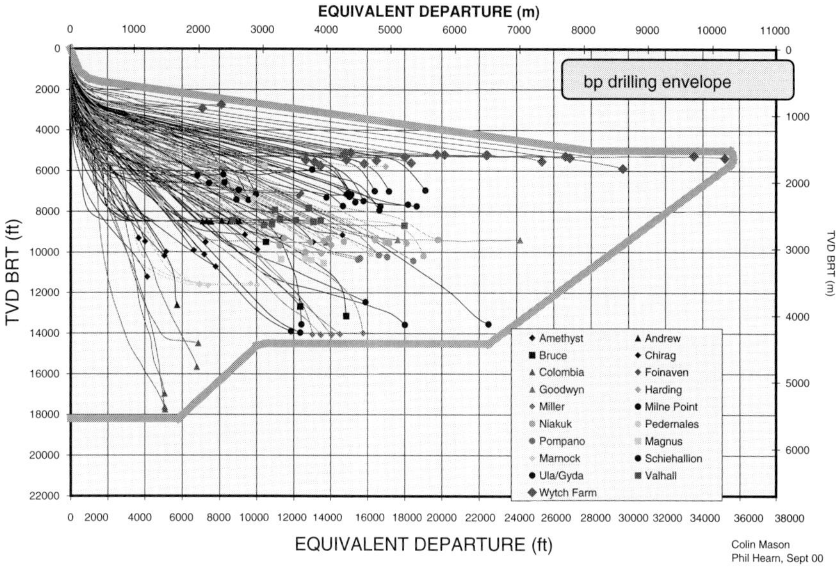

FIGURE 5.3 TVD versus horizontal displacement.

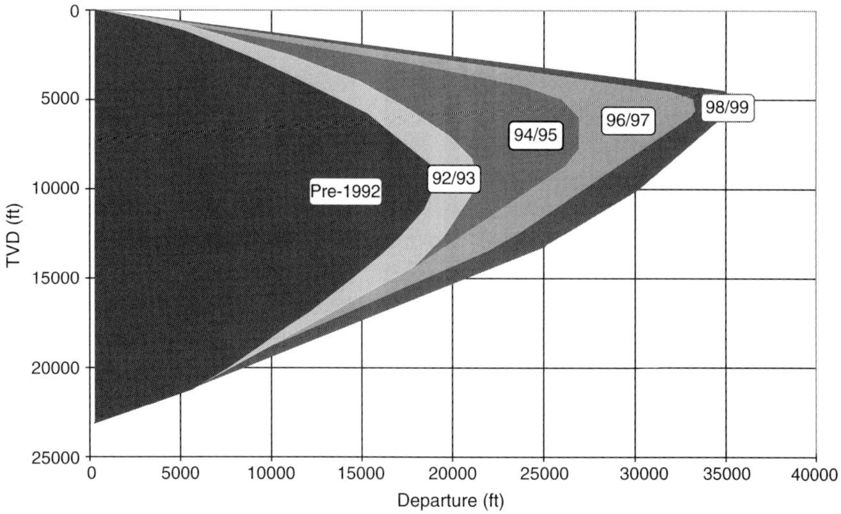

FIGURE 5.4 Extended-Reach Drilling–Evolution in the 1990s.

The key issues are torque, drag, and buckling; hole cleaning; ECDs; rig capability; survey accuracy and target definition; wellbore stability; differential sticking and stuck pipe; well control; casing wear; bit type; BHA type; logistics; and costs.

TABLE 5.1 Well Profiles

Well Type	Angle Build Rate (°/100 ft)	Feet Drilled to Turn 90°	Curve Radius (ft)	Maximum Lateral Length (ft)	Special Tools Required (?)
Long radius	2–8	1125–4500	1000–3000	8000	No
Medium radius	8–20	450–1125	700–1250	8000	No
Short radius	150–300	30–60	20–40	1000	Yes

Horizontal Wells Horizontal wells are defined according the rate of build from vertical to horizontal. Choosing one well profile instead of another is based on numerous criteria, including the location of the production zone in relation to the well's surface location, the length of horizontal lateral needed, minimizing cost for footage drilled, and maximizing value. Table 5.1 provides a summary of typical ranges of application of each of these technologies, although it seems that as soon as numbers are written, they become obsolete. Product and tool selection must be suited to and optimized for the respective footages, hours, hole angles, and rate of change of hole angles. Large-radius horizontal wells essentially use the same build rates to build the curve as ERD wells use. However, drill collar weight is placed high enough in the drill string so that the collars do not enter the curved build section. In this way, the weight is most effective axially along the drill string, causing less hole drag. Buckling is still a significant planning need. Casing wear is a lesser concern. Mud systems for horizontal laterals normally possess very low low-end viscosity, which maximizes turbulence when pumping, but solids will fall out of solution when pumping stops. Rotation of the drill string is still a key component to accomplish hole cleaning. Laterals are aligned to drill across any expected fractures to add most effectively to the effective permeability.

Deepwater Until as late as 10 years ago, anything off the continental shelf (500–600 ft) was considered deepwater. However, with recent advances in drilling, exploration, and production technologies, wells are being drilled and produced in depths of 10,000 ft of water. A generally accepted minimum depth of 1,000 ft is common today but will surely change along with the technology. Drilling costs are a major share of the total development cost. A deepwater well may also be an ERD or horizontal profile and therefore possess those sensitivities.

The key issues are rig-buoyed weight and variable deck load capacity; surface stack BOPs; shallow water flows; riser design; hole cleaning; mud temperature cooling; hydrates; low pore pressure; ability to predict pore pressure during planning and in real time; narrow range of pore pressure to fracture gradient; ECD control; wellbore stability; trouble cost; BHA type; logistics; and time cost sensitivity.

New technologies for reducing finding costs include dual density and gradient drilling, composite risers, expandable casing, casing drilling, hole-enlarging devices, and surface stack BOPs.

5.4.3 Real Time Optimization Practices

Prudent use of real time information and outcomes during drilling facilitate improved optimization. Real-time decisions must be allowed versus the well plan. This does not, however, mean that all decision-making should be placed on one person.

Drilling Mud Optimum mud properties control the borehole effectively but minimize the negative effect on instantaneous ROP. Water loss, spurt loss, and low-gravity solids (i.e., drilled solids) directly reduce instantaneous ROP.

Directional Drilling Techniques Unnecessary "drilling on the line" (especially on ERD wells) that is dictated by the customer or desired by the directional driller will add unnecessary doglegs, hole drag, time and costs, and hole-cleaning problems to a well. Meaningful criteria specific to the particular well plan should be developed.

Evaluation of the Last Bit Information can be applied to the next bit run if the formation and the application method are substantially similar for the upcoming need. Bit BHA for the next run may change.

5.4.4 Drill-off Tests

Drill-off tests establish the preferred operational parameters in an effort to optimize ROP and bit life. The objective is not necessarily to maximize ROP, because bit life may be severely shortened. The objective is to produce the highest ROP with an acceptable bit and bearing wear rate, if applicable. This is accomplished by using only the magnitudes of rpm and WOB that cause a reasonably sufficient increase in ROP to justify the bit wear that will be caused. New parameters are needed when the formation changes throughout the drilling interval. Sometimes, the ROP is very sensitive to the WOB, and it is important to keep the WOB replaced as it drills off.

Once the WOB and rpm constraints are understood for the downhole tools and possible drill string vibration, the process for performing a drill-off test is as follows:

- Select a WOB believed to be reasonable and then vary the rpm up to the maximum, identifying the lowest rpm that produced the highest ROP.
- Select a reasonable rpm and then vary the WOB through its range, identifying the lowest WOB that produced the highest ROP.

5.4.5 Downhole Vibration

The effects of downhole vibration can be disastrous, and recognition, mitigation, and planning for vibration can be critical for success. Drill string failure, downhole tool (DHT) failures, and bit destruction are possible. Less than catastrophic outcomes include reduction in ROP, bit, DHT, and drill string life. A formula for drill string harmonics rule of thumb was removed from the API RP7G recommendations during the 1998 revision because of the inability to accurately predict vibration occurrence. Vibration observed at the surface is not always a model of what is occurring downhole.

Mitigation is accomplished by modifying operational parameters or the drill string and BHA and bit type selection, individually or in combination (e.g., using a downhole motor or turbine). Rig instrumentation or real-time downhole vibration monitoring can be added to recognize drilling vibration. The first step of mitigation operationally involves increasing the rpm. If this does not stop the event, WOB can be lessened. The last choice is to stop and then restart drilling. Damage can be caused very quickly. Quick recognition and mitigation is imperative.

5.4.6 Trendology

Drilling must use trend analysis because changes during drilling (over time) tell stories that can and should be used to prevent nonproductive time and to optimize drilling practices. Tabular data or graphical representations can be used to assess the data. The following relationships are predictable, and any unexpected changes are indicators of possible problems.

Standpipe Pressure (SPP) versus Strokes Per Minute (SPM) An unexpected change may be caused by pump liner failure, drill string and tool washouts, lost nozzles in drill bits, plugged tools, or downhole motor failure.

Torque and Drag Torque and drag change during the operation. Tracking of these criteria can indicate the borehole condition and hole cleaning. The measurement criteria are pick-up and slack-off weights and free rotating torque. When plotted against time, unexpected and rapid changes are indicators of potential problems. High-angle wells, which typically involve higher daily spread rates and higher risks of occurrence are prime candidates for this level of effort, as are wells with expected hole stability problems.

ROP Changes in ROP during a bit run should correlate to changes in formation properties (e.g., type, hardness, characteristics), mud properties, or operating parameters (e.g., WOB, rpm, flow rate). An increase in pore pressure can cause a safety incident or borehole instability.

Mud The mud weight, spurt loss, water loss, and drilled solids all directly affect ROP, hole stability, hole washouts, and future cementing effectiveness.

5.5 POST-RUN EVALUATION

Continuous improvement demands understanding to what extent each variable influences the outcome and how to apply the knowledge. The post-run evaluation process completes the cycle shown in Figure 5.1. This should be a continuous improvement cycle.

The data acquired from the well just drilled become the input data for subsequent well planning and drilling stages. These data have become useful information for achieving new learning and greater optimization.

The key criteria that affected the outcome should be the primary focus of evaluation. The data will be used in the specialized "tools" used by operator and service company engineers. End of well (EOW) summary reports are created as a method to communicate the lessons learned.

References

[1] The Hon Lord Cullen, "The Public Inquiry into the Piper Alfa Disaster," London, HMSO, 1990.

[2] Comer, P. J., Fitt, J. S., and Ostebo, R., "A Driller's HAZOP Method," SPE Paper #15867, 1986.

[3] Murchison W. J., Murchison Drilling Schools, Inc., "Drilling Technology for the Man on the Rig," 1998.

[4] Devereux, S., "Practical Well Planning and Drilling Manual," PennWell Publishing Company, Tulsa, OK, 1998.

[5] Perrin, V. P., Mensa-Wilmot, G., and Alexander, W. L., "Drilling Index: A New Approach to Bit Performance Evaluation," SPE/IADC Paper #37595, 1997.

[6] Wolfson, L., Mensa-Wilmot, G., and Coolidge, R., "Systematic Approach Enhances Drilling Optimization and PDC Bit Performance in North Slope ERD Program" SPE Paper #50557, 1998.

[7] Fear, M. J., Meany, N. C., and Evans, J. M., "An Expert System for Drill Bit Selection," SPE/IADC Paper #27470, 1994.

[8] "Standard DS-1, Drill Stem Design and Inspection," 2nd Edition, T. H. Hill Associates, Inc., Texas, Houston, March 1998.

[9] API RP7G, Recommended Practice for Drill Stem Design and Operating Limits, 16th Edition, 1998.

[10] Shanks, E., Schroeder, J., Ambrose, W., and Steddum, R., "Surface BOP for Deepwater Moderate Environment Drilling Operations from a Floating Drilling Unit," Offshore Technology Conference #14265, 2002.

[11] Krepp, T. A., and Richardson, B., "Step Improvements Made in Timor Sea Drilling Performance," World Oil, May 1997.

[12] Payne, M. L., Cocking, D. A., Hatch, A. J., "Critical Technologies for Success in Extended Reach Drilling, SPE #28293, 1994.

[13] Cocking, D. A., Bezant, P. N., and Tooms, P. J., "Pushing the ERD Envelope at Wytch Farm," SPE/IADC Paper #37618, 1997.

[14] Mims, M. G., Krepp, A. N., and Williams, H. A., "Drilling Design and Implementation for Extended Reach and Complex Wells," K and M Technology Group, LLC, 1999.
[15] Chitwood, J. E., and Hunter, W. A., "Well Drilling Completion and Maintenance Technology Gaps," Offshore Technology Conference #13090, 2001.
[16] Fear, M. J., "How to Improve Rate of Penetration in Field Operations," IADC/SPE #35107, 1996.
[17] Aldred, W. D., and Sheppard, M. C., "Drillstring Vibrations: A New Generation Mechanism and Control Strategies," SPE #24582, 1992.
[18] Belaskie, J. P., Dunn, M. D., and Choo, D. K., "Distinct Applications of MWD, Weight on Bit, and Torque," SPE #19968, 1990.

Further Reading

Maidla, E. E., and Ohara, S., "Field Verification of Drilling Models and Computerized Selection of Drill Bit, WOB, and Drillstring Rotation," SPE #19130, 1989.
Hill, T. H., Guild, G. J., and Summers, M. A., "Designing and Qualifying Drill Strings For Extended Reach Drilling," SPE #29349, 1995.
Saleh, S. T., and Mitchell, B. J., "Wellbore Drillstring Mechanical and Hydraulic Interaction," SPE #18792, 1989.
Brett, J. F., Beckett, A. D., Holt, C. A., and Smith, D. L., "Uses and Limitations of Drillstring Tension and Torque Models for Monitoring Hole Conditions," SPE #16664, 1988.
Besaisow, A. A., Jan, Y. M., and Schuh, F. J., "Development of a Surface Drillstring Vibration Measurement System," SPE #14327, 1985.
Kriesels, P. C., Huneidi, I., Owoeye, O. O., and Hartmann, R. A., "Cost Savings through an Integrated Approach to Drillstring Vibration Control," SPE #57555, 1999.
Burgess, T. M., McDaniel, G. L., and Das, P. K. "Improving BHA Tool Reliability With Drillstring Vibration Models: Field Experience and Limitations," SPE #16109, 1987.

CHAPTER

6

Well Pressure Control

OUTLINE

6.1 INTRODUCTION

Basically all formations penetrated during drilling are porous and permeable to some degree. Fluids contained in pore spaces are under pressure that is overbalanced by the drilling fluid pressure in the well bore. The borehole pressure is equal to the hydrostatic pressure plus the friction pressure loss in the annulus. If for some reason the borehole pressure falls below the formation fluid pressure, the formation fluids can enter the well. Such an event is known as a *kick*. This name is associated with a rather sudden flowrate increase observed at the surface.

A formation fluid influx (a kick) may result from one of the following reasons:

- abnormally high formation pressure is encountered
- lost circulation

- mud weight too low
- swabbing in during tripping operations
- not filling up the hole while pulling out the drillstring
- recirculating gas or oil cut mud.

If a kick is not controlled properly, a blowout will occur. A blowout may develop for one or more of the following causes:

- lack of analysis of data obtained from offset wells
- lack or misunderstanding of data during drilling
- malfunction or even lack of adequate well control equipment

6.2 SURFACE EQUIPMENT

A formation gas or fluid kick can be efficiently and safely controlled if the proper equipment is installed at the surface. One of several possible arrangement of pressure control equipment is shown in Figure 6.1. The blowout preventer (BOP) stack consists of a spherical preventer (i.e., Hydril) and ram type BOPs with blind rams in one and pipe rams in another with a drilling spool placed in the stack.

A spherical preventer contains a packing element that seals the space around the outside of the drill pipe. This preventer is not designed to shut off the well when the drill pipe is out of the hole. The spherical preventer allows stripping operations and some limited pipe rotation.

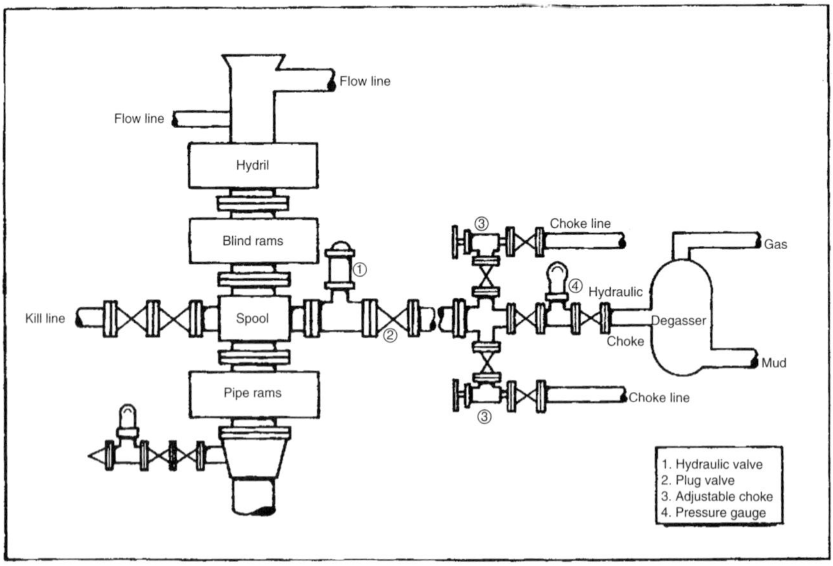

FIGURE 6.1 Pressure control equipment [1].

Hydril Corporation, Shaffer, and other manufactures provide several models with differing packing element designs for specific types of service. The ram type preventer uses two concentric halves to close and seal around the pipe, called pipe rams or blind rams, which seal against the opposing half when there is no pipe in the hole. Some pipe rams will only seal on a single size pipe; 5 in. pipe rams only seal around 5 in. drill pipe. There are also variable bore rams, which cover a specific size range such as $3\frac{1}{2}$ in. to 5 in. that seal on any size pipe in their range.

Care must be taken before closing the blind rams. If pipe is in the hole and the blind rams are closed, the pipe may be damaged or cut. A special type of blind rams that will sever the pipe are called shear blind rams. These rams will seal against themselves when there is no pipe in the hole, or, in the case of pipe in the hole, the rams will first shear the pipe and then continue to close until they seal the well.

A drilling spool is the element of the BOP stack to which choke and kill lines are attached. The pressure rating of the drilling spool and its side outlets should be consistent with BOP stack. The kill line allows pumping mud into the annulus of the well in the case that is required. The choke line side is connected to a manifold to enable circulation of drilling and formation fluids out of the hole in a controlled manner.

A degasser is installed on the mud return line to remove any small amounts of entrained gas in the returning drilling fluids. Samples of gas are analyzed using the gas chromatograph.

If for some reason the well cannot be shut in, and thus prevents implementation of regular kick killing procedure, a diverter type stack is used rather, the BOP stack described above. The diverter stack is furnished with a blow-down line to allow the well to vent wellbore gas or fluids a safe distance away from the rig. Figure 6.2 shows a diverter stack arrangement.

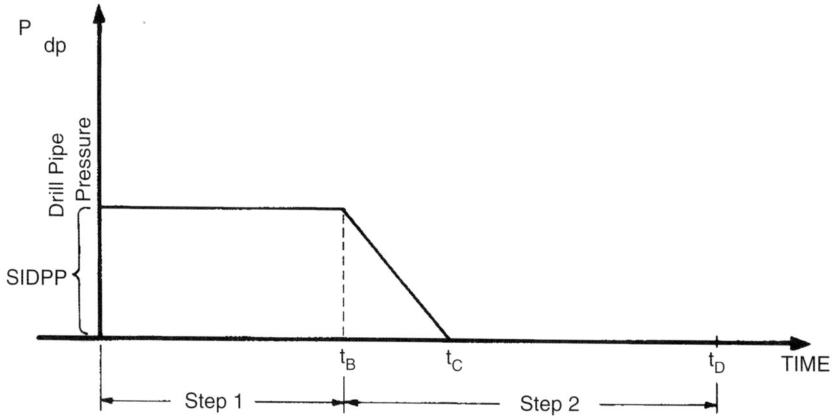

FIGURE 6.2 Driller's method—Schematic diagram of drill pipe pressure vs. time.

6.3 WHEN AND HOW TO CLOSE THE WELL

While drilling, there are certain warning signals that, if properly analyzed, can lead to early detection of gas or formation fluid entry into the wellbore.

1. *Drilling break.* A relatively sudden increase in the drilling rate is called a drilling break. The drilling break may occur due to a decrease in the difference between borehole pressure and formation pressure. When a drilling break is observed, the pumps should be stopped and the well watched for flow at the mud line. If the well does not flow, it probably means that the overbalance is not lost or simply that a softer formation has be encountered.
2. *Decrease in pump pressure.* When less dense formation fluid enters the borehole, the hydrostatic head in the annulus is decreased. Although reduction in pump pressure may be caused by several other factors, drilling personnel should consider a formation fluid influx into the wellbore as one possible cause. The pumps should be stopped and the return flow mud line watched carefully.
3. *Increase in pit level.* This is a definite signal of formation fluid invasion into the wellbore. The well must be shut in as soon as possible.
4. *Gas-cut mud.* When drilling through gas-bearing formations, small quantities of gas occur in the cuttings. As these cuttings are circulated up, the annulus, the gas expands. The resulting reduction in mud weight is observed at surface. Stopping the pumps and observing the mud return line help determine whether the overbalance is lost.

If the kick is gained while tripping, the only warning signal we have is an increase in fluid volume at the surface (pit gain). Once it is determined that the pressure overbalance is lost, the well must be closed as quickly as possible. The sequence of operations in closing a well is as follows:

1. Shut off the mud pumps.
2. Raise the Kelly above the BOP stack.
3. Open the choke line
4. Close the spherical preventer.
5. Close the choke slowly.
6. Record the pit level increase.
7. Record the stabilized pressure on the drill pipe (Stand Pipe) and annulus pressure gauges.
8. Notify the company personnel.
9. Prepare the kill procedure.

If the well kicks while tripping, the sequence of necessary steps can be given below:

1. Close the safety valve (Kelly cock) on the drill pipe.
2. Pick up and install the Kelly or top drive.
3. Open the safety valve (Kelly cock).
4. Open the choke line.
5. Close the annular (spherical) preventer.
6. Record the pit gain along with the shut in drill pipe pressure (SIDPP) and shut in casing pressure (SICP).
7. Notify the company personnel.
8. Prepare the kill procedure.

Depending on the type of drilling rig and company policy, this sequence of operations may be changed.

6.4 GAS-CUT MUD

A gas-cut mud is a warning sign of possible formation fluid influx, although it is not necessarily a serious problem. Due to gas expandability, it usually gives the appearance of being more serious than it actually is.

The bottomhole hydrostatic head of gas-cut mud and the expected pit gain can be calculated from the following equations:

$$P_h = P_s + \frac{\gamma_s H}{(1 - a_s)} - \frac{P_s a_s}{(1 - a_s)} \ln\left(\frac{P_h}{P_s}\right) \tag{6.1}$$

$$V_p = \frac{A_n P_s a_s}{\bar{\gamma}_s} \ln\left(\frac{P_h}{P_s}\right) \tag{6.2}$$

where

$$\gamma_s = \frac{P_s M a_s}{z R T_{av}} + \gamma_m (1 - a_s) \tag{6.3}$$

where P_h = hydrostatic pressure of gas-cut mud in lb/ft^2, abs
P_s = casing pressure at the surface in lb/ft^2, abs
H = vertical hole depth with the gas-cut mud in ft
a_s = gas concentration in mud (ratio of gas total fluid volume)
$\bar{\gamma}_s$ = gas-cut mud specific weight at surface in lb/ft^3
A_n = cross-sectional area of annulus in ft^2
γ_m = original mud specific weight in lb/ft^3
z = the average gas compressibility factor
M = gas molecular mass

T_{av} = the average fluid temperature in °R
R = universal gas constant in lb · ft

Example 6.1

Calculate the expected reduction in bottomhole pressure and pit gain for the data as given below:

Hole size = $9\frac{5}{8}$ in.
Hole depth = 10,000 ft
Drilling rate = 60 ft/hr
Original mud density = 10 lb/gal
Mud flowrate = 350 gal/min
Formation pore pressure gradient = 0.47 psi/ft
Porosity = 30%
Water saturation = 25%
Gas saturation = 75%
Gas specific gravity = 0.6 (air specific gravity is 1.0)
Surface temperature = 540°R
Temperature gradient = 0.01°F/ft

Solution

Gas volumetric rate entering the annulus

$$\dot{V} = \frac{\pi(9.625)^2}{4 \times 144} \times 60 \times \frac{1}{60} \times 0.3 \times 0.75$$
$$= 0.1136\,\text{ft}^3\,\text{min} = 0.85\,\text{gal/min}$$

Gas volumetric flowrate at surface (at constant temperature) is

$$\dot{V} = \frac{0.85 \times 4700}{14.7} = 272.6\,\text{gal/min}$$

The surface pressure is assumed to be 14.7 psia. Gas concentration as surface is

$$a_s = \frac{272.6}{272.6 + 350} = 0.4378$$

Mud weight at surface (assume z = 1.0) is

$$\gamma_s = \frac{14.7 \times 144 \times 0.6 \times 29 \times 0.4378}{1{,}544 \times 540} + 10 \times 7.48(1 - 0.4378)$$
$$= 42.07\,\text{lb/ft}^3, \text{ or } 5.62\,\text{lb/gal}$$

Hydrostatic pressure of gas-cut mud is

$$p_h = 14.7 + \frac{0.052 \times 5.62 \times 9{,}038}{(1-0.4378)} - \frac{14.7 \times 0.4378}{1-0.4378} \ln\left(\frac{p_h}{14.7}\right)$$

Solving the above yields

$$P_h = 4646\,\text{psia}$$

Hydrostatic head at the bottom of the hole is

$$P_{bh} = 0.052 \times 10(10{,}000 - 9{,}038) + 4{,}646 = 5{,}146\,\text{psia}$$

Since the hydrostatic pressure of the original mud is 5,214.7 psia, the reduction in the hydrostatic pressure is about 69 psi. Because the pore pressure at the vertical depth of 10,000 ft is 4,700 psi, the hydrostatic pressure of the gas-cut mud is sufficient to prevent any formation fluid kick into the hole.

6.5 THE CLOSED WELL

Upon shutting in the well, the pressure builds up both on the drillpipe and casing sides. The rate of pressure buildup and time required for stabilization depend upon formation fluid type, formation properties, initial differential pressure and drilling fluid properties. In Reference [2], technique is provided for determining the shut-in pressures if the drillpipe pressure is recorded as a function of time. Here we assume that after a relatively short time the conditions are stabilized. At this time we record the shut-in drillpipe pressure (SIDPP) and the shut-in casing pressure (SICP). A small difference between their pressures indicates liquid kick (oil, saltwater) while a large difference is evidence of gas influx. This is true for the same kick size (pit gain).

Assuming the formation fluid does not enter the drillpipe, we know that the SIDPP plus the hydrostatic head of the drilling fluid inside the pipe equals the pressure of the kick fluid (formation pressure). The formation pressure is also equal to the SICP plus the hydrostatic head of the original mud, plus the hydrostatic head of the kick fluid in the annulus.

Thus

$$P_p = 0.052\,\bar{\gamma}_m H + \text{SIDPP} \tag{6.4}$$

$$= 0.052\,\bar{\gamma}_m (H - L) + 0.052\,\bar{\gamma}_f L + \text{SICP} \tag{6.5}$$

where H = vertical hole depth in ft

$\bar{\gamma}_m$ = original mud specific weight lin lb/gal

$\bar{\gamma}_f$ = formation fluid specific weight in lb/gal
L = vertical length of the kick in ft
p_p = formation pore pressure in psi

If the hole is vertical the kick length can be calculated as

$$L = \frac{V}{VC_{dc}} \tag{6.6}$$

or

$$L = L_{dc} = \frac{V - L_{dc}VC_{dc}}{VC_{dp}} \tag{6.7}$$

where V = pit gain in bbl
VC_{dc} = annulus volume capacity opposite the drill collars in bbl/ft
L_{dc} = length of drill collar in ft
VC_{dp} = annulus volume capacity opposite the drillpipe in bbl/ft

Equation 6.6 is applicable if the kick length is shorter than the drill collars while Equation 6.7 is used if the kick fluid column is longer than the drill collars.

From Equations 6.4 and 6.5 it is found that the required mud weight increase can be calculated from

$$\Delta\gamma_m = \frac{19.23 \times SIDPP}{H} \tag{6.8}$$

The formation fluid density can be obtained from

$$\bar{\gamma}_f = \bar{\gamma}_m - \frac{SICP - SIDPP}{0.052H} \tag{6.9}$$

6.6 KICK CONTROL PROCEDURES

There are several techniques available for kick control (kick-killing procedures). In this section only three methods will be addressed.

1. *Driller's method.* First the kick fluid is circulated out of the hole and then the drilling fluid density is raised up to the proper density (kill mud density) to replace the original mud. An alternate name for this procedure is the two circulation method.
2. *Engineer's method.* The drilling fluid is weighted up to kill density while the formation fluid is being circulated out of the hole. Sometimes this technique is known as the *one circulation method.*
3. *Volumetric method.* This method is applied if the drillstring is off the bottom.

The guiding principle of all these techniques is that bottomhole pressure is held constant and slightly above the formation pressure at any stage of the process. To choose the most suitable technique one ought to consider (a) complexity of the method, (b) drilling crew experience and training, (c) maximum expected surface and borehole pressure and (d) time needed to reestablish pressure overbalance and resume normal drilling operations.

6.6.1 Driller's Method

The driller's method of controlling a kick is accomplished in two main steps:

Step 1. The well is circulated at half the normal pump speed while keeping the drillpipe pressure constant (see Figure 6.2). This is accomplished by adjusting the choke on the mud line so that the bottomhole pressure is constant and above the formation fluid pressure. To maintain a constant bottomhole pressure the formation fluid is allowed to expand, which usually results in a noticeable increase in casing pressure. This step is completed when the formation fluid is out of the hole. At this time casing pressure should be equal to the initial SIDPP if the well could be shut in.

Step 2. When the formation fluid is out of the hole, a kill mud is circulated down the drillpipe. To obtain constant bottomhole pressure, the casing pressure is kept constant (Figure 6.3) while the drillpipe pressure drops. Once the kill mud reaches the bottom of the hole the control moves

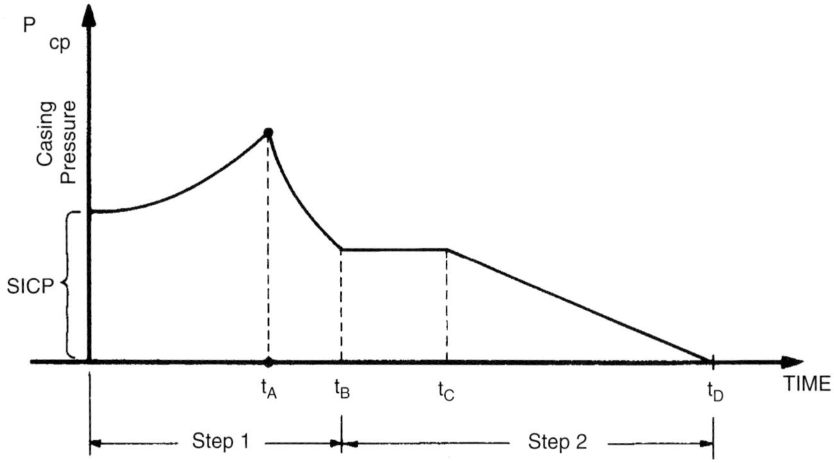

FIGURE 6.3 Driller's method—Schematic diagram of casing pressure vs. time. t_A = kick fluid out the top of the hole; t_B = kick fluid out of the hole; t_C = kill mud at the bottom of the hole; t_D = killing procedure completed.

back to the drillpipe side. The drillpipe pressure is maintained constant (almost constant) while the new mud fills the annulus.

Example 6.2

Consider the following data:

Vertical hole depth = 10,000 ft
Hole diameter = $8\frac{3}{4}$ in.
Drillpipe diameter = $4\frac{1}{2}$ in.
Mud density = 12 lb/gal
Yield point = 12 lb/100 ft^2
Circulating pressure at reduced speed = 800 psi
Shut-in drillpipe pressure = 300 psi

Calculate the following:

1. Required mud weight to restore the safe overbalance
2. Initial circulating pressure (ICP)
3. Final circulating pressure (FCP)
4. Specific weight of formation fluid

Solution

Mud weight increase to balance formation pressure is

$$\Delta\bar{\gamma}_m = \frac{20 \times 300}{10,000} = 0.6\,\text{lb/gal}$$

Trip margin $\Delta\bar{\gamma}_t$ is

$$\Delta\bar{\gamma}_t = \frac{yp}{6(D_h - D_p)} = \frac{12}{6(8.75 - 4.5)} = 0.47\,\text{lb/gal}$$

Consequently the required mud weight is

$$\bar{\gamma}_m = 12 + 0.6 + 0.47 = 13.1\,\text{lb/gal}$$

Initial circulating pressure is

$$\text{ICP} = 800 + 300 = 1,100\,\text{psi}$$

Final circulating pressure is the system pressure loss corrected for new mud weight. This is

$$\text{FCP} \cong 800\left(\frac{13.1}{12}\right) = 873\,\text{psi}$$

6.6.2 Engineer's Method

This method consists of four phases.

Phase 1. During this phase the drilling fluid, weighted to the desired density, is placed in the drillpipe. When the drillpipe is filled with heavier mud, the standpipe pressure is gradually reduced. The expected drillpipe pressure versus the number of pump strokes (or time) must be prepared in advance. Only by pumping with a constant number of strokes and simultaneously maintaining the standpipe pressure in accordance with the schedule can one keep the bottomhole pressure constant and above the formation pressure. The annulus pressure at the surface generally rises due to formation fluid expansion, although for some formation fluid the casing pressure may decrease. This depends on phase behavior of the formation fluid and irregularities in the hole geometry.

Phase 2. This phase is initiated when the kill mud begins filling the annulus and is finished when the formation fluid reaches the choke. The standpipe pressure remains essentially constant by proper adjustment of the choke.

Phase 3. The formation fluid is circulated out of the hole whole heavier mud fills the annulus. Again the choke operator maintains the drill pipe pressure constant and constant pumping speed.

Phase 4. During this phase the original mud that follows the kick fluid is circulated out of the hole and a kill mud fills up the annulus. The choke is opened more and more to keep the drill pipe pressure constant. At the end of this phase and safe pressure overbalance is restored.

A qualitative relationship between the drillpipe pressure, casing pressure and circulating time is shown is Figure 6.4 and 6.5, respectively.

6.6.3 Volumetric Method

This method can be used if the kick is taken during tripping up the hole with the bit far from the bottom of the hole. Again the constant bottomhole pressure principle is used to control the situation.

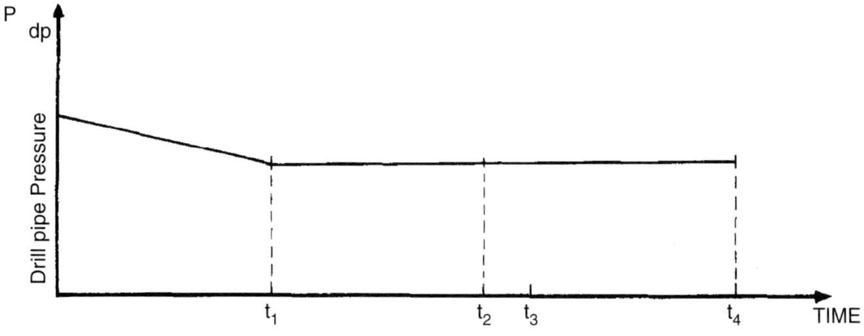

FIGURE 6.4 Engineer's method—drill pipe pressure vs. time.

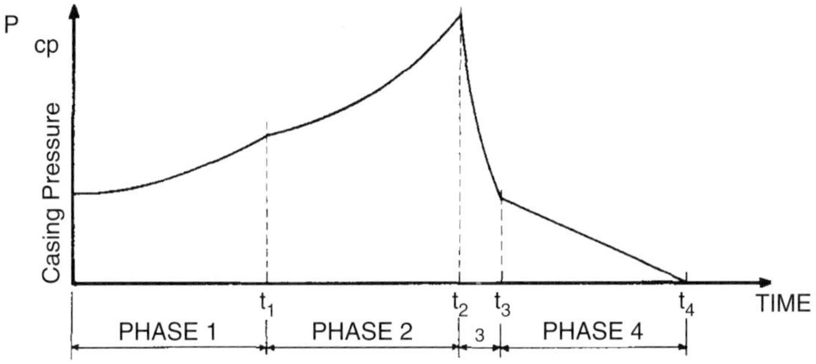

FIGURE 6.5 Engineer's method—casing pressure vs. time; t_1 = kill mud at the bottom of the hole; t_2 = formation fluid reaches the choke; t_3 = kick fluid out of the hole; t_4 = pressure overbalance is restored.

The fundamental principle of this method is equating the pit volume change with the corresponding change in annulus pressure. We write

$$\Delta p_s = \frac{\Delta V_p}{AV} 0.052\,\bar{\gamma}_m \tag{6.10}$$

where Δ_{p_s} = change in surface pressure at casing side in psi
ΔV_p = change in pit volume (volume gained) in bbl
$\bar{\gamma}_m$ = mud specific weight in lb/gal
AV = volume capacity of annulus in bbl/ft

Note that the term $(0.052\gamma_m/AV)$ expresses the expected increase of casing pressure when 1 bbl of pit gain is recorded.

The magnitude of the casing pressure during kick control is

$$P = SICP + \frac{\Delta V_p}{AV} 0.052\,\bar{\gamma}_m \tag{6.11}$$

When the pit volume stabilizes, there is equilibrium in the annulus and the kick fluid is out of the hole.

6.7 MAXIMUM CASING PRESSURE

Determination of the maximum expected casing pressure is required for selection of the kick control technique. If the driller's method is used for kick control, the maximum casing pressure P_{cmax} (psi) is calculated assuming gas influx into the hole. This is

$$P_{c\,max} = \frac{SIDPP}{2} + \left[\left(\frac{SIDPP}{2}\right)^2 + 0.052\,\bar{\gamma}_m + \frac{p_p \Delta V_p T_2 z_2}{z_1 T_1 AV}\right]^{0.5} \tag{6.12}$$

where SIDPP = shut-in drillpipe pressure in psi
$\bar{\gamma}_m$ = original mud density specific weight in lb/gal
P_p = formation pressure
ΔV_p = pit gain in bbl
T_2 = gas temperature at surface in °R
T_1 = gas temperature at the bottom of the hole in °R
z_1 = gas compressibility factor at surface conditions
z_2 = gas compressibility factor at the bottom of the hole
AV = volume capacity of annulus in bbl/ft

For the engineer's method the maximum expected casing pressure can be calculated from

$$p_{c\,max} = 0.026(\bar{\gamma}_k - \bar{\gamma}_m)L_m + \left([0.026(\bar{\gamma}_k - \bar{\gamma}_m)L_m]^2 + 0.052\,\bar{\gamma}_k \frac{p_p \Delta V_p z_2 T_2}{z_1 T_1 AV}\right)^{0.5} \tag{6.13}$$

where $\bar{\gamma}_k$ = kill mud density
L_m = length of original mud in the annulus at the moment the bubble reaches the top of the hole

Example 6.3

Calculate the maximum expected casing pressure for the driller's and engineer's techniques of kick control for the data as below:

Vertical hole depth = 12,500 ft
Original mud weight = 12.0 lb/gal
Shut-in drillpipe pressure = 260 psi
Pit gain = 30 bbl
Volume of mud inside the drillstring = 175 bbl
Annular volume capacity = 0.13 bbl/ft
Gas compressibility ratio $z_1/z_2 = 1.35$
Gas temperature at the surface = 120°F
Well-bore temperature gradient = 1.2°F/100 ft
Surface temperature = 80°F

Solution

Required mud weight for killing the well is

$$\bar{\gamma}_k = 12 + \frac{260}{0.052 \times 12{,}500} + 0.5 = 12.9\,\text{lb/gal}$$

For trip margin, a 0.5 lb/gal was added arbitrarily. Bottomhole temperature is

$$T_1 = 80 + 1.2 \times 125 = 230°\text{F}$$

Formation pore pressure is

$$p_p = 260 + 0.052 \times 12 \times 12{,}500 = 8{,}060\,\text{psi}$$

For the driller's method

$$p_{cmax} = \frac{260}{2} + \left[\left(\frac{260}{2}\right)^2 + 0.052 \times 12 \frac{8{,}060 \times 30 \times 1.35 \times 580}{690 \times 0.13}\right]^{0.5}$$

$$= 1{,}285\,\text{psi}$$

For the engineer's method

$$p_{cmax} = 0.026(12.9 - 12.0)\frac{175}{0.13} + \left[(0.026 \times 0.9 \times 1{,}346.1)^2 + 0.052 \times 12.9 \frac{8{,}060 \times 30 \times 1.35 \times 580}{690 \times 0.13}\right]^{0.5}$$

$$= 1{,}222\ \text{psi}$$

The engineer's method normally results in lower maximum surface pressure than the driller's method. For last field-type calculations, the following equation for the engineer's method can be used:

$$p_{cmax} = 200\left(\frac{p_p \Delta V_p \bar{\gamma}_k}{AV}\right)^{0.5} \quad (6.14)$$

All symbols in Equation 6.14 are the same as those used above.

6.8 MAXIMUM BOREHOLE PRESSURE

When the top of the gas bubble just reaches the casing setting depth, the open part of the hole is exposed to the highest pressure. If this pressure is less than the formation fraction pressure, the kick can be circulated out of the hole safely without danger of an underground blowout.

If the driller's method is used, this maximum pressure p_{bm} can be obtained from

$$p_{bm}^2 - (p_p - 0.052 \times \bar{\gamma}_m \times H + 0.052\,\bar{\gamma}_m D_1)p_{bm} - 0.052\,\bar{\gamma}_m \frac{\Delta V_p p_p}{AV} = 0 \quad (6.15)$$

Example 6.4

The data for this example are the same as the data used in Example 6.3.

Casing setting depth = 9,200 ft
Formation fracturing pressure at casing shoe = 7,550 psi

It is required to determine whether or not the formation at the casing shoe will break down while circulating the kick out of the hole. The annulus volume capacity in an open hole is 0.046 bbl/ft.

Solution

Using Equation 6.15 we write

$$p_{bm}^2 - (8{,}060 - 0.052 \times 12 \times 12{,}500 + 0.052 \times 12 \times 7{,}550)p_{bm}$$
$$- 0.052 \times 12 \times \frac{30 \times 8{,}060}{0.046} = 0$$

Solving the above equation yields $p_{bm} = 5{,}561$ psi. Since the maximum expected pressure in an open hole is less than the formation fracture pressure, the kick can be safely circulated out of the hole.

References

[1] Moore, P. L., *Drilling Practices Manual,* The Petroleum Publishing Co., Tulsa, Oklahoma, 1974.

[2] Miska, S., F. Luis, and A. Schofer-Serini, "Analysis of the Inflow and Pressure Buildup Under Impending Blowout Conditions," *Journal of Energy Resources Technology,* March 1992.

Further Reading

Rehm, B., *Pressure Control in Drilling*. The Petroleum Publishing Co., Tulsa, Oklahoma, 1976.

CHAPTER

7

Fishing Operations and Equipment

OUTLINE

A fish is a part of the drill string that separates from the upper remaining portion of the drill string while the drill string is in the well. This can result from the drill string failing mechanically, or from the lower portion of the drill string becoming stuck or otherwise becoming disconnected from drill string upper portion. Such an event will instigate an operation to free and retrieve the lower portion (or fish) from the well with a strengthened specialized string. Junk is usually described as small items of non-drillable metals that fall or are left behind in the borehole during the drilling, completion, or workover operations. These non-drillable items must be retrieved before operations can be continued [1–3]. The process of removing a fish or junk from the borehole is called *fishing*.

It is important to remove the fish or junk from the well as quickly as possible. The longer these items remain in a borehole, the more difficult these parts will be to retrieve. Further, if the fish or junk is in an open hole section of a well the more problems there will be with borehole stability. There is an important tradeoff that must be considered during any fishing operation. Although the actual cost of a fishing operation is normally small compared to the cost of the drilling rig and other investments in support of the overall drilling operation, if a fish or junk cannot be removed from the borehole in a timely fashion, it may be necessary to sidetrack (directionally drill around the obstruction) or drill another borehole. Thus, the economics of the fishing operation and the other incurred costs at the well site must be carefully and continuously assessed while the fishing operation is underway. It is very important to know when to terminate the fishing operation and get on with the primary objective of drilling a well. Equation 7.1 can be used to determine the number of days that should be allowed for a fishing operation.

The number of days, D (days), is

$$D = \frac{V + C_s}{R + C_d} \tag{7.1}$$

where

V is the replacement value of the fish (dollars or other money),
C_s is the estimated cost of the sidetrack or the cost of restarting the well (dollars),
R_f is the cost per day of the fishing tool and services (dollars/day),
C_d is the cost per day of the drilling rig (and appropriate support) (dollars/day).

7.1 CAUSES AND PREVENTION

There are a number of causes for fishing operations. Many of the problems that lead to fishing operations can be prevented by careful operational planning and being very watchful as drilling operations progress for indications of possible future borehole troubles [4]. The major causes of a fishing or junk retrieval operation are

1. *Differential Pressure Sticking*. It is estimated that the cost of stuck pipe in deep oil and gas wells can be approximately 25% of the overall budget. Therefore, a large portion of the fishing tools available were developed for the recovery of stuck pipe. This condition occurs when a portion of the drill string becomes stuck against wall of an open hole section of the borehole. This is due to smooth surface of the drill string (usually the drill

collars) has become embedded in the filter cake on the wall. Differential sticking is possible in most borehole operations where drilling mud is being used as the circulation fluid. Underbalanced operations generally avoid this differential pressure sticking. Sticking occurs when the pressure exerted by the mud column is greater than the pressure of the formation fluids. Normally the drill string is differentially stuck when

   a. The drill string cannot be rotated, raised or lowered, but circulating pressure is normal.
   b. The drill collars are opposite a permeable formation.
   c. Sticking was instantaneous when the pipe was stationary after drilling at a higher than normal penetration rate. In some cases a differentially stuck string or bottomhole assembly may be freed by reducing the mud weight.

   This will reduce the differential pressure between the column of mud and the permeable zone. However, this procedure should not be used if well control is a problem.

2. *Under Gauge Borehole*. Mud filter cake can build excessively across a low pressure permeable formation when the circulation rate is low, water loss is very high, and there is an extended period between trips. Under these conditions, a drill string or logging tools can become stuck in the under gauge borehole and/or filter cake. Filter cake build up is usually slow and appears as drag on the multi-channel recorder, or as an under gauge hole on a caliper survey.
3. *Key seats*. Key seats develop where there is a sudden change in hole deviation or above a washout in a deviated hole. Doglegs above the drill collars are subject to erosion or wear by the drill pipe on the high side of the dogleg. Continuous rotation can slowly cut a groove into the dogleg forming what is known as a key seat. The drill pipe body and tool joints wear a groove in the formation approximately the same diameter as the tool joints. The wear is confined to a narrow groove, because high tension in the drill pipe prevents side ways movement. During a trip out of the hole, the BHA may be pulled into these grooves and the grooves may be too small to allow the BHA to pass through. In this situation no attempt should be made to jar the collars through a key seat. A possible solution to this problem would be to circulate and rotate the drill string and move the string in small increments up through the key seat. All tight spots (over pull and depth of the over pull) should be noted and recorded on the IADC daily report and the drilling recorder. A tight spot that occurs on two successive trips out of the hole with over pull on the second trip greater than the first, is an indication of a key seat forming. A key seat wiper or string reamer should be run on the third trip.
4. *Tapered hole*. Abrasive hole sections will tend to dull bits and thereby reduce bit and stabilizer gauge. Attempting to maximize the length of a

bit run in an abrasive formation may prove to be a false economy since an under gauge hole will likely lead to a reaming operation. If the driller fails to ream a new bit to bottom when this situation exists, the bit may jam in the under gauge hole. Proper grading and gauging of the old bit prior to running a new one will prevent this problem. Since this type of sticking usually occurs while tripping in the hole. In such situations, the string should be jarred up immediately. This will usually free the drill string. If jarring does not free the string, it will be necessary to make a back-off and wash over. Wash over procedures will be discussed below.

5. *Object along side the drill string*. Occasionally, an object such as a wrench, bolt, slip or tong part, or a hammer will fall into the hole along side the drill string. Except when the string can be pushed or pulled around the object or the object can be pushed into the wall of the hole, serious fishing problems can develop. This is especially true when the string gets jammed to one side of the hole in a cased hole. A visual check of all hand and other surface tools is required to see if anything is missing. Never leave the hole unprotected or leave loose objects lying around the rotary area. Jarring might free the string, if not, a short wash over is required using an internal spear to catch the string when it falls.
6. *Inadequate hole cleaning occurs as a result of*
   - **a.** A drill string washout above the bit.
   - **b.** Low circulation rate in a large hole with an unweighted mud system.
   - **c.** Sloughing shale.
   - **d.** Gravel bed in the shallow portion of the hole.
   - **e.** Partial returns.

Indications of sticking due to inadequate hole cleaning are:

a. A significant change in the amount of returns across the shaker before sticking.
b. A decrease in pump pressure or increase in pump strokes followed by an increase in drag while picking up on the pipe (a washout in the drill string).
c. An increase in pump pressure and drag.
d. The inability to circulate if the pipe sticks.
e. Frequent bridges on trips.

Even with the challenging wells being drilled today the incident of parted strings occur less often than decades ago. Improved maintenance, inspection procedures, monitoring systems, materials and coatings are all contributing to a reduction in this fishing operations. The biggest challenge when fishing a parted string is in the interpretation of the condition of the top of the fish. A string may part due to any of the following reasons:

a. A twist off after the drill string has become stuck.
b. A washout.
c. A back lash and subsequent unscrewing of the string.
d. Junk wearing through a tubular.
e. Metal fatigue in the string.

When working shallow without the benefit of a torque limit switch and if the string becomes stuck, the torque can build up very rapidly in the string and cause a twist off. If a work string is in poor condition a twist off can occur at any depth (with or without a torque limit switch). A twist off can be the most difficult type of fishing job due to the possible condition of the top of the fish. Turbulent flow of the circulating drilling fluid can damage a connection and cause a washout in the metal of the connection itself. If such a washout is not detected, the drill string can be weakened in the washed out area resulting in the failure of the string component. A washout in a tubular can in turn washout the formation. This would reduce the annular velocity in the washout area which in turn world diminish hole cleaning. Any time there is a drop in the standpipe pressure that cannot be explained, the string should be pulled immediately.

A string that alternately sticks then releases while drilling forward can result in a build up of torque which, when released, rotates the lower portion of the string at an accelerated rate. The inertia of the lower portion of the string can make the string back off. In a gauged area of the hole, the string could be screwed back together, but in a washed out area it will be necessary to run special tools to engage the fish. This can also occur off bottom while torque is in the string. Junk pushed into a soft formation can later damage tubulars rotating against the junk. It is always better to remove the junk than to push it to the side. Metal fatigue can cause a string to fail under normal operating parameters. Fatigue can be reduced by establishing the working life of string components and replacing them at the appropriate respective time intervals.

In general, parted strings are easier to fish than stuck pipe. However, if a fish is in an open hole, the likelihood of recovery a fish will diminishes with time. If a fish has a connection is facing up, a screw-in assembly with jars should be run. If the fish top cannot be screwed into, an overshot with a jarring assembly should be run. Different types of fishing and jarring assemblies are shown in Figure 7.1. The condition of the bottom of the string pulled after a string parts should give an indication of the condition of the top of the fish and, thus, determine what tools should be used for the fishing job. The piece of fish pulled out of the hole should be a reverse mirror image of what the top of the fish looks like.

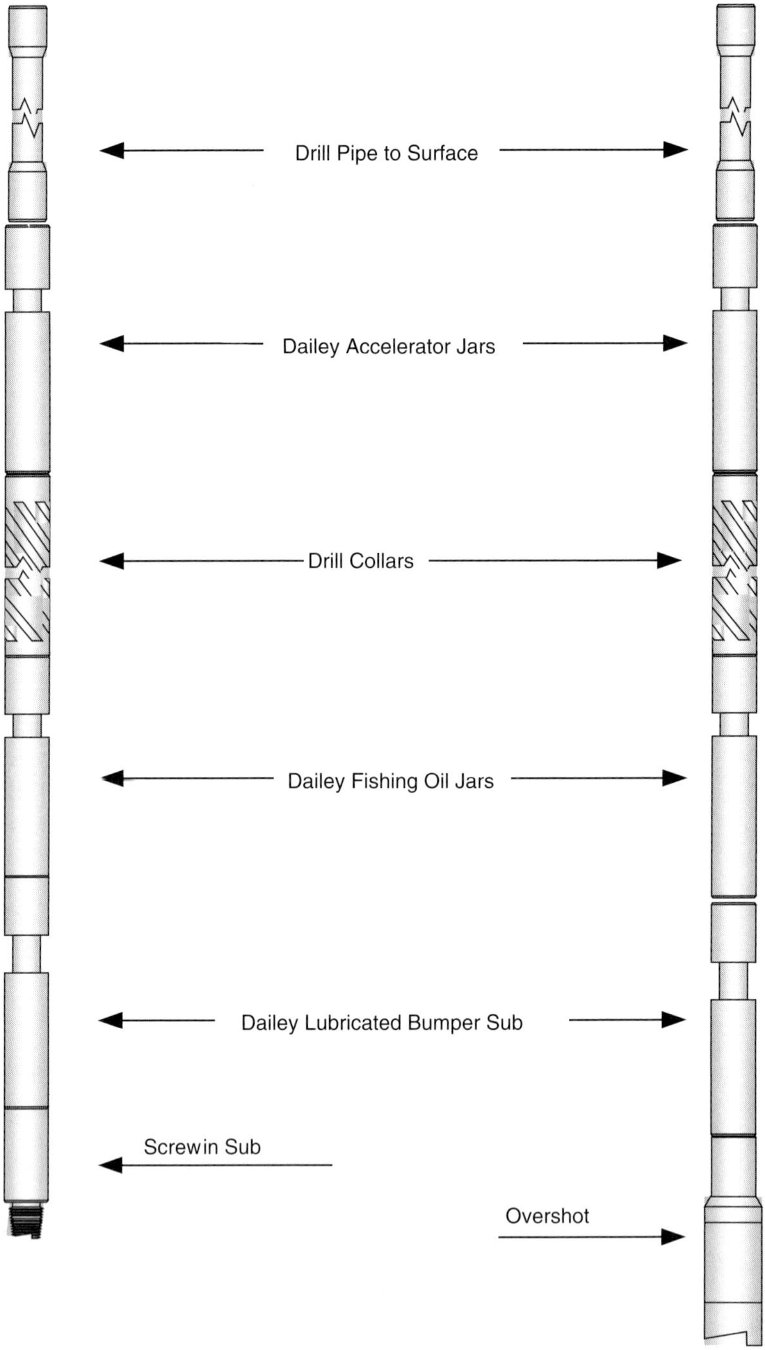

FIGURE 7.1 Screw in Sub and Overshot Jarring Assemblies. (Courtesy Dailey Fishing Tools)

7.2 PIPE RECOVERY AND FREE POINT

When a drill string or other tubular becomes stuck in a borehole it is very important that the depth where the pipe is stuck be determined. In most cases this can be accomplished via a simple calculation. The depth to where the drill string is free and where sticking commences is called the free point [2]. This free point can be calculated using measurements taken on the rig floor.

The length of the drill string from the surface to the free point is L_f. This will be less than the overall length of the drill string D. To obtain the L_f, the following procedure is used:

1. An upward force F_1 is applied to the top of the drill string via the draw-works. The force should be slightly greater than the total buoyant weight of the drill string. This ensures that the entire drill string is in tension.
2. With this tension on the drill string, a reference mark is make on the drill string exposed at the top of the string.
3. A greater upward force F_2 is then applied to the drill string. This causes the free portion of the drill string to elastically stretch by an amount ΔL. The stretch (or elastic displacement) is measured by the movement of the original reference mark. The magnitude of F_2 must be limited by the yield stress, or elastic limit stress of the specific pipe steel grade.

Knowing the stretch ΔL and the forces applied F_1 and F_2, Hooke's law, the length to the free point is

$$L_f = EA\frac{\Delta L}{(F_2 - F_1)} \tag{7.2}$$

where

- E is the Elastic Modulus (Young's Modulus) of steel (i.e., 29×10^6 psi, 200 GPa),
- A is the cross-sectional area of the pipe body (ft^2, m^2),
- ΔL is the stretch distance (ft, m),
- F_1 is the force to place the entire drill string in tension (lb, N),
- F_2 is a force greater than F_1 but less that the force limited by the yield stress of the pipe grade (lb, N).

Example 7.1

A drill string API $3\frac{1}{2}$ in., 13.3 lb/ft nominal weight, Grade E75 drill pipe is stuck in a 12,000 ft borehole. The driller places 150,000 lb of tension on the top of the drill string above the normal weight of the string and makes a mark on the string at top of the rotary table. The driller then increases the tension to 210,000 lb beyond the weight of the drill string. This later tension will give a maximum stress in the drill pipe body that is less than the 75,000 psi yield stress of Grade E75. The original mark on the drill string shows that the free portion of the drill string has stretched 4 ft due to the additional tension placed on the drill string by the drawworks.

The cross-sectional area of the drill pipe body is approximately

$$A = \frac{13.3}{490}$$
$$A = 0.271\,\text{ft}^2$$

where the specific weight of steel is 490 lb/ft^3. Equation 7.1 is

$$L_f = \frac{(29 \times 10^6)(144)(0.0271)(4.0)}{(210000 - 150000)}$$
$$L_f = 7805\,\text{ft}$$

There is a special downhole tool that can be used to determine the location where the drill string (or other tubulars) is stuck. This is called a free point tool. Prior to using the free point tool must be calibrated in the free drill string (free pipe). This calibration is accomplished using an anchor system in the free string prior to normal operations being initiated. Once the tool is calibrated, the calibration should not be changed during the course of the operation. The free point tool is calibrated so that 100 units of movement is defined as completely free pipe. When the tool enters an increased cross-sectional area (increased wall thickness) such as Hevi-Wate drill pipe (HWDP), the scale will change so that 70 units of movement is accepted to be completely free pipe (the calibration of the free point tool has not changed). As the tool goes from HWDP into drill collars, there is another scale change. In the drill collars, 40 to 50 units of movement indicates completely free drill collars. A unit of movement in a free point application is a nondefined reference term.

There are four types of anchoring systems used in free point tools today. The oldest is the bow spring anchor. This design has been in use since the late 1950s. The others are the permanent magnet anchoring system, the electromagnet anchoring system, and the motorized anchoring system. All free point tools measure movement between the two anchors (average about 17/1000 of 1 in. in movement). All free point tools read movement between the two anchors in the form of tension on the pipe, compression on the pipe, right hand torque on the pipe, and the newest tools, including the read left hand torque. It is critical that both torque and stretch readings be measured and compared. If the two readings differ, the torque reading is generally more reliable.

The interval between free pipe and stuck pipe is called the transition area. The interval is a length of the borehole. The length of the transition area is an important clue in helping to identify why the pipe is stuck. For example, in free pointing a stuck packer, the transition area would be extremely short if the pipe is free from the surface to the top of the packer. If the pipe were stuck due to fill on top of the packer, the transition are would be spread out over a longer interval. In an open hole a keyseat would result in a very short transition area while differential sticking would result in the transition area being spread out over a much longer interval. When parting the pipe it is important to leave some free pipe looking up (one or two joints of free pipe is recommended). Back-off at least 100 ft. below a casing seat or up inside the casing. Do not back-off in a washed out section or immediately below a drop in hole angle. Always make the back-off at the depth that will facilitate the best chance of future engagement to the fish.

7.3 PARTING THE PIPE

There are seven options to be considered prior to parting the pipe. These are chemical cut, jet cutter, internal mechanical cutter, outside mechanical cutter, multi-string cutter, severing tool, and washover back-off safety joint/washover procedures.

There are five requirements for a back-off to be successful:

1. Free pipe: the connection to be backed off must be free.
2. Torque: the correct amount of left hand torque is required.
3. Weight: the connection being shot must be at neutral weight.
4. Shot placement: the short must be fired across the connection.
5. Shot: use the proper size string shot/prima cord.

7.3.1 Chemical Cut

The first chemical cutter was developed by McCullough Tool Company and used in the field in 1957. Today there are several manufactures of chemical cutters. The chemical cutter is lowered inside the pipe (that is to be cut) to a depth of one or two joints above the stuck point. A collar locator is used to correlate depths. Chemical cuts do not require that the pipe be torqued up. This affords a safer operation and is recommended in bad strings of pipe. Sometimes pipe will rotate freely, even though it cannot be pulled from the hole. This makes it impossible to back off the pipe. The chemical cutter utilizes a blast of powerful acid (at high speed and temperature) to make a smooth cut without flare or distortion to the OD or ID of the pipe. It will not damage the outer string of casing or tubing making for easy engagement of the pipe being cut.

7.3.2 Jet Cutter

The shaped explosive charges using parabolic geometry were developed after World War I to penetrate thick steel armor. This shaped change was adapted to fit in casing or tubing and became production jet perforating changes that replaced the earlier bullet perforators. Further improvements in the technology allowed the shaped charge concept to be used in a 360° circle design that can be used to completely sever a steel tube. Advantages of the jet cutter are that the jet cutter does not have mechanical slips to set so the condition of the tubular being cut, or what the ID is coated with, has little bearing on the operation of the cutting. Jet cutters are shorter in length than a chemical cutter and greater size ranges are available. The disadvantage of a jet cutter is that the pipe being cut will be deformed and must be dressed off before fishing. Also, adjacent strings could be damaged in multiple tubing completed wells.

7.3.3 Internal Mechanical Cutter

The mechanical cutter usually cannot compete with cutters that can be run on wire line due to the cost of trip time. An internal mechanical cutter is shown in Figure 7.2. However, if large OD tubulars are being cut a mechanical cutter can be cost effective due to the high cost of large OD chemical and jet cutters. Also a mechanical cutter can be run in conjunction with a spear which allows cutting and retrieving in a single trip. The mechanical cutter is an option that will has merit in several situations. These are shallow depth cuts, large OD tubular cuts, the need to cut and retrieve in a single trip, and in well conditions too adverse for wireline conveyed cutters. The mechanical cutter is lowered into the hole on a tubular string to the point where the cut is to be made. At this point, right hand rotation will allow the friction assembly to unscrew from the mandrel and a gradual lowering of the tool permits the cone to be driven through the slips thereby anchoring the tool in the pipe. As the slips firmly engage, the wedge block forces the knives outward. This action is continued until the pipe is cut. With the cut complete the pipe is raised, the slips disengage, the knives retracted and the friction assembly returns automatically to the running in position. A unique feature of this tool is the automatic nut which allows the resetting and disengaging of the tool frequently without coming out of the hole.

7.3.4 Outside Mechanical Cutter

Figure 7.3 shows an external mechanical cutter. Washing over the stuck pipe is done with washover pipe and a rotary shoe slightly larger than the cutter to make a gauge run. When washing over the desired section is completed, the washover pipe is pulled out of the hole and the rotary shoe is replaced with the cutter. The cutter and washpipe are then lowered over the stuck pipe. To operate, the cutter is slowly raised until the dog assembly engages the joint. The string is then lowered slightly to reduce excess pressure on the knives as the cut is started. Rotating the cutter to the right starts the cut. A slight upward pull and slow uniform rotation is

FIGURE 7.2 Type "C" Internal Mechanical Cutter. (Courtesy Weatherford International)

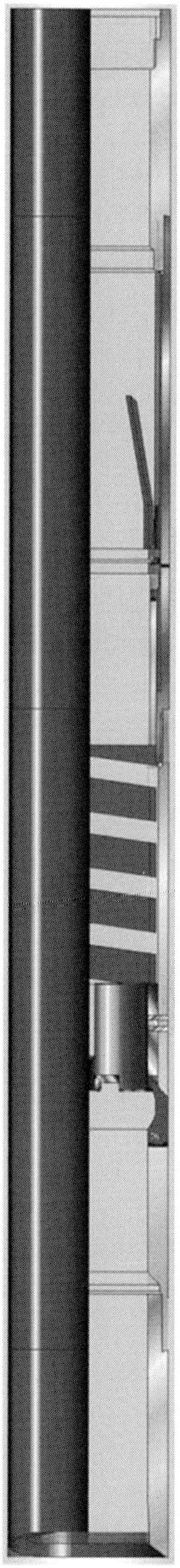

FIGURE 7.3 Type "A" Outside Mechanical Cutter. (Courtesy Weatherford International)

maintained while the cut is being made. When the cut is completed, the string is raised, bringing with it the cut off section of pipe which is held in the cutter and washpipe. At the surface the cut off section is stripped out of the washpipe and the process is repeated.

7.3.5 Multi-String Cutter

Figure 7.4 shows a multi-string cutting assembly. The multi-string cutter is lowered into the hole on a working string. This cutting assembly is used for cutting multiple string of casing cemented together. There are three or four bladed cutter designs and both are hydraulically actuated. The tool is run to the desired depth and rotation is begun. As hydraulic pressure increases the knives are forced into the cutting position and the cut is made. In making the multiple cut, a series of different length knives are used in succession, each enlarging the window cut by the previous set. The tool is easily retracted by easing the hydraulic pressure and picking up on the string. Knives can be quickly exchanged from the rig floor. A well stabilized cutter will do a better downhole cutting job. A space out assembly is required on floating rigs so that the cutter can be accurately repositioned after each knife change. In making single or multipe cuts in

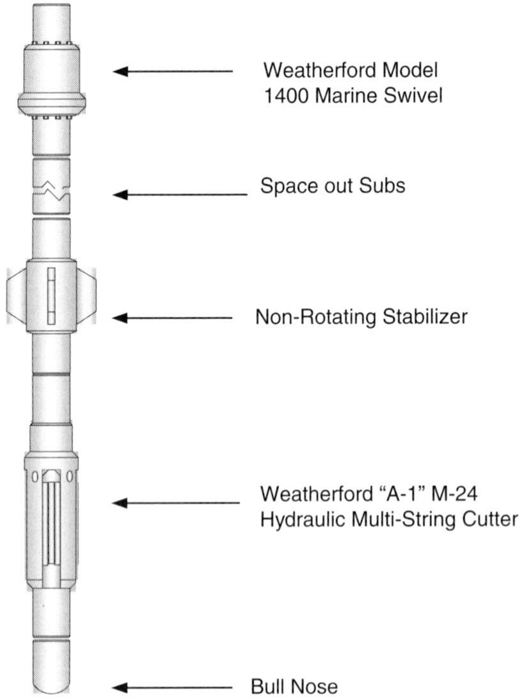

FIGURE 7.4 Multi-String Cutting Assembly. (Courtesy Weatherford International)

large pipe (20 in.) or larger, a replaceable blade stabilizer is used since cutter stabilization is most important for efficient cutting action. On floating rigs, a marine swivel also must be used to stabilize the cutting depth and land properly in the wellhead.

7.3.6 Severing Tool

The severing tool is an option for parting pipe, drill collars, casing or other tubing. This tool is so powerful it will often eliminate many fishing options. The 3 in. OD severing tool is rated to sever up to an 11 in. OD drill collar. The explosive material in a severing tool is C-4. If attempting to sever anything other than drill collars, it is recommended that the charge be placed in a tool joint or coupling. The severing tool needs metal mass in order to shatter the drill collar (or other thick walled tubular). If the tool is fired in the pipe body or casing the severing tool may balloon and rail split the pipe wall rather than part it. When the severing tool is fired across a connection on drill pipe or casing, even if the pipe does not part, the connection is usually damaged enough so that the pipe can be pulled into.

7.3.7 Washover Back-off Safety Joint/Washover Procedures

The washover back-off safety joint is used to cross the wash pipe back to the jarring assembly and allow for washing over, screwing into a fish, and making a string shot back off in one trip. Should the washpipe become stuck, the washover safety joint provides a means to back off at the top of the washpipe with considerably less torque than is required to back off the work string. Due to the high risk of sticking the washpipe, a washover safety joint should be used any time a washover assembly is run in open hole. A fish can be screwed into with the washover safety joint, but several factors should be considered before this is done.

A washover assembly is shown in Figure 7.5. A typical washover assembly should consist of the following tools from the bottom up:

- A rotary shoe
- Washpipe
- Washover safety joint
- Bumper jars
- Oil jars
- Drill collars
- Accelerator jar
- Drill pipe or work string

A jarring assembly should always be run with washpipe in case the washpipe becomes stuck. The washover assembly is run into the hole to within 10 ft to the top of the fish. Circulation is established and the washpipe is lowered slowly over the top of the fish. The free portion of the fish

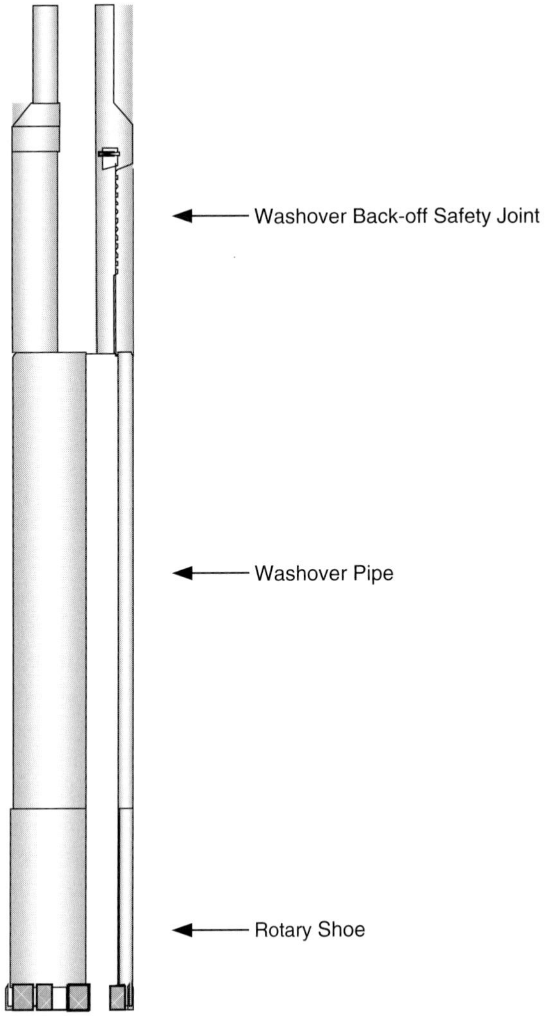

FIGURE 7.5 Washover Assembly. (Courtesy Weatherford International)

standing above the stuck point acts as a guide for the washpipe. Once over the fish, free torque, circulating pressure and all other parameters should be recorded. Use rotary speeds of 40 to 50 rpm in soft formations and higher rotary speeds for harder formation. As with all milling applications, the surface speed should be calculated to insure the correct carbide dressed shoe feed rate. For crushed carbide the recommended feed rate is 150 to 200 surface feet per minute. During this operation, torque and circulating pressure should be closely monitored. The OD of the washpipe in relation to the hole circumference makes sticking the washpipe a real possibility. The string should not be allowed to be static for any period of time. If the torque becomes too great, it is necessary to come out of the hole and lay down about half of the washpipe. If the hole is crooked and the location of the back off was poorly chosen, the top of the fish may be laying under a bend. Shorting of the washpipe may allow it to conform to the curvature of the hole more easily.

When the entire length of the washpipe had washed over the fish, the washpipe can be pulled from the hole and a jarring assembly run back to recover the fish. Special tools exist that allow the washing over, screwing in, and recovering of the fish in one trip. Use of these tools is very dependent on hole conditions and require a seasoned fishing tool operator. If the fish has been engaged but will not pull, a back-off should be made and the washover operation repeated to recover another section of the fish.

When a fish has been washed over and recovered in one trip, the fish must be stripped out of the washpipe. This can be a long and difficult operation. The drill pipe and part of the fishing string is pulled, broken out and set back until the washpipe is in the rotary table and the slips are set around it. The washpipe safety joint with the fish attached is then broken out and raised a few feet. A split slip holder or bowl is then screwed into the threads on the top of the washpipe and split slips are set around the fish to hold it. The fish is stripped out of the washpipe with the elevators. The washpipe can be pulled to change the rotary shoe if necessary. The whole washover process is repeated until all of the fish is recovered. When the last section of the fish is retrieved, the fish is stripped until the bit or stabilizer is near the bottom of the rotary shoe. The fish and washpipe must be pulled, suspended and broken out together. This double stripping job takes about one hour per 100 ft of washpipe. There are tools such as the drill collar spear that can be used to lower the fish and suspend it in the bottom joint of washpipe so that the washpipe can be pulled first, saving a double stripping job.

7.4 JARS, BUMPER SUBS AND INTENSIFIERS

Jars, Bumper Subs and Intensifiers are basic components in a fishing assembly. Improved design, new materials, and sophisticated computer programs for jar placement are continually improving jar overall jar performance. The fishing bumper sub is engineered to withstand sustained bumping loads and displacements during fishing, drilling, and workover operations. These tools are designed to permit a 10 to 60 in. vertical strokes downward. The stroke is always available in the tool, whether rotating or not, but the ease of the stroke may be affected by friction within the bumper sub itself. The lubricated bumper sub is better suited for high or sustained circulating pressures. In all bumper subs, an adequate striking surface (to produce impact) must be provided at the limits of the free movement of the sub mechanism. Specially designed splines, torque keys, or torque slippers provide a source of continuous torque transmission. In addition the bores permit full circulation at all times.

By providing an immediate bumping action these tools can help to free drill pipe, drill collars, bits and other tools that have become stuck, lodged or keyseated. To achieve a good impact with a bumper sub, the drill string stretch must be utilized to provide the speed for the impact. This requires dropping the drill or working string until the jar is approximately 50% closed; the string is stopped abruptly at the surface and the downward momentum continues down, supplying the impact. This is a delicate operation and requires a little practice. Another function of the bumper sub is to release engaging tools and provide a "soft touch" when attempting to get over the top of a fish. Whenever a releasing type of fishing tool is to be used, a bumper sub should be run just above it to aid in its release.

The fishing jar is a straight pull operated jar that employs elementary principles of hydraulics and mechanics. Fishing jars can be mechanically or hydraulically actuated. Hydraulic fishing jars are most common. Hydraulic jars require no setting or adjustments before going in the hole, or after the fish has been engaged. The intensity of the jarring blow may be controlled within a wide range of impact load applied by the amount of over pull applied to the string before the jar is tripped. As pull is applied to the jar, oil is forced from one side of the inner body to the other through some type of orifice slots that will meter volume of fluid into storage chambers. This allows for ample time to pull the required stretch in the string.

To minimize the dulling effect on a grapple, hydraulic oil jars should be closed or reset using only sufficient weight to overcome friction in the inner body. During resetting, oil is forced in the opposite direction, unimpeded form one cavity to the other. Some jars require that the resetting be done slowly to optimize the life of the jar. It is possible to jar with torque in the string though it is not recommended. In most cases, the friction created by the torque will slow down the upward movement, reducing the impact loads. When used in fishing operations, an oil jar should be installed below a string of drill collars. For maximum effectiveness, a jar intensifier should be used in the string.

7.4.1 Drill Collars in a Jarring Assembly

In a jarring assembly, drill collars provide the mass to deliver the impact. The drill collars can be compared to the weight of head of a conventional hammer. A light hammerhead can deliver a fast impact blow, but the lack of mass reduces inertia, which means the impact is quickly diminished. A heavy hammerhead delivers a slow impact blow, and the high mass increases inertia, increasing the effectiveness of the impact. Too many drill collars can have the same effect as too-heavy a hammerhead. Just as a hammer could be too heavy to swing, the friction produced between the borehole wall and a long string of drill collars could diminish the impact force at the fish.

7.4.2 Fluid Accelerator or Intensifier

During conventional jarring operations with either mechanical or hydraulic jars, the intensity of the blow struck is a function of, and proportional to the accelerated movement of the drill string above the jar. This accelerated movement will often be diminished by friction between the string and the wall of the hole. In such cases much of the energy will be lost. Also, at very shallow depths, lack of stretch in the running string causes significant loss in the effectiveness of the acceleration.

The intensifier provides the means to store the required energy immediately above the jar and drill collars. This is done to effectively offset the friction loss of stretch or drag on the string. An important secondary contribution of the intensifier is the use of its contained hydraulic fluid or gas to cushion the shock of the string as it rebounds after each jarring stroke. This reduces the tendency to cause shock damage to the tool and string. Use of the intensifier allows fewer drill collars to be used in specific cases than would otherwise be possible. This is particularly true at shallow operating depths where excessive numbers of drill collars are sometimes used to produce great mass in place of available stretch. However, the use of too many drill collars can be damaging to the tools and the string. The stretch created when over-pulling the drill string provides the energy required to actuate the fishing jar. Because of this energy, oil jars will work without intensifiers but less effectively. Illustrations of Jars, Bumper Subs and Intensifiers are shown in Figure 7.6.

7.5 ATTACHMENT DEVICES

An attachment or engaging device provides a means to reconnect to the equipment (fish) left in a wall. Engaging devices should have as many as possible of the following features:

1. There should be a means to release the engaging tool from the fish.
2. Full and positive circulation through the fish should be possible.
3. There should be an unobstructed ID that will allow wire line access to the fish.
4. The device should have the highest tensile and torsion yield possible.
5. The device should be able to withstand up and down jarring impact.
6. The tool should allow the application of torque through the work string and into the fish when applying overpull or set down weight or when the engaging tool is at the neutral weight.

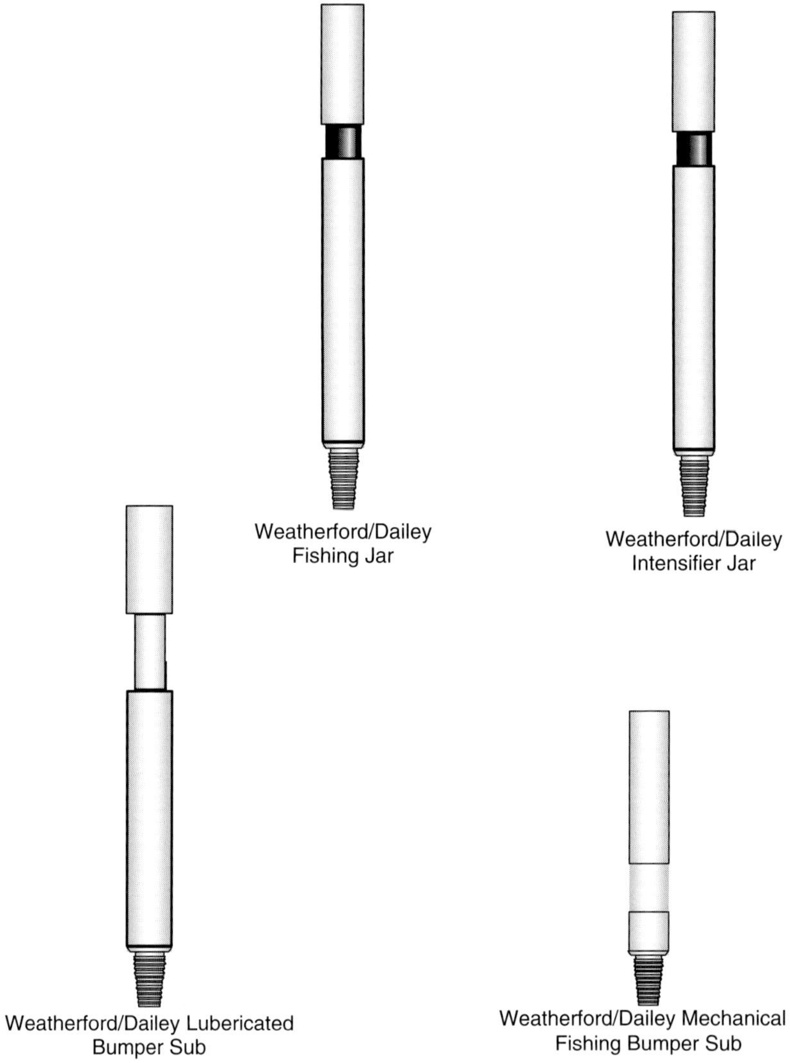

FIGURE 7.6 Jars, Bumper Subs and Intensifiers. (Courtesy Weatherford International)

In all cases, the preferred method of engaging a fish is screwing in the with a screw in sub, joint of drill pipe, or a drill collar. This will meet all of the above requirements as well as providing the simplest and most fail safe means of solid engagement. Once screwed back together, the screw in point will have as high or higher mechanical properties as the fish below it. Screwing in will allow pump pressures equal to that of the string to be applied to the fish. Other than the jars, the ID of the string will not have changed after screwing in.

Screwing in is the only option when down jars will be used due to the releasing mechanism of most engaging tools. Most engaging tools require over pull to maintain a good bite on the fish, which makes it very difficult to work torque down to the fish. The only drawback to screwing in is that a wire line string shot back-off becomes the surest method of backing off at the screw in point.

7.5.1 Cutlip Screw-in Sub

When a fish is partially buried in the wall or partially behind a ledge or even against the high side of the hole, a cut lip sub can be used to pull the connection over to align with the screw-in the assembly. Due to the cutting of the threaded area, this sub is not API approved. If a fish is in a gauged section of a hole in medium to hard rock formations, with low borehole deviation (~5° or less), a joint of drill pipe, or a drill collar may be used to screw into the fish. A cutlip screw-in sub is shown in Figure 7.7.

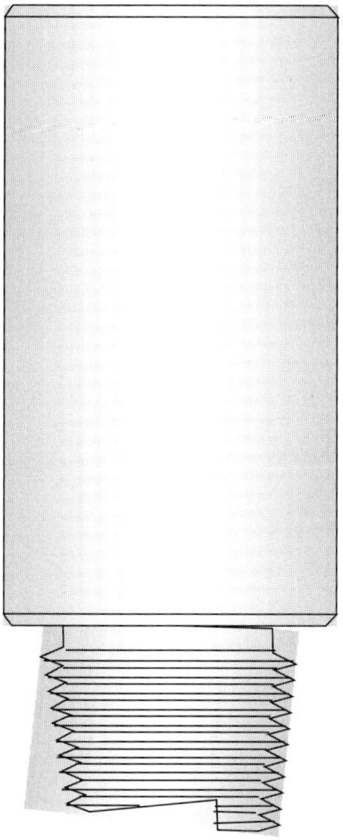

FIGURE 7.7 Cutlip Screw in Sub. (Courtesy Weatherford International)

7.5.2 Skirted Screw-in Assembly

A skirted screw-in assembly gives an operator all the hole sweeping advantages of an overshot with the strength, torque and circulating ability of a screw-in sub. The skirt can be of an OD that will not allow the skirted assembly to pass the fish. The cut lip makes it possible to pull the fish inside the skirt.

Skirted assemblies are most often made up from a washpipe triple connection bushing and a blank rotary shoe. The shoe can then be cut into the desired shape, cut lip or wallhook. A Weatherford skirted cutlip screw-in assembly is shown in Figure 7.8.

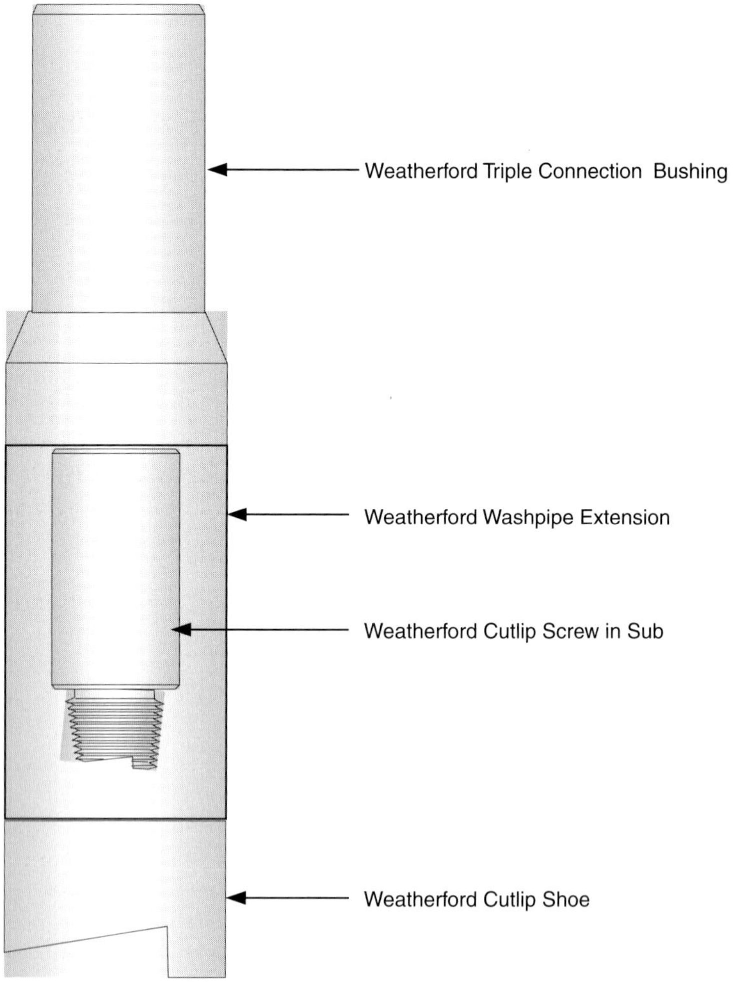

FIGURE 7.8 Skirted Cutlip Screw in Assembly. (Courtesy Weatherford International)

7.5.3 External Engaging Devices

The second choice for engaging a fish is with an external catch device (when it meets the strength requirements). An external engaging device usually will not restrict the ID of the fishing assembly, thus, leaving wireline options open and allowing full circulation. An external catch device can also be of sufficient OD to eliminate the possibility of passing the fish.

7.5.4 Series 150 Releasing and Circulating Overshot

The most-popular external catch device is the Series 150 releasing and circulating overshot. The tool was invented over 60 years ago and has not changed from the original design. This overshot provides the strongest means available to externally engage, pack off, and pull a fish when screwing in is not possible. Its basic simplicity, rugged construction, and availability make this tool the standard for all external catch fishing tools.

The Series 150 is designed to engage, pack off, and grapple (and retrieve) a specific size of tubing, pipe, coupling, tool joint, drill collar or smooth OD tool. Through the installation of the proper parts, these tools may be adapted to engage and pack off a wide range of diameters for each overshot size. Series 150 overshots are available in Full Strength or Slim Hole types. The overshot consists of three outside components: the top sub, bowl and guide. The basic overshot may be addressed with either of two sets of internal parts, depending on whether the fish to be retrieved is near the maximum catch size for the particular overshot. If the fish diameter is near the maximum retrieval for the overshot, a spiral grapple spiral grapple control, and type A packer are used. If the fish is a tubing collar, a type D collar pack-off assembly will be used. If the fish diameter is below the maximum retrieval size (by approximately $\frac{1}{2}$ in.), then basket grapple and mill control packer are used. Both the spiral and basket grapple are the engaging mechanism. The spiral control and basket grapple control locks the torque in the grapple but still allows the grapple to move up and down inside the bowl.

The bowl of the overshot is designed with a helical tapered spiral section within its inside diameter. The gripping mechanism (the grapple) is fitted into this section. When an upward pull is exerted against a fish, a contraction strain is spread evenly over a long section of the fish. The catch range is from $\frac{1}{32}$ to $\frac{1}{16}$ in. above and below the specified catch range. The overshot is engaged by slacking off over the fish, then a straight pull will engage the overshot. As long as there is an over pull, either right or left hand torque may be applied. To release the Bowen overshot, the string is bumped or jarred down on to break the freeze between the overshot bowl and the grapple. Once the freeze has been broken, rotation to the right while slowly picking up on the string will walk the grapple off the fish. A Series 150 releasing and circulating overshot is shown in Figure 7.9.

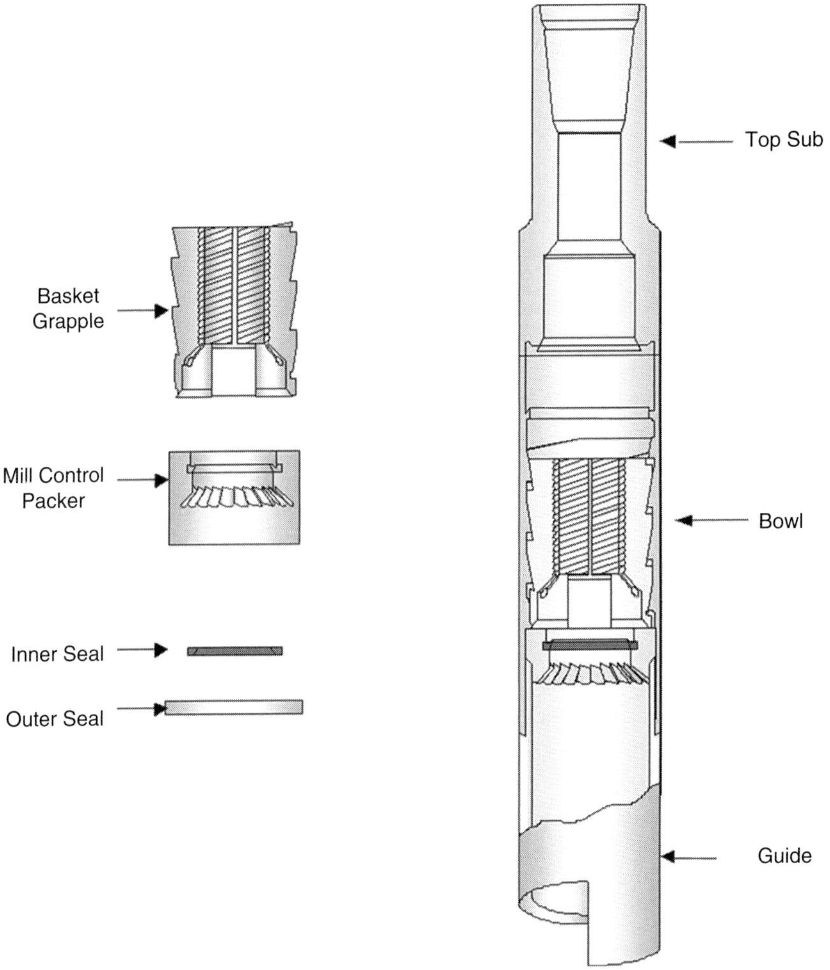

FIGURE 7.9 Series "150" Releasing and Circulating Overshot. (Courtesy Weatherford International)

Tables 7.1 and 7.2 give specific restrictions for overshot sizes in openhole and cased hole environments, respectively.

7.5.5 High-Pressure Pack-Off

The mill control and Type A packers used with the overshot have a limited pressure rating. If high pressures will be required or there are well control concerns, it is advisable to run a high pressure pack-off. The

TABLE 7.1 Recommended Drill Pipe and Drill Collars For Open Hole and Cased Hole

API Casing		Drill Pipe Recommended In Casing		Drill Pipe Recommended In Open Hole		Drill Collars Recommended In Cased Open Hole		Washpipe Recommended For Fishing			Overshot Recommended For Fishing	
Size & Wt.	Bit Size	DP Size	TJ OD	DP Size	Max TJ OD	Min OD	Max OD	OD; Conn	Weight	I.D.	OD	Max Catch
$4\frac{1}{2}''$ 9.50#–11.6	$3\frac{7}{8}''$	$2\frac{7}{8}''$ PAC	$3\frac{1}{8}''$	$2\frac{7}{8}''$ PAC	$3\frac{1}{8}''$	$2\frac{7}{8}''$	$3\frac{1}{16}''$	$3\frac{3}{4}''$ FJWP	9.35#	3.250″	$3\frac{3}{4}''$	$3\frac{1}{16}''$
$4\frac{1}{2}''$ 13.50#	$3\frac{3}{4}''$	$2\frac{7}{8}''$ PAC	$3\frac{1}{8}''$	$2\frac{7}{8}''$ PAC	$2\frac{1}{2}''$	$2\frac{7}{8}''$	$3\frac{1}{16}''$	*$3\frac{3}{4}''$ FJWP	9.35#	3.250″	$3\frac{3}{4}''$	$3\frac{1}{16}''$
$4\frac{1}{2}''$ 15.10#	$3\frac{5}{8}''$	$2\frac{3}{8}''$ WFJ	$2\frac{1}{2}''$	$2\frac{3}{8}''$ WFJ	$2\frac{3}{8}''$	$2\frac{1}{4}''$	$2\frac{1}{2}''$	$3\frac{1}{2}''$ FJWP	8.81#	2.992″	$3\frac{3}{8}''$	$2\frac{1}{2}''$
5″ 11.50#–15#	$4\frac{1}{4}''$	$2\frac{7}{8}''$ PAC	$3\frac{1}{8}''$	$2\frac{7}{8}''$ PAC	$3\frac{1}{8}''$	3″	$3\frac{1}{8}''$	4″ FJWP	11.34#	3.428″	$4\frac{1}{8}''$	$3\frac{1}{8}''$
5″ 18#	$4\frac{1}{8}''$	$2\frac{7}{8}''$ PAC, $2\frac{3}{8}''$ SLH90	$3\frac{1}{8}''$	$2\frac{7}{8}''$ PAC, $2\frac{3}{8}''$ SLH90	$3\frac{1}{8}''$	3″	$3\frac{1}{8}''$	$3\frac{3}{4}''$ FJWP	9.35#	3.250″	$3\frac{7}{8}''$	$3\frac{1}{8}''$
5″ 29.3#–24.2#	4″–$3\frac{7}{8}''$	$2\frac{7}{8}''$ PAC, $2\frac{3}{8}''$ SLH90	$3\frac{1}{8}''$	$2\frac{7}{8}''$ PAC, $2\frac{3}{8}''$ SLH90	$3\frac{1}{8}''$	$2\frac{7}{8}''$	$3\frac{1}{16}''$	$3\frac{3}{4}''$ FJWP	9.35#	3.250″	$3\frac{3}{4}''$	$3\frac{1}{16}''$
$5\frac{1}{2}''$ 13#–17#	$4\frac{3}{4}''$	$2\frac{7}{8}''$ RPO	$3\frac{7}{8}''$	$2\frac{7}{8}''$ RPO	$3\frac{7}{8}''$	$3\frac{1}{4}''$	$3\frac{1}{2}''$	$4\frac{1}{2}''$ FJWP	14.98#	3.826″	$4\frac{11}{16}''$	$3\frac{21}{32}''$
$5\frac{1}{2}''$ 13#–17#	$4\frac{3}{4}''$	$2\frac{7}{8}''$ SH	$3\frac{5}{8}''$	$2\frac{7}{8}''$ SH	$3\frac{5}{8}''$	$3\frac{1}{4}''$	$3\frac{1}{2}''$	$4\frac{1}{2}''$ FJWP	14.98#	3.826″	$4\frac{11}{16}''$	$3\frac{21}{32}''$
$5\frac{1}{2}''$ 20#	$4\frac{5}{8}''$	$2\frac{7}{8}''$ PAC	$3\frac{1}{8}''$	$2\frac{7}{8}''$ PAC	$3\frac{1}{8}''$	$3\frac{1}{4}''$	$3\frac{1}{2}''$	$4\frac{3}{8}''$ FJWP	13.58#	3.749″	$4\frac{9}{16}''$	$3\frac{21}{32}''$
$5\frac{1}{2}''$ 20#	$4\frac{5}{8}''$	$2\frac{7}{8}''$ RPO	$3\frac{7}{8}''$	$2\frac{7}{8}''$ RPO	$3\frac{7}{8}''$	$3\frac{1}{4}''$	$3\frac{1}{2}''$	$4\frac{3}{8}''$ FJWP	13.58#	3.749″	$4\frac{9}{16}''$	$3\frac{21}{32}''$
$5\frac{1}{2}''$ 23#–26#	$4\frac{1}{2}''$	$2\frac{7}{8}''$ IF	$3\frac{3}{8}''$	$2\frac{7}{8}''$ IF	$3\frac{3}{8}''$	3″	$3\frac{1}{4}''$	$4\frac{3}{8}''$ FJWP	13.58#	3.749″	$4\frac{3}{8}''$	$3\frac{3}{8}''$
$5\frac{1}{2}''$ 23#–26#	$4\frac{1}{2}''$	$2\frac{7}{8}''$ SH	$3\frac{1}{8}''$	$2\frac{7}{8}''$ SH	$3\frac{1}{8}''$	3″	$3\frac{1}{4}''$	$4\frac{3}{8}''$ FJWP	13.58#	3.749″	$4\frac{3}{8}''$	$3\frac{3}{8}''$
$6\frac{5}{8}''$ 32#	$5\frac{3}{8}''$	$2\frac{7}{8}''$ RPO	$3\frac{7}{8}''$	$2\frac{7}{8}''$ RPO	$3\frac{7}{8}''$	$3\frac{1}{2}''$	$4\frac{1}{8}''$	5″ FJWP	17.93#	4.276″	$5\frac{1}{8}''$	$4\frac{1}{4}''$
$6\frac{5}{8}''$ 32#	$5\frac{3}{8}''$	$2\frac{7}{8}''$ IF	$4\frac{1}{8}''$	$2\frac{7}{8}''$ IF	$4\frac{1}{8}''$	$3\frac{1}{2}''$	$4\frac{1}{8}''$	5″ FJWP	17.93#	4.276″	$5\frac{1}{8}''$	$4\frac{1}{4}''$
$6\frac{5}{8}''$ 32#	$5\frac{3}{8}''$	$3\frac{1}{2}''$ SH	$4\frac{1}{8}''$	$3\frac{1}{2}''$ SH	$4\frac{1}{8}''$	$3\frac{1}{2}''$	$4\frac{1}{8}''$	5″ FJWP	17.93#	4.276″	$5\frac{1}{8}''$	$4\frac{1}{4}''$
7″ 17#–23#–26#	$6\frac{1}{4}''$–$6\frac{1}{8}''$	$3\frac{1}{2}''$ IF, 4″ SH	$4\frac{3}{4}''$	$3\frac{1}{2}''$ IF, 4″ SH	$4\frac{3}{4}''$	$4\frac{9}{16}''$	$4\frac{3}{4}''$	$5\frac{3}{4}''$ FJWP	21.53#	5.000″	$5\frac{3}{4}''$	$4\frac{3}{4}''$
7″ 17#–23#–26#	$6\frac{1}{4}''$–$6\frac{1}{8}''$	$3\frac{1}{2}''$ IF, 4″ SH	$4\frac{3}{4}''$–$4\frac{7}{8}''$	$3\frac{1}{2}''$ IF, 4″ SH	$4\frac{7}{8}''$	$4\frac{9}{16}''$	5″	6″ FJWP	22.81#	5.240″	$5\frac{7}{8}''$	5″
7″ 29#–32#	6″	$3\frac{1}{2}''$ IF, 4″ SH	$4\frac{3}{4}''$	$3\frac{1}{2}''$ IF, 4″ SH	$4\frac{3}{4}''$	$4\frac{9}{16}''$	$4\frac{3}{4}''$	$5\frac{3}{4}''$ FJWP	21.53#	5.000″	$5\frac{3}{4}''$	$4\frac{3}{4}''$
7″ 35#	$5\frac{7}{8}''$	$3\frac{1}{2}''$ IF, 4″ SH	$4\frac{3}{4}''$	$3\frac{1}{2}''$ IF, 4″ SH	$4\frac{3}{4}''$	$4\frac{1}{4}''$	$4\frac{1}{2}''$	$5\frac{1}{2}''$ FJWP	16.87#	4.892″	$5\frac{3}{4}''$	$4\frac{3}{4}''$
7″ 38#	$5\frac{3}{4}''$	$3\frac{1}{2}''$ SH	$4\frac{1}{8}''$	$3\frac{1}{2}''$ SH	$4\frac{1}{8}''$	$4\frac{1}{8}''$	$4\frac{1}{4}''$	$5\frac{3}{8}''$ FJWP	18.93#	4.670″	$5\frac{9}{16}''$	$4\frac{1}{4}''$
$7\frac{5}{8}''$ 20#–33.7#	$6\frac{3}{4}''$–$6\frac{5}{8}''$	$3\frac{1}{2}''$ IF, 4″ SH	$4\frac{3}{4}''$–$4\frac{7}{8}''$	$3\frac{1}{2}''$ IF, 4″ SH	$4\frac{7}{8}''$	$4\frac{9}{16}''$	$5\frac{1}{4}''$	$6\frac{3}{8}''$ FJWP	24.03#	5.625″	$6\frac{3}{8}''$	$5\frac{1}{4}''$
$7\frac{5}{8}''$ 20#–33.7#	$6\frac{3}{4}''$–$6\frac{5}{8}''$	4″ FH	$5\frac{1}{4}''$	4″ FH	$5\frac{1}{4}''$	$4\frac{9}{16}''$	$5\frac{1}{4}''$	$6\frac{3}{8}''$ FJWP	24.03#	5.265″	$6\frac{3}{8}''$	$5\frac{1}{4}''$
$7\frac{5}{8}''$ 39#	$6\frac{1}{2}''$–$6\frac{3}{8}''$	$3\frac{1}{2}''$ IF, 4″ SH	$4\frac{3}{4}''$–$4\frac{7}{8}''$	$3\frac{1}{2}''$ IF, 4″ SH	$4\frac{7}{8}''$	$4\frac{9}{16}''$	5″	6″ FJWP	22.81#	5.240″	$5\frac{7}{8}''$	5″
$8\frac{5}{8}''$ 24#–40#	$7\frac{7}{8}''$ – $\frac{7}{58}''$	$4\frac{1}{2}''$ XH	$6\frac{1}{4}''$	$4\frac{1}{2}''$ XH	$6\frac{1}{4}''$	6″	$6\frac{1}{4}''$	$7\frac{3}{8}''$ FJWP	28.04#	6.625″	$7\frac{3}{8}''$	$6\frac{1}{4}''$
$8\frac{5}{8}''$ 44#–49#	$7\frac{1}{2}''$–$7\frac{3}{8}''$	4″ FH	$5\frac{1}{4}''$	4″ FH	$5\frac{1}{4}''$	$5\frac{1}{2}''$	6″	7″ FJWP	25.66#	6.276″	$7\frac{1}{8}''$	6″
$9\frac{5}{8}''$ 29.3# 36#	$8\frac{3}{4}''$	$4\frac{1}{2}''$ XH, 5″ XH	$6\frac{1}{4}''$–$6\frac{5}{8}''$	$4\frac{1}{2}''$ XH, 5″ XH	$6\frac{5}{8}''$	6″	7″	$8\frac{1}{8}''$ FJWP	35.92#	7.250″	$8\frac{1}{8}''$	7″
$9\frac{5}{8}''$ 40#–43.5#	$8\frac{5}{8}''$	$4\frac{1}{2}''$ XH, 5″ XH	$6\frac{1}{4}''$–$6\frac{5}{8}''$	$4\frac{1}{2}''$ XH, 5″ XH	$6\frac{5}{8}''$	6″	7″	$8\frac{1}{8}''$ FJWP	35.92#	7.250″	$8\frac{1}{8}''$	7″
9.5/8″ 47#	$8\frac{1}{2}''$	$4\frac{1}{2}''$ XH, 5″ XH	$6\frac{1}{4}''$–$6\frac{5}{8}''$	$4\frac{1}{2}''$ XH, 5″ XH	$6\frac{5}{8}''$	6″	7″	$8\frac{1}{8}''$ FJWP	35.92#	7.250″	$8\frac{1}{8}''$	7″
$9\frac{5}{8}''$ 53.5#	$8\frac{3}{8}''$	$4\frac{1}{2}''$ XH, 5″ XH	$6\frac{1}{4}''$–$6\frac{5}{8}''$	$4\frac{1}{2}''$ XH, 5″ XH	$6\frac{5}{8}''$	6″	7″	$8\frac{1}{8}''$ FJWP	35.92#	7.250″	$8\frac{1}{8}''$	7″

Note: The chart above is a guideline only and should be used as a rule of thumb; the final decision is that of customer; 4″ pin-up drill pipe is an option inside 7″ casing, discuss with fishing tool manager for guidance.

TABLE 7.2 Cased Hole Fishing: Recommended Overshot/Washpipe/Cutter

Casing Size-Wt.	Pipe Size-Wt.	Connection Type O.D.	Overshot Size-Type	Maximum Catch	Washpipe Size	Maximum Washover	External Cutter Type/Size	Maximum Cutter Clearance
$4\frac{1}{2}''$ 9.50–11.60#	$2\frac{3}{8}''$ 4.70#	8RD, $3\frac{1}{16}''$	$3\frac{3}{4}''$ S.H.	$3\frac{1}{16}''$	$3\frac{3}{4}''$	$3\frac{1}{8}''$	Bowen Mech. $3\frac{7}{8}''$	$3\frac{1}{16}''$
$4\frac{1}{2}''$ 13.50#	$2\frac{3}{8}''$ 4.70#	8RD, $3\frac{1}{16}''$	$3\frac{3}{4}''$ S.H.	$3\frac{1}{16}''$	$3\frac{3}{4}''$	$3\frac{1}{8}''$	Turn Down $3\frac{7}{8}''$ ($3\frac{3}{4}''$)	$3\frac{1}{16}''$
$4\frac{1}{2}''$ 15.10#	$2\frac{3}{8}''$ 4.70#	8RD, $3\frac{1}{16}''$	$3\frac{5}{8}''$ X.S.H.	$2\frac{7}{8}''$	$3\frac{1}{2}''$	$2\frac{5}{8}''$	None	—
$5''$ 11.50–23.20#	$2\frac{3}{8}''$ 4.70#	8RD, $3\frac{1}{16}''$	$3\frac{3}{4}''$ S.H.	$3\frac{1}{16}''$	$3\frac{3}{4}''$	$3\frac{1}{8}''$	Bowen Mech. $3\frac{7}{8}''$	$3\frac{1}{16}''$
$5\frac{1}{2}''$ 14.17#	$2\frac{3}{8}''$ 4.70#	8RD, $3\frac{1}{16}''$	$4\frac{11}{16}''$ F.S.	$3\frac{21}{32}''$	$4\frac{1}{2}''$	$3\frac{3}{4}''$	H.E. $4\frac{9}{16}''$	$3\frac{3}{8}''$
$5\frac{1}{2}''$ 14.17#	$2\frac{7}{8}''$ 6.50#	8RD, $3\frac{21}{32}''$	$4\frac{11}{16}''$ F.S.	$3\frac{21}{32}''$	$4\frac{1}{2}''$	$3\frac{3}{4}''$	Bowen $4\frac{11}{16}''$	$3\frac{3}{4}''$
$5\frac{1}{2}''$ 20.00#	$2\frac{3}{8}''$ 4.70#	8RD, $3\frac{1}{16}''$	$4\frac{9}{16}''$ S.H.	$3\frac{21}{32}''$	$4\frac{3}{8}''$	$3\frac{5}{8}''$	Bowen Hyd. $4\frac{9}{16}''$	$3\frac{9}{16}''$
$5\frac{1}{2}''$ 20.00#	$2\frac{7}{8}''$ 6.50#	8RD, $3\frac{21}{32}''$	$4\frac{9}{16}''$ S.H.	$3\frac{21}{32}''$	$4\frac{3}{8}''$	$3\frac{5}{8}''$	Bowen Hyd. $4\frac{9}{16}''$	$3\frac{9}{16}''$
$5\frac{1}{2}''$ 23.00#	$2\frac{3}{8}''$ 4.70#	8RD, $3\frac{1}{16}''$	$4\frac{1}{4}''$ F.S.	$3\frac{1}{8}''$	$4\frac{3}{8}''$	$3\frac{5}{8}''$	Bowen Mech. $3\frac{7}{8}''$	$3\frac{1}{16}''$
$5\frac{1}{2}''$ 23.00#	$2\frac{7}{8}''$ 6.50#	8RD, $3\frac{21}{32}''$	$4\frac{3}{8}''$ S.H.	$3\frac{1}{2}''$	$4\frac{3}{8}''$	$3\frac{5}{8}''$	None	—
$7''$ 20–38#	$2\frac{7}{8}''$ 6.50#	8RD, $3\frac{21}{32}''$	$5\frac{3}{4}''$ F.S.	$4\frac{3}{4}''$	$5\frac{1}{2}''$	$4\frac{3}{4}''$	H.E. $5\frac{9}{16}''$	$4''$
$7''$ 20–38$''$	$3\frac{1}{2}''$ 9.30#	8RD, $4\frac{1}{2}''$	$5\frac{3}{4}''$ F.S.	$4\frac{3}{4}''$	$5\frac{1}{2}''$	$4\frac{3}{4}''$	Bowen Mech. $5\frac{7}{8}''$	$4\frac{1}{2}''$
$7\frac{5}{8}''$ 26.40–45.30#	$3\frac{1}{2}''$ 9.30#	8RD, $4\frac{1}{2}''$	$5\frac{3}{4}''$ F.S.	$4\frac{3}{4}''$	$5\frac{1}{2}''$	$4\frac{3}{4}''$	Bowen Mech. $5\frac{7}{8}''$	$4\frac{1}{2}''$
$7\frac{5}{8}''$ 26.40–45.30#	$3\frac{1}{2}''$ 13.30#	IF, $4\frac{3}{4}''$	$5\frac{3}{4}''$ F.S.	$4\frac{3}{4}''$	$5\frac{1}{2}''$	$4\frac{3}{4}''$	Bowen Mech. $6\frac{1}{4}''$	$4\frac{3}{4}''$
$8\frac{5}{8}''$ 24–40#	$4''$ 14.00#	FH, $5\frac{1}{4}''$	$7\frac{3}{8}''$ F.S.	$5\frac{3}{4}''$	$7\frac{3}{8}''$	$6\frac{1}{2}''$	Bowen Mech $7\frac{5}{8}''$	$6\frac{1}{4}''$
$8\frac{5}{8}''$ 24–40#	$4\frac{1}{2}''$ 16.60#	XH, $6\frac{1}{4}''$	$7\frac{3}{8}''$ S.H.	$6\frac{1}{4}''$	$7\frac{3}{8}''$	$6\frac{1}{2}''$	Bowen Mech $7\frac{5}{8}''$	$6\frac{1}{4}''$
$8\frac{5}{8}''$ 44–49#	$4''$ 14.00#	FH, $5\frac{1}{4}''$	$7\frac{3}{8}''$ S.H.	$6\frac{1}{4}''$	$7''$	$6\frac{1}{8}''$	None	—
$8\frac{5}{8}''$ 44–49#	$4\frac{1}{2}''$ 16.60#	XH, $6\frac{1}{4}''$	$7\frac{3}{8}''$ S.H.	$6\frac{1}{4}''$	$7\frac{3}{8}''$	$6\frac{1}{2}''$	None	—
$9\frac{5}{8}''$ 36–53.50#	$4\frac{1}{2}''$ 16.60#	XH, $6\frac{1}{4}''$	$8\frac{1}{8}''$ F.S.	$6\frac{5}{8}''$	$8\frac{1}{8}''$	$7\frac{1}{16}''$	Bowen Mech. $8\frac{1}{8}''$	$6\frac{1}{2}''$

Tubing Dimensions	Tube O.D.	Upset O.D.	Coupling O.D.
$2\frac{3}{8}''$ 4.70# 8RD EUE	$2\frac{3}{8}''$	$2\frac{19}{32}''$	$3\frac{1}{16}''$
$2\frac{7}{8}''$ 6.50# 8RD EUE	$2\frac{7}{8}''$	$3\frac{3}{32}''$	$3\frac{21}{32}''$
$3\frac{1}{2}''$ 9.30# 8RD EUE	$3\frac{1}{2}''$	$3\frac{3}{4}''$	$4\frac{1}{2}''$
$4''$ 11.00# 8RD EUE	$4''$	$4\frac{1}{4}''$	$5''$

pack-off is run directly above the overshot bowl. Chevron packing is used for sealing. This type of assembly can be used as a permanent tubing patch. The pack-off is good for approximately 5,000 psi.

7.5.6 Oversize Cutlip Guide

The oversize guide is used when the fish OD in relation to the hole ID makes it possible for the overshot to bypass the fish. The oversize guide may be used to better utilize inventory by keeping a smaller selection of overshots. But care should be taken not to sacrifice strength to do this. An oversize cutlip guide is shown in Figure 7.10.

7.5.7 Wallhook Guide

The wallhook guide is used with a bent joint of pipe or hydraulic knuckle joint to sweep a washed out section of hole when a fish cannot be pulled into the overshot with a standard cutlip guide. The assembly is made up so the lip of the wallhook has the maximum possible reach. Once the overshot has passed the top of the fish, the string is slowly rotated until torque indicates the wallhook is madeup to the fish. The string is then elevated until the torque is lost. The fish should be in a position that it can be engaged by slacking off on the overshot. A wallhook guide is shown in Figure 7.10.

7.5.8 Hollow Mill Container and Hollow Mill

The hollow mill container with a hollow mill insert is used to dress off the top of a fish that has been damaged. This devise is used to engage the

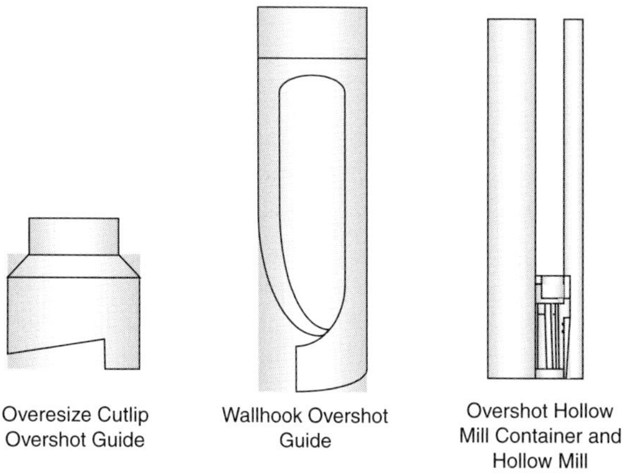

FIGURE 7.10 Oversize Cutlip Guide, Wallhook Guide and Hollow Mill Container with Hollow Mill. (Courtesy Weatherford International)

fish in one trip. This accessory is most often used when a fish top is flared after a jet cut or in the event a twist off has occurred. This assembly is run directly below the overshot bowl. A hollow mill container and hollow mill device is shown in Figure 7.10.

7.5.9 Bowen Series 70 Short Catch Overshot

Series 70 short catch overshots are specifically designed to engage the exposed portion of a fish too short to be caught with a conventional overshots. The short catch overshot consists of a top sub, a bowl, a basket grapple and a basket grapple control. The grapple is inserted in the top end of the overshot and is positioned at the extreme lowered end of the bowl. Spiral grapples, mill control packers and Type A packers cannot be used with a short catch overshot. A Series 70 short catch overshot is shown in Figure 7.11.

7.5.10 Internal Engaging Devices

If the fish must be engaged internally, releasable spears are used. Spears often restrict the ID of the fishing assembly. There is no pack off system built into the tool but certain accessories may be used to accomplish circulation. The Itco Type Bowen releasing spear is designed to assure positive engagement with a fish. Built to withstand severe jarring and pulling strains, it engages the fish over a large area without damage to the fish. If the fish cannot be pulled, the spear may be released. The Itco type Bowen releasing spear consists of a mandrel, grapple, and a locking ring and nut. The mandrel may be a shoulder type or flush type. The flexible one-piece grapple has an internal helix which mates with the helix on the mandrel. The tang of the grapple rests against a stop on the mandrel when the spear is in the engaged position. A Bowen Itco releasing spear is shown in Figure 7.12.

7.5.11 Box Taps and Taper Taps

Box taps are used to engage the outside diameter of a fish, usually in a situation where conventional releasing overshots would not be a feasible option. They are designed to tap threads into the steel tubing (or other steel downhole tools) so that the fish can be retrieved. Standard box taps are made with an integral guide on the lower end to insure proper engagement to the fish even when the fish is not centralized in the hole.

Taper taps are used to engage the inside of a fish where conventional releasing spears would not be feasible. Like the box tap, the taper tap

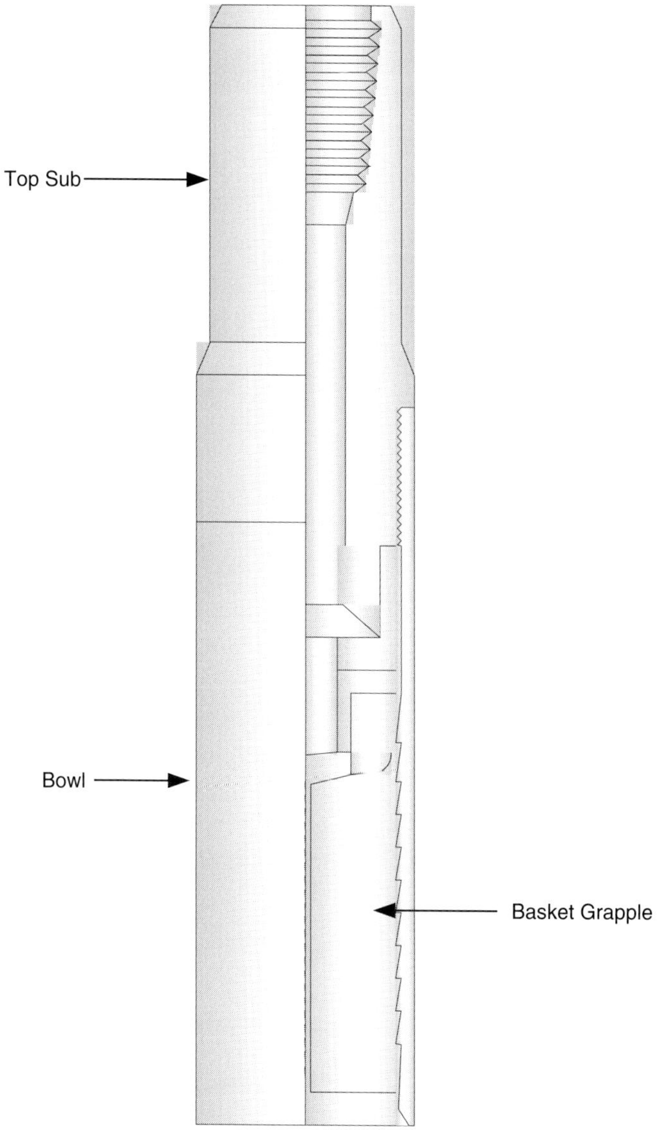

FIGURE 7.11 Series "70" Short Catch Overshot. (Courtesy Bowell Tool Co.)

is designed to cut threads where no threads existed. Standard taps are machined from high quality heat treated steel, with buttress type threads that are carborized for hardness. They can be ordered with right or left hand wickers. Box and taper taps can be used to fish tubulars, packers, or other

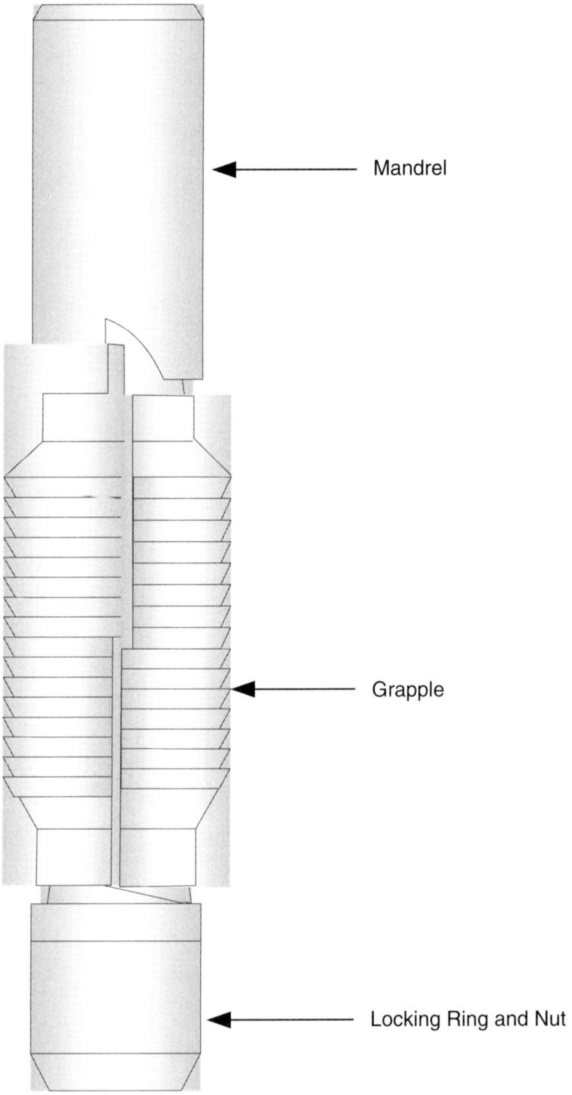

FIGURE 7.12 ITCO Type Releasing Spear. (Courtesy Bowell Tool Co.)

types of downhole equipment that cannot be fished by conventional means. They have the widest catch range of any fisting tool. Since all taps are non-releasable, a safety joint should always be run just above them. A box tap and taper tap are shown in Figure 7.13.

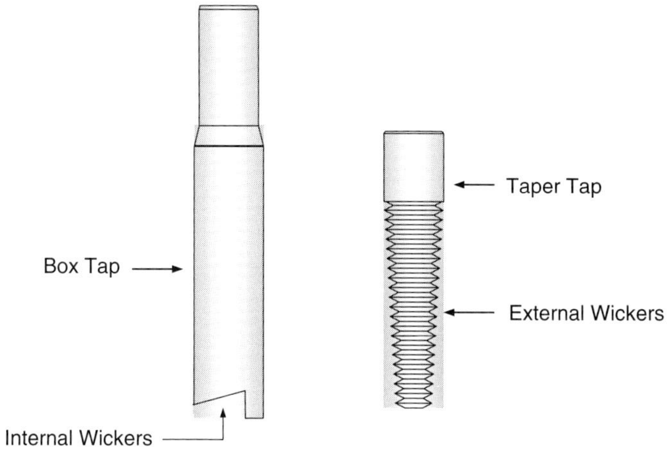

FIGURE 7.13 Box Tap and Taper Tap. (Courtesy Weatherford International)

7.6 FISHING FOR JUNK

Junk is all that other devices that may fall into the hole or become detached in the hole that are not tubulars. The type of junk retrieval device to be used in the fishing operation will depend on the properties of the junk (to be milled up or recovered). There are many tools designed to recover junk, and often an operational practice preference will determine which tool is used.

7.6.1 Poor Boy Junk Basket

One of the simplest fishing tools is the poor boy junk basket. It is run in the hole, circulation established, then the tool is slowly rotated down over the fish. If the basket is nearly hole size, its finger like catchers will gather junk toward the center of hole. When weight is applied the basket will bend inward to trap the junk inside. It is most effective for a small solid mass lying loose on bottom, such as a single bit cone. A poor boy junk basket is shown in Figure 7.14.

7.6.2 Boot Basket

A boot basket is run just above a bit or mill to collect a variety of small pieces of junk. The boot basket design traps junk by producing a sudden decrease in annular velocity when cuttings (with entrained junk) pass the larger OD of the boot and reach the smaller OD of the mandrel and top connection. A boot basket is shown in Figure 7.14.

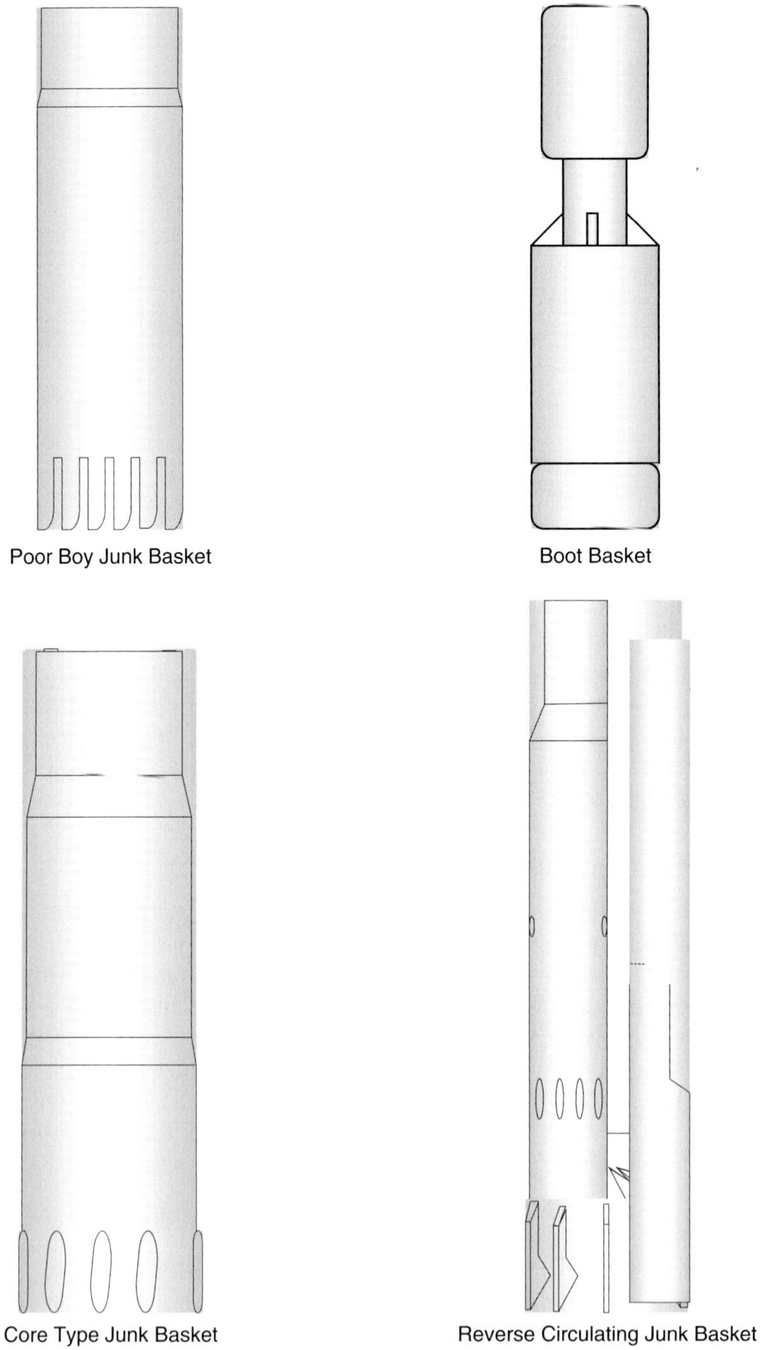

FIGURE 7.14 Poor Boy, Boot, Core Type and Reverse Circulating Junk Baskets. (Courtesy Weatherford International)

7.6.3 Core Type Junk Basket

The core-type junk basket is used to retrieve junk such as bit cones that may or may not be embedded in the wall or bottom of the rock formation. A milling shoe is made up on the bottom of the tool. After it is run nearly to bottom, the mud is circulated at reduced pressure and the tool is slowly rotated and lowered to make contact with the junk. Weight is gradually increased on the shoe as it cuts away the edges of the junk and the formation. The junk is forced into the barrel and a short core is cut. Rotation and circulation are stopped, torque is released from the drill string, and the working string is raised to break the core. Catchers inside the basket hold the core and the junk as it is tripped out of the hole. A core type junk basket is shown in Figure 7.14.

7.6.4 Jet Powered Junk Baskets and Reverse Circulating Junk Baskets

One of the most common types of junk baskets is the venturi jet basket. The basket is run into the hole and fluid is circulated to clean all cuttings and debris off the junk. A ball is then dropped down the pipe and seated in a valve to produce reverse circulation. The circulating fluid is then jetted outward and downward against the full circumference of the hole where it is deflected in a manner that directs all annular flow below this point into the shoe. Along with drilling fluid (mud), small pieces of junk are sucked up into the barrel where pieces are caught by catcher fingers. A reverse circulating junk basket is shown in Figure 7.14.

7.6.5 Hydrostatic Junk Baskets

The hydrostatic junk basket derives its actuating forces from circulating fluid pressure differential. The tool and drill string are lowered into the hole nearly empty of fluid. Once on bottom with the junk basket over the junk, application of weight on the string (downward) trips a valve allowing drilling fluids under great hydrostatic pressure to enter the empty drill string carrying debris (junk) with it. The effect is the same as lowering an empty straw, its top covered with a finger, into a glass of water. When the finger is removed, water rushes into the straw. The hydrostatic junk retriever can be opened and closed many times before being removed from the hole.

7.6.6 Milling Tools

Junk mills are designed to mill bit cones, reamer blades, or any other junk which may obstruct the wellbore. Many milling tools are fabricated for a specific milling operation. Some mills are manufactured from a solid piece of AISI 4140 heat-treated steel. Others are fabricated by welding the blades and stabilizer pads on to a simple tubular body. Although these tools are quite simple in appearance, the junk mill requires thought and experience to obtain the desired junk removal results. These tools should be designed to be versatile and to withstand spudding, heavy weights, and fast rotation. Factors that affect milling rates and the design of the mill are the type of fish, being milled, the stability of the fish, and its hardness.

7.6.7 Mill Designs

The concave design centers the mill over the fish and is less likely to sidetrack. The cone buster junk mill is a very popular type of mill. It is used to mill loose junk, or dress off the top of a fish. The Type P mill is a very aggressive mill. It has a large throat (rather than smaller water courses). The bladed junk mill is used to mill up packers and will cut faster through sand, cement or fill. String mills are used to enlarge windows in a whipstock operation, or to mill out a collapsed casing. Tapered mills are used to dress liner tops, smooth out perforations, and are sometimes run with pilot mills to stabilize and keep the casing clear. Pilot mills are used to mill up casing, drill pipe or liners. They perform better if the fish is stabilized (to reduce vibration and chatter). Pilot milling should be treated as a machining process. Shock subs are usually run directly above the mill to reduce vibration. A section mill is used to cut windows in the casing to facilitate a sidetrack operation. They can also be used to mill out internal casing patches or to cut pipe. Junk, string, taper, pilot, and section mills are shown in Figure 7.15.

7.6.8 Impression Block

One method that is sometimes used to access the condition of the top of the fish is to run an Impression block. A typical impression block consists of a block of lead that is molded into a circular steel body. The block is made up on drill collars and run into the hole until it just above the fish. Circulation is started to wash all fill off the top of the fish so that a good impression can be obtained. The block is lowered gently to touch the fish and weight is applied. The top of the fish indents the bottom of the soft lead, leaving an impression that can be examined and measured at the surface.

7.6.9 Fishing Magnets

Ferrous metallic junk can often be retrieved using a fishing magnet that is made up to the bottom of a working string. This tool is a powerful permanent magnet that has ports for circulation and a nonmagnetic brass sleeve housing to prevent junk from clinging to the side of the magnet. A fishing magnet is lowered into the hole with circulation to wash cuttings off the

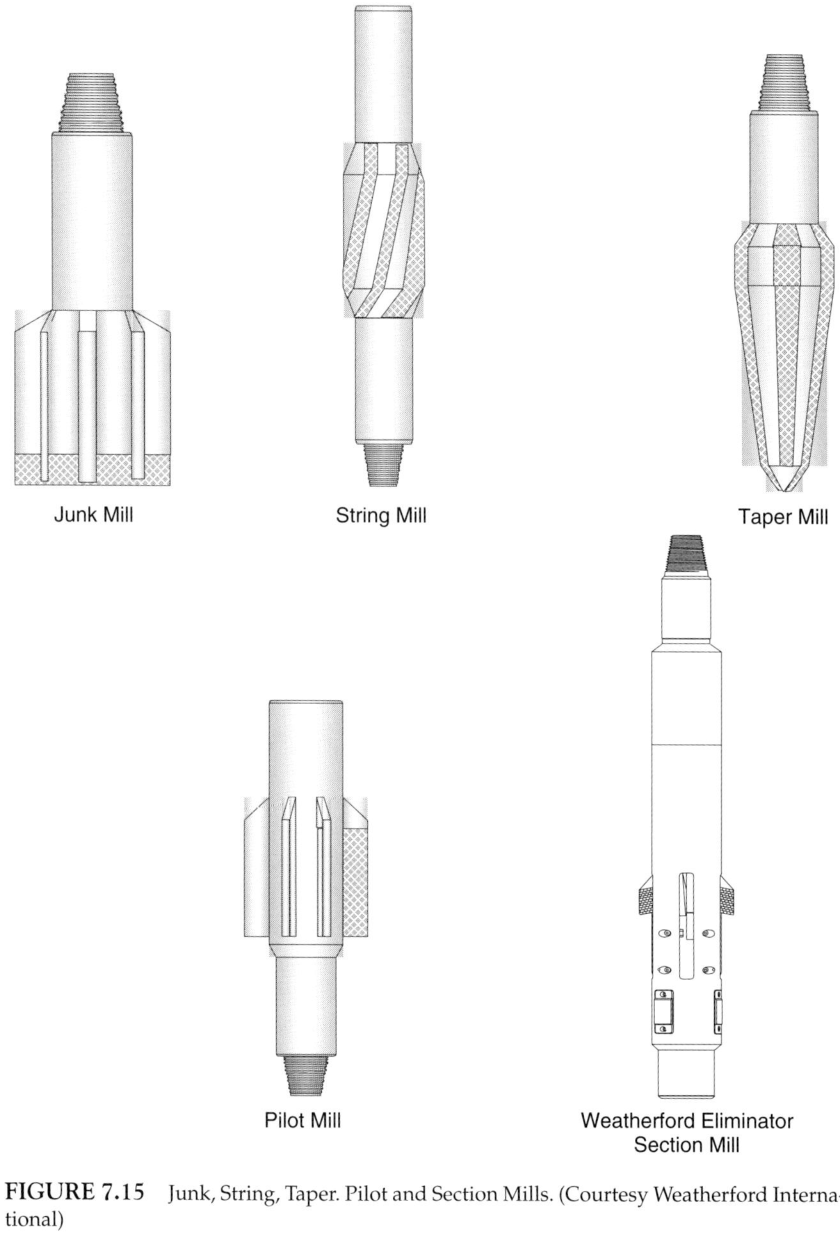

FIGURE 7.15 Junk, String, Taper. Pilot and Section Mills. (Courtesy Weatherford International)

top of the fish. A cut lip skirt on the bottom helps to rake the junk into the face of the magnet for engagement and also prevents the junk from being knocked off while pulling out of the hole. Do not run excessive weight on a fishing magnet; it is not a drilling tool.

7.6.10 Junk Shot

When a large or oddly shaped piece of junk is in the wellbore and it cannot be retrieved using regular junk baskets, a junk shot may be used to break piece into smaller pieces. This junk shot tool contains a shaped charge. This tool must be run on collars and drill pipe to keep the force of its explosion directed downhole (at the junk). The tool is lowered to just above the junk and circulation is established to wash all fill off the junk. When the shaped charge is fired, its downward-directed blow breaks up the junk so that it can be finished with more conventional means.

7.7 ABANDONMENT

Well abandonment is a process to permanently shut-in a previously producing well in a safe and environmentally responsible way. This process is applied to both land and offshore wells. Obviously, offshore wells are the most complicated to permanently abandon. To abandon a well the casing is generally cut below the mud line. In the United States and its territorial waters, the depth below the mud line is defined in U.S. Federal Government regulations [5]. Other countries and regions of the world often use similar but different regulation sources. This type of operation was at one time the hardest perform. But with today's new recovery systems and new casing cutting technologies NPT has been greatly reduced, saving the operators thousands of dollars in rig time. Some companies have advanced wellhead recovery systems available on the market. An example of such tool systems is the Weatherford M.O.S.T. Tool System that cuts and retrieves the casing, wellhead, and the temporary guide base in a single trip. These advanced systems when used with a downhole mud motor for cutting can eliminate wellhead damage and confirm cuts without tripping pipe. There are two different types of tools; the deepwater version, and the standard version. The deepwater versions utilize a mud motor and can be used at all water depths and water currents. The standard version has three options for cutting the casing; rotary table, top drive, or a mud motor. When using the rotary table or top drive for cutting a marine swivel must be used. The rotary table or top drive options of cutting the pipe are used primarily for shallow water operations (less than 2,000 ft and water currents less than 3 knots). Using the standard version with a mud motor increases cutting efficiencies (all water depths and water currents). The deepwater and standard versions (using a marine swivel) are shown in Figures 7.16 and 7.17, respectively.

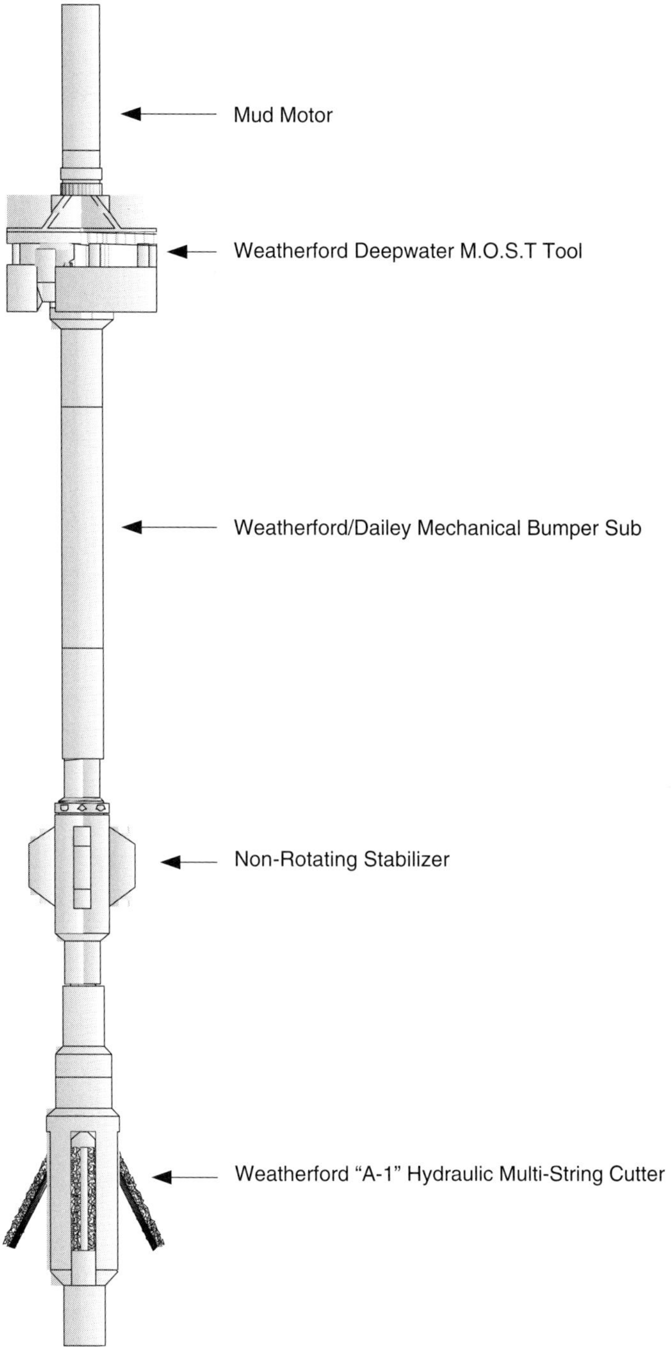

FIGURE 7.16 Deepwater MOST Tool. (Courtesy of Weatherford International)

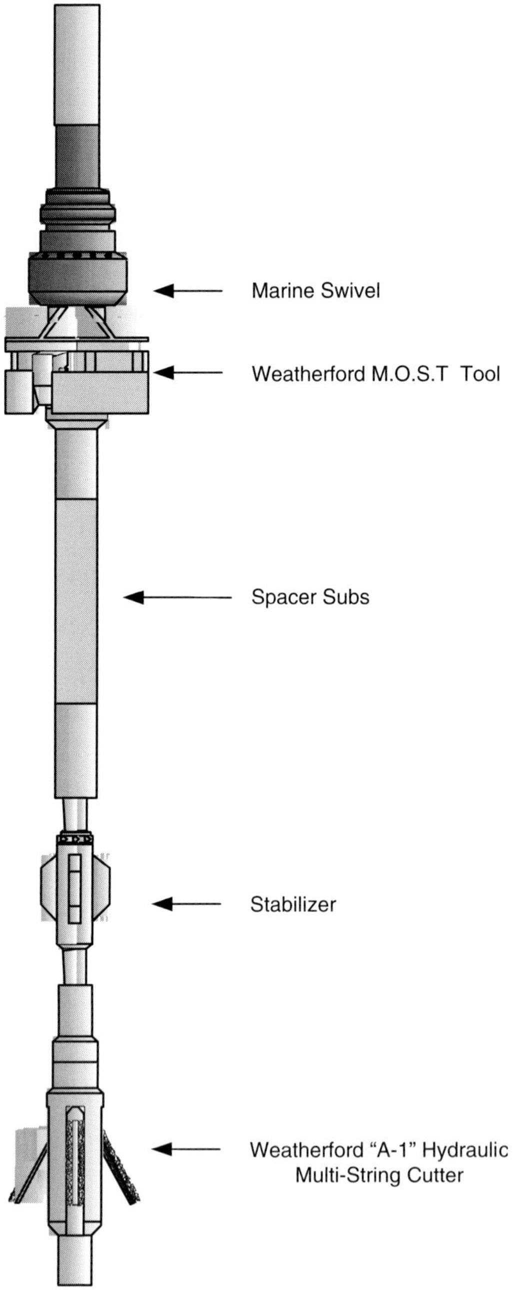

FIGURE 7.17 Tool with Marine Swivel. (Courtesy Weatherford International)

7.8 WIRELINES

Wirelines are utilized in a variety of operations in drilling and completions and production operations. During exploration and development drilling operations wireline logging operations, retrieval of bottomhole samples, and placement and retrieval of downhole tools are vital to the assessment of new reservoirs and the continued operation of producing reservoirs. The wireline subject has been concentrated for convenience in this Fishing and Abandonment section.

7.8.1 Wireline Construction

There are two basic categories of wirelines. All wirelines are anchored at one end at the surface with a wireline head that can be mounted on the drilling rig floor or on some other auxiliary equipment (e.g., logging truck bed) at the operational site. The free end of the wireline carries a load of some type of downhole tool (a deadload or tool for providing a downhole actuating force).

The first category of wirelines are made up of simple armored cables of various diameters and strengths that have no electrical conductor lines integrated into the overall construction. These wirelines are used to carry loads (e.g., special downhole tools) or otherwise provide actuating forces inside the wellbore. The second category of wirelines are an integrated construction with armored cables (for load carrying) surrounding a variety of electrical conductor lines.

The standard load carrying armored cables are special galvanized improved plow steel wires. The wires of the standard cables are coated with zinc. The tensile strength of each wire used in the standard cables are in the range of 270,000 to 300,000 psi. These standard load carrying cables are not recommended for operations in H_2S and CO_2 environments.

For operations in H_2S and CO_2 it is necessary to use load carrying armored cables constructed of stainless steel wires. Stainless steel wires have tensile strengths that are slightly below those of the standard wires discussed above (e.g., range of 240,000 to 270,000 psi). Thus, armored cables constructed with stainless steel are rated at slightly lower breaking strengths than the standard armored cables discussed above. Usually H_2S and CO_2 are accompanied by higher downhole temperatures. Higher operating borehole temperatures will result in decreased breaking strength for all armored cables.

All mechanical properties are for room temperature conditions.

7.8.2 Electrical Conductors

Most wireline manufactures rate their conductors cables at DC rather than AC. Rating a conductor cables for maximum AC voltage presents difficulties. AC voltage ratings are dependent on frequency and actual wave shape. Such an AC voltage can be characterized by an equivalent DC voltage that will provide the same amount of heating. In many applications, AC power is transmitted as alternating voltages and currents in a since or cosine wave variation in time. Also, 60 hertz is a common power frequency. If this is the method of transmission, then a 707-volt sine wave has 1000 volts peak voltage. Therefore, a 707-volt average voltage at a frequency of 60 hertz would correspond to 1000 volts DC. At higher frequencies there can be more stress on the dielectric due to heating. Generally, the higher the frequency, the lower the breakdown voltage.

All electrical data for resistance and capacitance are nominal values and have been corrected to 20°C.

Proper selection of armored cable materials and conductor insulation materials can allow wirelines to operate up to temperature of 600°F (316°C).

7.8.3 Simple Armored Wirelines

This category of wirelines is composed of a variety of armored wirelines that are used in numerous downhole, rig site and/or wellhead operations. These wirelines vary in diameters from $\frac{3}{16}$ in. (4.80 mm) to as high as $\frac{9}{16}$ in. (14.30 mm).

Table 7.3 gives the typical construction, dimensions, and mechanical properties for sand lines.

Table 7.4 gives the typical construction, dimensions, and mechanical properties for Swablines.

Table 7.5 gives the typical construction, dimensions, and mechanical properties of a variety of utility wirelines used in wellhead operations.

TABLE 7.3 Sand Lines

Type (Nominal Diameter)	$\frac{7}{16}''$ (11.11 mm)	$\frac{1}{2}''$ (12.70 mm)	$\frac{9}{16}''$ OS (14.29 mm)
CONSTRUCTION	6 × 7 (6/1)	6 × 7 (6/1)	6 × 7 (6/1)
Polypropylene fiber diameter	0.219″(5.56 mm)	0.250″(6.35 mm)	0.290″(7.40 mm)
Stranded wire diameter	0.142″(3.61 mm)	0.164″(4.17 mm)	0.190″(4.83 mm)
Center wire diameter	0.050″(1.27 mm)	0.056″(1.42 mm)	0.066″(1.68 mm)
Outer wire diameter	0.046″(1.17 mm)	0.054″(1.37 mm)	0.062″(1.57 mm)
Steel wire diameter	0.438″(11.11 mm)	0.500″(12.7 mm)	0.582″(14.8 mm)
MECHANICAL CHARACTERISTICS			
Breaking strength	16.5 Klb (7.5 Kkg)	22.3 Klb (10.1 Kkg)	29.8 Klb (13.5 Kkg)
Breaking strength	17.4 Klb (7.9 Kkg)	23.6 Klb (10.7 Kkg)	31.3 Klb (14.2 Kkg)
Weight	277.5 lb/Kft (413 kg/km)	375.6 lb/Kft (559 kg/km)	500.6 lb/Kft (745 kg/km)
Diameter tolerance	+4% – 0%	+4% – 0%	+4% – 0%
Stretch coefficient	ft/Kft/Klb	ft/Kft/Klb	ft/Kft/Klb

(Courtesy of Camesa, Inc.)

TABLE 7.4 Swabline

Type (Nominal diameter)	$\frac{1}{4}''$ (6.35 mm)	$\frac{5}{16}''$ (7.94 mm)	$\frac{7}{16}''$ (11.11 mm)
CONSTRUCTION (Special)	1 × 19 (12/6/1)	1 × 19 (12/6/1)	1 × 19 (12/6/1)
Inner Layer – Left			
Outside diameter	0.154″(3.91 mm)	0.196″(4.98 mm)	0.244″(6.2 mm)
Center wire diameter	0.058″(1.47 mm)	0.072″(1.83 mm)	0.102″(2.59 mm)
Number of outer diameter	6	6	6
Outer wire diameter	0.048″(1.22 mm)	0.062″(1.57 mm)	0.084″(2.13 mm)
Outer Layer – Right			
Outside diameter	0.250″(6.35 mm)	0.317″(8.05 mm)	0.438″(11.1 mm)
Number wires	12	12	12
Wire diameter	0.048″(1.22 mm)	0.062″(1.57 mm)	0.084″(2.13 mm)
MECHANICAL CHARACTERISTICS			
Breaking strength	8.4 Klb (3.8 Kkg)	13.9 Klb (6.3 Kkg)	25.6 Klb (11.6 Kkg)
Weight	124.3 lb/Kft (185 kg/km)	207.6 lb/Kft (309 kg/km)	381.7 lb/Kft (568 kg/km)
Diameter tolerance	+2% – 0%	+2% – 0%	+2% – 0%
Stretch coefficient	1.70 ft/Kft/Klb	1.10 ft/Kft/Klb	0.60 ft/Kft/Klb

Type (Nominal diameter)	$\frac{3}{16}''$ (4.76 mm)	$\frac{7}{32}''$ (5.56 mm)	$\frac{1}{4}''$ (6.35 mm)
CONSTRUCTION (Special)	1 × 16 (9/6/1)	1 × 16 (9/6/1)	1 × 16 (9/6/1)
Inner Layer – Left			
Outside diameter	0.100″(2.54 mm)	0.118″(3.0 mm)	—
Center wire diameter	0.036″(0.91 mm)	0.042″(1.07 mm)	—
Number of outer diameter	6	6	—
Outer wire diameter	0.032″(0.81 mm)	0.038″(.97 mm)	—
Outer Layer – Right			
Outside diameter	0.188″(4.78 mm)	0.224″(5.69 mm)	—
Number wires	9	9	—
Wire diameter	0.044″(1.12 mm)	0.054″(1.37 mm)	—
MECHANICAL CHARACTERISTICS			
Breaking strength	4.6 Klb (2.1 Kkg)	6.8 Klb (3.1 Kkg)	—
Weight	70.6 lb/Kft (105 kg/km)	104.2 lb/Kft (155 kg/km)	— —
Diameter tolerance	+2% – 0%	+2% – 0%	—
Stretch coefficient	2.85 ft/Kft/Klb	1.80 ft/Kft/Klb	—

(Courtesy of Camesa, Inc.)

Most manufactures of wirelines will construct specially designed armored wirelines for unique downhole operations.

7.8.4 Armored Wirelines with Electrical Conductors

This category of wirelines is composed of a variety of armored wirelines that surround electrical conductors.

TABLE 7.5 Utility Lines

Type (Nominal diameter)	$\frac{3}{16}''$ **(4.76 mm)**	$\frac{7}{32}''$ **(5.56 mm)**	$\frac{1}{14}''$ **(6.35 mm)**
CONSTRUCTION (Normal)	1 × 19 (9/9/1)	1 × 19 (9/9/1)	1 × 19 (9/9/1)
Inner Layer – Left			
Outside diameter	0.112″(2.84 mm)	0.130″(3.30 mm)	0.152″(3.86 mm)
Centre wire diameter	0.056″(1.42 mm)	0.066″(1.68 mm)	0.076″(1.93 mm)
Number of outer wires	9	9	9
Outer wire diameter	0.028″(.711 mm)	0.032″(.81 mm)	0.038″(0.97 mm)
Outer Layer – Right			
Outside diameter	0.188″(4.78 mm)	0.219″(5.56 mm)	0.250″(6.35 mm)
Number wires	9	9	9
Wire diameter	0.050″(1.27 mm)	0.058″(1.47 mm)	0.066″(1.68 mm)
MECHANICAL CHARACTERISTICS			
Breaking strength	6.4 Klb (2.9 Kkg)	8.6 Klb (3.9 Kkg)	11.4 Klb (5.2 Kkg)
Weight	88.6 lb/Kft (131.8 kg/km)	127.7 lb/Kft (190 kg/km)	159 lb/Kft (237 kg/km)
Diameter tolerance	+2% – 0%	+2% – 0%	+2% – 0%
Stretch coefficient	ft/Kft/Klb	ft/Kft/Klb	ft/Kft/Klb
Type (Nominal diameter)	$\frac{3}{16}''$	$\frac{7}{32}''$	$\frac{1}{4}''$
CONSTRUCTION (Special)	1 × 19 (9/9/1)	1 × 19 (9/9/1)	1 × 19 (9/9/1)
Inner Layer – Left			
Outside diameter	0.108″(2.74 mm)	0.118″(3. 0 mm)	—
Centre wire diameter	0.052″(1.32 mm)	0.060″(1.52 mm)	—
Number of outer wires	9	9	—
Outer wire diameter	0.026″(0.66 mm)	0.029″(0.74 mm)	—
Outer Layer – Right			
Outside diameter	0.188″(4.78 mm)	0.219″(5.56 mm)	—
Number wires	9	9	—
Wire diameter	0.049″(1.24 mm)	0.057″(1.45 mm)	—
MECHANICAL CHARACTERISTICS			
Breaking strength	5.9 Klb (2.7 Kkg)	7.9 Klb (3.6 Kkg)	—
Weight	84 lb/Kft (125 kg/km)	111.5 lb/Kft (166 kg/km)	— —
Diameter tolerance	+2% – 0%	+2% – 0%	—
Stretch coefficient	1.51 ft/Kft/Klb	ft/Kft/Klb	—

(Courtesy of Camesa, Inc.)

Table 7.6 shows the $\frac{1}{10}$ths of an inch (2.54 mm) diameter monoconductor wireline. The table gives typical construction, dimensions, mechanical properties, and electrical properties.

Table 7.7 shows the $\frac{7}{32}$ths of an inch (5.69 mm) diameter three-conductor wireline. The table gives typical construction, dimensions, mechanical properties, and electrical properties.

TABLE 7.6 1/10″ (2.54 mm) Monoconductor 1N10

PROPERTIES		
Cable diameter	0.101″ + 0.004″ − 0.002″	(2.56 mm + 0.10 mm − 0.05 mm)
Minimum sheave diameter	6″	(16 cm)
Cable stretch coefficient	13.1 ft/Kft/Klb	(14.72 m/Km/5KN)
ELECTRICAL		
Maximum conductor voltage	300 VDC	
Conductor AWG rating	24	
Minimum insulation resistance	1,500 MegaΩ/Kft at 500 VDC	(457 MegaΩ/Km at 500 VDC)
Armor electrical resistance	22 Ω/Kft	(72.2 Ω/Km)
MECHANICAL		
Cable breaking strength		
Ends fixed	1,000 lb	(4.4 KN) Nominal
Ends free	670 lb	(3.0 KN) Nominal
Maximum suggested working tension	500 lb	(2.2 KN)
Number and size of wires		
Inner armor	12 × 0.0140″	(0.356 mm)
Outer armor	18 × 0.0140″	(0.356 mm)
Average wire braking strength		
Inner armor	42 lb	(0.19 KN)
Outer armor	42 lb	(0.19 KN)

(Continued)

TABLE 7.6 (*Continued*)

Cable Type	Core Description							Cable Weight		
	Temp Rating	Plastic Insulation	Type Thickness	Copper Construction	Res Typical	Cap. Typical	O.D. Each	In Air	In H_2O	Spec. Gravity
	°F		in.	in.	Ω/Kft	pf/ft	in.	lb/Kft		
	(°C)		(mm)	(mm)	(Ω/Km)	(pf/m)	(mm)	(Kg/Km)		
1N10RP	300	Poly	0.012	7 × 0.0085	21.0	51	0.049	19	16	5.42
	(149)		(0.305)	(7 × 0.216)	(69.0)	(167)	(1.244)	(28)	(24)	

The armor wires are high tensile, galvanized extra improved plow steel (GEIPS), and coated with anti-corrosion compound for protection during shipping and storing. Wires are preformed and cables are post tensioned.
Core assembly – Copper strand consists of six wires around one center wire. Conductor resistance is measured at 68 °F. Voids in the copper strand are filled with a water-blocking agent to reduce water and gas migration.
SUPERSEAL – a special pressure seal agent, is applied between armor layers.
The temperature rating assumes a normal gradient for both temperature and weight.
(Courtesy of Camesa, Inc.)

TABLE 7.7 7/32″ (5.69 mm) 3-CONDUCTOR 3H22

PROPERTIES		
Cable diameter	0.224″ + 0.005″ − 0.002″	(5.69 mm + 0.13 mm − 0.05 mm)
Minimum sheave diameter	13″	(33 cm)
Cable stretch coefficient (nominal)	2.50 ft/Kft/Klb	(2.81 m/Km/5KN)
ELECTRICAL		
Maximum conductor voltage	300 VDC	
Conductor AWG rating	23	
Minimum insulation resistance	1,500 MegΩ/Kft at 500 VDC	(457 MegΩ/Km at 500 VDC)
Armor electrical resistance	4.80 Ω/Kft	(15.7 Ω/Km)
MECHANICAL		
Cable breaking strength		
Ends fixed	4,700 lb	(20.9 KN) Nominal
Ends free	2,880 lb	(12.8 KN) Nominal
Maximum suggested working tension	2,350 lb	(10.5 KN)
Number and size of wires		
Inner armor	18 × 0.0220″	(0.559 mm)
Outer armor	18 × 0.0310″	(0.787 mm)
Average wire braking strength		
Inner armor	103 lb	(0.46 KN)
Outer armor	204 lb	(0.91 KN)

(Continued)

TABLE 7.7 (*Continued*)

Cable Type		Core Description							Cable Weight		
	Temp Rating	Plastic Type	Insulation Thickness	Copper Construction	Res Typical	Cap. Typical	OD Each	Jacket Type	In Air	In H_2O	Spec. Gravity
	°F (°C)		in. (mm)	in. (mm)	Ω/Kft (Ω/Km)	pf/ft (pf/m)	in. (mm)		lb/Kft (Kg/Km)		
3H22RPP	300 (149)	Poly	0.011 (0.279)	7 × 0.0085 (7 × 0.216)	22.5 (73.8)	36 (118)	0.047 (1.194)	Poly	81 (120)	69 (103)	6.91
3H22RXZ	420 (216)	Camtane	0.011 (0.279)	7 × 0.0085 (7 × 0.216)	22.5 (73.8)	35 (115)	0.047 (1.194)	Tefzel	82 (122)	72 (104)	6.82
3H22RTZ	500 (260)	Teflon	0.011 (0.279)	7 × 0.0085 (7 × 0.216)	22.5 (73.8)	40 (131)	0.047 (1.194)	Tefzel	83 (124)	71 (106)	6.81
3H22RTA	550 (288)	Teflon	0.011 (0.279)	7 × 0.0085 (7 × 0.216)	22.5 (73.8)	36 (118)	0.047 (1.194)	PFA	83 (124)	71 (106)	6.81

The armor wires are high tensile. Galvanized extra improved plow steel (GEEIPS), and coated with anti-corrosion compound for protection during shipping and storing. Wires are preformed and cables are post tensioned.
Core assembly – Copper strand consists of six wires around one center wire. Voids in the copper strand are with a water-blocking agent to reduce water and gas migration. Conductor resistance is measured at 68 °F.
The temperature rating assumes a normal gradient for both temperature and weight.
(Courtesy of Camesa, Inc.)

Table 7.8 shows the $\frac{5}{16}$th of an inch (8.26 mm) diameter seven-conductor wireline. The table gives typical construction, dimensions, mechanical properties, and electrical properties.

Table 7.9 shows the $\frac{3}{16}$ths of an inch (4.80 mm) diameter four-conductor wireline. The table gives typical construction, dimensions, mechanical properties, and electrical properties.

Most manufactures of wirelines will construct specially designed armored electrical conductor wirelines for unique downhole operations.

7.8.5 Wireline Operating and Breaking Strengths

During field operations the line tensions should be no greater than one half of the wireline breaking strength. This will insure that no permanent damage is done to the wireline.

In the tables above the breaking strength for each wireline is given. For stainless steel armored cables or other special construction the manufacturer should be consulted for breaking strength.

7.8.6 Wireline Stretching

Many field operations using wirelines requires that the wireline stretch be determined. One common need is the necessity of determine the true depth of the particular logging tool in the well. The true depth can be determined by adding the amount of cable stretch ΔL to the measured wireline length L in the well. The amount of wireline stretch is determined from

$$DL = \left(\frac{KL}{2}\right)[(T + W_t) - 2T_m] \tag{7.3}$$

where

- L is the length of wireline in the well (in units of 1,000 ft)
- T is the surface tension (in units of 1,000 lb)
- W_t is the effective weight (in the well fluid) of any downhole tool on the free end of the wireline (in units of 1,000 lb)
- T_m is the marked cable tension (usually 1,000 lb)
- K is the elastic stretch coefficient (ft/1,000 ft/1,000 lb)

Example 7.2

Determine the stretch of 10,000 ft of a monoconductor 7H32RZ wireline with a 300 = lb tool (weight in drilling mud used). The stretch coefficient is 1.8 ft/kft/klb. The wireline has been marked 1000 lb tension. The wirelines

TABLE 7.8 5/16″ (8.26 mm) 7-Conductor 7H32

PROPERTIES		
Cable diameter	0.325″ + 0.005″ – 0.002″	(8.26 mm + 0.13 mm – 0.05 mm)
Minimum sheave diameter	18″	(45 cm)
Cable stretch coefficient (nominal)	1.8 ft/Kft/Klb	(2.02 m/Km/5KN)
ELECTRICAL		
Maximum conductor voltage	1000 VDC	
Conductor AWG rating	22	
Minimum insulation resistance	1,500 MegΩ/Kft at 500 VDC	(457 MegΩ/Km at 500 VDC)
Armor electrical resistance	2.3 Ω/Kft	(7.5 Ω/Km)
MECHANICAL		
Cable breaking strength		
Ends fixed	9,500 lb	(42.3 KN) Nominal
Ends free	5,700 lb	(25.4 KN) Nominal
Maximum suggested working tension	4,750 lb	(21.1 KN)
Number and size of wires		
Inner armor	18 × 0.0320″	(0.813 mm)
Outer armor	18 × 0.0445″	(1.130 mm)
Average wire braking strength		
Inner armor	217 lb	(0.97 KN)
Outer armor	420 lb	(1.87 KN)

Cable Type				Core Description					Cable Weight		
	Temp Rating	Plastic Type	Insulation Thickness	Copper Construction	Res Typical	Cap. Typical	OD Each	Tape Type	In Air	In H_2O	Spec. Gravity
	°F (°C)		in. (mm)	in. (mm)	Ω/Kft (Ω/Km)	pf/ft (pf/m)	in. (mm)		lb/Kft (Kg/Km)		
7H32RP	300 (149)	Poly	0.013 (0.330)	7 × 0.0100 (7 × 0.254)	15.8 (51.8)	55 (180)	0.056 (1.422)	Dacron	183 (272)	152 (226)	5.86
7H32RZ	420 (216)	Tefzel	0.013 (0.330)	7 × 0.0100 (7 × 0.254)	15.8 (51.8)	67 (220)	0.056 (1.422)	Nomex	188 (280)	157 (234)	6.03
7H32RA	500 (260)	PFA	0.013 (0.330)	7 × 0.0100 (7 × 0.254)	15.8 (51.8)	58 (190)	0.056 (1.422)	Nomex	190 (283)	159 (237)	6.08

The armor wires are high tensile, galvanized extra improved plow steel (GEEIPS), and coated with anti-corrosion compound for protection during shipping and storing. Wires are preformed and cables are post tensioned.
Core assembly – Conductors are bound with conductive tape and voids are filled with conductive paste and string.
Conductors are "Water Blocked" to reduce water and gas migration. Conductor resistance is measured at 68°F.
The temperature rating assumes a normal gradient for both temperature and weight.
Center conductor construction is 7 × 0.0100″. The typical capacitance is decreased by approximately 5 to 10% in comparison to the outer conductors.

TABLE 7.9 3/16″ (4.80 mm) 4-Conductor 4H18

PROPERTIES		
Cable diameter	0.189″ + 0.004″ – 0.002″	(4.80 mm + 0.10 mm – 0.05 mm)
Minimum sheave diameter	10″	(25 cm)
Cable stretch coefficient (nominal)	4.25 ft/Kft/Klb	(4.78 m/Km/5KN)
ELECTRICAL		
Maximum conductor voltage	300 VDC	
Conductor AWG rating	23	
Minimum insulation resistance	1,500 MegΩ/Kft at 500 VDC	(457 MegΩ/Km at 500 VDC)
Armor electrical resistance	6.70 Ω/Kft	(22.0 Ω/Km)
MECHANICAL		
Cable breaking strength		
Ends fixed	3,100 lb	(13.8 KN) Nominal
Ends free	1,900 lb	(8.5 KN) Nominal
Maximum suggested working tension	1,550 lb	(6.9 KN)
Number and size of wires		
Inner armor	18 × 0.0185″	(0.470 mm)
Outer armor	18 × 0.0248″	(0.630 mm)
Average wire breaking strength		
Inner armor	72 lb	(0.32 KN)
Outer armor	130 lb	(0.58 KN)

Cable Type		Core Description							Cable Weight		
	Temp Rating	Plastic Type	Insulation Thickness	Copper Construction	Res Typical	Cap. Typical	OD Each	Jacket Type	In Air	In H_2O	Spec. Gravity
	°F		in.	in.	W/Kft	pf/ft	in.		lb/Kft		
	(°C)		(mm)	(mm)	(W/Km)	(pf/m)	(mm)		(Kg/Km)		
4H18RPP	300	Poly	0.0075	7 × 0.0085	22.5	36	0.047	Poly	84	69	5.73
	(149)		(0.191)	(7 × 0.216)	(73.8)	(118)	(1.194)		(124)	(103)	
4H18RXZ	420	Camtane	0.0075	7 × 0.0085	22.5	35	0.047	Tefzel	84	69	5.73
	(216)		(0.191)	(7 × 0.216)	(73.8)	(115)	(1.194)		(124)	(103)	
4H18RTZ	500	Teflon	0.0075	7 × 0.0085	22.5	40	0.047	Tefzel	86	71	5.85
	(260)		(0.191)	(7 × 0.216)	(73.8)	(131)	(1.194)		(127)	(105)	
4H18RTA	550	Teflon	0.0075	7 × 0.0085	22.5	36	0.047	PFA	88	73	5.94
	(288)		(0.191)	(7 × 0.216)	(73.8)	(103)	(1.194)		(130)	(107)	

The armor wires are high tensile, galvanized extra improved plow steel (GEEIPS), and coated with anti-corrosion compound for protection during shipping and storing. Wires are preformed and cables are post tensioned.
Core assembly – Copper strand consists of six wires around one center wire. Voids in the copper strand are with a water-blocking agent to reduce water and gas migration. Conductor resistance is measured at 68 °F.
The temperature rating assumes a normal gradient for both temperature and weight.
(Courtesy of Camesa, Inc.)

tension measured at the surface is 5000 lbs. Using Equation 7.3,

$$\Delta L = \left[\frac{1.8(10.0)}{2}\right][(5.0 + 0.3) - 2(1.0)]$$

$$\Delta L = 29.7\,\text{ft}$$

Thus, the true depth as measured by wireline would be 10,000 ft + 29.7 ft = 10,029.7 ft.

References

[1] Main, W.C., "Detection of Incipient Drill-pipe Failures," *API Drilling and Production Practices*, 1949.
[2] Moore, E.E., "Fishing and Freeing Stuck Drill Pipe," *The Petroleum Engineer*, April 1956.
[3] Gatlin, C., *Petroleum Engineering: Drilling and Well Completions*, Prentice-Hall, 1960.
[4] Short, J.A., *Fishing and Casing Repair*, PennWell, 1981.
[5] "Well Abandonment and Inactive Well Practices for U. S. Exploration and Production Operations, Environment Guidance Document," *API Bulletin E3*, 1st Edition, January 1993.

Further Reading

McCray, A.W., and Cole, F.W., *Oil Well Drilling Technology*, University of Oklahoma Press, Norman, Oklahoma, 1959.

CHAPTER

8

Casing and Casing String Design

OUTLINE

8.1 TYPES OF CASING

Based on the primary function of the casing string, there are five types of casing to be distinguished.

Stove or Surface Casing The stovepipe is usually driven to sufficient depth (15–60 ft) to protect loose surface formation and to enable circulation of the drilling fluid. This pipe is sometimes cemented in predrilled holes.

Conductor String This string acts as a guide for the remaining casing strings into the hole. The purpose of the conductor string is also to cover unconsolidated formations and to seal off overpressured formations. The conductor string is the first string that is always cemented to the top and equipped with casing head and blowout prevention (BOP) equipment.

Surface Casing This is set deeply enough to protect the borehole from caving-in in loose formations frequently encountered at shallow depths, and protects the freshwater sands from contamination while subsequently drilling a deeper hole. In case the conductor string has not been set, the surface casing is fitted with casing head and BOP.

Intermediate Casing Also called protection string, this is usually set in the transition zone before abnormally high formation pressure is encountered, to protect weak formations or to case off loss-of-circulation zones. Depending upon geological conditions, the well may contain two or even three intermediate strings. Production string (oil string) is the string through which the well is produced.

Intermediate or production string can be set a liner string. The liner string extends from the bottom of the hole upward to a point about 150–250 ft above the lower end of the upper string.

Casing Program Design Casing program design is accomplished by two steps. In the first step, the casing sizes and corresponding bit sizes should be determined. In the second step, the setting depth of the individual casing strings ought to be evaluated. Before starting the casing program design, the designer ought to know the following basic information:

- The purpose of the well (exploratory or development drilling)
- Geological cross-sections that should consist of type of formations, expected hole problems, pore and formation's fracture pressure, number and depth of water, oil, gas horizons
- Available rock bits, reamer shoes and casing sizes
- Load capacity of a derrick and mast if the type of rig has already been selected

Before starting the design, it must be assumed that the production casing size and depth of the well has been established by the petroleum engineer in cooperation with a geologist, so that the hole size (rock bit diameter) for the casing may be selected. Considering the diameter of the hole, a sufficient clearance beyond the coupling outside diameter must be provided to allow for mud cake and also for a good cementing job. Field experience shows that the casing clearance should range from about 1.0 in. to 3.5 in. Larger casing sizes require greater value of casing clearance. Once the hole size for production string has been selected, the smallest casing through which a given bit will pass is next determined. The bit diameter should be a little less (0.05 in.) than casing drift diameter. After choosing the casing with appropriate drift diameter, the outside coupling diameter of this casing may be found. Next, the appropriate size of the bit should be determined and the procedure repeated.

Expandable casing technology, expandable drill bits, under-reamers and other tools for optimizations of borehole and/or string designs, are not covered in this section.

Example 8.1

The production casing string for a certain well is to consist of 5-in. casing. Determine casing and corresponding bit sizes for the intermediate, surface and conductor string. Take casing data and bit sizes from Table 8.1.

Solution

For production hole, select a $6\frac{3}{4}$-in. rock bit. Therefore, the casing clearance $= 6.75 - 5.563 = 1.187$ in.

For intermediate string, select a 7 5/8-in. casing, assuming that wall thickness that corresponds to drift diameters of 6.640 or smaller will not be used. For the 7 5/8-in. intermediate string, use a 9 7/8-in. bit. The casing clearance $= 9.875 - 8.5 = 1.375$ in.

For surface string, select a $10\frac{3}{4}$-in. casing. Note that only unit weights corresponding to drift diameters of 10.036 and 9.894 in. can be used. For the $10\frac{3}{4}$-in. casing, use a $13\frac{3}{4}$–in. bit, so the casing clearance $= 13.75 - 11.75 = 2.0$ in.

For conductor string, select 16 in. casing; the bit size will then be 20 in. and the casing clearance $= 20 - 17 = 3$ in.

Having defined bit and casing string sizes, the setting depths of the individual strings should be determined.

The operation of setting is governed by the principle according to which casing should be placed as deep as possible. However, the designer must remember to ensure the safety of the drilling crew from possible blowout, and to maintain the hole stability, well completion aspects (formation damage) and state regulations.

In general, casing should be set

- Where drilling fluid could contaminate freshwater that might be used for drinking or other household purposes
- Where unstable formations are likely to cave or slough into the borehole
- Where loss of circulation may result in blowout
- Where drilling fluid may severely damage production horizon

Currently, a graphical method of casing setting depth determination is used. The method is based on the principle according to which the borehole pressure should always be greater than pore pressure and less than fracture pressure. (Drilling with borehole pressures lower than pore pressure requires the use of under-balanced drilling technologies not covered in this chapter.)

For practical purposes, a safety margin for reasonable kick conditions should be imposed (Figure 8.1). Even when borehole pressure is adjusted correctly, problems may arise from the contact between the drilling fluid and the formation. It depends upon the type of drilling fluid and formation, but in general, the more time spent drilling in an open hole, the

TABLE 8.1 Casing Dimensions for Rock Bit Selection

API Casing Data (Bit Sizes and Clearances)						
Casing Specifications					Recommended Max. Bit Size	
Casing Size D.D. (inches)	Casing Coupling D.D. (inches)	Nominal Weight lb./ft.	Inside Diameter I.D. (inches)	API Drift I.D. (inches)	Roller Cone Bit Size D.D. (inches)	Fixed Cutter Bit Size D.D. (inches)
4.500	5.000	9.50	4.090	3.965	$3\frac{7}{8}$	$3\frac{7}{8}$
4.500	5.000	10.50	4.052	3.927	$3\frac{7}{8}$	$3\frac{7}{8}$
4.500	5.000	11.60	4.000	3.875	$3\frac{7}{8}$	$3\frac{7}{8}$
4.500	5.000	13.50	3.920	3.795	$3\frac{3}{4}$	$3\frac{3}{4}$
5.000	5.563	11.50	4.560	4.435	$3\frac{7}{8}$	$3\frac{7}{8}$
5.000	5.563	13.00	4.494	4.369	$3\frac{7}{8}$	$3\frac{7}{8}$
5.000	5.563	15 00	4.408	4.283	$3\frac{7}{8}$	$3\frac{7}{8}$
5.000	5.563	18.00	4.276	4.151	$3\frac{7}{8}$	$3\frac{7}{8}$
5.500	6.050	14.00	5.012	4.887	$4\frac{3}{4}$	$4\frac{3}{4}$
5.500	6.050	15.50	4.950	4.825	$4\frac{3}{4}$	$4\frac{3}{4}$
5.500	6.050	17.00	4.992	4.767	$4\frac{3}{4}$	$4\frac{3}{4}$
5.500	6.050	20.00	4.778		$4\frac{1}{2}$	$4\frac{1}{2}$
5.500	6.050	23.00	4.670	4.545	$3\frac{7}{8}$	$4\frac{1}{2}$
6.625	7.390	20.00	6.049	5.924	$5\frac{7}{8}$	$5\frac{7}{8}$
6.625	7.390	24.00	5.921	5.798	$4\frac{3}{4}$	$4\frac{3}{4}$
6.625	7.390	28.00	5.791	5.666	$4\frac{3}{4}$	$4\frac{3}{4}$

6.625	7.390	32.00	5.675	5.550	$4\frac{3}{4}$	$4\frac{3}{4}$
7.000	7.656	17.00	6.538	6.413	$6\frac{1}{4}$	$6\frac{1}{4}$
7.000	7.656	20.00	6.456	6.331	$6\frac{1}{4}$	$6\frac{1}{4}$
7.000	7.656	23.00	6.366	6.241	$6\frac{1}{8}$	$6\frac{1}{8}$
7.000	7.656	26.00	6.276	6.151	$6\frac{1}{8}$	$6\frac{1}{8}$
7.000	7.656	29.00	6.184	6.059	6	6
7.000	7.656	32.00	6.094	5.969	$5\frac{7}{8}$	$5\frac{7}{8}$
7.000	7.656	35.00	6.004	5.879	$5\frac{7}{8}$	$5\frac{7}{8}$
7.000	7.656	38.00	5.920	5.795	$4\frac{3}{4}$	$4\frac{3}{4}$
7.625	8.500	20.00	7.125	7.000	$6\frac{3}{4}$	$6\frac{3}{4}$
7.625	8.500	24.00	7.025	6.900	$6\frac{3}{4}$	$6\frac{3}{4}$
7.625	8.500	26.40	6.969	6.844	$6\frac{3}{4}$	$6\frac{3}{4}$
7.625	8.500	29.70	6.875	6.750	$6\frac{3}{4}$	$6\frac{3}{4}$
7.625	8.500	33.70	6.765	6.640	$6\frac{1}{2}$	$6\frac{1}{2}$
7.625	8.500	39.00	6.625	6.500	$6\frac{1}{2}$	$6\frac{1}{2}$
8.625	9.625	24.00	8.097	7.972	$7\frac{7}{8}$	$7\frac{7}{8}$
8.625	9.625	28.00	8.017	7.892	$7\frac{7}{8}$	$7\frac{7}{8}$
8.625	9.625	32.00	7.921	7.795	$6\frac{3}{4}$	$6\frac{3}{4}$
8.625	9.625	36.00	7.825	7.700	$6\frac{3}{4}$	$6\frac{3}{4}$

(Continued)

TABLE 8.1 (*Continued*)

API Casing Data (Bit Sizes and Clearances)						
Casing Specifications					Recommended Max. Bit Size	
Casing Size D.D. (inches)	Casing Coupling D.D. (inches)	Nominal Weight lb./ft.	Inside Diameter I.D. (inches)	API Drift I.D. (inches)	Roller Cone Bit Size D.D. (inches)	Fixed Cutter Bit Size D.D. (inches)
8.625	9.625	40.00	7.725	7.600	$6\frac{3}{4}$	$6\frac{3}{4}$
8.625	9.625	44.00	7.625	7.500	$6\frac{3}{4}$	$6\frac{3}{4}$
8.625	9.625	49.00	7.511	7.386	$6\frac{3}{4}$	$6\frac{3}{4}$
9.625	10.625	29.30	9.063	8.907	$8\frac{3}{4}$	$8\frac{3}{4}$
9.625	10.625	32.30	9.001	8.845	$8\frac{3}{4}$	$8\frac{3}{4}$
9.625	10.625	36.00	8.921	8.765	$8\frac{3}{4}$	$8\frac{3}{4}$
9.625	10.625	40.00	8.835	8.679	$8\frac{1}{2}$	$8\frac{1}{2}$
9.625	10.625	43.50	8.755	8.599	$8\frac{1}{2}$	$8\frac{1}{2}$
9.625	10.625	47.00	8.681	8.525	$8\frac{1}{2}$	$8\frac{1}{2}$
9.625	10.625	53.50	8.535	8.379	$8\frac{3}{8}$	$8\frac{3}{8}$
10.750	11.750	32.75	10.192	10.038	$9\frac{7}{8}$	$9\frac{7}{8}$
10.750	11.750	40.50	10.050	9.894	$9\frac{7}{8}$	$9\frac{7}{8}$
10.750	11.750	45.50	9.950	9.794	$9\frac{1}{2}$	$9\frac{1}{2}$
10.750	11.750	51.00	9.850	9.694	$9\frac{1}{2}$	$9\frac{1}{2}$
10.750	11.750	55.50	9.760	9.604	$9\frac{1}{2}$	$9\frac{1}{2}$
10.750	11.750	80.70	9.660	9.504	$9\frac{1}{2}$	$9\frac{1}{2}$
10.750	11.750	65.70	9.560	9.404	$8\frac{3}{4}$	$8\frac{3}{4}$

11.750	12.750	42.00	11.084	10.928	$10\frac{5}{8}$	$10\frac{5}{8}$
11.750	12.750	47.00	11.000	10.844	$10\frac{5}{8}$	$10\frac{5}{8}$
11.750	12.750	54.00	10.880	10.724	$10\frac{5}{8}$	$10\frac{5}{8}$
11.750	12.750	60.00	10.772	10.616	$9\frac{7}{8}$	$9\frac{7}{8}$
13.375	14.375	48.00	12.715	12.559	$12\frac{1}{4}$	$12\frac{1}{4}$
13.375	14.375	54.50	12.615	12.459	$12\frac{1}{4}$	$12\frac{1}{4}$
13.375	14.375	61.00	12.515	12.359	$12\frac{1}{4}$	$12\frac{1}{4}$
13.375	14.375	68.00	12.415	12.259	$12\frac{1}{4}$	$12\frac{1}{4}$
13.375	14.375	72.00	12.347	12.191	11	$10\frac{5}{8}$
16.000	17.000	65.00	15.250	15.062	$14\frac{3}{4}$	$14\frac{3}{4}$
16.000	17.000	75.00	15.124	14.936	$14\frac{3}{4}$	$14\frac{3}{4}$
16.000	17.000	84.00	15.010	14.822	$14\frac{3}{4}$	$14\frac{3}{4}$
18.625	20.000	87.50	17.755	17.567	$17\frac{1}{2}$	$17\frac{1}{2}$
20.000	21.000	94.00	19.124	18.936	$17\frac{1}{2}$	$17\frac{1}{2}$
20.000	21.000	106.50	19.000	18.812	$17\frac{1}{2}$	$17\frac{1}{2}$
20.000	21.000	133.00	18.730	18.542	$17\frac{1}{2}$	$17\frac{1}{2}$
20.000	21.000	169.00	18.376	18.188	$17\frac{1}{2}$	$17\frac{1}{2}$

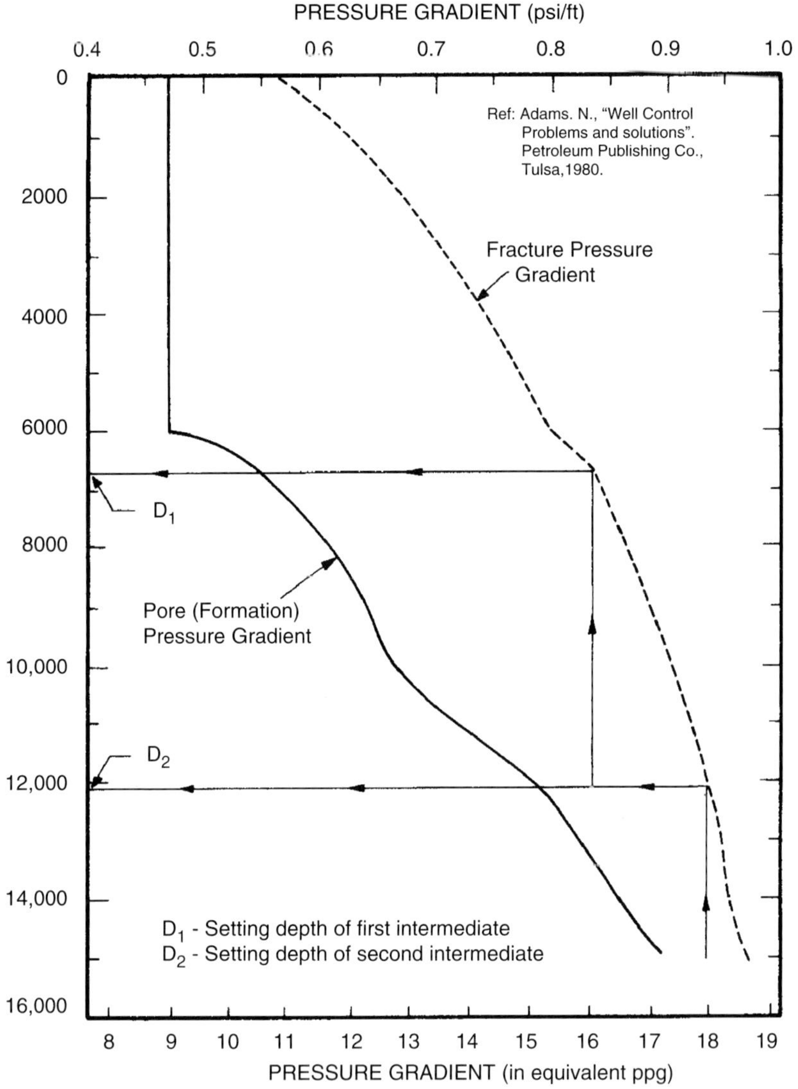

FIGURE 8.1 Well planning.

greater the possibility of formation caving or sloughing into the borehole. Formation instability may lead to expensive work in the borehole, which influences the time and cost of the drilling operation. To arrest or reduce this problem, special treatment drilling fluids might be used, but these special drilling fluids are expensive. Therefore, the casing and drilling fluid programs depend on each other, and solving the issue of correct casing setting depth evaluation is a rather complicated, optimizing problem.

Example 8.2

Suppose that in some area the expected formation pressure gradient is 0.65 psi/ft and formation fracture pressure gradient is 0.85 psi/ft. A gas-bearing formation is expected at depth of 15,000 ft. Assume that a gas kick occurs that, to be removed from the hole, induces a surface pressure of 2,000 psi. The first intermediate casing is set at a depth of 7,200 ft. Determine the setting depth for the second intermediate casing string if required in given conditions. Assume drilling fluid pressure gradient = 0.65 psi/ft.

Solution

The formation fracture pressure line is

$$P_f = (0.85)(D)$$

The borehole pressure line is

$$P_{bh} = 2{,}000 + (0.65)(D)$$

So

$$(0.85)(D) = 2{,}000 + (0.65)(D)$$

$$D = \frac{2{,}000}{0.2} = 10{,}000 \text{ ft.}$$

The second intermediate casing string is required and must be set at a depth of 10,000 ft.

8.2 CASING DATA

Special note: Nothing in this specification should be interpreted as indicating a preference by the committee for any material or process, nor as indicating equality between the various materials or processes. In the selection of materials and processes, the purchaser must be guided by his or her experience and by the service for which the pipe is intended.

Casing is classified according to its manner of manufacture, steel grade, dimensions, and weights, and the type of coupling.

The following include excerpts from, and references to API Specification 5CT, Seventh Edition, October 1, 2001.

8.2.1 Process of Manufacture

Casing and liners shall be seamless or electric welded as defined below and as specified on the purchase.

a. Seamless pipe is defined a wrought steel tubular product made without a welded seam. It is manufactured by hot working steel or, if necessary, by subsequently cold finishing the hot-worked tubular product to produce the desired shape, dimension and properties.

b. Electric-welded pipe is defined as pipe having one longitudinal seam formed by electric-flash welding or electric-resistance welding, without the addition of extraneous metal. The weld seam of electric-welded pipe shall be heat treated after welding to a minimum temperature of 1,000°F (580°C), or processed in such a manner that no untempered martensite remains.
c. Pup-joints shall be made from standard casing or tubing or by machining thick-wall casing, tubing or bar stock. Cold-drawn tubular products without appropriate heat treatment are not acceptable.
d. Casing and tubing accessories shall be seamless and made from standard casing or tubing, or by machining thick-wall casing, tubing or mechanical tubes, or bar stock or hot forgings.
e. Electric-welded Grade P110 pipe and Grade Q125 casing shall be provided only when the supplementary requirement in API-Specification 5CT Section A.5 (SR11) is specified on the purchase agreement.
f. Grade Q125 upset casing shall be provided only when the supplementary requirement in API-Specification 5CT Section A.4 (SR10) is specified on the purchase agreement.

Further information about heat treatment requirements for the manufacturing process of various grades of pipe steel, as well as straightening methods and traceability requirements are further described in API-Specification 5CT Sections 6.2 through 6.4.2.

8.2.2 Material Requirements (Section 7, API Specification 5CT)

Tensile Properties Product shall conform to the tensile requirements specified in Table 8.2. The tensile properties of upset casing and tubing, except elongation of the upset ends, shall comply with the requirements given for the pipe body. In case of dispute, the properties (except elongation) of the upset shall be determined from a tensile test specimen cut from the upset. A record of such tests shall be available to the purchaser.

Yield Strength The yield strength shall be the tensile stress required to produce the elongation under load specified in Table 8.2 as determined by an extensometer.

Tensile Tests Tensile properties shall be determined by tests on longitudinal specimens conforming to the requirements of Paragraph 10.4 in API Specification 5CT, and ASTM A370: Mechanical Testing of Steel Products, Supplement II, Steel Tubular Products.

TABLE 8.2 Tensile and Hardness Requirements

Group	Grade	Type	Total Elongation Under Load %	Yield Strength ksi min.	Yield Strength ksi max.	Tensile Strength min. ksi	Hardenss[a] max. HRC	Hardenss[a] max. HBW/HBS	Specified Wall Thickness in	Allowable Hardness Variation[b] HRC
1	2	3	4	5	6	7	8	9	10	11
1	H40		0.5	40	80	60				
	J55		0.5	55	80	75				
	K55		0.5	55	80	95				
	N80	1	0.5	80	110	100				
	N80	Q	0.5	80	110	100				
2	M65		0.5	65	85	85	22	235		
	L80	1	0.5	80	95	95	23	241		
	L80	9Cr	0.5	80	95	95	23	241		
	L80	13Cr	0.5	80	95	95	23	241		
	C90	1.2	0.5	90	105	100	25.4	255	≤ 0.500	3.0
	C90	11	0.5	90	105	100	25.4	255	0.501 to 0.749	4.0
	C90	1.2	0.5	90	105	100	25.4	255	0.750 to 0.999	5.0
	C90	1.2	0.5	90	105	100	25.4	255	≥ 1.000	6.0
	C95		0.5	95	110	105				
	T95	1.2	0.5	95	110	105	25.4	255	≤ 0.500	3.0
	T95	1.2	0.5	95	110	105	25.4	255	0.500 to 0.749	4.0
	T95	1.2	0.5	95	110	105	25.4	255	0.750 to 0.999	5.0
	T95	1.2	0.5	95	110	105	25.4	255	≥ 1.000	6.0
3	P110		0.6	110	140	125				
4	Q125		0.65	125	150	135	b		≤ 0.500	3.0
	Q125		0.65	125	150	135	b		0.500 to 0.749	4.0
	Q125		0.65	125	150	135	b		≥ 0.750	5.0

[a] In case of dispute, laboratory Rockwell C hardness testing shall be used as the referee method.

[b] No hardness limits are specified, but the maximum variation is restricted as a manufacturing control in accordance with section 7.8 and 7.9 of API Spec. 5CT.

Source: From API Specification 5CT, page 164.

Further Inspection and Testing Requirements for the manufacturing process are described in API Specification 5CT, Sections 7 and 10, and includes such as chemical composition, flattening test of welded pipe, hardness, impact testing and energy absorption, grain size determination, hardenability, Sulfide stress cracking test, metallographic evaluation, hydrostatic testing, dimensional tests and more.

8.2.3 Dimensions, masses, tolerances (section 8, API Specification 5CT)

Labels and Sizes In the dimensional tables from API-Spec 5CT, pipe is designated by labels, and by size (outside diameter). The outside diameter size of external-upset pipe is the outside diameter of the body of the pipe, not the upset portion.

Dimensions and Masses Pipe shall be furnished in the sizes, wall thickness and masses (as shown in Tables 8.3, 8.4) as specified on the purchase agreement except Grades C90, T95 and Q125 which may be furnished in other sizes, masses and wall thickness as agreed between purchaser and manufacturer. All dimensions shown without tolerances are related to the basis for design and are not subject to measurement to determine acceptance or rejection of product.

Diameter The outside diameter shall be within the tolerances specified below. For threaded pipe, the outside diameter at the threaded ends shall be such that the thread length, L_4, and the full-crest thread length, L_C, are within the tolerances and dimensions specified in API Spec 5B. For pipe furnished non-upset and plain-end and which is specified on the purchase agreement for the manufacture of pup-joints, the non-upset plain-end tolerances shall apply to the full length. (Inside diameter, d, is governed by the outside diameter and mass tolerances)

Tolerances The following tolerances apply to the outside diameter, D, of pipe:

Label 1	Tolerance on outside diameter, D
$<4-1/2$	±0.031 in
$\geq 4-1/2$	+1% D to − 0.5% D

For upset pipe, the following tolerances apply to the outside diameter of the pipe body immediately behind the upset for a distance of approximately 5.0 in. for sizes 5-1/2 in. and smaller, and a distance approximately equal

TABLE 8.3 Dimensions and Masses for Round Thread, Buttress Thread and Extreme-Line Casing

Labels[a]		Outside Diameter	Nominal linear Mass T&C[b,c]	Wall Thickness	Inside Diameter	Drift Diameter	Calculated mass[c]						
							Plain End	e_w Mass Gain or Loss Due to End Finishing[d] lb					
								Round Thread		Buttress Thread		Extreme-Line	
1	2	D in.	lb/ft	t in.	d in.	in.	W_{pe} lb/ft	Short	Long	Reg. OD	SCC	Stand'd	Optional
1	2	3	4	5	6	7	8	9	10	11	12	13	14
4½	9.50	4.500	9.50	0.205	4.090	3.965	9.41	4.20	—	—	—	—	—
4½	10.50	4.500	10.50	0.224	4.052	3.927	10.24	3.80	—	5.00	2.56	—	—
4½	11.60	4.500	11.60	0.250	4.000	3.875	11.36	3.40	3.80	4.60	2.16	—	—
4½	13.50	4.500	13.50	0.290	3.920	3.795	13.05	—	3.20	4.00	1.56	—	—
4½	15.10	4.500	15.10	0.337	3.826	3.701	15.00	—	2.80	3.20	0.76	—	—
5	11.50	5.000	11.50	0.220	4.560	4.435	11.24	5.40	—	—	—	—	—
5	13.00	5.000	13.00	0.253	4.494	4.369	12.84	4.80	5.80	6.60	2.42	—	—
5	15.00	5.000	15.00	0.296	4.408	4.283	14.88	4.20	5.20	5.80	1.62	4.60	—
5	18.00	5.000	18.00	0.362	4.276	4.151	17.95	—	4.20	4.40	0.22	1.40	—
5	21.40	5.000	21.40	0.437	4.126	4.001	21.32	—	2.95	2.46	−1.72	—	—
5	23.20	5.000	23.20	0.478	4.044	3.919	23.11	—	2.30	2.05	−2.09	—	—
5	24.10	5.000	24.10	0.500	4.000	3.875	24.05	—	1.95	1.24	−2.94	—	—
5½	14.00	5.500	14.00	0.244	5.012	4.887	13.71	5.40	—	—	—	—	—
5½	15.50	5.500	15.50	0.275	4.950	4.825	15.36	4.80	5.80	6.40	2.10	5.80	4.20
5½	17.00	5.500	17.00	0.304	4.892	4.767	16.89	4.40	5.40	5.80	1.50	4.80	3.20
5½	20.00	5.500	20.00	0.361	4.778	4.653	19.83	—	4.40	4.60	0.30	1.40	−0.20

(Continued)

TABLE 8.3 (*Continued*)

Labels[a]		Outside Diameter	Nominal linear Mass T&C[b,c]	Wall Thickness	Inside Diameter	Drift Diameter	Calculated mass[c]						
							Plain End	e_W Mass Gain or Loss Due to End Finishing[d] lb					
								Round Thread		Buttress Thread		Extreme-Line	
1	2	D in.	lb/ft	t in.	d in.	in.	W_{pe} lb/ft	Short	Long	Reg. OD	SCC	Stand'd	Optional
1	2	3	4	5	6	7	8	9	10	11	12	13	14
$5\frac{1}{2}$	23.00	5.500	23.00	0.415	4.670	4.545	22.56	—	3.20	3.40	– 0.90	0.00	– 1.60
$5\frac{1}{2}$	26.80	5.500	26.80	0.500	4.500	4.375	26.72	—	—	—	—	—	—
$5\frac{1}{2}$	29.70	5.500	29.70	0.562	4.376	4.251	29.67	—	—	—	—	—	—
$5\frac{1}{2}$	32.60	5.500	32.60	0.625	4.250	4.125	32.57	—	—	—	—	—	—
$5\frac{1}{2}$	35.30	5.500	35.30	0.687	4.126	4.001	35.35	—	—	—	—	—	—
$5\frac{1}{2}$	38.00	5.500	38.00	0.750	4.000	3.875	38.08	—	—	—	—	—	—
$5\frac{1}{2}$	40.50	5.500	40.50	0.812	3.876	3.751	40.69	—	—	—	—	—	—
$5\frac{1}{2}$	43.10	5.500	43.10	0.875	3.750	3.625	43.26	—	—	—	—	—	—
$6\frac{5}{8}$	20.00	6.625	20.00	0.288	6.049	5.924	19.51	11.00	13.60	14.40	2.38	—	—
$6\frac{5}{8}$	24.00	6.625	24.00	0.352	5.921	5.796	23.60	9.60	12.00	12.60	0.58	3.40	1.80
$6\frac{5}{8}$	28.00	6.625	28.00	0.417	5.791	5.666	27.67	—	10.20	10.60	– 1.42	0.20	– 1.40
$6\frac{5}{8}$	32.00	6.625	32.00	0.475	5.675	5.550	31.23	—	8.80	9.00	– 3.02	– 1.40	– 3.00
7	17.00	7.000	17.00	0.231	6.538	6.413	16.72	10.00	—	—	—	—	—
7	20.00	7.000	20.00	0.272	6.456	6.331	19.56	9.40	—	—	—	—	—

7	23.00	7.000	23.00	0.317	6.366	6.250[e]	22.65	8.00	10.40	11.00	1.60	6.00	4.20
7	23.00	7.000	23.00	0.317	6.366	6.241	22.65	8.00	10.40	11.00	1.60	6.00	4.20
7	26.00	7.000	26.00	0.362	6.276	6.151	25.69	7.20	9.40	9.60	0.20	2.80	1.00
7	29.00	7.000	29.00	0.408	6.184	6.059	28.75	—	8.00	8.20	−1.20	0.60	−1.20
7	32.00	7.000	32.00	0.453	6.094	6.000[e]	31.70	—	6.60	6.80	−2.60	−0.60	−2.40
7	32.00	7.000	32.00	0.453	6.094	5.969	31.70	—	6.60	6.80	−2.60	−0.60	−2.40
7	35.00	7.000	35.00	0.498	6.004	5.879	34.61	—	5.60	5.60	−3.80	1.00	−1.80
7	38.00	7.000	38.00	0.540	5.920	5.795	37.29	—	4.40	4.20	−5.20	−0.20	−3.00
7	42.70	7.000	42.70	0.626	5.750	5.625	42.65	—	—	—	—	—	—
7	46.40	7.000	46.40	0.687	5.625	5.500	46.36	—	—	—	—	—	—
7	50.10	7.000	50.10	0.750	5.500	5.375	50.11	—	—	—	—	—	—
7	53.60	7.000	53.60	0.812	5.376	5.251	53.71	—	—	—	—	—	—
7	57.10	7.000	57.10	0.875	5.250	5.125	57.29	—	—	—	—	—	—
$7\frac{5}{8}$	24.00	7.625	24.00	0.300	7.025	6.900	23.49	15.80	—	—	—	—	—
$7\frac{5}{8}$	26.20	7.625	26.40	0.328	6.969	6.844	25.59	15.20	19.00	20.60	6.21	6.40	4.00
$7\frac{5}{8}$	29.70	7.625	29.70	0.375	6.875	6.750	29.06	—	17.40	18.80	4.41	2.60	0.20
$7\frac{5}{8}$	33.70	7.625	33.70	0.430	6.765	6.640	33.07	—	15.80	17.00	2.61	0.00	2.40
$7\frac{5}{8}$	39.00	7.625	39.00	0.500	6.625	6.500	38.08	—	13.60	14.60	0.21	2.20	4.60
$7\frac{5}{8}$	42.80	7.625	42.80	0.562	6.501	6.376	42.43	—	12.01	11.39	3.06	—	—

(Continued)

TABLE 8.3 *(Continued)*

Labels[a] 1	Labels[a] 2	Outside Diameter D in.	Nominal linear Mass T&C[b,c] lb/ft	Wall Thickness t in.	Inside Diameter d in.	Drift Diameter in.	Calculated mass[c]: Plain End W_{pe} lb/ft	e_W Mass Gain or Loss Due to End Finishing[d] lb: Round Thread Short	Round Thread Long	Buttress Thread Reg. OD	Buttress Thread SCC	Extreme-Line Stand'd	Extreme-Line Optional
1	2	3	4	5	6	7	8	9	10	11	12	13	14
$7\frac{5}{8}$	45.30	7.625	45.30	0.595	6.435	6.310	44.71	—	11.04	11.04	3.36	—	—
$7\frac{5}{8}$	47.10	7.625	47.10	0.625	6.375	6.250	46.77	—	10.16	9.23	5.17	—	—
$7\frac{5}{8}$	51.20	7.625	51.20	0.687	6.251	6.126	50.95	—	—	—	—	—	—
$7\frac{5}{8}$	55.30	7.625	55.30	0.750	6.125	6.000	55.12	—	—	—	—	—	—
$7\frac{3}{4}$	46.10	7.750	46.10	0.595	6.560	6.500[e]	45.54	—	—	—	—	—	—
$7\frac{3}{4}$	46.10	7.750	46.10	0.595	6.560	6.435	45.54	—	—	—	—	—	—
$8\frac{5}{8}$	24.00	8.625	24.00	0.264	8.097	7.972	23.60	23.60	—	—	—	—	—
$8\frac{5}{8}$	28.00	8.625	28.00	0.304	8.017	7.892	27.04	22.20	—	—	—	—	—
$8\frac{5}{8}$	32.00	8.625	32.00	0.352	7.921	7.875[e]	31.13	20.80	27.60	28.30	6.03	13.20	8.80
$8\frac{5}{8}$	32.00	8.625	32.00	0.352	7.921	7.796	31.13	20.80	27.60	28.20	6.03	13.20	8.80
$8\frac{5}{8}$	36.00	8.625	36.00	0.400	7.825	7.700	35.17	19.40	25.60	26.20	4.03	7.60	4.20
$8\frac{5}{8}$	40.00	8.625	40.00	0.450	7.725	7.625[e]	39.33	—	23.80	24.20	2.03	4.00	0.60
$8\frac{5}{8}$	40.00	8.625	40.00	0.450	7.725	7.600	39.33	—	23.80	24.20	2.03	4.00	0.60
$8\frac{5}{8}$	44.00	8.625	44.00	0.500	7.625	7.500	43.43	—	21.80	22.20	0.03	1.60	−1.80

$8\frac{5}{8}$	49.00	8.625	49.00	0.557	7.511	7.286	48.04	—	19.60	19.80	– 2.37	– 0.80	– 4.20
$9\frac{5}{8}$	32.30	9.625	32.30	0.312	9.001	8.845	31.06	24.40	—	—	—	—	—
$9\frac{5}{8}$	36.00	9.625	36.00	0.352	8.921	8.765	34.89	23.00	32.00	31.00	6.48	—	—
$9\frac{5}{8}$	40.00	9.625	40.00	0.395	8.835	8.750[e]	38.97	21.40	30.00	29.00	4.48	10.60	7.20
$9\frac{5}{8}$	40.00	9.625	40.00	0.395	8.835	8.679	38.97	21.40	30.00	29.00	4.48	10.60	7.20
$9\frac{5}{8}$	43.50	9.625	43.50	0.435	8.755	8.599	42.73	—	28.20	27.20	2.68	5.40	2.00
$9\frac{5}{8}$	47.00	9.625	47.00	0.472	8.681	8.525	46.18	—	26.60	25.60	1.08	2.20	– 1.20
$9\frac{5}{8}$	53.50	9.625	53.50	0.545	8.535	8.500[e]	52.90	—	23.40	22.40	– 2.12	– 1.20	– 4.60
$9\frac{5}{8}$	53.50	9.625	53.50	0.545	8.535	8.379	52.90	—	23.40	22.40	– 2.12	– 1.20	– 4.60
$9\frac{5}{8}$	58.40	9.625	58.40	0.595	8.435	8.375[e]	57.44	—	21.50	20.13	– 4.40	—	—
$9\frac{5}{8}$	58.40	9.625	58.40	0.595	8.435	8.279	57.44	—	21.50	20.13	– 4.40	—	—
$9\frac{5}{8}$	59.40	9.625	59.40	0.609	8.407	8.251	58.70	—	—	—	—	—	—
$9\frac{5}{8}$	64.90	9.625	64.90	0.672	8.281	8.125	64.32	—	—	—	—	—	—
$9\frac{5}{8}$	70.30	9.625	70.30	0.734	8.157	8.001	69.76	—	—	—	—	—	—
$9\frac{5}{8}$	75.60	9.625	75.60	0.797	8.031	7.875	75.21	—	—	—	—	—	—
$10\frac{3}{4}$	32.75	10.750	32.75	0.279	10.192	10.036	31.23	29.00	—	—	—	—	—
$10\frac{3}{4}$	40.50	10.750	40.50	0.350	10.050	9.894	38.91	26.40	—	34.40	7.21	—	—
$10\frac{3}{4}$	45.50	10.750	45.50	0.400	9.950	9.875[e]	44.26	24.40	—	31.80	4.61	21.20	—
$10\frac{3}{4}$	45.50	10.750	45.50	0.400	9.950	9.794	44.26	24.40	—	31.80	4.61	21.20	—

(Continued)

TABLE 8.3 (Continued)

Labels[a] 1	Labels[a] 2	Outside Diameter D in.	Nominal linear Mass T&C[b,c] lb/ft	Wall Thickness t in.	Inside Diameter d in.	Drift Diameter in.	Plain End W_{pe} lb/ft	Calculated mass[c]: e_w Mass Gain or Loss Due to End Finishing[d] lb: Round Thread Short	Round Thread Long	Buttress Thread Reg. OD	Buttress Thread SCC	Extreme-Line Stand'd	Extreme-Line Optional
1	2	3	4	5	6	7	8	9	10	11	12	13	14
$10\frac{3}{4}$	51.00	10.750	51.00	0.450	9.850	9.694	49.55	22.60	—	29.40	2.21	18.40	—
$10\frac{3}{4}$	55.50	10.750	55.50	0.495	9.760	9.625[e]	54.26	20.80	—	27.00	−0.19	15.80	—
$10\frac{3}{4}$	55.50	10.750	55.50	0.495	9.760	9.604	54.26	20.80	—	27.00	−0.19	15.80	—
$10\frac{3}{4}$	60.70	10.750	60.70	0.545	9.660	9.504	59.45	18.80	—	24.40	—	13.00	—
$10\frac{3}{4}$	65.70	10.750	65.70	0.595	9.560	9.404	64.59	16.80	—	22.00	—	—	—
$10\frac{3}{4}$	73.20	10.750	73.20	0.672	9.406	9.250	72.40	—	—	—	—	—	—
$10\frac{3}{4}$	79.20	10.750	79.20	0.734	9.282	9.126	78.59	—	—	—	—	—	—
$10\frac{3}{4}$	85.30	10.750	85.30	0.797	9.156	9.000	84.80	—	—	—	—	—	—
$11\frac{3}{4}$	42.00	11.750	42.00	0.333	11.084	11.000[e]	40.64	29.60	—	—	—	—	—
$11\frac{3}{4}$	42.00	11.750	42.00	0.333	11.084	10.928	40.64	29.60	—	—	—	—	—
$11\frac{3}{4}$	47.00	11.750	47.00	0.375	11.000	10.844	45.60	27.60	—	35.80	—	—	—
$11\frac{3}{4}$	54.00	11.750	54.00	0.435	10.880	10.724	52.62	25.00	—	32.40	—	—	—
$11\frac{3}{4}$	60.00	11.750	60.00	0.489	10.772	10.625[e]	58.87	22.60	—	29.60	—	—	—
$11\frac{3}{4}$	60.00	11.750	60.00	0.489	10.772	10.616	58.87	22.60	—	29.60	—	—	—
$11\frac{3}{4}$	65.00	11.750	65.00	0.534	10.682	10.625[e]	64.03	—	—	—	—	—	—
$11\frac{3}{4}$	65.00	11.750	65.00	0.534	10.682	10.526	64.03	—	—	—	—	—	—

$11\frac{3}{4}$	71.00	11.750	71.00	0.582	10.586	10.430	69.48	—	—	—	—	—	—
$13\frac{3}{8}$	48.00	13.375	48.00	0.330	12.715	12.559	46.02	33.20	—	—	—	—	—
$13\frac{3}{8}$	54.50	13.375	54.50	0.380	12.615	12.459	52.79	30.80	—	40.20	—	—	—
$13\frac{3}{8}$	61.00	13.375	61.00	0.430	12.515	12.359	59.50	28.40	—	36.80	—	—	—
$13\frac{3}{8}$	68.00	13.375	68.00	0.480	12.415	12.259	66.17	25.80	—	33.60	—	—	—
$13\frac{3}{8}$	72.00	13.375	72.00	0.514	12.347	12.250[e]	70.67	24.20	—	31.60	—	—	—
$13\frac{3}{8}$	72.00	13.375	72.00	0.514	12.347	12.191	70.67	24.20	—	31.60	—	—	—
16	65.00	16.000	65.00	0.375	15.250	15.062	62.64	42.60	—	—	—	—	—
16	75.00	16.000	75.00	0.438	15.124	14.936	72.86	38.20	—	45.60	—	—	—
16	84.00	16.000	84.00	0.495	15.010	14.822	82.05	34.20	—	39.60	—	—	—
16	109.00	16.000	109.00	0.656	14.688	14.500	107.60	—	—	—	—	—	—
$18\frac{5}{8}$	87.50	18.625	87.50	0.435	17.755	17.567	84.59	73.60	—	86.40	—	—	—
20	94.00	20.000	94.00	0.438	19.124	18.936	91.59	47.00	61.20	54.80	—	—	—
20	106.50	20.000	106.50	0.500	19.000	18.812	104.23	41.60	54.80	48.40	—	—	—
20	133.00	20.000	133.00	0.635	18.730	18.542	131.45	30.00	40.60	35.20	—	—	—

See Figures 8.2, 8.3, 8.4 and 8.5.

[a] Labels are for information and assistance in ordering.

[b] Nominal linear masses, threaded and coupled (col. 4) are shown for information only.

[c] The densities of martensitic chromium steels (L80 Types 9Cr and 13Cr) are less than those of carbon steels. The masses shown are therefore not accurate for martensitic chromium steels. A mass correction factor of 0.989 may be used.

[d] Mass gain or loss due to end finishing. See formula on page 406.

[e] Drift diameter for most common bit size. This drift diameter shall be specified on the purchase agreement and marked on the pipe. See 8.10 in API Spec. 5CT for drift requirements.

Source: From API Specification 5CT, page 177, 178.

TABLE 8.4 Extreme-Line Casing Upset End Dimensions and Masses

				Finished pin-and-box dimensions[a]								
Labels		Outside Diameter in.	Nominal Linear, Mass Upset and Threaded lb/ft	Pin-and-Box Outside Diameter (Turned) −0.010 +0.020 in.		Pin Inside Diameter (Bored) −0.015 +0.015 in.	Box Inside Diameter (Bored) −0 +0.030 in.	Pin-and-Box Made-up (Power-Tight)[b]			Drift Dia'r for Finished Upset member in.	Drift Diameter for Full Length Drifting Min. in.
1	2			Standard	Optional	Standard and Optional	Standard and Optional	Outside Diameter −0.010 +0.020 in		Inside Diameter −0.015 +0.015 in	Standard and Optional	Standard and Optional
		D		M	M	B	D	Standard[c]	Optional[c]	Standard and Optional		
1	2	3	4	5	6	7	8	9	10	11	12	13
5	15.00	5.000	15.00	5.360	—	4.208	4.235	5.360	—	4.198	4.183	4.151
5	18.00	5.000	18.00	5.360	—	4.208	4.235	5.360	—	4.198	4.183	4.151
$5\frac{1}{2}$	15.50	5.500	15.50	5.860	5.780	4.746	4.773	5.860	5.780	4.736	4.721	4.653
$5\frac{1}{2}$	17.00	5.500	17.00	5.860	5.780	4.711	4.738	5.860	5.780	4.701	4.686	4.65
$5\frac{1}{2}$	20.00	5.500	20.00	5.860	5.780	4.711	4.738	5.860	5.780	4.701	4.686	4.653
$5\frac{1}{2}$	23.00	5.500	23.00	5.860	5.780	4.619	4.647	5.860	5.780	4.610	4.595	4.545
$6\frac{5}{8}$	24.00	6.625	24.00	7.000	6.930	5.792	5.818	7.000	6.930	5.781	5.766	5.730
$6\frac{5}{8}$	28.00	6.625	28.00	7.000	6.930	5.741	5.768	7.000	6.930	5.731	5.716	5.666
$6\frac{5}{8}$	32.00	6.625	32.00	7.000	6.930	5.624	5.652	7.000	6.930	5.615	5.600	5.550
7	23.00	7.000	23.00	7.390	7.310	6.182	6.208	7.390	7.310	6.171	6.156	6.151
7	26.00	7.000	26.00	7.390	7.310	6.182	6.208	7.390	7.310	6.171	5.156	6.151
7	29.00	7.000	29.00	7.390	7.310	6.134	6.160	7.390	7.310	6.123	6.108	6.059

7	32.00	7.000	32.00	7.390	7.310	6.042	6.069	7.390	7.310	6.032	6.017	5.969
7	35.00	7.000	35.00	7.530	7.390	5.949	5.977	7.530	7.390	5.940	5.925	5.879
7	38.00	7.000	38.00	7.530	7.390	5.869	5.897	7.530	7.390	5.860	5.845	5.795
$7\frac{5}{8}$	26.40	7.625	26.40	8.010	7.920	6.782	6.807	8.010	7.920	6.770	6.755	6.750
$7\frac{5}{8}$	29.70	7.625	29.70	8.010	7.920	6.782	6.807	8.010	7.920	6.770	6.755	6.750
$7\frac{5}{8}$	33.70	7.625	33.70	8.010	7.920	6.716	6.742	8.010	7.920	6.705	6.690	6.640
$7\frac{5}{8}$	39.00	7.625	39.00	8.010	7.920	6.575	6.602	8.010	7.920	6.565	6.550	6.500
$8\frac{5}{8}$	32.00	8.625	32.00	9.120	9.030	7.737	7.762	9.120	9.030	7.725	7.710	7.700
$8\frac{5}{8}$	36.00	8.625	36.00	9.120	9.030	7.737	7.762	9.120	9.030	7.725	7.710	7.700
$8\frac{5}{8}$	40.00	8.625	40.00	9.120	9.030	7.674	7.700	9.120	9.030	7.663	7.648	7.600
$8\frac{5}{8}$	44.00	8.625	44.00	9.120	9.030	7.575	7.602	9.120	9.030	7.565	7.550	7.500
$8\frac{5}{8}$	49.00	8.625	49.00	9.120	9.030	7.460	7.488	9.120	9.030	7.451	7.436	7.386
$9\frac{5}{8}$	40.00	9.625	40.00	10.100	10.020	8.677	8.702	10.100	10.020	8.665	8.650	8.599
$9\frac{5}{8}$	43.50	9.625	43.50	10.100	10.020	8.677	8.702	10.100	10.020	8.665	8.650	8.599
$9\frac{5}{8}$	47.00	9.625	47.00	10.100	10.020	8.633	8.658	10.100	10.020	8.621	8.606	8.525
$9\frac{5}{8}$	53.50	9.625	53.50	10.100	10.020	8.485	8.512	10.100	10.020	8.475	8.460	8.379
$10\frac{3}{4}$	45.50	10.750	45.50	11.460	—	9.829	9.854	11.460	—	9.819	9.804	9.794
$10\frac{3}{4}$	51.00	10.750	51.00	11.460	—	9.729	9.754	11.460	—	9.719	9.704	9.694
$10\frac{3}{4}$	55.50	10.750	55.50	11.460	—	9.639	9.664	11.460	—	9.629	9.614	9.604
$10\frac{3}{4}$	60.70	10.750	60.70	11.460	—	9.539	9.564	11.460	—	9.529	9.514	9.504

See Table 8.3 and Figure 8.5.

[a] Labels are for information and assistance in ordering.

[b] Nominal linear masses, threaded and coupled (col. 4) are shown for information only.

[c] The densities of martensitic chromium steels (L80 Types 9Cr and 13Cr) are less than those of carbon steels. The masses shown are therefore not accurate for martensitic chromium steels. A mass correction factor of 0.989 may be used.

Source: From API Specification 5CT, page 180.

to the outside diameter for sizes larger than 5-1/2 in. Measurements shall be made with calipers or snap gauges.

Label 1	Tolerance on outside diameter, m_{eu} or L_0
≤3−1/2	+3/32 in. to −1/32 in.
>3−1/2 to ≤5	+7/64 in. to −0.75% D
>5 to ≤8−5/8	+1/8 in. to −0.75% D
>8−5/8	+5/32 in. to −0.75% D

Wall Thickness Each length of pipe shall be measured for conformance to wall thickness requirements. The wall thickness at any place shall not be less than the tabulated thickness, t, minus the permissible under-tolerance of 12.5%. Wall thickness measurements shall be made with a mechanical caliper, a go/no-go gauge or with a properly calibrated nondestructive testing device of appropriate accuracy. In case of dispute, the measurements determined by the mechanical caliper shall govern.

Mass Each length of casing shall be weighed separately. The pipe may be weighed plain-end, upset, non-upset, threaded, or threaded and coupled. Threaded-and-coupled pipe may be weighed with the couplings screwed on or without couplings, provided proper allowance is made for the mass of the coupling. Threaded-and-coupled pipe, integral-joint pipe, and pipe shipped without couplings shall be weighed with or without thread protectors if proper allowances are made for the mass of the thread protectors. The masses determined as described above shall conform to the calculated masses as specified herein (or adjusted calculated masses) for the end finish specified on the purchase agreement, within the following stipulated mass tolerances:

Amount	Tolerance
Single lengths	+6.5% to −3.5%
Carload 18 144 kg (40,000 lb) or more	−1.75%
Carload less than 18 144 kg (40,000 lb)	−3.5%
Order items 18 144 kg (40,000 lb) or more	−1.75%
Order item less than 18 144 kg (40,000 lb)	−3.5%

The calculated masses shall be determined in accordance with the following formula:

$$m_L = (m_{pe} L) + e_m$$

where

m_L is the calculated mass of a piece of pipe of length L, in pounds (kilograms)
m_{pe} in the plain-end mass, in pounds per foot (kilograms per meter)
L length of pipe, including end finish (see note on length determination) in ft (meters)
e_m is the mass gain or loss due to end-finishing, in pounds (kilograms). For plain-end non-upset pipe, $e_m = 0$

Note: The densities of martensitic chromium steels (L80 Types 9Cr and 13Cr) are less than those of carbon steels. The masses shown are therefore not accurate for martensitic chromium steels. A mass correction factor of 0.989 may be used.

Length Pipe shall be furnished in range lengths conforming to Table 8.5, as specified on the purchase order. When pipe is furnished with threads and couplings, the length shall be measured to the outer face of the coupling, or if measured without couplings proper allowance shall be made to include the length of coupling. The extreme-line casing and integral joint tubing lengths shall be measured to the outer face of the box end. For pup joints and connectors, the length shall be measured from end to end.

Casing Jointers If so specified on the purchase order for round-thread casing only, jointers (two pieces coupled to make a standard length) may be furnished to a maximum of 5% of the order, but no length used in making a jointer shall be less than 5 ft.

Coupling API standards established three types of threaded joints:

1. Coupling joints with rounded thread (Figures 8.2 and 8.3) (long or short)
2. Coupling joints with asymmetrical trapezoidal thread buttress (Figure 8.4)
3. Extreme-line casing with trapezoidal thread without coupling (Figure 8.5)

There are also many non-API joints, like Hydril "CTS," Hydril "Super FJ-P," Armco SEAL-LOC, Mannesmann metal-to-metal seal casing and others.

The following are excerpts both from API Specification 5CT, Seventh Edition, October 2001 and API RP 5B1, Fifth Edition, October 1999.

8.2.4 Elements of Threads

Threaded connections are complicated mechanisms consisting of many elements that must interact in prescribed fashion to perform a useful function. Each of these elements of a thread may be gauged individually as

TABLE 8.5 Round-Thread Casing Coupling Dimensions, Masses and Tolerances

Label 1	Size[a] Outside diameter D in.	Outside diameter W in.	Minimum length in Short N_L	Minimum length in Long N_L	Diameter of recess Q^b in.	Width of bearing face b in.	Mass lb Short	Mass lb Long
1	2	3	4	5	6	7	8	9
$4\frac{1}{2}$	4.500	5.000	$6\frac{1}{4}$	7	$4\frac{19}{32}$	$\frac{5}{32}$	7.98	9.16
5	5.000	5.563	$6\frac{1}{2}$	$7\frac{3}{4}$	$5\frac{3}{32}$	$\frac{3}{16}$	10.27	12.68
$5\frac{1}{2}$	5.500	6.050	$6\frac{3}{4}$	8	$5\frac{19}{32}$	$\frac{1}{8}$	11.54	14.15
$6\frac{5}{8}$	6.625	7.390	$7\frac{1}{4}$	$8\frac{3}{4}$	$6\frac{23}{32}$	$\frac{1}{4}$	20.11	25.01
7	7.000	7.656	$7\frac{1}{4}$	9	$7\frac{3}{32}$	$\frac{3}{16}$	18.49	23.87
$7\frac{5}{8}$	7.625	8.500	$7\frac{1}{2}$	$9\frac{1}{4}$	$7\frac{25}{32}$	$\frac{7}{32}$	27.11	34.46
$8\frac{5}{8}$	8.625	9.625	$7\frac{3}{4}$	10	$8\frac{25}{32}$	$\frac{1}{4}$	35.79	47.77
$9\frac{5}{8}$	9.625	10.625	$7\frac{3}{4}$	$10\frac{1}{2}$	$9\frac{25}{32}$	$\frac{1}{4}$	39.75	56.11
$10\frac{3}{4}$	10.750	11.750	8	—	$10\frac{29}{32}$	$\frac{1}{4}$	45.81	—
$11\frac{3}{4}$	11.750	12.750	8	—	$11\frac{29}{32}$	$\frac{1}{4}$	49.91	—
$13\frac{3}{8}$	13.375	14.375	8	—	$13\frac{17}{32}$	$\frac{7}{32}$	56.57	—
16	16.000	17.000	9	—	$16\frac{7}{32}$	$\frac{7}{32}$	76.96	—
$18\frac{5}{8}$	18.625	20.000	9	—	$18\frac{27}{32}$	$\frac{7}{32}$	119.07	—
20	20.000	21.000	9	$11\frac{1}{2}$	$20\frac{7}{32}$	$\frac{7}{32}$	95.73	126.87

Tolerance on outside diameter W, ±1% but not greater than ±1/8 in. Groups 1, 2 and 3.
Tolerance on outside diameter W, ±1% but not greater than +1/8 in., Group 4.
[a] The size designation for the coupling is the same as the size designation for the pipe on which the coupling is used.
[b] Tolerance on diameter of recess, Q, for all groups is 0 to +0.031 in.
Source: From API Specification 5CT, page 184. See also Figures 8.2 and 8.5.

described in API RP 5B1, 5th Edition, October 1999. The thread elements are defined as

1. *Thread height or depth.* The thread height or depth is the distance between the threaded crest and the thread root normal to the axis of the thread (Figure 8.6).
2. *Lead.* For pipe thread inspection purposes, lead is defined as the distance from a point on a thread to a corresponding point on the adjacent thread measured parallel to the thread axis (Figure 8.6).
3. *Taper.* Taper is the change of diameter of a thread expressed in in./ft of thread length.

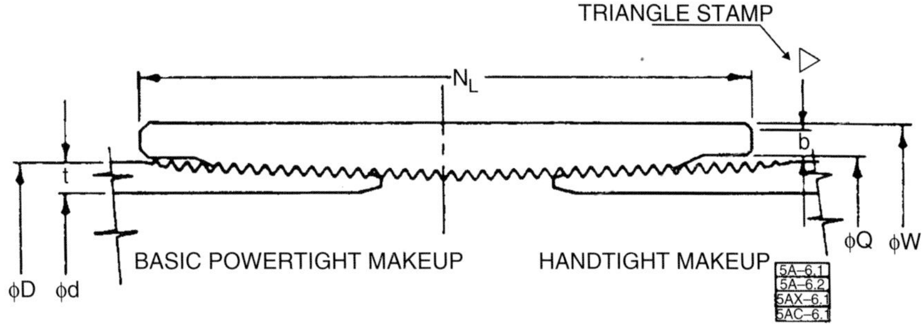

FIGURE 8.2 D.1—Short round thread casing and coupling. From API Spec 5CT 7th Edition, October 2001, p. 136.

See Table 8.3 for pipe dimensions, Table 8.5 for coupling dimensions and API Spec 5B for L4.

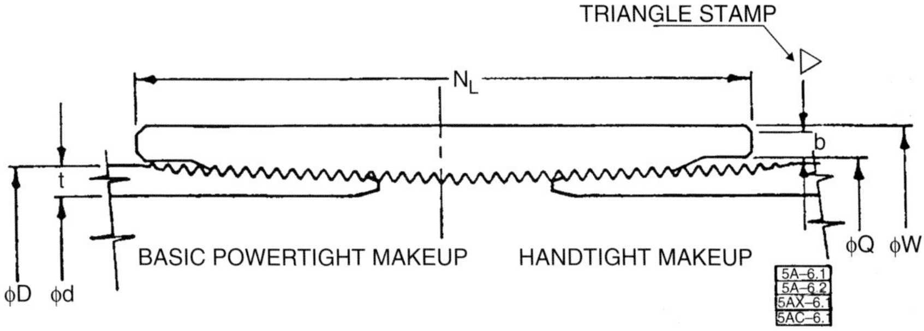

FIGURE 8.3 Long round thread casing and coupling. From API Spec 5CT 7th Edition, October 2001, p. 136.

See Table 8.3 for pipe dimensions, Table 8.5 for coupling dimensions and API Spec 5B for L4.

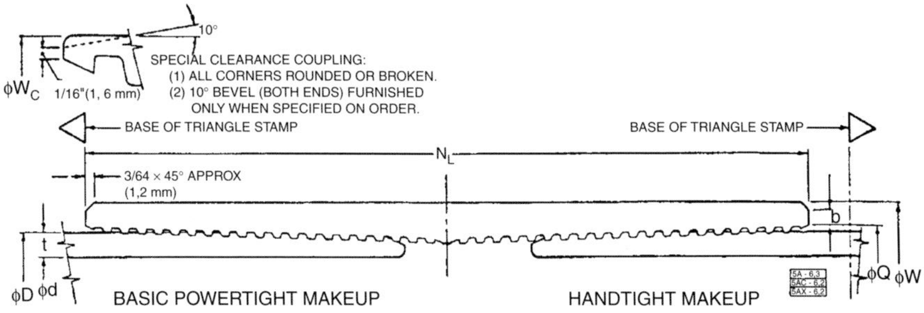

FIGURE 8.4 Buttress thread casing and coupling. From API Spec 5CT 7th Edition, October 2001, p. 137.

See Table 8.3 for pipe dimensions, Table 8.6 for coupling dimensions and API Spec 5B for L4.

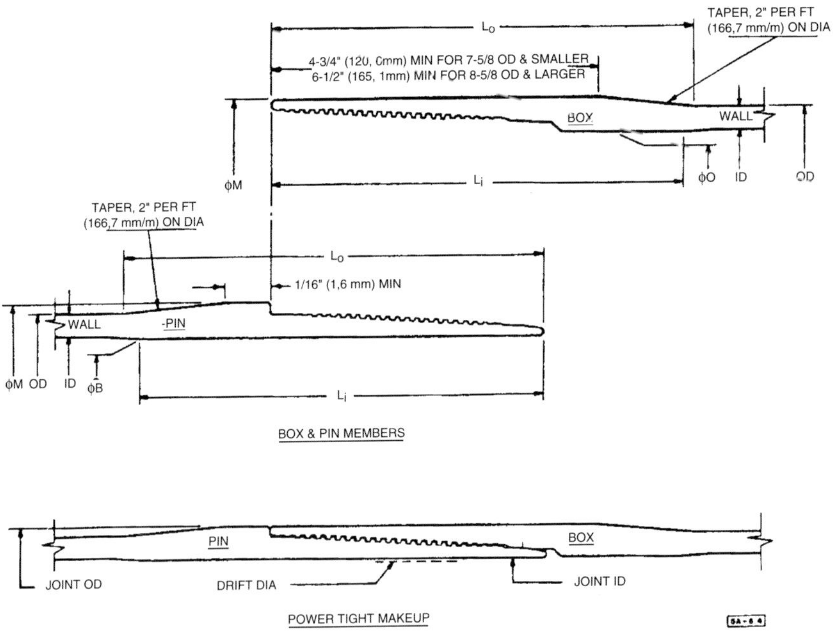

Size	Length of Upset					
	Pin † Min.		Box† Min.		Pin or Box Max	
in. OD	in. L_i	mm	in. L_i	mm	in. L_o	mm
5	$6\frac{5}{8}$	168,3	7	177,8	8	203,2
$5\frac{1}{2}$	$6\frac{5}{8}$	168,3	7	177,8	8	203,2
$6\frac{5}{8}$	$6\frac{5}{8}$	168,3	7	177,8	8	203,2
7	$6\frac{5}{8}$	168,3	7	177,8	8*	203,2*
$7\frac{5}{8}$	$6\frac{5}{8}$	168,3	7	177,8	8	203,2
$8\frac{5}{8}$	8	203,3	$8\frac{3}{4}$	222,2	11	279,4
$9\frac{5}{8}$	8	203,2	$8\frac{3}{4}$	222,2	11	279,4
$10\frac{3}{4}$	8	203,2	$8\frac{3}{4}$	222,2	$12\frac{3}{4}$	323,8

†L_i is the minimum length from end of pipe of the machined diameter B on pin, or machined diameter D plus length of thread on box, to the beginning of the internal upset runout.

*L_o shall be 9 in. (228,6 mm) max. for 7 in. — 35 lb/ft and 7 in. — 38 lb/ft casing.

FIGURE 8.5 Extreme-line casing. Taken from API Spec 5CT Seventh Edition, October 201, p. 141, 142.

See Table 8.4 for pipe dimensions and API Spec 5B for thread details.

Round Threads (Figure 8.6) The purpose of round top (crest) and round bottom (root) is

A. To improve the resistance of the threads from galling in make-up.
B. To provide a controlled clearance between make-up thread crest and root for foreign particles or contaminants.
C. To make the crest less susceptible to harmful damage from minor scratches or dents. If sufficient interference is applied during makeup,

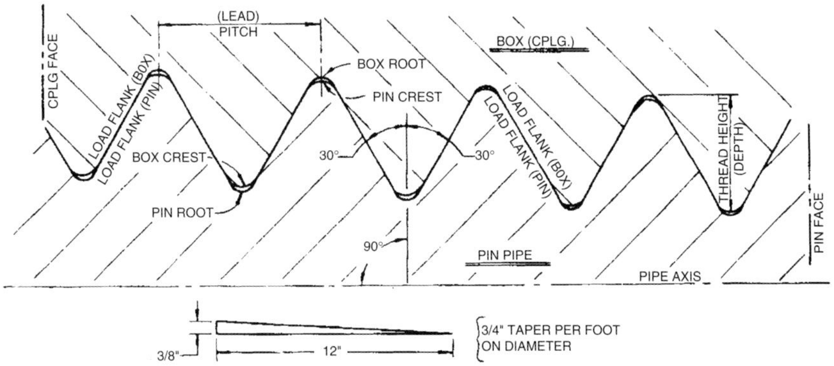

FIGURE 8.6 Round thread casing and tubing thread configuration. From API RP 5B1, 5th Edition, October 199, p. 6.

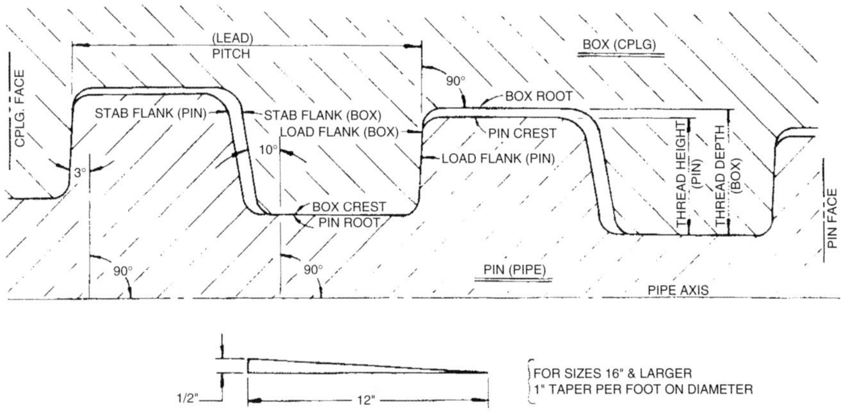

FIGURE 8.7 Buttress thread configuration for 16-in. OD and larger casing. From API RP 5B1, 5th Edition, October 199, p. 7.

the leak path through the connection should be through the annular clearance between mated crest and roots. Proper thread compound is necessary to ensure leak resistance.

Buttress Threads (Figure 8.7 and 8.8) Buttress threads are designed to resists high axial tension or compression loading in addition to offering resistance to leakage.

The 3° load flank offers resistance to disengagement under high axial tension loading, while the 10° stub flank offers resistance to high axial compression loading. In any event, leak resistance is again accomplished with use of proper thread compound and/or thread coating agents. Leak

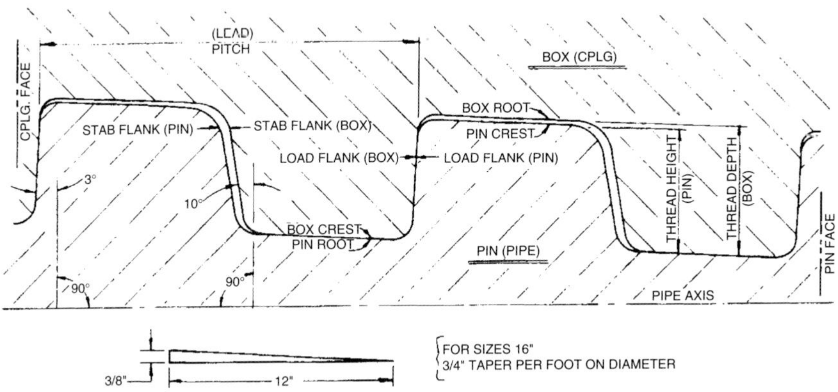

FIGURE 8.8 Buttress thread configuration for 13 3/8-in. OD and smaller casing. From API RP 5B1, 5th Edition, October 199, p. 7.

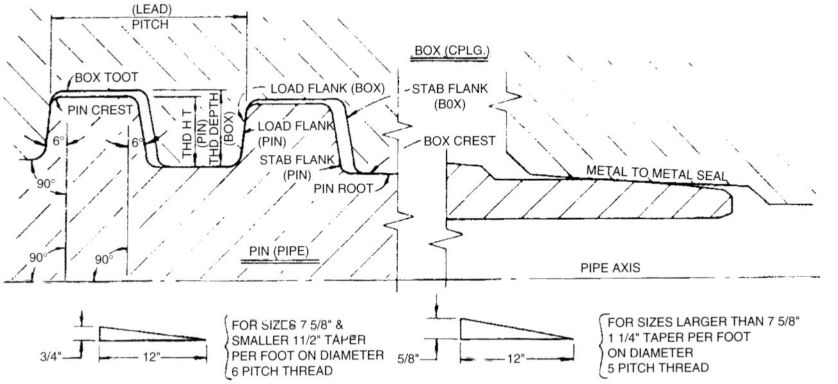

FIGURE 8.9 Extreme-line casing thread configuration. From API RP 5B1, 5th Edition, October 199, p. 7.

resistance is controlled by proper assembly (interference) within the perfect thread only.

8.2.5 Extreme-Line Casing (Integral Connection)

Extreme-line casing in all sizes uses a modified acme type thread having a 12° included angle between stub and load flanks, and all threads have crests and roots flat and parallel to the axis (Figure 8.9). For all sizes, the threads are not intended to be leak resistant when made up. Threads are used purely as a mechanical means to hold the joint members together during axial tension loading. The connection uses upset pipe ends for pin and box members that are an integral part of the pipe body. Axial compression load resistance is primarily offered by external shouldering to the connection or makeup.

Leak resistance is obtained on makeup by interference of metal-to-metal seal between a long radius curved seal surface on the pin member engaging a conical metal seal surface of the box member (Figures 8.5 and 8.9).

Thread compound is not necessarily a critical agent to ensure leak resistance, but instead is used primarily as an antigalling or antiseizure agent.

Material Couplings for pipe (both casing and tubing) shall be seamless and, unless otherwise specified on the purchase order, shall be of the same grade as the pipe. Exceptions are stipulated in Sections 9.2 and 9.3 in API Spec 5CT, for grades H-40, J-55, K-55, M65, N-80, and P-110, where couplings of stipulated higher grades are acceptable when specified on the purchase agreement.

When couplings are electroplated, the electroplating process should be controlled to minimize hydrogen absorption.

Note: Most buttress thread couplings will not develop the highest minimum joint strength unless couplings of the next higher order are specified (see API Specification 5CT for more detailed information).

Physical Properties Couplings shall conform to the mechanical requirements specified in clauses 7 and 10 in API Spec 5CT, including the frequency of testing, re-test provision, etc. A record of these tests shall be open to inspection by the purchaser. Tensile tests shall be made on each heat of steel from which couplings are produced.

Dimensions and Tolerances Couplings shall conform to the dimensions and tolerances shown in Tables 8.5 and 8.6. Unless otherwise specified, threaded and coupled casing and tubing shall be furnished with regular couplings.

Note: Couplings inspection procedures are described by API RP 5B1, 5th Edition, October 1999.

8.2.6 Thread Protectors

Design The entity performing the threading shall apply external and internal thread protectors of such design, material and mechanical strength to protect the thread and end of the pipe from damage under normal handling and transportation. External thread protectors shall cover the full length of the thread on the pipe, and internal thread protectors shall cover the equivalent total pipe thread length of the internal thread. Thread protectors shall be of such design and material to inhibit infiltration of dust and water to the threads during transportation and normal storage period. Normal storage period shall be considered as approximately one year. The

TABLE 8.6 Buttress-Thread Casing Coupling Dimensions, Masses and Tolerances

Label 1	Size[a] Outside diameter D in.	Outside diameter in Regular W in.	Special clearance W_C	Minimum length N_L in.	Diameter of counterbore Q^b in.	Width of bearing face B in.	Mass lb Regular	Special clearance
1	2	3	4	5	6	7	8	9
$4\frac{1}{2}$	4.500	5.000	4.875	$8\frac{7}{8}$	4.640	$\frac{1}{8}$	10.12	7.68
5	5.000	5.563	5.375	$9\frac{1}{8}$	5.140	$\frac{5}{32}$	13.00	8.82
$5\frac{1}{2}$	5.500	6.050	5.875	$9\frac{1}{4}$	5.640	$\frac{5}{32}$	14.15	9.85
$6\frac{5}{8}$	6.625	7.390	7.000	$9\frac{5}{8}$	6.765	$\frac{1}{4}$	24.49	12.46
7	7.000	7.656	7.375	10	7.140	$\frac{7}{32}$	23.24	13.84
$7\frac{5}{8}$	7.625	8.500	8.125	$10\frac{3}{8}$	7.765	$\frac{5}{16}$	34.88	20.47
$8\frac{5}{8}$	8.625	9.625	9.125	$10\frac{5}{8}$	8.765	$\frac{3}{8}$	45.99	23.80
$9\frac{5}{8}$	9.625	10.625	10.125	$10\frac{5}{8}$	9.765	$\frac{3}{8}$	51.05	26.49
$10\frac{3}{4}$	10.750	11.750	11.250	$10\frac{5}{8}$	10.890	$\frac{3}{8}$	56.74	29.52
$11\frac{3}{4}$	11.750	12.750	—	$10\frac{5}{8}$	11.890	$\frac{3}{8}$	61.80	—
$13\frac{3}{8}$	13.375	14.375	—	$10\frac{5}{8}$	13.515	$\frac{3}{8}$	70.03	—
16	16.000	17.000	—	$10\frac{5}{8}$	16.154	$\frac{3}{8}$	88.81	—
$18\frac{5}{8}$	18.625	20.000	—	$10\frac{5}{8}$	18.779	$\frac{3}{8}$	138.18	—
20	20.000	21.000	—	$10\frac{5}{8}$	20.154	$\frac{3}{8}$	110.45	—

See also Figure 8.6.
Tolerance on outside diameter W, ± 1%.
[a]The size designation for the coupling is the same as the size designation for the pipe on which the coupling is used.
Source: From API Specification 5CT, page 185.

thread forms in protectors shall be such that the product threads are not damaged by the protectors. Thread protectors are not required for pup-joints and accessories provided they are packaged in a manner that protects the threads.

Material Protector material shall contain no compounds capable of causing corrosion or promoting adherence of the protectors to the threads and shall be suitable for service temperatures from −50°F to +150°F (−46°C to +66°C).

Note. Bare steel protectors shall not be used on Grade L80 Types 9Cr and 13Cr tubulars.

Minimum Performance Properties of Casing Results of years of field experience have revealed that to reduce the risk of failure, the minimum yield strength should be used instead of average yield strength to determine the performance properties of casing.

Values for collapse resistance, internal yield pressure, pipe body, and joint strength for steel grades as in Table 8.2 are given in Table 8.7. Table 8.7 is directly taken from API Bulletin 5C2, 21st Edition, October 1999. Formulas and procedures for calculating the values in Table 8.8 are given in API Bulletin 5C3 6th Edition, October 1, 1994 [152] and are as follows:

Collapse Pressure (Section 2 of API Bulletin 5C3, 6th edition, October 1, 1994).

Yield Strength Collapse Pressure Formula The yield strength collapse pressure is not a true collapse pressure but rather the external pressure P_{yp} that generates minimum yield stress Y_p on the inside of the wall of a tube as calculated by

$$P_{yp} = 2Y_p \left[\frac{(D/t) - 1}{(D/t)^2} \right] \tag{8.1}$$

The applicable D/t ratios for field strength collapse pressure are shown in Table 8.8.

Plastic Collapse Pressure Formula The minimum collapse pressure for the plastic range of collapse is calculated by

$$P_p = Y_p \left[\frac{A}{D/t} - B \right] - C \tag{8.2}$$

The formula for minimum plastic collapse pressure is applicable for D/t values as shown in Table 8.8. The factors A, B and C are given in Table 8.9.

TABLE 8.7 Minimum Performance of Casing

1 Size: Outside Diameter in. D	2 Nominal Weight, Threads and Coupling lb per ft	3 Grade	4 Wall Thickness in. t	5 Inside Diameter in. d	6 Threaded and Coupled: Drift Diameter in.	7 Threaded and Coupled: Outside Diameter of Coupling in. W	8 Threaded and Coupled: Outside Diameter Special Clearance Coupling in. W_e	9 Extreme Line: Drift Diameter in.	10 Extreme Line: Outside Diameter of Box Powertight in. M	11 Collapse Resistance psi	12 Pipe Body Yield Strength 1000 lbs
4½	9.50	H-40	0.205	4.090	3.965	5.000	—	—	—	2,700	111
	9.50	J-55	0.205	4.090	3.965	5.000	—	—	—	3,310	152
	10.50	J-55	0.224	4.052	3.927	5.000	4.875	—	—	4,010	163
	11.60	J-55	0.250	4.000	3.875	5.000	4.875	—	—	4,960	184
	9.50	K-55	0.205	4.090	3.965	5.000	—	—	—	3,310	152
	10.50	K-55	0.224	4.052	3.927	5.000	4.875	—	—	4,010	165
	11.60	K-55	0.250	4.000	3.875	5.000	4.875	—	—	4,960	184
	11.60	C-75	0.250	4.000	3.875	5.000	4.875	—	—	6,130	250
	13.50	C-75	0.290	3.920	3.795	5.000	4.875	—	—	8,170	288
	11.60	L-80	0.250	4.000	3.875	5.000	4.875	—	—	6,350	267
	13.50	L-80	0.290	3.920	3.795	5.000	4.875	—	—	8,540	307
	11.60	N-80	0.250	4.000	3.875	5.000	4.875	—	—	6,350	267
	13.50	N-80	0.290	3.920	3.795	5.000	4.875	—	—	8,540	307
	11.60	C-95	0.250	4.000	3.875	5.000	4.875	—	—	7,010	317
	13.50	C-95	0.290	3.920	3.795	5.000	4.875	—	—	9,650	364
	11.60	P-110	0.250	4.000	3.875	5.000	4.875	—	—	7,560	367
	13.50	P-110	0.290	3.920	3.795	5.000	4.875	—	—	10,670	422
	15.10	P-110	0.337	3.826	3.701	5.000	4.875	—	—	14,320	485
5	11.50	J-55	0.220	4.560	4.435	5.563	—	—	—	3,060	182
	13.00	J-55	0.253	4.494	4.369	5.563	5.375	—	—	4,110	208
	15.00	J-55	0.296	4.408	4.283	5.563	5.375	4.151	5.360	5,550	241
	11.50	K-55	0.220	4.560	4.435	5.563	—	—	—	3,060	182
	13.00	K-55	0.253	4.494	4.369	5.563	5.375	—	—	4,140	208
	15.00	K-55	0.296	4.408	4.283	5.563	5.375	4.151	5.360	5,550	241
	15.00	C-75	0.296	4.408	4.283	5.563	5.375	4.151	5.360	6,970	328
	18.00	C-75	0.362	4.276	4.151	5.563	5.375	4.151	5.360	10,000	396
	21.40	C-75	0.437	4.126	4.001	5.563	5.375	—	—	11,060	470
	24.10	C-75	0.500	4.000	3.875	5.563	5.375	—	—	13,500	530
	15.00	L-80	0.296	4.408	4.283	5.563	5.375	4.151	5.360	7,250	350
	18.00	L-80	0.362	4.276	4.151	5.563	5.375	4.151	5.360	10,490	422
	21.40	L-80	0.437	4.126	4.001	5.563	5.375	—	—	12,760	501
	24.10	L-80	0.500	4.000	3.875	5.563	5.375	—	—	14,400	566
	15.00	N-80	0.296	4.408	4.283	5.563	5.375	4.151	5.360	7,230	350
	18.00	N-80	0.362	4.276	4.151	5.563	5.375	4.151	5.360	10,400	422
	21.40	N-80	0.437	4.126	4.001	5.563	5.375	—	—	12,760	501
	24.10	N-80	0.500	4.000	3.873	5.563	5.375	—	—	14,400	566
	15.00	C-95	0.296	4.408	4.283	5.563	5.375	4.151	5.360	8,090	415
	18.00	C-95	0.362	4.276	4.151	5.563	5.375	4.151	5.360	12,010	501
	21.40	C-95	0.437	4.126	4.001	5.563	5.375	—	—	15,150	595
	24.10	C-95	0.500	4.000	3.875	5.563	5.375	—	—	17,100	672
	15.00	P-110	0.206	4.408	5.283	5.563	5.375	4.151	5.360	8,830	481
	18.00	P-110	0.362	4.276	4.151	5.563	5.375	4.151	5.360	13,450	580
	21.40	P-110	0.437	4.126	4.001	5.563	5.375	—	—	17,550	689
	24.10	P-110	0.500	4.000	3.875	5.563	5.375	—	—	10,800	778

1 Size: Outside Diameter in. D	2 Nominal Weight lb per ft	3 Grade	13 Internal Yield Pressure at Minimum Yield, psi: Plain End or Extreme Line	14 Round Thread: Short	15 Round Thread: Long	16 Buttress Thread, Regular Coupling: Same Grade	17 Buttress Thread, Regular Coupling: Higher Grader	18 Buttress Thread, Special Clearance Coupling: Same Grade	19 Buttress Thread, Special Clearance Coupling: Higher Grade	20 ★ Joint Strength — 1,000 lbs, Threaded and Coupled, Round Thread: Short	21 Round Thread: Long	22 Buttress Thread: Regular Coupling	23 Buttress Thread: Regular Coupling Higher Grade:	24 Buttress Thread: Special Clearance Coupling	25 Buttress Thread: Special Clearance Coupling Higher Grade‡	26 Extreme Line: Standard Joint	27 Extreme Line: Optional Joint
4½	9.50	H-40	3,190	3,190	—	—	—	—	—	77	—	—	—	—	—	—	—
	9.50	J-55	4,380	4,380	—	—	—	—	—	101	—	—	—	—	—	—	—
	10.50	J-55	4,790	4,790	—	4,790	4,790	4,790	4,790	132	—	203	203	203	203	—	—
	11.60	J-55	5,350	5,350	5,350	5,350	5,350	5,350	5,350	154	162	225	225	225	225	—	—
	9.50	K-55	4,380	4,380	—	—	—	—	—	112	—	—	—	—	—	—	—
	10.50	K-55	4,790	4,790	—	4,790	4,790	4,790	4,790	146	—	249	249	249	249	—	—
	11.60	K-55	5,350	5,350	5,350	5,350	5,350	5,350	5,350	170	180	277	277	277	277	—	—
	11.60	C-75	7,290	—	7,290	7,290	—	7,290	—	—	212	288	—	288	—	—	—
	13.50	C-75	8,460	—	8,460	8,460	—	7,490	—	—	257	331	—	320	—	—	—
	11.60	L-80	7,780	—	7,780	7,780	7,780	7,780	7,780	—	212	291	—	291	—	—	—
	13.50	L-80	9,020	—	9,020	9,020	9,020	7,990	9,020	—	257	334	—	320	—	—	—
	11.60	N-80	7,780	—	7,780	7,780	7,780	7,780	7,780	—	223	304	304	304	304	—	—
	13.50	N-80	9,020	—	9,020	9,020	9,020	7,990	9,020	—	270	349	349	337	349	—	—
	11.60	C-95	9,240	—	9,240	9,240	—	9,240	—	—	284	325	325	325	—	—	—
	13.50	C-95	10,710	—	10,710	10,710	—	9,490	—	—	284	374	374	353	—	—	—
	11.60	P-110	10,690	—	10,690	10,690	10,690	10,690	10,690	—	279	385	385	385	385	—	—
	13.50	P-110	12,410	—	12,410	12,410	12,410	10,990	12,410	—	338	443	443	421	443	—	—
	15.10	P-110	14,420	—	14,420	13,460	14,420	10,990	14,310	—	406	509	509	421	509	—	—
5	11.50	J-55	4,240	4,240	—	—	—	—	—	133	—	—	—	—	—	—	—
	13.00	J-55	4,870	4,870	4,870	4,870	4,870	4,870	4,870	169	182	252	252	252	252	—	—
	15.00	J-55	5,700	5,700	5,700	5,700	5,700	5,130	5,700	207	223	293	293	287	293	328	—
	11.50	K-55	4,240	4,240	—	—	—	—	—	147	—	—	—	—	—	—	—
	13.00	K-55	4,870	4,870	4,870	4,870	4,870	4,870	4,870	186	201	309	309	309	309	—	—
	15.00	K-55	5,700	5,700	5,700	5,700	5,700	5,130	5,700	228	246	359	359	359	359	416	—
	15.00	C-75	7,770	—	7,770	7,770	—	6,990	—	—	295	375	—	364	—	416	—
	18.00	C-75	9,500	—	9,500	9,290	—	6,990	—	—	376	452	—	364	—	446	—
	21.40	C-75	11,470	—	10,140	9,290	—	6,990	—	—	466	510	—	364	—	—	—
	24.10	C-75	13,130	—	10,140	9,920	—	6,990	—	—	538	510	—	364	—	—	—
	15.00	L-80	8,290	—	8,290	8,290	8,290	7,460	8,290	—	295	379	—	364	—	416	—
	18.00	L-80	10,140	—	10,140	9,910	10,140	7,460	10,140	—	376	457	—	364	—	446	—
	21.40	L-80	12,240	—	10,810	9,910	—	7,460	—	—	466	510	—	364	—	—	—
	24.10	L-80	14,000	—	10,810	9,910	—	7,460	—	—	538	510	—	364	—	—	—
	15.00	N-80	8,290	—	8,290	8,290	8,290	7,460	8,290	—	311	396	396	383	396	437	—
	18.00	N-80	10,140	—	10,140	9,910	10,140	7,460	10,140	—	396	477	477	383	477	469	—
	21.40	N-80	12,240	—	10,810	9,910	12,240	7,460	10,250	—	490	537	566	383	479	—	—
	24.10	N-80	14,000	—	10,810	9,910	13,620	7,460	10,250	—	567	537	639	383	479	—	—
	15.00	C-95	9,840	—	9,840	9,840	—	8,850	—	—	326	424	—	402	—	459	—
	18.00	C-95	12,040	—	12,040	11,770	—	8,850	—	—	416	512	—	402	—	493	—
	21.40	C-95	14,530	—	12,840	11,770	—	8,850	—	—	515	563	—	402	—	—	—
	24.10	C-95	16,630	—	12,850	11,770	—	8,850	—	—	505	563	—	402	—	—	—
	15.00	P-110	11,400	—	11,400	11,400	11,400	10,250	11,400	—	388	503	503	479	503	547	—
	18.00	P-110	13,940	—	13,940	13,620	13,910	10,250	13,940	—	495	606	606	479	606	587	—
	21.40	P-110	16,820	—	14,870	13,620	16,820	10,250	13,980	—	613	671	720	479	613	—	—
	24.10	P-110	19,250	—	14,870	13,620	18,580	10,250	13,080	—	708	671	812	479	613	—	—

5½	14.00	H-40	0.244	5.012	4.887	6.050	—	—	—	2,630	161	3,110	3,110	—	—	—	—	—	130	—	—	—	—	—	—	—
	14.00	J-55	0.244	5.012	4.887	6.050	—	—	—	3,120	222	4,270	4,270	—	—	—	—	—	172	—	—	—	—	—	—	—
	15.50	J-55	0.273	4.950	4.825	6.050	5.875	4.653	5.860	4,040	248	4,810	4,810	4,810	4,810	4,810	4,730	4,810	202	217	300	300	300	300	339	339
	17.00	J-55	0.304	4.892	4.767	6.050	5.875	4.653	5.860	4,910	273	5,320	5,320	5,320	5,320	5,320	4,730	5,320	229	247	329	329	318	329	372	372
	14.00	K-55	0.244	5.012	4.887	6.050	—	—	—	3,120	222	4,270	4,270	—	—	—	—	—	189	—	—	—	—	—	—	—
	15.50	K-55	0.275	4.050	4.825	6.050	5.875	4.653	5.860	4,040	248	4,810	4,810	4,810	4,810	4,810	4,730	4,810	222	239	366	366	366	366	429	429
	17.00	K-55	0.304	4.892	4.767	6.050	5.875	4.653	5.860	4,910	273	5,320	5,320	5,320	5,320	5,320	4,730	5,320	252	272	402	402	402	402	471	471
	17.00	C-75	0.304	4.892	4.767	6.050	5.875	4.653	5.860	6,070	372	7,250	—	7,250	7,250	—	6,450	—	—	327	423	—	403	—	471	471
	20.00	C-75	0.361	4.778	4.653	6.050	5.875	4.653	5.860	8,410	437	8,610	—	8,610	8,430	—	6,450	—	—	403	497	—	403	—	497	479
	23.00	C-75	0.415	4.670	4.545	6.050	5.875	4.545	5.860	10,460	497	9,900	—	9,260	8,430	—	6,450	—	—	473	550	—	403	—	549	479
	17.00	L-80	0.304	4.892	4.767	6.050	5.875	4.653	5.860	6,280	367	7,740	—	7,740	7,740	7,740	6,880	7,740	—	338	428	—	403	—	471	471
	20.00	L-80	0.361	4.778	4.653	6.050	5.875	4.653	5.860	8,830	466	9,190	—	9,190	8,990	9,190	6,880	9,190	—	416	503	—	403	—	497	479
	23.00	L-80	0.415	4.670	4.545	6.050	5.875	4.545	5.860	11,160	530	10,560	—	9,880	8,990	10,560	6,880	9,460	—	489	550	—	403	—	549	479
	17.00	N-80	0.304	4.892	4.767	6.050	5.875	4.653	5.860	6,280	397	7,740	—	7,740	7,740	7,740	6,880	7,740	—	348	446	446	424	446	496	496
	20.00	N-80	0.361	4.778	4.653	6.050	5.875	4.653	5.860	8,830	466	9,190	—	9,190	8,990	9,190	6,880	9,190	—	428	524	524	424	524	523	504
	23.00	N-80	0.415	4.670	4.545	6.050	5.875	4.545	5.860	11,160	530	10,560	—	9,880	8,990	10,560	6,880	9,460	—	502	579	596	424	530	577	504
	17.00	C-95	0.304	4.892	4.767	6.050	5.875	4.653	5.860	8,930	471	9,190	—	9,190	9,190	—	8,170	—	—	374	480	—	445	—	521	521
	20.00	C-95	0.361	4.778	4.653	6.050	5.875	4.653	5.860	10,000	554	10,910	—	10,910	10,680	—	8,170	—	—	460	563	—	445	—	549	530
	23.00	C-95	0.415	4.070	4.515	6.050	5.875	4.545	5.860	12,920	630	12,540	—	11,730	10,680	—	8,170	—	—	540	608	—	445	—	606	530
	17.00	P-110	0.304	4.892	4.767	6.050	5.875	4.653	5.860	7,460	546	10,640	—	10,640	10,640	10,640	9,460	10,640	—	445	568	568	530	568	620	620
	20.00	P-110	0.361	4.778	4.653	6.050	5.875	4.653	5.860	11,080	641	12,640	—	12,640	12,360	12,640	9,460	12,640	—	548	667	667	530	667	654	630
	23.00	P-110	0.415	4.070	4.515	6.050	5.875	4.545	5.860	14,520	729	14,520	—	13,580	12,360	14,520	9,460	12,890	—	643	724	759	530	668	732	630
6	20.00	H-40	0.288	6.019	5.924	7.390	—	—	—	2,520	229	3,040	3,040	—	—	—	—	—	184	—	—	—	—	—	—	—
	20.00	J-55	0.288	6.019	5.924	7.234	7.000	—	—	2,970	315	4,180	4,180	4,180	4,180	4,180	4,060	4,180	245	266	374	374	374	374	—	—
	28.00	J-55	0.352	5.921	5.756	7.390	7.000	5.730	7.000	4,560	382	5,110	5,110	5,110	5,110	5,110	4,060	5,110	314	340	453	453	390	453	477	477
	20.00	K-55	0.288	6.019	5.924	7.234	—	—	—	2,970	315	4,180	4,180	4,180	4,180	4,180	4,060	4,180	267	290	453	453	453	453	—	—
	28.00	K-55	0.352	5.921	5.756	7.390	7.000	5.730	7.000	4,560	382	5,110	5,110	5,110	5,110	5,110	4,060	5,110	342	372	548	548	494	520	605	605
	24.00	C-75	0.352	5.921	5.796	7.390	7.000	5.730	7.000	5,570	520	6,970	—	6,970	6,970	—	5,540	—	—	453	583	—	494	—	605	605
	28.00	C-75	0.417	5.741	5.666	7.390	7.000	5.666	7.000	7,830	610	8,260	—	8,260	8,260	—	5,540	—	—	552	683	—	494	—	648	644
	32.00	C-75	0.475	5.675	5.550	7.390	7.000	5.550	7.000	9,830	688	9,410	—	9,410	9,200	—	5,540	—	—	638	771	—	494	—	717	644
	24.00	L-80	0.352	5.921	5.796	7.390	7.000	5.730	7.000	5,760	555	7,440	—	7,440	7,440	—	5,910	—	—	473	592	—	494	—	605	605
	28.00	L-80	0.417	5.741	5.666	7.390	7.000	5.666	7.000	8,170	651	8,810	—	8,810	8,810	—	5,910	—	—	576	693	—	494	—	648	644
	32.00	L-80	0.475	5.675	5.550	7.390	7.000	5.550	7.000	10,320	734	10,040	—	10,040	9,820	—	5,910	—	—	666	783	—	494	—	717	644
	24.00	N-80	0.352	5.921	5.796	7.390	7.000	5.730	7.000	5,760	555	7,440	—	7,440	7,440	7,440	5,910	7,440	—	481	615	615	520	615	637	637
	28.00	N-80	0.417	5.741	5.666	7.390	7.000	5.666	7.000	8,170	651	8,810	—	8,810	8,810	8,810	5,910	8,120	—	586	721	721	520	650	682	678
	32.00	N-80	0.475	5.675	5.550	7.390	7.000	5.550	7.000	10,320	734	10,040	—	10,040	9,820	10,040	5,910	8,120	—	677	814	814	520	650	756	678
	24.00	C-95	0.352	5.921	5.796	7.390	7.000	5.730	7.000	6,290	659	8,830	—	8,830	8,830	—	7,020	—	—	545	605	—	546	—	688	608
	28.00	C-95	0.417	5.741	5.666	7.390	7.000	5.666	7.000	9,200	773	10,460	—	10,460	10,460	—	7,020	—	—	605	780	—	546	—	716	712
	32.00	C-95	0.475	5.675	5.550	7.390	7.000	5.550	7.000	11,800	872	11,920	—	11,920	11,660	—	7,020	—	—	769	880	—	546	—	793	712
	24.00	P-110	0.352	5.921	5.796	7.390	7.000	5.730	7.000	6,710	763	10,230	—	10,230	10,230	10,230	8,120	10,230	—	641	786	786	650	786	796	796
	28.00	P-110	0.417	5.741	5.666	7.390	7.000	5.666	7.000	10,140	895	12,120	—	12,120	12,120	12,120	8,120	11,080	—	781	922	922	650	832	852	848
	32.00	P-110	0.475	5.675	5.550	7.390	7.000	5.550	7.000	13,200	1009	13,800	—	13,800	13,500	13,800	8,120	11,080	—	904	1,040	1,040	650	832	944	848

(Continued)

TABLE 8.7 (Continued)

1	2	3	4	5	6	7	8	9	10	11	12	13	14	15	16	17	18	19	20	21	22	23	24	25	26	27
																			★ Joint Strength — 1,000 lbs							
					Threaded and Coupled			Extreme Line				Internal Yield Pressure at Minimum Yield, psi							Threaded and Coupled						Extreme Line	
													Round Thread		Buttress Thread				Round Thread		Buttress Thread					
															Regular Coupling		Special Clearance Coupling									
Size: Outside Diameter in. *D*	Nominal Weight, Threads and Coupling lb per ft	Grade	Wall Thickness in. *t*	Inside Diameter in. *d*	Drift Diameter in.	Outside Diameter of Coupling in. *W*	Outside Diameter Special Clearance Coupling in. W_c	Drift Diameter in.	Outside Diameter of Box Power-tight in. *M*	Collapse Resistance psi	Pipe Body Yield Strength 1000 lbs	Plain End or Extreme Line	Short	Long	Same Grade	Higher Grade	Same Grade	Higher Grade	Short	Long	Regular Coupling	Regular Coupling Higher Grade:	Special Clearance Coupling	Special Clearance Coupling Higher Grade‡	Standard Joint	Optional Joint
7	17.00	H-40	0.231	6.578	6.413	7.656	—	—	—	1,450	196	2,310	2,310	—	—	—	—	—	122	—	—	—	—	—	—	—
	20.00	H-40	0.272	6.456	6.331	7.656	—	—	—	1,980	200	2,720	2,720	—	—	—	—	—	176	—	—	—	—	—	—	—
	20.00	J-55	0.272	6.456	6.331	7.656	—	—	—	2,270	316	3,740	3,740	—	—	—	—	—	234	—	—	—	—	—	—	—
	23.00	J-55	0.317	6.346	6.241	7.656	7.375	6.151	7.390	3,270	366	4,360	4,360	4,360	4,360	4,360	3,950	4,360	284	313	432	432	421	432	499	499
	26.00	J-55	0.362	6.276	6.131	7.656	7.375	6.151	7.390	4,320	415	4,980	4,980	4,980	4,980	4,980	3,950	4,980	334	367	490	490	421	490	506	506
	20.00	K-55	0.272	6.456	6.331	7.656	—	—	—	2,270	316	3,740	3,740	—	—	—	—	—	254	—	—	—	—	—	—	—
	23.00	K-55	0.317	6.368	6.241	7.656	7.375	6.151	7.390	3,270	364	4,360	4,360	4,360	4,360	4,350	3,950	4,360	309	341	522	522	522	522	532	632
	26.00	K-55	0.362	6.276	6.135	7.656	7.375	6.151	7.390	4,320	415	4,980	4,980	4,980	4,980	4,980	3,950	4,980	364	401	592	592	533	561	541	641
	23.00	C-75	0.317	6.366	6.211	7.656	7.375	6.151	7.390	3,770	499	5,940	—	5,940	5,940	—	5,380	—	—	416	557	—	533	—	685	632
	26.00	C-75	0.362	6.276	6.151	7.656	7.375	6.151	7.390	5,250	566	6,790	—	6,790	6,790	—	8,380	—	—	489	631	—	533	—	611	641
	29.00	C-75	0.408	6.184	6.059	7.656	7.375	6.059	7.390	6,760	634	7,650	—	7,650	7,650	—	8,380	—	—	562	707	—	533	—	695	674
	32.00	C-75	0.458	6.004	5.069	7.656	7.375	5.969	7.390	8,250	699	8,490	—	8,490	7,930	—	5,380	—	—	633	779	—	533	—	681	674
	35.00	C-75	0.498	6.004	5.879	7.656	7.375	5.879	7.530	9,710	763	9,340	—	8,660	7,930	—	8,380	—	—	701	833	—	533	—	856	781
	38.00	C-75	0.540	5.920	5.795	7.656	7.375	5.795	7.530	10,680	822	10,120	—	8,660	7,930	—	8,380	—	—	767	833	—	533	—	917	781
	23.00	L-80	0.317	6.366	6.241	7.656	7.375	6.151	7.390	3,830	532	6,340	—	6,340	6,340	8,346	5,740	6,340	—	435	565	—	533	—	632	632
	26.00	L-80	0.362	6.276	6.151	7.656	7.375	6.151	7.390	5,410	604	7,240	—	7,240	7,210	7,210	5,740	7,210	—	511	641	—	533	—	641	641
	29.00	L-80	0.408	6.184	6.059	7.656	7.375	6.059	7.390	7,020	676	8,160	—	8,160	8,160	8,160	5,740	7,890	—	587	718	—	533	—	685	674
	32.00	L-80	0.483	6.094	5.368	7.656	7.375	5.969	7.390	8,600	745	9,060	—	9,060	8,160	9,060	5,740	7,890	—	661	791	—	533	—	761	674
	35.00	L-80	0.498	6.004	5.879	7.656	7.375	5.879	7.530	10,180	814	9,900	—	9,240	8,460	9,960	5,740	7,890	—	734	833	—	533	—	856	761
	38.00	L-80	0.590	5.920	5.793	7.656	7.375	6.795	7.530	11,390	877	10,800	—	9,240	8,460	10,800	5,740	7,890	—	801	833	—	533	—	917	761
	23.00	N-80	0.317	6.366	6.241	7.656	7.375	6.151	7.390	3,330	532	6,340	—	6,340	6,340	6,340	5,740	6,340	—	442	588	588	561	588	666	666
	26.00	N-80	0.362	6.276	6.151	7.656	7.375	6.151	7.390	5,410	604	7,240	—	7,240	7,240	7,240	5,740	7,240	—	519	667	667	561	667	675	675
	29.00	N-80	0.408	6.184	6.059	7.656	7.375	6.059	7.390	7,020	676	8,160	—	8,160	8,160	8,160	5,740	7,890	—	507	745	746	561	702	721	709
	32.00	N-80	0.453	6,036	5.069	7.656	7.375	5.969	7.390	8,600	745	9,060	—	9,060	8,460	9,060	5,740	7,890	—	672	823	823	561	702	801	709
	35.00	N-80	0.498	6,006	5.879	7.656	7.375	5.879	7.530	10,180	814	9,960	—	9,240	8,460	9,960	5,740	7,890	—	746	876	898	561	702	805	801
	38.00	N-80	0.540	5.920	5.795	7.656	7.375	5.795	7.530	11,330	877	10,800	—	9,240	8,460	10,800	5,740	7,890	—	814	876	968	561	702	895	801
	23.00	C-95	0.317	6.366	6.711	7.656	7.375	6.151	7.390	4,150	632	7,530	—	7,530	7,530	—	6,810	—	—	605	636	—	589	—	699	699
	26.00	C-95	0.362	6.274	6.151	7.656	7.375	6.151	7.390	5,870	717	8,600	—	8,600	8,500	—	6,810	—	—	593	722	—	589	—	709	709
	29.00	C-95	0.408	6.184	6.059	7.656	7.375	6.059	7.390	7,820	803	9,690	—	9,690	9,690	—	6,810	—	—	683	808	—	889	—	757	744
	32.00	C-95	0.453	6.094	5.969	7.656	7.375	5.959	7.390	9,730	885	10,760	—	10,760	10,050	—	6,810	—	—	768	891	—	589	—	841	744
	35.00	C-95	0.498	6.001	5.879	7.656	7.375	5.879	7.530	11,640	966	11,830	—	10,970	10,050	—	6,810	—	—	853	920	—	589	—	940	841
	38.00	C-95	0.540	5.820	5.795	7.656	7.375	5.795	7.530	13,429	1041	12,820	—	10,970	10,050	—	6,810	—	—	931	920	—	589	—	1,013	841
	26.00	P-110	0.362	6.276	6.151	7.656	7.375	6.151	7.390	6,210	830	9,960	—	9,960	9,960	9,960	7,890	9,960	—	693	853	853	702	853	844	844
	29.00	P-110	0.408	4.184	6.059	7.656	7.375	6.059	7.390	8,310	929	11,220	—	11,220	11,220	11,220	7,890	10,760	—	797	955	955	702	898	902	884
	32.00	P-110	0.453	6.091	5.969	7.656	7.375	5.959	7.390	10,760	1025	12,460	—	12,460	11,640	12,460	7,890	10,760	—	897	1,053	1,053	702	898	1,002	886
	35.00	P-110	0.498	6.004	5.879	7.656	7.375	5.879	7.530	13,010	1119	13,700	—	12,700	11,640	13,700	7,890	10,760	—	996	1,096	1,150	702	898	1,118	1,002
	38.00	P-110	0.540	8.920	5.795	7.656	7.375	5.795	7.530	15,110	1205	14,850	—	12,700	11,640	14,850	7,890	10,760	—	1,087	1,096	1,239	702	898	1,207	1,002
7⅝	24.00	H-40	0.300	7.025	6.900	8.500	—	—	—	2,040	276	2,750	2,750	—	—	—	—	—	212	—	—	—	—	—	—	—
	26.40	J-55	0.328	6.969	6.844	8.500	8.125	6.750	8.010	2,890	414	4,140	4,140	4,140	4,140	4,140	4,140	4,140	316	346	483	483	483	483	558	553
	26.40	K-55	0.328	6.969	6.844	8.500	8.125	6.750	8.010	2,890	414	4,140	4,140	4,140	4,140	4,140	4,140	4,140	342	377	581	581	581	581	700	700
	26.40	C-75	0.328	6.969	6.814	8.500	8.125	6.750	8.010	3,280	564	5,650	—	5,650	5,650	—	5,650	—	—	461	624	—	624	—	700	700
	29.70	C-75	0.375	6.875	6.750	8.500	8.125	6.750	8.010	4,670	611	6,450	—	6,450	6,450	—	6,140	—	—	542	709	—	709	—	700	700
	33.70	C-75	0.430	6.755	6.610	8.500	8.125	6.640	8.010	6,320	725	7,400	—	7,400	7,400	—	6,140	—	—	635	806	—	735	—	766	744
	39.00	C-75	0.500	6.625	6.500	8.500	8.125	6.500	8.010	8,430	839	8,610	—	8,610	8,610	—	6,140	—	—	751	929	—	735	—	851	744
	42.80	C-75	0.562	6.301	6.374	8.500	8.125	—	—	10,240	231	9,670	—	9,670	9,190	—	6,140	—	—	852	1,035	—	735	—	—	—
	47.10	C-75	0.628	6.373	6.250	8.500	8.125	—	—	11,200	1031	10,760	—	9,840	9,190	—	6,140	—	—	953	1,140	—	735	—	—	—
	26.40	L-80	0.328	6.969	6.844	8.500	8.125	6.750	8.010	3,400	602	6,020	—	6,020	6,020	6.020	6,020	6,020	—	482	635	—	635	—	700	700
	29.70	L-80	0.375	6.875	6.730	8.500	8.125	6.750	8.010	4,700	683	6,890	—	6,890	6,890	6,890	6,550	6,890	—	566	721	—	721	—	700	700
	33.70	L-80	0.400	6.765	6.620	8.500	8.125	6.610	8.010	6,560	778	7,900	—	7,900	7,900	7,900	6,550	7,900	—	664	820	—	735	—	766	744
	39.00	L-80	0.500	6.825	6.500	8.500	8.125	6.500	8.010	3.810	895	9,180	—	9,180	9,180	9,180	6,550	9,000	—	786	945	—	735	—	851	744
	42.00	L-80	0.562	6.301	6.374	8.500	8.125	—	—	10,810	928	10,320	—	10,320	9,790	—	6,550	—	—	892	1,053	—	735	—	—	—
	47.10	L-80	0.625	6.375	6.250	8.500	8.125	—	—	12,040	1100	11,480	—	10,490	9,970	—	6,550	—	—	997	1,160	—	735	—	—	—
	26.40	N-80	0.328	6.969	6.844	8.500	8.125	6.750	8.010	3,400	602	6,020	—	6,020	6,020	6,020	6,020	6,020	—	490	659	659	659	659	737	737
	29.70	N-80	0.375	6.875	6.750	8.500	8.125	6.750	8.010	4,700	683	6,890	—	6,890	6,890	6,890	6,550	6,890	—	575	749	749	749	749	737	737
	33.70	N-80	0.430	6.765	4.810	8.500	8.125	6.640	8.010	6,560	778	7,900	—	7,900	7,900	7,900	6,550	7,000	—	674	852	852	773	852	806	784
	39.00	N-80	0.500	6.625	6.500	8.500	8.125	6.500	8.010	5,810	695	9,180	—	9,180	9,180	9,180	6,550	9,000	—	798	981	981	773	967	896	784
	42.80	N-80	0.562	6.301	6.776	8.300	8.125	—	—	10,810	998	10,320	—	10,320	9,700	10,320	6,550	9,000	—	905	1,003	1,093	773	967	—	—
	47.10	N-80	0.623	6.375	6.230	8.500	8.125	—	—	12,010	1100	11,480	—	10,490	9,790	11,480	6,550	9,000	—	1,013	1,205	1,204	773	967	—	—
	26.40	C-95	0.328	6.969	6.814	8.500	8.125	6.750	8.010	3,710	714	7,150	—	7,150	7,150	—	7,150	—	—	560	716	—	716	—	774	774
	29.70	C-95	0.375	6.875	6.750	8.500	8.125	6.720	8.010	5,120	811	8,180	—	8,180	8,180	—	7,780	—	—	659	813	—	812	—	774	774
	33.70	C-95	0.400	6.765	6.610	8.500	8.125	6.640	8.010	7,260	923	9,380	—	9,380	9,380	—	7,780	—	—	772	925	—	812	—	846	823
	38.90	C-95	0.500	6.625	6.500	8.500	8.125	6.500	8.010	9,389	1061	10,900	—	10,900	10,900	—	7,780	—	—	914	1,065	—	812	—	941	823
	42.80	C-95	0.562	6.501	6.725	8.500	8.125	—	—	12,400	1185	12,250	—	12,250	11,620	—	7,780	—	—	1,037	1,187	—	812	—	—	—
	47.10	C-95	0.623	6.375	6.250	8.500	8.125	—	—	14,300	1396	13,630	—	12,460	11,620	—	7,780	—	—	1,159	1,300	—	812	—	—	—

	29.70	P-110	0.375	6.875	6.750	8.500	8.125	6.750	8.010	5,340	940	9,470	—	9,470	9,470	9,470	9,000	9,470	—	769	960	960	960	960	922	922
	33.70	P-110	0.430	6.755	6.410	8.500	8.125	6.600	8.010	7,950	1069	10,860	—	10,860	10,860	10,860	9,000	10,860	—	901	1,093	1,093	967	1,093	1,008	979
	39.00	P-110	0.500	6.425	6.500	8.500	8.125	6.500	8.010	11,140	1231	12,620	—	12,620	12,620	12,620	9,000	12,280	—	1,066	1,258	1,258	967	1,237	1,120	979
	42.80	P-110	0.542	6.501	6.376	8.500	8.125	—	—	13,910	1372	14,190	—	14,190	13,460	14,100	9,000	12,280	—	1,210	1,402	1,402	967	1,237	—	—
	47.10	P-110	0.625	6.375	6.250	8.500	8.125	—	—	16,330	1512	15,780	—	14,430	13,460	15,780	9,000	12,280	—	1,353	1,545	1,545	967	1,237	—	—
8⅝	28.00	H-40	0.304	8.017	7.892	9.625	—	—	—	1,640	318	2,470	2,470	—	—	—	—	—	233	—	—	—	—	—	—	—
	32.00	H-40	0.352	7.921	7.796	9.625	—	—	—	2,210	366	2,860	2,860	—	—	—	—	—	279	—	—	—	—	—	—	—
	24.00	J-55	0.264	8.097	7.972	9.625	—	—	—	1,370	381	2,950	2,950	—	—	—	—	—	244	—	—	—	—	—	—	—
	32.00	J-55	0.352	7.921	7.796	9.625	9.125	7.700	9.120	2,530	503	3,930	3,930	3,930	3,930	3,930	3,930	3,930	372	417	579	579	579	579	686	686
	36.00	J-55	0.400	7.825	7.700	9.625	9.125	7.700	9.120	3,450	568	4,460	4,460	4,460	4,460	4,460	4,060	4,460	434	486	654	654	654	654	688	688
	24.00	K-55	0.264	8.097	7.972	9.625	—	—	—	1,370	381	2,950	2,950	—	—	—	—	—	263	—	—	—	—	—	—	—
	32.00	K-55	0.352	7.921	7.796	9.625	9.125	7.700	9.120	2,530	503	3,930	3,930	3,930	3,930	3,930	3,930	3,930	402	452	690	690	690	690	869	869
	36.00	K-55	0.400	7.825	7.700	9.625	9.125	7.700	9.120	3,450	568	4,460	4,460	4,460	4,460	4,460	4,060	4,460	468	526	780	780	780	780	871	871
	36.00	C-75	0.400	7.825	7.700	9.625	9.125	7.700	9.120	4,020	775	6,090	—	6,090	6,090	—	5,530	—	—	648	847	—	839	—	871	871
	40.00	C-75	0.450	7.725	7.600	9.625	9.125	7.600	9.120	5,350	867	6,850	—	6,850	6,850	—	5,530	—	—	742	947	—	839	—	942	886
	44.00	C-75	0.500	7.625	7.300	9.625	9.125	7.500	9.120	6,680	957	7,610	—	7,610	7,610	—	5,530	—	—	834	1,046	—	839	—	1,007	886
	49.00	C-75	0.557	7.511	7.386	9.625	9.125	7.386	9.120	8,200	1059	8,480	—	8,480	8,480	—	5,530	—	—	939	1,157	—	839	—	1,007	886
	36.00	L-80	0.400	7.825	7.700	9.625	9.125	7.700	9.120	4,100	827	6,490	—	6,490	6,490	—	5,900	—	—	678	864	—	839	—	871	871
	40.00	L-80	0.450	7.725	7.600	9.625	9.125	7.600	9.120	5,520	925	7,300	—	7,300	7,300	—	5,900	—	—	776	966	—	839	—	942	886
	44.00	L-80	0.500	7.625	7.500	9.625	9.125	7.500	9.120	6,950	1021	8,120	—	8,120	8,120	—	5,900	—	—	874	1,066	—	839	—	1,007	886
	49.00	L-80	0.557	7.511	7.386	9.625	9.125	7.386	9.120	8,570	1129	9,040	—	9,040	9,040	—	5,900	—	—	983	1,180	—	839	—	1,007	886
	36.00	N-80	0.400	7.825	7.700	9.625	9.125	7.700	9.120	4,100	827	6,490	—	6,490	6,490	6,490	5,900	6,490	—	688	895	895	883	895	917	917
	40.00	N-80	0.450	7.725	7.600	9.625	9.125	7.600	9.120	5,520	925	7,300	—	7,300	7,300	7,300	5,900	7,300	—	788	1,001	1,001	883	1,001	992	932
	44.00	N-80	0.500	7.625	7.500	9.625	9.125	7.500	9.120	6,950	1021	8,120	—	8,120	8,120	8,120	5,900	8,110	—	887	1,105	1,105	883	1,103	1,060	932
	49.00	N-80	0.557	7.511	7.386	9.626	9.125	7.386	9.120	8,570	1129	9,040	—	9,040	9,040	9,040	5,900	8,110	—	997	1,222	1,222	883	1,103	1,060	932
	36.00	C-95	0.400	7.825	7.706	9.625	9.125	7.700	9.120	4,360	982	7,710	—	7,710	7,710	—	7,010	—	—	789	976	—	927	—	963	963
	40.00	C-95	0.450	7.725	7.600	9.625	9.125	7.600	9.120	6,010	1098	8,670	—	8,670	8,670	—	7,010	—	—	904	1,092	—	927	—	1,042	979
	44.00	C-95	0.500	7.625	7.500	9.625	9.125	7.500	9.120	7,730	1212	9,640	—	9,640	9,640	—	7,010	—	—	1,017	1,206	—	927	—	1,113	979
	49.00	C-95	0.557	7.511	7.386	9.625	9.125	7.386	9.120	9,690	1341	10,740	—	10,740	10,740	—	7,010	—	—	1,144	1,334	—	927	—	1,113	979
	40.00	P-110	0.450	7.725	7.600	9.625	9.125	7.600	9.120	6,380	1271	10,040	—	10,040	10,040	10,040	8,110	10,040	—	1,055	1,288	1,288	1,103	1,288	1,240	1,165
	44.00	P-110	0.500	7.625	7.500	9.625	9.125	7.500	9.120	8,400	1404	11,160	—	11,160	11,160	11,160	8,110	11,060	—	1,186	1,423	1,423	1,103	1,412	1,326	1,165
	49.00	P-110	0.557	7.511	7.386	9.625	9.125	7.386	9.120	10,720	1553	12,430	—	12,430	12,430	12,430	8,110	11,060	—	1,335	1,574	1,574	1,103	1,412	1,326	1,165
9⅝	32.30	H-40	0.312	9.001	8.845	10.625	—	—	—	1,400	365	2,270	2,270	—	—	—	—	—	254	—	—	—	—	—	—	—
	36.00	H-40	0.262	8.921	8.765	10.625	—	—	—	1,740	410	2,560	2,560	—	—	—	—	—	294	—	—	—	—	—	—	—
	36.00	J-55	0.352	8.921	8.765	10.625	10.125	—	—	2,020	569	3,520	3,520	3,520	3,520	3,520	3,520	3,520	394	453	639	639	639	639	—	—
	40.00	J-55	0.395	8.835	8.679	10.625	10.125	8.599	10.100	2,570	630	3,950	3,950	3,950	3,950	3,950	9,660	3,950	455	520	714	714	714	714	770	770
	36.00	K-55	0.352	8.921	8.765	10.625	10.125	—	—	2,020	564	3,520	3,520	3,520	3,520	3,520	3,520	3,520	425	489	755	755	755	755	—	—
	40.00	K-55	0.395	8.835	8.679	10.625	10.125	8.599	10.100	2,570	630	3,950	3,950	3,950	3,950	3,950	3,660	3,950	486	561	843	843	843	843	975	975
	40.00	C-75	0.395	8.835	8.679	10.625	10.125	8.599	10.100	2,980	859	5,390	—	5,390	5,390	—	4,990	—	—	694	926	—	926	—	975	975
	43.50	C-75	0.435	8.755	8.599	10.625	10.125	8.599	10.100	3,750	942	5,930	—	5,930	5,930	—	4,990	—	—	776	1,016	—	934	—	975	975
	47.00	C-75	0.472	8.681	8.525	10.625	10.125	8.525	10.100	4,630	1018	6,440	—	6,440	6,440	—	4,990	—	—	852	1,098	—	934	—	1,032	1,032
	53.50	C-75	0.545	8.535	8.579	10.625	10.125	8.379	10.100	6,380	1166	7,430	—	7,430	7,430	—	4,990	—	—	999	1,257	—	934	—	1,175	1,053
	40.00	L-80	0.395	8.835	8.679	10.625	10.125	8.599	10.100	3,090	916	5,750	—	5,750	5,750	—	5,320	—	—	727	947	—	934	—	975	975
	43.50	L-80	0.435	8.755	8.599	10.625	10.125	8.599	10.100	3,810	1005	6,330	—	6,330	6,330	—	5,320	—	—	813	1,038	—	934	—	975	975
	47.00	L-80	0.472	8.681	8.525	10.625	10.125	8.525	10.100	4,750	1086	6,870	—	6,870	6,870	—	5,320	—	—	893	1,122	—	934	—	1,032	1,032
	53.50	L-80	0.545	8.535	8.379	10.625	10.125	8.379	10.100	6,620	1244	7,930	—	7,930	7,930	—	5,320	—	—	1,047	1,286	—	934	—	1,173	1,053
	40.00	N-80	0.395	8.835	8.679	10.625	10.125	8.539	10.100	3,090	916	5,750	—	5,750	5,750	5,750	5,320	5,750	—	737	979	979	979	979	1,027	1,027
	43.50	N-80	0.435	8.755	8.599	10.625	10.125	8.590	10.100	3,610	1005	6,330	—	6,330	6,330	6,330	5,320	6,330	—	825	1,074	1,074	983	1,074	1,027	1,027
	47.00	N-80	0.472	8.681	8.525	10.625	10.125	8.325	10.100	4,750	1086	6,870	—	6,870	6,870	6,870	5,320	6,870	—	905	1,161	1,161	983	1,161	1,086	1,086
	53.50	N-80	0.545	8.535	8.379	10.625	10.125	8.379	10.100	6,620	1244	7,930	—	7,930	7,930	7,930	5,320	7,310	—	1,062	1,329	1,329	983	1,229	1,235	1,109
	40.00	C-95	0.395	8.835	8.679	10.625	10.125	8.599	10.100	3,330	1088	6,820	—	6,820	6,820	—	6,310	—	—	847	1,074	—	1,032	—	1,078	1,078
	43.50	C-95	0.435	8.755	8.599	10.625	10.125	8.599	10.100	4,130	1193	7,510	—	7,510	7,510	—	6,310	—	—	948	1,178	—	1,032	—	1,078	1,078
	47.00	C-95	0.472	8.681	8.525	10.625	10.125	8.525	10.100	5,080	1239	8,150	—	8,150	8,150	—	6,310	—	—	1,010	1,273	—	1,032	—	1,141	1,141
	53.50	C-95	0.545	8.535	8.379	10.625	10.125	8.379	10.100	7,230	1477	9,410	—	9,410	9,410	—	6,310	—	—	1,220	1,458	—	1,032	—	1,297	1,164
	43.50	P-110	0.435	8.755	8.599	10.625	10.125	8.599	10.100	4,430	1381	8,700	—	8,700	8,700	8,700	7,310	8,700	—	1,106	1,388	1,388	1,229	1,388	1,283	1,283
	47.00	P-110	0.472	8.681	8.525	10.625	10.125	8.525	10.100	5,310	1403	9,440	—	9,440	9,440	9,440	7,310	9,440	—	1,213	1,500	1,500	1,229	1,500	1,358	1,358
	53.50	P-110	0.545	8.535	8.379	10.625	10.125	8.379	10.100	7,930	1710	10,900	—	10,900	10,900	10,900	7,310	9,970	—	1,422	1,718	1,718	1,229	1,573	1,544	1,386

(Continued)

TABLE 8.7 (Continued)

1	2	3	4	5	6	7	8	9	10	11	12	13	14	15	16	17	18	19	20	21	22	23	24	25	26	27
					Threaded and Coupled			Extreme Line				Internal Yield Pressure at Minimum Yield, psi							★ Joint Strength — 1,000 lbs							
													Round Thread		Buttress Thread				Threaded and Coupled						Extreme Line	
															Regular Coupling		Special Clearance Coupling		Round Thread		Buttress Thread					
Size: Outside Diameter in. D	Nominal Weight, Threads and Coupling lb per ft	Grade	Wall Thickness in. t	Inside Diameter in. d	Drift Diameter in.	Outside Diameter of Coupling in. W	Outside Diameter Special Clearance Coupling in. W_e	Drift Diameter in.	Outside Diameter of Box Powertight in. M	Collapse Resistance psi	Pipe Body Yield Strength 1000 lbs	Plain End or Extreme Line	Short	Long	Same Grade	Higher Grade	Same Grade	Higher Grade	Short	Long	Regular Coupling	Regular Coupling Higher Grade:	Special Clearance Coupling	Special Clearance Coupling Higher Grade‡	Standard Joint	Optional Joint
10¾	32.75	H-40	0.279	10.192	10.036	11.750	—	—	—	880	367	1,820	1,820	—	—	—	—	—	205	—	—	—	—	—	—	—
	40.50	H-40	0.350	10.050	9.894	11.750	—	—	—	1,420	457	2,280	2,280	—	—	—	—	—	314	—	—	—	—	—	—	—
	40.50	J-55	0.350	10.050	9.894	11.750	11.250	—	—	1,580	629	3,130	3,130	—	3,130	3,130	3,130	3,130	420	—	700	700	700	700	—	—
	45.50	J-55	0.400	9.950	9.794	11.750	11.250	9.794	11.460	2,090	715	3,580	3,580	—	3,580	3,580	3,290	3,580	493	—	796	796	796	796	975	—
	51.00	J-55	0.450	9.850	9.694	11.750	11.250	9.694	11.460	2,700	801	4,030	4,030	—	4,030	4,030	3,290	4,030	565	—	891	891	822	891	1,092	—
	40.50	K-55	0.350	10.050	9.894	11.750	11.250	—	—	1,580	629	3,130	3,130	—	3,130	3,130	3,130	3,130	450	—	819	819	819	819	—	—
	45.50	K-55	0.400	9.950	9.794	11.750	11.250	9.794	11.460	2,090	715	3,580	3,580	—	3,580	3,580	3,200	3,580	528	—	931	931	931	931	1,236	—
	51.00	K-55	0.450	9.850	9.694	11.750	11.250	9.694	11.460	2,700	801	4,030	4,030	—	4,030	4,030	3,290	4,030	606	—	1,043	1,043	1,041	1,043	1,383	—
	51.00	C-75	0.450	9.850	9.694	11.750	11.250	9.694	11.460	3,100	1092	5,490	5,490	—	5,490	—	4,490	—	756	—	1,160	—	1,041	—	1,383	—
	55.50	C-75	0.495	9.760	9.604	11.750	11.250	9,604	11.460	3,950	1196	6,040	6,040	—	6,040	—	4,490	—	843	—	1,271	—	1,041	—	1,515	—
	51.00	L-80	0.450	9.850	9.694	11.750	11.250	9.694	11.460	3,220	1165	5,860	5,860	—	5,860	—	4,790	—	794	—	1,190	—	1,041	—	1,383	—
	55.50	L-80	0.495	9.760	9.604	11.750	11.250	9.604	11.460	4,020	1276	6,450	6,450	—	6,450	—	4,790	—	884	—	1,303	—	1,041	—	1,515	—
	51.00	N-80	0.450	9.850	9.694	11.750	11.250	9.694	11.460	3,220	1165	5,860	5,860	—	5,860	5,860	4,790	5,860	804	—	1,228	1,228	1,096	1,228	1,456	—
	55.50	N-80	0.495	9.760	9.604	11.750	11.250	9.604	11.460	4,020	1276	6,450	6,450	—	6,450	6,450	4,790	6,450	895	—	1,345	1,345	1,096	1,345	1,595	—
	51.00	C-95	0.450	9.850	9.694	11.750	11.250	9.694	11.460	3,490	1383	6,960	6,960	—	6,960	—	5,680	—	927	—	1,354	—	1,151	—	1,529	—
	55.50	C-95	0.495	9.760	9.604	11.750	11.250	9.604	11.460	4,300	1515	7,660	7,660	—	7,660	—	5,680	—	1,032	—	1,483	—	1,151	—	1,675	—
	51.00	P-110	0.450	9.850	9.694	11.750	11.250	9.694	11.460	3,670	1602	8,060	8,060	—	8,060	8,060	6,580	8,060	1,080	—	1,594	1,594	1,370	1,594	1,820	—
	55.50	P-110	0.495	9.760	9.604	11.750	11.250	9.604	11.460	4,630	1754	8,860	8,860	—	8,860	8,860	6,580	8,860	1,203	—	1,745	1,745	1,370	1,745	1,993	—
	60.70	P-110	0.545	9.660	9.504	11.750	11.250	9.504	11.460	5,860	1922	9,760	9,760	—	9,760	9,760	6,580	8,970	1,338	—	1,912	1,912	1,370	1,754	2,000	—
	65.70	P-110	0.595	9.560	9.404	11.750	11.250	—	—	7,490	2088	10,650	10,650	—	10,650	10,650	6,580	8,970	1,472	—	2,077	2,077	1,370	1,754	—	—
11¾	42.00	H-40	0.333	11.084	10.928	12.750	—	—	—	1,070	478	1,980	1,980	—	—	—	—	—	307	—	—	—	—	—	—	—
	47.00	J-55	0.375	11.000	10.844	12.750	—	—	—	1,510	737	3,070	3,070	—	3,070	3,070	—	—	477	—	807	807	—	—	—	—
	54.00	J-55	0.435	10.880	10.724	12.750	—	—	—	2,070	850	3,560	3,560	—	3,560	3,560	—	—	568	—	931	931	—	—	—	—
	60.00	J-55	0.489	10.772	10.616	12.750	—	—	—	2,660	952	4,010	4,010	—	4,010	4,010	—	—	649	—	1,042	1,042	—	—	—	—
	47.00	K-55	0.375	11.000	10.844	12.750	—	—	—	1,510	737	3,070	3,070	—	3,070	3,070	—	—	509	—	935	935	—	—	—	—
	54.00	K-55	0.435	10.880	10.724	12.750	—	—	—	2,070	850	3,560	3,560	—	3,560	3,560	—	—	606	—	1,079	1,079	—	—	—	—
	60.00	K-55	0.489	10.772	10.616	12.750	—	—	—	2,660	952	4,010	4,010	—	4,010	4,010	—	—	693	—	1,208	1,208	—	—	—	—
	60.00	C-75	0.489	10.772	10.616	12.750	—	—	—	3,070	1298	5,460	5,460	—	5,460	—	—	—	869	—	1,361	—	—	—	—	—
	60.00	L-80	0.489	10.772	10.616	12.750	—	—	—	3,180	1384	5,830	5,830	—	5,830	—	—	—	913	—	1,399	—	—	—	—	—
	60.00	N-80	0.489	10.772	10.616	12.750	—	—	—	3,180	1384	5,830	5,830	—	5,830	—	—	—	924	—	1,440	1,440	—	—	—	—
	60.00	C-95	0.489	10.772	10.616	12.750	—	—	—	3,440	1644	6,920	6,920	—	6,920	—	—	—	1,066	—	1,596	—	—	—	—	—

13⅜	48.00	H-40	0.330	12.715	12.559	14.375	—	—	—	770	541	1,730	1,730	—	—	—	—	—	322	—	—	—	—	—	—	—
	54.50	J-55	0.380	12.615	12.459	14.375	—	—	—	1,130	853	2,730	2,730	—	2,730	2,730	—	—	514	—	909	909	—	—	—	—
	61.00	J-55	0.430	12.515	12.359	14.375	—	—	—	1,540	962	3,090	3,090	—	3,090	3,090	—	—	595	—	1,025	1,025	—	—	—	—
	68.00	J-55	0.480	12.415	12.259	14.375	—	—	—	1,950	1069	3,450	3,450	—	3,450	3,450	—	—	675	—	1,140	1,140	—	—	—	—
	54.50	K-55	0.380	12.615	12.459	14.375	—	—	—	1,130	853	2,730	2,730	—	2,730	2,730	—	—	547	—	1,038	1,038	—	—	—	—
	61.00	K-55	0.430	12.515	12.359	14.375	—	—	—	1,540	962	3,090	3,090	—	3,090	3,090	—	—	633	—	1,169	1,169	—	—	—	—
	68.00	K-55	0.480	12.415	12.259	14.375	—	—	—	1,950	1069	3,450	3,450	—	3,450	3,450	—	—	718	—	1,300	1,300	—	—	—	—
	68.00	C-75	0.480	12.415	12.259	14.375	—	—	—	2,220	1458	4,710	4,710	—	4,710	—	—	—	905	—	1,496	—	—	—	—	—
	72.00	C-75	0.514	12.347	12.191	14.375	—	—	—	2,590	1558	5,040	5,040	—	5,040	—	—	—	978	—	1,598	—	—	—	—	—
	68.00	L-80	0.480	12.415	12.259	14.375	—	—	—	2,260	1556	5,020	5,020	—	5,020	—	—	—	952	—	1,545	—	—	—	—	—
	72.00	L-80	0.514	12.347	12.191	14.375	—	—	—	2,670	1661	5,380	5,380	—	5,380	—	—	—	1,029	—	1,650	—	—	—	—	—
	68.00	N-80	0.480	12.415	12.259	14.375	—	—	—	2,260	1556	5,020	5,020	—	5,020	—	—	—	963	—	1,585	1,585	—	—	—	—
	72.00	N-80	0.514	12.347	12.191	14.375	—	—	—	2,670	1661	5,380	5,380	—	5,380	—	—	—	1,040	—	1,693	1,693	—	—	—	—
	68.00	C-95	0.480	12.415	12.259	14.375	—	—	—	2,320	1847	5,970	5,970	—	5,970	—	—	—	1,114	—	1,772	—	—	—	—	—
	72.00	C-95	0.514	12.347	12.191	14.375	—	—	—	2,820	1973	6,390	6,390	—	6,390	—	—	—	1,204	—	1,893	—	—	—	—	—
16	65.00	H-40	0.375	15.250	15.062	17.000	—	—	—	670	736	1,640	1,640	—	—	—	—	—	439	—	—	—	—	—	—	—
	75.00	J-55	0.438	15.124	14.936	17.000	—	—	—	1,020	1178	2,630	2,630	—	2,630	2,630	—	—	710	—	1,200	1,200	—	—	—	—
	84.00	J-55	0.495	15.016	14.822	17.000	—	—	—	1,410	1326	2,980	2,980	—	2,980	2,980	—	—	817	—	1,351	1,351	—	—	—	—
	75.00	K-55	0.438	15.124	14.936	17.000	—	—	—	1,020	1178	2,630	2,630	—	2,630	2,630	—	—	752	—	1,331	1,331	—	—	—	—
	84.00	K-55	0.495	15.010	14.822	17.000	—	—	—	1,410	1326	2,980	2,980	—	2,980	2,980	—	—	865	—	1,499	1,499	—	—	—	—
18⅝	87.50	H-40	0.435	17.755	17.567	20.000	—	—	—	*630	994	1,630	1,630	—	—	—	—	—	559	—	—	—	—	—	—	—
	87.50	J-55	0.435	17.755	17.567	20.000	—	—	—	*630	1367	2,250	2,250	—	—	—	—	—	754	—	1,329	1,329	—	—	—	—
	87.50	K-55	0.435	17.755	17.567	20.000	—	—	—	*630	1367	2,250	2,250	—	—	—	—	—	794	—	1,427	1,427	—	—	—	—
20	94.00	H-40	0.438	19.124	18.936	21.000	—	—	—	*520	1077	1,530	1,530	—	—	—	—	—	581	—	—	—	—	—	—	—
	94.00	J-55	0.438	19.124	18.936	21.000	—	—	—	*520	1480	2,110	2,110	2,110	—	—	—	—	784	907	1,402	1,402	—	—	—	—
	106.50	J-55	0.500	19.000	18.812	21.000	—	—	—	*770	1685	2,410	2,410	2,410	—	—	—	—	913	1,057	1,596	1,596	—	—	—	—
	133.00	J-55	0.635	18.730	18.542	21.000	—	—	—	1,500	2125	3,060	3,060	3,060	—	—	—	—	1,192	1,380	2,012	2,012	—	—	—	—
	94.00	K-55	0.438	19.124	18.036	21.000	—	—	—	*520	1480	2,100	2,110	2,110	—	—	—	—	824	955	1,479	1,479	—	—	—	—
	106.50	K-55	0.500	19.000	18.812	21.000	—	—	—	*770	1685	2,410	2,410	2,410	—	—	—	—	960	1,113	1,683	1,683	—	—	—	—
	133.00	K-55	0.625	18.730	18.542	21.000	—	—	—	1,500	2125	3,060	3,060	3,060	—	—	—	—	1,253	1,453	2,123	2,123	—	—	—	—

‡ For P-110 casing the next higher grade is 150YS, a non-API steel grade having a minimum yield strength of 150,000 psi.

★ Some joint strengths listed in Col. 20 through 27 are greater than the corresponding pipe body yield strength listed in Col. 12.

* Collapse resistance values calculated by elastic formula.

Taken from API Bul 5C2, 17th Edition, March 1980.

TABLE 8.8 D/t Ranges for Collapse Pressures

Steel Grade	D/t range for formula (1)	D/t range for formula (2)	D/t range for formula (3)	D/t range for formula (4)
H-40	16.44 & less	16.44 to 26.62	26.62 to 42.70	42.70 & greater
J-K-55	14.80 & less	14.80 to 24.39	24.39 to 37.20	37.20 & greater
C-75	13.67 & less	13.67 to 23.09	23.03 to 32.05	32.05 & greater
L-N-80	13.38 & less	13.38 to 22.46	22.46 to 31.05	31.05 & greater
C-95	12.83 & less	12.83 to 21.21	21.21 to 28.25	28.25 & greater
P-105	12.56 & less	12.56 to 20.66	20.66 to 26.88	26.88 & greater
P-110	12.42 & less	12.42 to 20.29	20.29 to 26.20	26.20 & greater

TABLE 8.9 Factors for Collapse Pressure Formulas

	Formula Factor				
Steel Grade	A	B	C	F	G
H-40	2.950	0.0463	755	2.047	0.03125
J-K-55	2.990	0.0541	1205	1.990	0.03360
C-75	3.060	0.0642	1805	1.985	0.0417
L-N-80	3.070	0.0667	1955	1.998	0.0434
C-95	3.125	0.0745	2405	2.047	0.0490
P-105	3.162	0.0795	2700	2.052	0.0515
P-110	3.180	0.0820	2855	2.075	0.0535

Transition Collapse Pressure Formula The minimum collapse pressure for the plastic to elastic transition zone P_T is calculated by

$$P_T = Y_p \left| \frac{F}{D/t} - G \right| \tag{8.3}$$

The factors F and G and applicable D/t range for the transition collapse pressure formula are shown in Tables 8.8 and 8.9, respectively.

Elastic Collapse Pressure Formula The minimum collapse pressure of the elastic range of collapse is calculated by

$$P_E = \frac{46.95 \times 10^6}{(D/t)((D/t)-1)^2} \tag{8.4}$$

Collapse Pressure under Axial Tension Stress The reduced minimum collapse pressure caused by the action of axial tension stress is calculated by

$$P_{CA} = P_{CO}\left[\sqrt{1-0.75[(S_A+P_i)/Y_p]^2}\right] - 0.5(S_A+P_i/Y_p) \tag{8.5}$$

Equation 8.5 is not applicable if P_{CO} is calculated from the elastic collapse formula.

Symbols in Equations 8.1 to 8.5 are as follows:

D = nominal outside diameter in in.
t = nominal wall thickness in in.
Y_p = minimum yield strength of pipe in psi
P_{Yp} = minimum yield strength collapse pressure in psi
P_p = minimum plastic collapse pressure in psi
P_T = minimum plastic/elastic transition collapse pressure in psi
P_E = minimum elastic collapse pressure in psi
P_{CA} = minimum collapse pressure under axial tension stress in psi
P_{CO} = minimum collapse pressure without axial tension stress in psi
S_A = axial tension stress in psi
P_i = internal pressure in psi

Internal Yield Pressure for Pipe Internal yield pressure for pipe is calculated from Equation 8.6. The factor 0.875 appearing in the formula 8.6 allows for minimum wall thickness.

$$P_i = 0.875\left[\frac{2Y_p t}{D}\right] \tag{8.6}$$

where

P_i = minimum internal yield pressure in psi
Y_p = minimum yield strength in psi
t = nominal wall thickness in in.
D = nominal outside diameter in inches.

Internal Yield Pressure for Couplings Internal yield pressure for threaded and coupled pipe is the same as for plain end pipe, except where a lower pressure is required to avoid leakage due to insufficient coupling strength. The lower pressure is based on

$$P = Y_c\left[\frac{W - d_1}{W}\right] \tag{8.7}$$

where

P = minimum internal yield pressure in psi
Y_c = minimum yield strength at coupling in psi
W = nominal outside diameter of coupling
d_1 = diameter of the root of the coupling thread at the end of the pipe in the powertight position (see API Bulletin 5C3, 6th Edition, October 1994).

Pipe Body Yield Strength (Section 2 of API 5C3) Pipe body yield strength is the axial load required to yield the pipe. It is taken as the product of the cross sectional area and the specified minimum yield strength for the particular grade of pipe. Values for pipe body yield strength were calculated by means of the following formula:

$$P_Y = 0.7854(D^2 - d^2)Y_p \tag{8.8}$$

where

P_Y = pipe body yield strength in psi
Y_p = minimum yield strength
D = specified outside diameter in in.
d = specified inside diameter in in.

8.2.7 Joint Strength (Section 9 of API 5C3)

Round Thread Casing Joint Strength Round thread casing joint strength is calculated from formulas 8.9 and 8.10. The lesser of the values obtained from the two formulas governs. Formulas 8.9 and 8.10 apply both to short and long threads and couplings. Formula 8.9 is for minimum strength of a joint failing by fracture, and formula 8.10 for minimum strength of a joint failing by thread jumpout or pullout.
The fracture strength is

$$P_j = 0.95 A_{jp} U_p \tag{8.9}$$

The pullout strength is

$$P_j = 0.95 A_{jp} L \left[\frac{0.74 D^{-0.59} U_p}{0.5L + 0.14D} + \frac{Y_p}{L + 0.14D} \right] \tag{8.10}$$

where

P_j = minimum joint strength in lb
A_{jp} = cross - sectional area of the pipe wall under the last perfect thread in in.^{2}
$= 0.7854(D - 0.1425)^2 - d^2)$ for eight round threads
D = nominal outside diameter of pipe in in.
d = nominal inside diameter of pipe in in.
L = engaged thread length in in.
$= L_4 - M$ for nominal makeup, API Spec 5B
Y_p = minimum yield strength of pipe in psi
U_p = minimum ultimate strength of pipe in psi

Buttress Thread Casing Joint Strength Buttress thread casing joint strength is calculated from formulas 8.11 and 8.12. The lesser of the values obtained from the two formulas governs.

Pipe thread strength is

$$P_j = 0.95\,A_p U_p |1.008 - 0.0396(1.083 - Y_p/U_p)D| \tag{8.11}$$

Casing thread strength is

$$P_j = 0.95\,A_c U_c \tag{8.12}$$

where

P_j = minimum joint strength in lb
Y_p = minimum yield strength of pipe in lb
U_p = minimum ultimate strength of pipe in psi
U_c = minimum ultimate strength of coupling in psi
A_p = cross-sectional area of plain end pipe in in.^{2}
$= 0.7854(D^2 - d^2)$
A_c = cross-sectional area of coupling in in.^{2}
$= 0.7854(W^2 - d_i^2)$
D = outside diameter of pipe in in.
W = outside diameter of coupling in in.
d = inside diameter of pipe in in.
d_i = diameter of the root of the coupling thread at the end of the pipe in the powertight position

Extreme-line Casing Joint Strength Extreme-line casing joint strength is calculated from

$$P_j = A_{cr} U_p \tag{8.13}$$

where

P_j = minimum joint strength in lb
A_{cr} = critical section area of box, pin or pipe, whichever is least, in in.^{2} (see API Bulletin 5C3)
U_p = specified minimum ultimate strength in psi (Table 8.9)

8.3 COMBINATION CASING STRINGS

The term *combination casing string* is generally applied to a casing string that is composed of more than one weight per foot, or more than one grade of steel, or both.

8.3.1 Design Consideration

Solving the problem of casing string design for known type and size of casing string relies on selection of the most economical grades and weights of casing that will withstand, without failure, the loads to which the casing will be subjected throughout the life of the well.

There are various established methods of designing a technically satisfactory combination casing string. The difference between these methods rely on different design models, different values of the safety factors and different sequences of calculations. There are no commonly accepted methods of combination string design nor accepted values for the safety factors. Some suggestions are offered below; however, the decision is left to the person responsible for the design.

In general, the following loads must be considered: tension, collapse, burst and compression. The reasonably worst working conditions ought to be assumed.

Collapse The casing must be designed against collapse to withstand the hydrostatic pressure of the fluid behind the casing at any depth, decreased by anticipated pressure inside the casing at the corresponding level. Usually, the maximum collapse pressure to be imposed on the casing is considered to be the hydrostatic pressure of the heaviest mud used to drill to the lending depth of the casing string, acting on empty string. Depending upon design model, it is recommended to use a design factor of 1.0 to 1.2. For example, if it is known that casing will never be empty inside, this fact should be considered for collapse pressure evaluation and selection of the magnitude of safety factor.

Burst Casing must be designed to resist expected burst pressure at any depth. In burst pressure consideration, it is suggested to consider different design models depending upon the type of casing string.

Conductor String It is assumed that the external pressure is zero. In any case, the maximum expected internal pressure cannot be greater than fracture pressure at the open hole below the conductor casing shoe; usually, it is the first formation right below the casing shoe. If this pressure is not known (in exploratory drilling), the burst pressure of gas equivalent to 0.9 or 1.0 psi/ft can be assumed. Hydrostatic head due to gas is neglected. For example, if the setting depth of conductor string is 1,000 ft, then the maximum expected burst pressure is even along the string and equal to (1,100)(1.0) = 1,100 psi; safety factor = 1.1 to 1.15.

8.3.2 Surface and Intermediate Strings

It is suggested to evaluate the burst load based on the internal pressure expected, reduced by the external pressure of the drilling fluid outside the string. Internal pressure is based on the expected bottomhole pressure of the next string with the hole being evacuated from drilling fluid up to a minimum of 50%. In exploratory wells, a reasonable assumption of expected formation pore pressure gradient is required.

Example 8.3

Evaluate an expected burst pressure acting on surface casing string in exploratory drilling if setting depth of the next string is 11,000 ft.

Solution

Step 1. Internal pressure in the borehole. Because the next string is set at 11,000 ft, the formation pore pressure gradient is assumed to be 0.65 psi/ft. Thus, the bottomhole pressure (at a depth of 11,000 ft) is (11,000)(0.65) = 7,150 psi. Assume 50% of evacuation; thus, (11,000)(0.5) = 5,500 ft.

Note: It is assumed that below 5,500 ft, the hole is filled with mud, which exerts a pressure gradient of 0.65 psi/ft. The hole above 5,500 ft is filled with gas, the weight of which is ignored.
The internal pressure at a depth of 5,500 ft is (5,500)(0.65) = 3,575 psi. Since the weight of gas is ignored for this type of string, the internal pressure at the top of the hole is also 3,575 psi.

Step 2. External pressure. It is assumed that there is drilling fluid with specific gravity of 1.2 outside the casing. Thus, the external pressure at surface = 0.0 psi and at 5,000 ft is (5,000)(1.2)(8.34) (0.052) ≅ 2,600 psi.

Step 3. Burst load (Pb). The burst pressure is equal to internal pressure reduced by the external pressure of the drilling fluid outside the casing. Therefore,

$$\text{at surface: } P_b = 3{,}575 - 0.0 = 3{,}573 \text{ psi}$$

$$\text{at 5,000 ft: } P_b = 3{,}575 - 2{,}600 = 987 \text{ psi}$$

The burst pressure line equation is as below:

$$P_b = 3{,}575 - 0.52(D) \text{ (psi)}$$

$$D = \text{depth at the hole (ft) (from 0 to 5,000 ft)}$$

Note: For practical purposes, a graphic solution is very advisable.

8.3.3 Production String

In exploratory drilling, it assumed that internal pressure acting on the casing is reduced by external saltwater pressure gradient of about 0.5 psi/ft. Internal pressure is based on expected gas pressure gradient. For long strings, the weight of gas is not ignored.

8.3.4 Tension Load

The maximum tensile load acting on the casing string is often considered as the static weight of the casing as measured in air.

Casing must be designed to satisfy these equations:

$$P_j = (W)(N_j) \quad (8.14)$$

$$P_y = (W)(N_p) \quad (8.15)$$

where

P_j = casing joint strength in lb
P_y = pipe body strength in lb
W = weight (mass) of casing suspended below the cross-section under consideration in lb
N_j = safety factor for joint
N_p = safety factor for pipe body

Safety factors (N_j, N_p) of 1.6 to 2.0, are used and should be applied to the minimum joint tensile strength or the minimum pipe body tensile strength, whichever is the smallest.

8.3.5 Compression Load

Under certain conditions, casing can be subjected to the compression load, such as if the weight of the inner strings (conductor or surface casing string) is transferred to the outer string or if the portion of the casing weight is slacked off on the bottom of the hole. This load may result in casing failure and, therefore, must also be considered.

It should be pointed out that hydrostatic pressure does not produce an effective compression and, therefore, is not considered. If the casing is suspended at the top of the hole that is filled with fluid, then the only effect of hydrostatic pressure is reduction of casing weight per foot and the string is effectively under tension.

An example is offered on the following pages of the design procedure to be followed in designing a combination string.

Example 8.4

Design a combination casing string if data are as below:

Type of well: exploration well
Type of casing: production for testing purposes
Casing setting depth: 12,000 ft
Casing size: 7 in.

Design model:

Collapse: Assumed external fluid pressure gradient of 0.52 psi/ft and casing empty inside. Safety factor for collapse = 1.0. Reduction of collapse pressure resistance due to the axial load is considered.

Burst: Assumed external pressure gradient of saltwater = 0.465 psi/ft and formation pore (gas) pressure gradient = 0.65 psi/ft. Gas weight is neglected. Safety factor = 1.1.
Tension: Casing suspended at the surface. Weight reduction due to buoyancy effect is ignored. Safety factor = 1.6.
Compression: casing not subjected to compression load.

The selected coupling should be long with round thread. The available casing grade is N-80 and unit weights as given below.

Steel Grade	Unit weight (lb/ft)	Cross-sectional area (in.)	Collapse pressure resistance	Burst pressure resistance	Joint strength (10^3 lb)	Pipe body strength (10^3 lb)
N-80	26.0	7.548	5,410	7,240	519	604
N-80	26.0	8.451	7,020	8,160	597	676
N-80	26.0	9.315	8,600	9,060	672	745

Solution

Part 1. Consider collapse pressure and tension load.

Step 1. Determine the lightest weight of casing to resist collapse pressure for a setting depth of 12,000 ft. Because the maximum collapse pressure is (12,000)(0.52) = 6,240 psi, select N-80, 29-lb/ft casing with collapse pressure resistance of 7,020 psi. (Note: assumed safety factor for collapse = 1.0.) This is Section 1.

Step 2. The next section (above section 1) is to consist of the next lighter casing, i.e., N-80, 26-lb/ft. This is Section 2. Neglecting the effect of the axial load due to the weight of Section 1 suspended below it, the setting depth of Section 2 is

$$D_2' = \frac{5{,}410}{(1.0)(0.52)} = 10{,}403\,\text{ft}$$

Under this assumption, the mass of Section 1 is

$$W_1' = (12{,}000 - 10{,}403)(29) = 46{,}313\,\text{lb}$$

For this axial load, the reduced minimum collapse pressure resistance of Section 2 can be calculated from formula 8.5.

(*Note*: internal pressure P_i for considered case = 0.)

$$P_{CA}(2) = 5{,}410\left[\sqrt{1-0.75\left(\frac{6{,}136}{80{,}000}\right)^2} - (0.5)\frac{6{,}136}{80{,}000}\right] = 5{,}190\,\text{psi}$$

(*Note*: $S_A = W'/A = 46{,}313/7.548 = 6{,}136$ psi)

Step 3. Using the obtained reduced minimum collapse pressure resistance of Section 2, calculate the setting depth of this section:

$$D_2'' = \frac{5{,}190}{(1.0)(0.52)} = 9{,}980\,\text{ft}$$

For obtained setting depth of Section 2, the weight of Section 1 is

$$W_1'' = (12{,}000 - 9{,}980)(29) = 58{,}580\,\text{lb}$$

and corresponding reduced minimum collapse pressure resistance of Section 2 is

$$P_{CA}'' = 5{,}410\left[\sqrt{1-0.75\left(\frac{7{,}761}{80{,}000}\right)^2} - (0.5)\frac{7{,}761}{80{,}000}\right] = 5{,}128\,\text{psi}$$

Step 4. The third assumed setting depth of Section 2 is usually taken as a correct setting depth, i.e.,

$$D_2'' = \frac{5{,}128}{(1.0)(0.52)} = 9{,}861\,\text{ft}$$

Then, the length and weight of Section 2 is

$$L_1 = 12{,}000 - 9{,}861 = 2{,}128\text{ ft} \quad \text{and} \quad W_1 = 62{,}015\text{ lb}$$

(*Note*: If the next lighter casing were available, then Steps 2 through 4 must be repeated for this casing and that would be Section 3, etc.).

The maximum length of Section 2 is limited by coupling load capacity and is calculated below:

$$L_{2\max} = \frac{519{,}000 - (62{,}015)(1.6)}{(1.6)(26)} = 10{,}090\text{ ft}$$

which is greater than its setting depth (9,861 ft). So, Section 2 extends to the top of the hole. (*Note*: If Section 2 would not cover the entire length of the hole, then the next stronger casing should be applied. That would be Section 3. The setting depth of Section 3 is governed by joint strength, not by collapse pressure.)

Part 2. Check casing string on burst pressure obtained in Part 1 and make necessary corrections.

Step 1. Determine external pressure.
At top: 0.0 psi
At bottom: (12,000)(0.465) = 5,580 psi.

Step 2. Determine internal pressure
At bottom: (12,000)(0.65) = 7,800 psi.
At top: 7,800 psi (weight of gas is ignored).

Step 3. Determine burst pressure.
At top: 7,800 psi.
At bottom: 7,800 − 5,580 = 2,220 psi.
Burst pressure line equation is

$$P_b = 7{,}800 - \frac{7{,}800 - 2{,}200}{12{,}000}(D)$$

$$= 7{,}800 - (0.456)(D)$$

$$D = \text{hole depth in ft}$$

Graphical solution is presented in Figure 8.10.

Step 4. It is apparent that Section 1 is capable of withstanding the expected burst pressure.

Step 5. Section 2 can withstand the expected burst pressure up to the depth calculated below:

$$\frac{7{,}240}{1.1} = 7{,}800 - (0.456)(D)$$

$$D = \frac{7{,}800 - 6{,}581}{0.456} = 2{,}673\ \text{ft}$$

Therefore, the length and weight of Section 2 is

$$9{,}861 - 2{,}673 = 7{,}188\ \text{ft}$$

$$W_2 = (7{,}188)(26) = 186{,}888\ \text{lb}$$

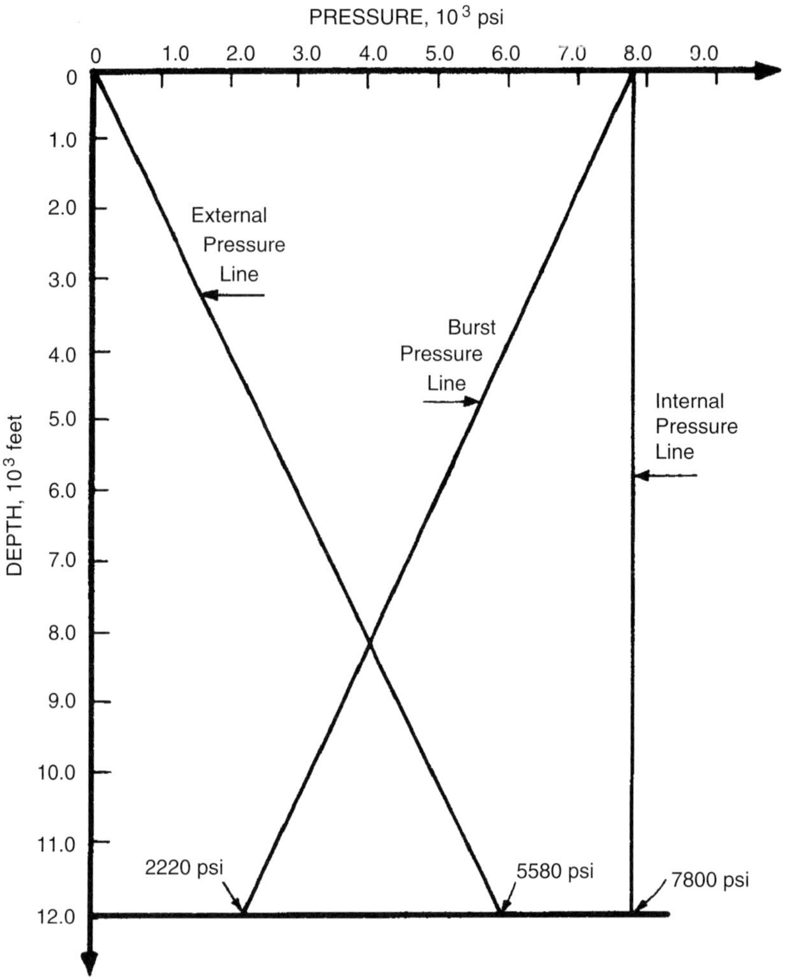

FIGURE 8.10 Graphical representation of burst load determination.

To cover the upper part of the hole (2,673 ft), stronger casing must be used.

Step 6. Take the next stronger casing, i.e., N-80, 29 lb/ft with burst pressure resistance of (8,160)/(1.1) = 7,418 psi.
This is Section 3. This casing can be used up to the hole depth of

$$7{,}418 = 7{,}800 - (0.456)(D)$$

$$D = \frac{7{,}800 - 7{,}418}{0.456} = 837 \text{ ft}$$

Then, the length and weight of Section 3 is 2,673 − 837 = 1,836 ft

$$W_3 = (1{,}836)(26) = 53{,}244\,\text{lb}$$

To cover the remaining 837 ft of the hole, stronger casing must be used.

Step 7. The next stronger casing is N-80, 32 lb/ft with burst pressure resistance of (9,060)/(1.1) = 8,236 psi. This is Section 4. N-80, 32 lb/ft casing is strong enough to cover the remaining part of the hole (*Note*: 8,236 psi > 7,800 psi). Therefore, the length and weight of Section 4 is 837 − 0.0 = 837 ft and $W_4 = 26{,}784$ lb.

Part 3. Check casing string on tension obtained in Part 2 and, if necessary, make corrections to satisfy the required magnitude of safety factor.

Step 1. Maximum length of Section 3 due to its joint strength is

$$L_3 = \frac{597{,}000 - (1.6)(62{,}015 + 186{,}888)}{(1.6)(29)}$$

$$= 4{,}283\,\text{ft}$$

Since the length of Section 3 is 1,836, the safety factor is even greater than required.

Step 2. Maximum length of Section 4 due to its joint strength is

$$L_4 = \frac{672{,}000 - (1.6)(62{,}015 + 186{,}888 + 26{,}784)}{(1.6)(32)}$$

$$= 4{,}509\,\text{ft}$$

Because L_4 is greater than the length obtained in Part 2 (837), it may be concluded that the requirements for tension are satisfied.

A summary of the results obtained is presented in the table below and in Figure 8.11.

Section no.	Setting depth (ft)	Length	Weight (lb)	Grade	Unit weight	Coupling
4	837	837	26,784	N-80	32	Long with round thread
3	2,637	1,836	53,244	N-80	29	Long with round thread
2	9,861	7,188	186,888	N-80	26	Long with round thread
1	12,000	2,138	62,015	N-80	29	Long with round thread

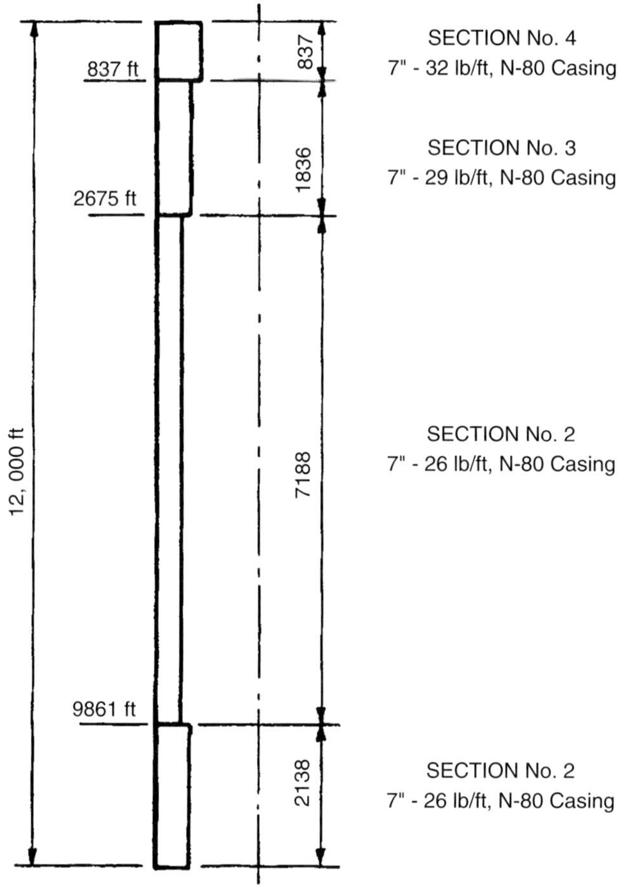

FIGURE 8.11 Schematic diagram of combination casing string.

8.4 RUNNING AND PULLING CASING

The following excerpts are taken from API Recommended Practice 5C1, "Care and Use of Casing and Tubing," 18th Edition, May 1999 (Section 4, Running and Pulling Casing).

8.4.1 Preparation and Inspection Before Running

4.1.1. New casing is delivered free of injurious defects as defined in API Specification 5CT and within the practical limits of the inspection procedures therein prescribed. Some users have found that, for a limited number of critical well applications, these procedures do not result in casing sufficiently free of defects to meet their needs for such critical applications. Various nondestructive inspection services have been employed by users to ensure that the desired quality of casing is being run. In view of this practice, it is suggested that the individual user:

a. Familiarize himself with inspection practices specified in the standards and employed by the respective mills, and with the definition of "injurious defect" contained in the standards.
b. Thoroughly evaluate any nondestructive inspection to be used by him on API tubular goods to assure himself that the inspection does in fact correctly locate and differentiate injurious defects from other variables that can be and frequently are sources of misleading "defect" signals with such inspection methods.

4.1.2. All casing, whether new, used, or reconditioned, should always be handled with thread protectors in place. Casing should be handled at all times on racks or on wooden or metal surfaces free of rocks, sand, or dirt other than normal drilling mud. When lengths of casing are inadvertently dragged in the dirt, the threads should be recleaned and serviced again as outlined in 4.1.7.

4.1.3. Slip elevators are recommended for long strings. Both spider and elevator slips should be clean and sharp and should fit properly. Slips should be extra long for heavy casing strings. The spider must be level.

Note: Slip and tong marks are injurious. Every possible effort should be made to keep such damage at a minimum by using proper up-to-date equipment.

4.1.4. If collar-pull elevators are used, the bearing surface should be carefully inspected for (a) uneven wear that may produce a side lift on the coupling with danger of jumping it off, and (b) uniform distribution of the load when applied over the bearing face of the coupling.

4.1.5. Spider and elevator slips should be examined and watched to see that all lower together. If they lower unevenly, there is danger of denting the pipe or badly slip-cutting it.

4.1.6. Care shall be exercised, particularly when running long casing strings, to ensure that the slip bushing or bowl is in good condition. Tongs may be sized to produce 1.5 percent of the calculated pullout strength (API Bulletin 5C3) with units changed to ft-lb (N·m) (150 percent of the guideline torque found in Table 1). Tongs should be examined for wear on hinge pins and hinge surfaces. The backup line attachment to the backup post should be corrected, if necessary, to be level with the tong in the backup position so as to avoid uneven load distribution on the gripping surfaces of the casing. The length of the backup line should be such as to cause minimum bending stresses on the casing and to allow full stroke movement of the makeup tong.

4.1.7. The following precautions should be taken in the preparation of casing threads for makeup in the casing strings:

a. Immediately before running, remove thread protectors from both field and coupling ends and clean the threads thoroughly, repeating as additional rows become uncovered.
b. Carefully inspect the threads. Those found damaged, even slightly, should be laid aside unless satisfactory, means are available for correcting thread damage.
c. The length of each piece of casing shall be measured prior to running. A steel tape calibrated in decimal feet (millimeters) to the nearest 0.01 feet (millimeters) should be used. The measurement should be made from the outermost face of the coupling or box to the position on the externally threaded end where the coupling or the box stops when the joint is made up power tight. On round-thread joints, this position is to the plane of the vanish point on the pipe; on buttress-thread casing, this position is to the base of the triangle stamp on the pipe; and on extreme line casing, this position is to the shoulder on the externally threaded end. The total of the individual lengths so measured will represent the unloaded length of the casing string. The actual length under tension in the hole can be obtained by consulting graphs that are prepared for this purpose and are available in most pipe handbooks.
d. Check each coupling for makeup. If the standoff is abnormally great, check the coupling for tightness. Tighten any loose couplings after thoroughly cleaning the threads and applying fresh compound over entire thread surfaces, and before pulling the pipe into the derrick.
e. Before stabbing, liberally apply thread compound to the entire internally and externally threaded areas. It is recommended that a thread compound that meets the performance objectives of API Bulletin 5A2 be used; however, in special cases where severe conditions are encountered, it is recommended that high-pressure silicone thread compounds as specified in API Bulletin 5A2 be used.
f. Place a clean thread protector on the field end of the pipe so that the thread will not be damaged while rolling pipe on the rack and pulling into the derrick. Several thread protectors may be cleaned and used repeatedly for this operation.
g. If a mixed string is to be run, check to determine that appropriate casing will be accessible on the pipe rack when required according to program.

h. Connectors used as tensile and lifting members should have their thread capacity carefully checked to ensure that the connector can safely support the load.
i. Care should be taken when making up pup joints and connectors to ensure that the mating threads are of the same size and type.

8.4.2 Drifting of Casing

4.2.1. It is recommended that each length of casing be drifted for its entire length just before running, with mandrels conforming to API Specification 5CT. Casing that will not pass the drill test should be laid aside.

4.2.2. Lower or roll each piece of casing carefully to the walk without dropping. Use rope snubber if necessary. Avoid hitting casing against any part of derrick or other equipment. Provide a hold-back rope at window. For mixed or unmarked strings, a drift or "jack rabbit" should be run through each length of casing when it is picked up from the catwalk and pulled onto the derrick floor to avoid running a heavier length or one with a lesser inside diameter than called for in the casing string.

8.4.3 Stabbing, Making Up, and Lowering

4.3.1. Do not remove thread protector from field end of casing until ready to stab.

4.3.2. If necessary, apply thread compound over the entire surface of threads just before stabbing. The brush or utensil used in applying thread compound should be kept free of foreign matter, and the compound should never be thinned.

4.3.3. In stabbing, lower casing carefully to avoid injuring threads. Stab vertically, preferably with the assistance of a man on the stabbing board. If the casing stand tilts to one side after stabbing, lift up, clean, and correct any damaged thread with a three-cornered file, then carefully remove any filings and reapply compound over the thread surface. After stabbing, the casing should be rotated very slowly at first to ensure that threads are engaging properly and not cross-threading. If spinning line is used, it should pull close to the coupling.

Note: Recommendations in 4.3.4 and 4.4.1 for casing makeup apply to the use of power tongs. For recommendations on makeup of casing with spinning lines and conventional tongs, see 4.4.2.

4.3.4. The use of power tongs for making up casing made desirable the establishment of recommended torque values for each size, weight, and grade of casing. Early studies and tests indicated that torque values are affected by a large number of variables, such as variations in taper, lead, thread height and thread form, surface finish, type of thread compound, length of thread, weight and grade of pipe, etc. In view of the number of variables and the extent that these variables, alone or in combination, could affect the relationship of torque values versus made-up position, it was evident that both applied torque and made-up position must be considered. Since the API joint pull-out strength formula in API Bulletin 5C2 contains several of the variables believed to affect torque, using a modified formula to establish torque values was investigated. Torque values obtained by taking 1 percent of the calculated pullout value were found to be generally comparable to values obtained by field makeup tests using API modified thread compound in accordance with API Bulletin 5A2. Compounds other than API modified thread compound may have other torque values. This procedure was therefore used to establish the makeup torque values listed in Table 1. All values are rounded to the nearest 10 ft-lb (10 N·m). These values shall be considered as a guide only, due to the very wide variations in torque requirements that can exist for a specific connection. Because of this, it is essential that torque be related to madeup position as outlined in 4.4.1. The torque values listed in Table 1 apply to casing with zinc-plated or phosphate-coated couplings. When making up connections with tin-plated couplings, 80 percent of the listed value can be used as a guide. The listed torque values are not applicable for making up couplings with PTFE (polytetrafluoroethylene) rings. When making up round thread connections with PTFE rings, 70 percent of the listed values are recommended. Buttress connections with PTFE seal rings may make up at torque values different from those normally observed on standard buttress threads.

Note: Thread galling of gall-prone materials (martensitic chromium steels, 9 Cr and 13 Cr) occurs during movement—stabbing or pulling and makeup or breakout. Galling resistance of threads is primarily controlled in two areas—surface preparation and finishing during manufacture and careful handling practices during running and pulling. Threads and lubricant must be clean. Assembly in the horizontal position should be avoided. Connections should be turned by hand to the hand-tight position before slowly power tightening. The procedure should be reversed for disassembly.

8.4.4 Field Makeup

4.4.1. The following practice is recommended for field makeup of casing:

a. For Round Thread, Sizes $4\frac{1}{2}$ through $1\frac{3}{8}$. OD

1. It is advisable when starting to run casing from each particular mill shipment to make up sufficient joints to determine the

torque necessary to provide proper makeup. See 4.4.2 for the proper number of turns beyond hand-tight position. These values may indicate that a departure from the values listed in Table 1 is advisable. If other values are chosen, the minimum torque should be not less than 75 percent of the value selected. The maximum torque should be not more than 125 percent of the selected torque.

2. The power tong should be provided with a reliable torque gauge of known accuracy. In the initial stages of makeup, any irregularities of makeup or in speed of makeup should be observed, since these may be indicative of crossed threads, dirty or damaged threads, or other unfavorable conditions. To prevent galling when making up connections in the field, the connections should be made up at a speed not to exceed 25 rpm.
3. Continue the makeup, observing both the torque gauge and the approximately position of the coupling face with respect to the thread vanish point position.
4. The torque values shown in Tables 1, 2, and 3 have been selected to give recommended makeup under normal conditions and should be considered as satisfactory providing the face of the coupling is flush with the thread vanish point or within two thread turns, plus or minus, of the thread vanish point.
5. If the makeup is such that the thread vanish point is buried two thread turns and 75 percent of the torque shown in Table 1 is not reached, the joint should be treated as a questionable joint as provided in 4.4.3.
6. If several threads remain exposed when the listed torque is reached, apply additional torque up to 125 percent of the value shown in Table 1. If the standoff (distance from face of coupling to the thread vanish point) is greater than three thread turns when this additional torque is reached, the joint should be treated as a questionable joint as provided in 4.4.3.

b. For buttress thread casing connections in sizes $4\frac{1}{2}$ through $13\frac{3}{8}$ OD, makeup torque values should be determined by carefully noting the torque required to make up each of several connections to the base of the triangle; then using the torque value thus established, make up the balance of the pipe of that particular weight and grade in the string.

c. For round thread and buttress thread, sizes 16, $18\frac{5}{8}$, and 20 outside diameter:

1. Makeup of sizes 16, $18\frac{5}{8}$, and 20 shall be to a position on each connection represented by the thread vanish point on 8-round thread and the base of the triangle on buttress thread using the minimum torque shown in Table 1 as a guide.

   On 8-round thread casing a $\frac{3}{8}$-inch (9.5-millimeter) equilateral triangle is die stamped at a distance of $L_1 + 1/16$ inch (1.6 millimeters) from each end. The base of the triangle will aid in locating the thread vanish point for basic power-tight makeup; however, the position of the coupling with respect to the base of the triangle shall not be a basis for acceptance or rejection of the product. Care shall be taken to avoid cross threading in starting these larger connections. The tongs selected should be capable of attaining high torques [50,000 ft-lb (67,800 N·m)] for the entire run. Anticipate that maximum torque values could be five times the minimum experienced in makeup to the recommended position.
2. Joints that are questionable as to their proper makeup in 4.4.1, item a.5 or a.6 should be unscrewed and laid down to determine the cause of improper makeup. Both the pipe thread and mating coupling thread should be inspected. Damaged threads or threads that do not comply with the specification should be repaired. If damaged or out-of-tolerance threads are not found to be the cause of improper makeup, then the makeup torque should be adjusted to obtain proper makeup (see 4.4.1, item a.1). It should be noted that a thread compound with a coefficient of friction substantially different from common values may be the cause of improper makeup.

4.4.2. When conventional tongs are used for casing makeup, tighten with tongs to proper degree of tightness. The joint should be made up beyond the hand-tight position at least three turns for sizes $4\frac{1}{2}$ through 7, and at least three and one-half turns for sizes $7\frac{5}{8}$ and larger, except $9\frac{5}{8}$, and $10\frac{3}{4}$ grade P110 and size 20 grade J55 and K55, which should be made up four turns beyond hand-tight position. When using a spinning line, it is necessary to compare hand tightness with spin-up tightness. In order to do this, make up the first few joints to the hand-tight position, then back off and spin up joints to the spin-up tight position. Compare relative position of these two makeups and use this information to determine when the joint is made up the recommended number of turns beyond hand tight.

4.4.3. Joints that are questionable as to their proper tightness should be unscrewed and the casing laid down for inspection and repair. When this is done, the mating coupling should be carefully inspected for damaged threads. Parted joints should never be reused without shopping or regauging, even though the joints may have little appearance of damage.

4.4.4. If casing has a tendency to wobble unduly at its upper end when making up, indicating that the thread may not be in line with the axis of the casing, the speed of rotation should be decreased to prevent galling of threads. If wobbling should persist despite reduced rotational speed, the casing should be laid down for inspection. Serious consideration should be given before using such casing in a position in the string where a heavy tensile load is imposed.

4.4.5. In making up the field joint, it is possible for the coupling to make up slightly on the mill end. This does not indicate that the coupling on the mill end is too loose but simply that the field end has reached the tightness with which the coupling was screwed on at the manufacturer's facility.

4.4.6. Casing strings should be picked up and lowered carefully and care exercised in setting slips to avoid shock loads. Dropping a string even a short distance may loosen couplings at the bottom of the string. Care should be exercised to prevent setting casing down on bottom or otherwise placing it in compression because of the danger of buckling, particularly in that part of the well where hole enlargement has occurred.

4.4.7. Definite instructions should be available as to the design of the casing string, including the proper location of the various grades of steel, weights of casing, and types of joint. Care should be exercised to run the string in exactly the order in which it was designed. If any length cannot be clearly identified, it should be laid aside until its grade, weight, or type of joint can be positively established.

4.4.8. To facilitate running and to ensure adequate hydrostatic head to contain reservoir pressures, the casing should be periodically filled with mud while being run. A number of things govern the frequency with which filling should be accomplished: weight of pipe in the hole, mud weight, reservoir pressure, etc. In most cases, filling every six to ten lengths should suffice. In no case should the hydrostatic balance of reservoir pressure be jeopardized by too infrequent filling. Filling should be done with mud of the proper weight, using a conveniently located hose of adequate size to expedite the filling operation. A quick opening and closing plug valve on the mud hose will facilitate the operation and prevent overflow. If rubber hose is used, it is recommended that the quickclosing valve be mounted where the hose is connected to the mud line, rather than at the outlet end of the hose. It is also recommended that at least one other discharge connection be left open on the mud system to prevent buildup of excessive pressure when the quick-closing valve is closed while the pump is still running. A cooper nipple at the end of the mud hose may be used to prevent damaging of the coupling threads during the filling operation.

Note: The foregoing mud fill-up practice will be unnecessary if automatic fill-up casing shoes and collars are used.

8.4.5 Casing Landing Procedure

Definite instructions should be provided for the proper string tension, also on the proper landing procedure after the cement has set. The purpose is to avoid critical stresses or excessive and unsafe tensile stresses at any time during the life of the well. In arriving at the proper tension and landing procedure, consideration should be given to all factors, such as well temperature and pressure, temperature developed due to cement hydration, mud temperature, and changes of temperature during producing operations. The adequacy of the original tension safety factor of the string as designed will influence the landing procedure and should be considered. If, however, after due consideration it is not considered necessary to develop special landing procedure instructions (and this probably applies to a very large majority of the wells drilled), then the procedure should be followed of landing the casing in the casing head at exactly the position in which it was hanging when the cement plug reached its lowest point or "as cemented."

8.4.6 Care of Casing in Hole

Drill pipe run inside casing should be equipped with suitable drill-pipe protectors.

8.4.7 Recovery of Casing

4.7.1. Breakout tongs should be positioned close to the coupling but not too close since a slight squashing effect where the tong dies contact the pipe surface cannot be avoided, especially if the joint is tight and/or the casing is light. Keeping a space of one-third to one-quarter of the diameter of the pipe between the tong and the coupling should normally prevent unnecessary friction in the threads. Hammering the coupling to break the joint is an injurious practice. If tapping is required, use the flat face, never the peen face of the hammer, and under no circumstances should a sledgehammer be used. Tap lightly near the middle and completely around the coupling, never near the end nor on opposite sides only.

4.7.2. Great care should be exercised to disengage all of the thread before lifting the casing out of the coupling. Do not jump casing out of the coupling.

4.7.3. All threads should be cleaned and lubricated or should be coated with a material that will minimize corrosion. Clean protectors should be placed on the tubing before it is laid down.
4.7.4. Before casing is stored or reused, pipe and thread should be inspected and defective joints marked for shopping and regauging.
4.7.5. When casing is being retrieved because of a casing failure, it is imperative to future prevention of such failures that a thorough metallurgical study be made. Every attempt should be made to retrieve the failed portion of the "as-failed" condition. When thorough metallurgical analysis reveals some facet of pipe quality to be involved in the failure, the results of the study should be reported to the API office.
4.7.6. Casing stacked in the derrick should be set on a firm wooden platform and without the bottom thread protector since the design of most protectors is not such as to support the joint or stand without damage to the field thread.

8.4.8 Causes of Casing Troubles

The more common causes of casing troubles are listed in 4.8.1 through 4.8.16.

4.8.1. Improper selection for depth and pressures encountered.
4.8.2. Insufficient inspection of each length of casing or of field-shop threads.
4.8.3. Abuse in mill, transportation, and field handling.
4.8.4. Nonobservance of good rules in running and pulling casing.
4.8.5. Improper cutting of field-shop threads.
4.8.6. The use of poorly manufactured couplings for replacements and additions.
4.8.7. Improper care in storage.
4.8.8. Excessive torquing of casing to force it through tight places in the hole.
4.8.9. Pulling too hard on a string (to free it). This may loosen the couplings at the top of the string. They should be retightened with tongs before finally setting the string.
4.8.10. Rotary drilling inside casing. Setting the casing with improper tension after cementing is one of the greatest contributing causes of such failures.
4.8.11. Drill-pipe wear while drilling inside casing is particularly significant in drifted holes. Excess doglegs in deviated holes, or occasionally in straight holes where corrective measures are taken, result in concentrated bending of the casing that in turn results in excess internal wear, particularly when the doglegs are high in the hole.
4.8.12. Wire-line cutting, by swabbing or cable-tool drilling.

4.8.13. Buckling of casing in an enlarged, washed-out uncemented cavity if too much tension is released in landing.
4.8.14. Dropping a string, even a very short distance.
4.8.15. Leaky joints, under external or internal pressure, are a common trouble, and may be due to the following:

a. Improper thread compound.
b. Undertonging.
c. Dirty threads.
d. Galled threads, due to dirt, careless stabbing, damaged threads, too rapid spinning, overtonging, or wobbling during spinning or tonging operations.
e. Improper cutting of field-shop threads.
f. Pulling too hard on string.
g. Dropping string.
h. Excessive making and breaking.
i. Tonging too high on casing, especially on breaking out. This gives a bending effect that tends to gall the threads.
j. Improper joint makeup at mill.
k. Casing ovality or out-of-roundness.
l. Improper landing practice, which produces stresses in the threaded joint in excess of the yield point.

4.8.16. **Corrosion.** Both the inside and outside of casing can be damaged by corrosion, which can be recognized by the presence of pits or holes in the pipe. Corrosion on the outside of casing can be caused by corrosive fluids or formations in contact with the casing or by stray electric current flowing out of the casing into the surrounding fluids or formations. Severe corrosion may also be caused by sulphate-reducing bacteria. Corrosion damage on the inside is usually caused by corrosive fluids produced from the well, but the damage can be increased by the abrasive effects of casing and tubing pumping equipment any by high fluid velocities such as those encountered in some gas-lifted wells. Internal corrosion might also be due to stray electrical currents (electrolysis) or to dissimilar metals in close contact (bimetallic galvanic corrosion).

Because corrosion may result from so many different conditions, no simple or universal remedy can be given for its control. Each corrosion problem must be treated as an individual case and a solution attempted in the light of the known corrosion factors and operating conditions. The condition of the casing can be determined by visual or optical-instrument inspections. Where these are not practical, a casing-caliper survey can be made to determine the condition of the inside surfaces. No tools have yet been designed for determining the condition of the outside of casing in a well. Internal casing caliper surveys indicate the extent, location, and severity of corrosion. On the basis of the industry's experience to date, the following practices and measures can be used to control corrosion of casing:

a. Where external casing corrosion is known to occur or stray electrical current surveys indicate that relatively high currents are entering the well, the following practices can be employed:

    1. Good cementing practices, including the use of centralizers, scratchers, and adequate amounts of cement to keep corrosive fluids from contact with the outside of the casing.
    2. Electrical insulation of flow lines from wells by the use of nonconducting flange assemblies to reduce or prevent electrical currents from entering the well.
    3. The use of highly alkaline mud or mud treated with a bactericide as a completion fluid will help alleviate corrosion caused by sulfate-reducing bacteria.
    4. A properly designed cathodic protection system similar to that used for line pipe, which can alleviate external casing corrosion. Protection criteria for casing differ somewhat from the criteria used for line pipe. Literature on external casing corrosion or persons competent in this field should be consulted for proper protection criteria.

b. Where internal corrosion is known to exist, the following practices can be employed:

    1. In flowing wells, packing the annulus with fresh water or low-salinity alkaline muds. (It may be preferable in some flowing wells to depend upon inhibitors to protect the inside of the casing and the tubing.)
    2. In pumping wells, avoiding the use of casing pumps. Ordinarily, pumping wells, should be tubed as close to bottom as practical, regardless of the position of the pump, to minimize the damage to the casing from corrosive fluids.
    3. Using inhibitors to protect the inside of the casing against corrosion.

c. To determine the value and effectiveness of the above practices and measures, cost and equipment-failure records can be compared before and after application of control measures. Inhibitor effectiveness may be checked also by means of caliper surveys, visual examinations of readily accessible pieces of equipment, and water analyses for iron content. Coupons may also be helpful in determining whether sufficient inhibitor is being used. When lacking previous experience with any of the above measures, they should be used cautiously and on a limited scale until appraised for the particular operating conditions.

d. In general, all new areas should be considered as being potentially corrosive and investigations should be initiated early in the life of a field, and repeated periodically, to detect and localize corrosion before it has done destructive damage. These investigations should cover: (1) a complete chemical analysis of the effluent water, including pH, iron, hydrogen sulfide, organic acids, and any other substances that influence or indicate the degree of corrosion. An analysis of the produced gas for carbon dioxide and hydrogen sulfide is also desirable; (2) corrosion rate tests by using coupons of the same materials as in the well; and (3) the use of caliper or optical instrument inspections. Where conditions favorable to corrosion exist, a qualified corrosion engineer should be consulted. Particular attention should be given to mitigation of corrosion where the probable life of subsurface equipment is less than the time expected to deplete a well.

e. When H_2S is present in the well fluids, casing of high yield strength may be subject to sulfide corrosion cracking. The concentration of H_2S necessary to cause cracking in different strength materials is not yet well defined. Literature on sulfide corrosion or persons competent in this field should be consulted.

CHAPTER

9

Well Cementing

OUTLINE

9.1 INTRODUCTION

Cementing the casing and *liner cementing* (denoted as *primary cementing*) are probably the most important operations in the development of an oil and gas well. The drilling group is usually responsible for cementing the casing and the liner. The quality of these cementing operations will affect the success of follow-on drilling, completion, production and workover efforts in the well [1–3].

In addition to primary cementing of the casing and liner, there are other important well cementing operations. These are *squeeze cementing* and *plug cementing*. Such operations are often called *secondary* or *remedial cementing* [2].

Well cementing materials vary from basic Portland cement used in civil engineering construction of all types, to highly sophisticated special-purpose resin-based or latex cements. The purpose of all of these cementing materials is to provide the well driller with a fluid state slurry of cement, water and additives that can be pumped to specific locations within the well. Once the slurry has reached its intended location in the well and a setup time has elapsed, the slurry material can become a nearly impermeable, durable solid material capable of bonding to rock and steel casing.

The most widely used cements for well cementing are the Portland-type cements. The civil engineering construction industry uses Portland cement and water slurries in conjunction with clean rock aggregate to form concrete. The composite material formed by the addition of rock aggregate forms a solid material that has a compressive strength that is significantly higher than the solid formed by the solidified cement and water slurry alone. The rock of the aggregate usually has a very high compressive strength (of the order of 5,000 to 20,000 psi). The cement itself will have a compressive strength of about 1,000 to 3,000 psi. Therefore, the rock of the aggregate forms a solid material that has a very high compressive strength of the order of 5,000 to 2,000 psi (345 to 1830 bar). The cement itself will have a compressive strength of about 1,000 to 3,000 psi (69 to 207 bar). Therefore, the rock aggregate together with the matrix of solid cement can form a high-strength composite concrete with compressive strengths of the order of 4,000 to 15,000 (276 to 1035 bar).

The well drilling industry does not generally use aggregate with the cement except for silica flour and Ottawa sand. This is mainly due to the tight spacing within a well that precludes the passage of the larger particles of aggregate through the system. Thus, the well drilling industry refers to this material as simply cement. The slurry pumped to wells is usually a slurry of cement and water with appropriate additives. Because of the lack of aggregate, the compressive strength of well cements are restricted to the order of 200 to about 3,000 psi (14 to 207 bar) [1].

Oil well compressive strengths in the range of 500 psi (35 bar) have long been considered acceptable. Cement mixture selection is based on a much broader criteria then compressive strength alone.

9.2 CHEMISTRY OF CEMENTS

Cement is made of calcareous and argillaceous rock materials that are usually obtained from quarries. The process of making cement requires that these raw rock materials be ground, mixed and subjected to high temperatures.

The calcareous materials contain calcium carbonate or calcium oxide. Typical raw calcareous materials are as follows:

Limestone. This is sedimentary rock that is formed by the accumulation of organic marine life remains (shells or coral). Its main component is calcium carbonate.

Cement rock. This is a sedimentary rock that has a similar composition as the industrially produced cement.

Chalk. This is a soft limestone composed mainly of marine shells.

Marl. This is loose or crumbly deposit that contains a substantial amount of calcium carbonate.

Alkali waste. This is a secondary source and is often obtained from the waste of chemical plants. Such material will contain calcium oxide and/or calcium carbonate.

The argillaceous materials contain clay or clay minerals. Typical raw argillaceous materials are as follows:

Clay. This material is found at the surface of the earth and often is the major component of soils. The material is plastic when wetted, but becomes hard and brittle when dried and heated. It is composed mainly of hydrous aluminum silicates as well as other minerals.

Shale. This sedimentary rock is formed by the consolidation of clay, mud and silt. It contains substantial amounts of hydrous aluminum silicates.

Slate. A dense fine-grained metamorphic rock containing mainly clay minerals. Slate is obtained from metamorphic shale.

Ash. This is a secondary source and is the by-product of coal combustion. It contains silicates.

There are two processes used to manufacture cements: the dry process and the wet process. The dry process is the least expensive of the two, but is the more difficult to control.

In the dry process the limestone and clay materials are crushed and stored in separate bins and their composition analyzed. After the composition is known, the contents of the bins are blended to achieve the desired ultimate cement characteristics. The blend is ground to a mesh size of 100–200. This small mesh size maximizes the contact between individual particles.

In the wet process the clay minerals are crushed and slurried with water to allow pebbles and other rock particles to settle out. The limestone is also crushed and slurried. Both materials are stored in separate bins and analyzed. Once the desired ultimate composition is determined, the slurry blend is ground and then partially dried out.

After the blends have been prepared (either in the dry or wet process), these materials are fed at a uniform rate into a long rotary kiln. The materials are gradually heated to a liquid state. At temperatures up to about 1,600°F the free water evaporates, the clay minerals dehydroxylate and crystallize, and $CaCO_3$ decomposes. At temperatures up to about 1600°F (870°C) the $CaCO_3$ and CaO react with aluminosilicates and the materials become liquids. Heating is continued to as high as 2800°F (1540°C).

When the kiln material is cooled it forms into crystallized clinkers. These are rather large irregular pieces of the solidified cement material. These clinkers are ground and a small amount of gypsum is added (usually about 1.5 to 3%). The gypsum prevents flash setting of the cement and also controls free CaO. This final cement product is sampled, analyzed and stored. The actual commercial cement is usually a blend of several different cements. This blending ensures a consistent product.

There are four chemical compounds that are identified as being the active components of cements.

Tricalcium aluminate ($3CaO \cdot Al_2O_3$) hydrates rapidly and contributes most to heat of hydration. This compound does not contribute greatly to the final strength of the set cement, but it sets rapidly and plays an important role in the early strength development. This setting time can be controlled by the addition of gypsum. The final hydrated product of tricalcium aluminate is readily attacked by sulfate waters. High-sulfate-resistant (HSR) cements have only a 3% or less content of this compound. High early strength cements have up to 15% of this compound.

Tricalcium silicate ($3CaO \cdot SiO_2$) is the major contributor to strength at all stages, but particularly during early stages of curing (up to 28 days). The average tricalcium silicate content is from 40% to maximum of 67%. The retared cements will contain from 40% to 45%. The high early strength cements will contain 60% to 67%.

Dicalcium silicate ($2CaO \cdot SiO_2$) is very important in the final strength of the cement. This compound hydrates very slowly. The average dicalcium silicate content is 25% to 35%.

Tetracalcium aluminoferrite ($4CaO \cdot Al_2O_3 \cdot FeO_3$) has little effect on the physical properties of the cement. For high-sulfate-resistant (HSR) cements, API specifications require that the sum of the tetracalcium aluminoferrite content plus twice the tricaclium aluminate may not exceed a maximum of 24%.

In addition to the four compounds discussed above, the final Portland cement may contain gypsum, alkali sulfates, magnesia, free lime and other components. These do not significantly affect the properties of the set cement, but they can influence rates of hydration, resistance to chemical attack and slurry properties.

When the water is added to the final dry cement material, the hydration of the cement begins immediately. The water is combined chemically with the cement material to eventually form a new immobile solid. As the cement hydrates, it will bond to the surrounding surfaces. This cement bonding is complex and depends on the type of surface to be cemented. Cement bonds to rock by a process of crystal growth. Cement bonds to the outside of a casing by filling in the pit spaces in the casing body [4].

9.3 CEMENTING PRINCIPLES

There are two basic oil well cementing activities: primary cementing and secondary cementing.

Primary cementing refers to the necessity to fix the steel casing or liner (which is placed in the drilled borehole) to the surrounding formations adjacent to the casing or liner. The purposes of primary cementing are the following:

1. Support vertical and radial loads applied to casing
2. Isolate porous formations from producing zone formations
3. Exclude unwanted subsurface fluids from the producing interval
4. Protect casing from corrosion
5. Resist chemical deterioration of cement
6. Confine abnormal formation pressures

When applied to the various casing or liner strings used in oil well completions, the specific purposes of each string are as follows:

- Conductor casing string is cemented to prevent the drilling fluid from escaping and circulating outside the casing.
- Surface casing string must be cemented to protect fresh-water formations near the surface and provide a structural connection between the casing and the subsurface competent rock formations. This subsurface structural connection will allow the blowout preventor to be affixed to the top of this casing to prevent high-pressure fluids from being vented to the surface. Further, this structural connection will give support for deeper run casing or liner strings.
- Intermediate casing strings are cemented to seal off abnormal pressure formations and cover both incompetent formations, which could cave or slough, and lost circulation formations.
- Production casing string is cemented to prevent the produced fluids from migrating to nonproducing formations and to exclude other fluids from the producing interval.

Cementing operations are carried out with surface equipment specially designed to carry out the primary and secondary cementing operations in oil wells. The key element in any well cementing operation is the recirculating blender (Figure 9.1). The recirculating blender has replaced the older jet mixing hopper. The blender provides a constant cement slurry specific weight that could never be achieved by the older equipment. A very careful control of the slurry weight is critical to a successful cementing operation. The recirculating blender is connected to a cement pump that in turn pumps the cement at low circulation rates (and high pressures if necessary) to the cementing head at the top of the casing string (see Figures 9.2 and 9.3 and Ref. [3]).

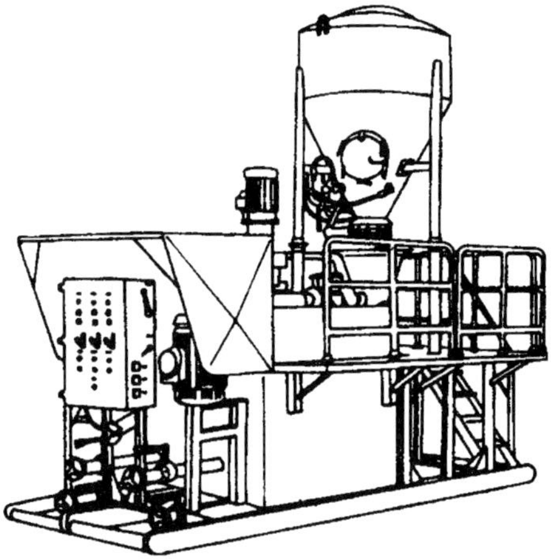

FIGURE 9.1 Recirculating blender [2].

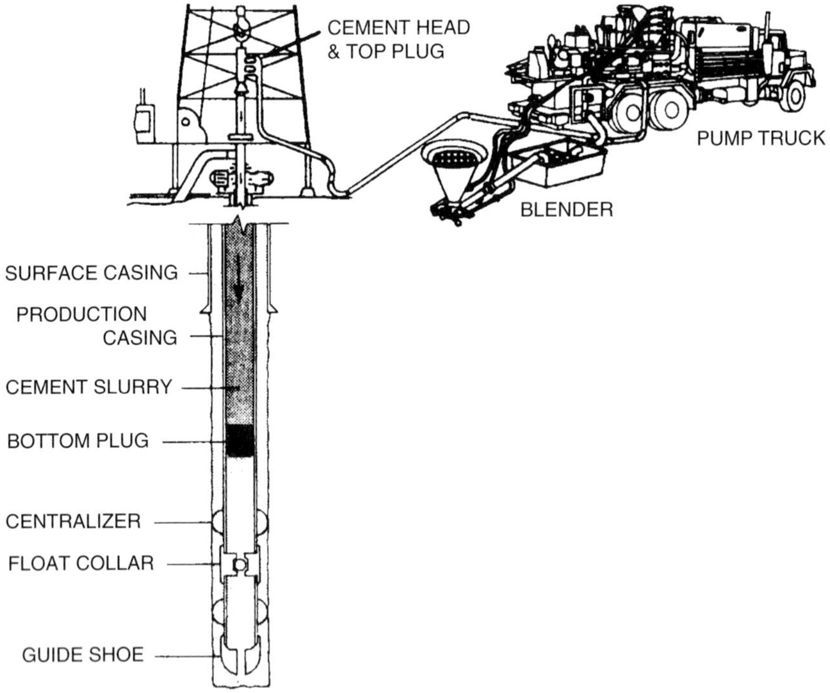

FIGURE 9.2 Blender, pump truck, cementing head and subsurface equipment [2].

The proper amount of water must be mixed with the dry cement product to ensure only sufficient water for hydration of the cement. Excess water above that needed for hydration will reduce the final strength of the set

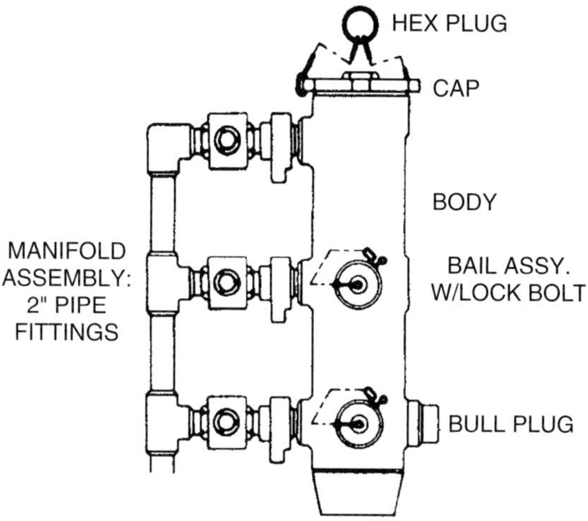

FIGURE 9.3 Cementing head [2].

cement and leave voids in the cement column that are filled with unset liquid. Insufficient water for proper hydration will leave voids filled with dry unset cement, or result in a slurry too viscous to pump.

The cement is usually dry mixed with the additives that are usually added for a particular cementing application. Often this mixing of additives is carried out at the service company central location. Depending on the application of the well cement, there are a variety of additives that can be used to design the cement slurry characteristics. These are accelerators, retarders, dispersants, extenders, weighting agents, gels, foamers and fluid-loss additives. With these additives the cement slurry and ultimately the set cement can be designed for the particular cementing operation. It is necessary that the engineer in charge of the well carry out the necessary engineering design of the slurries to be used in the well. In addition, the engineer should ensure that the work is carried out in accordance to the design specifications. This critical activity should not be left to the service company technicians.

9.4 STANDARDIZATION AND PROPERTIES OF CEMENTS

The American Petroleum Institute (API) has nine classes of well cements. These are as follows [5]:

Class A: Intended for use from surface to 6,000 ft (1,830 m) depth, when special properties are not required. Available only in ordinary type (similar to ASTM C 150, Type I[1]).

1. American Society for Testing and Materials (ASTM).

Class B: Intended for use from surface to 6,000 ft (1,830 m) depth, when conditions require moderate to high sulfate resistance. Available in both moderate (similar to ASTM C 150, Type II) and high-sulfate resistant types.

Class C: Intended for use from surface to 6,000 ft (1,830 m) depth, when conditions require high early strength. Available in ordinary and moderate (similar to ASTM C 150, Type III) and high-sulfate-resistant types.

Class D: Intended for use from 6,000 to 10,000 ft (1,830 to 3,050 m) depth, under conditions of moderately high temperatures and pressures. Available in both moderate and high-sulfate-resistant types.

Class E: Intended for use from 10,000 to 14,000 ft (3,050 to 4,270 m) depth, under conditions of high temperatures and pressures. Available in both moderate and high-sulfate-resistant types.

Class F: Intended for use from 10,000 to 16,000 ft (3,050 to 4,880 m) depth, under conditions of extremely high temperatures and pressures. Available in both moderate and high-sulfate-resistant types.

Class G: Intended for use from surface to 8,000 ft (2,440 m) depth as manufactured, or can be used with accelerators and retarders to cover a wide range of well depths and temperatures. No additions other than calcium sulfate or water, or both, shall be interground or blended with the clinker during manufacture of Class G well cement. Available in moderate and high-sulfate-resistant types.

Class H: Intended for use as a basic well cement from surface to 8,000 ft (2,440 m) depth as manufactured, and can be used with accelerators and retarders to cover a wide range of well depths and temperatures. No additions other than calcium sulfate or water, or both, shall be interground or blended with the clinker during manufacture of Class H well cement. Available in moderate and high-(tentative) sulfate-resistant types.

Class J: Intended for use as manufactured from 12,000 to 16,000 ft (3,660 to 4,880 m) depth under conditions of extremely high temperatures and pressures or can be used with accelerators and retarders to cover a range of well depths and temperatures. No additions of retarder other than calcium sulfate or blended with the clinker during manufacture of class J well cement.

The ASTM specifications provide for five types of Portland cements: Types I, II, III, IV and V; they are manufactured for use at atmospheric conditions [6]. The API Classes A, B and C correspond to ASTM Types I, II and III, respectively. The API Classes D, E, F, G, H and J are cements manufactured for use in deep wells and to be subject to a wide range of pressures and temperatures. These classes have no corresponding ASTM types.

Sulfate resistance is an extremely important property of well cements. Sulfate minerals are abundant in some underground formation waters that

TABLE 9.1 Properties of the Various Classes of API Cements

Cement Class	A	B	C	D	E	F	G	H
Specific Gravity	3.14	3.14	3.14	3.16	3.16	3.16	3.16	3.16
Surface Area (cm^2/gm)	1500	1600	2200	1200	1200	1200	1400	1200
Weight Per Sack, lb	94	94	94	94	94	94	94	94
Bulk Volume, ft^3/sk	1	1	1	1	1	1	1	1
Absolute Volume, gal/sk	3.59	3.59	3.55	3.57	3.57	3.57	3.62	3.57

can come into contact with set cement. The sulfate chemicals, which include magnesium and sodium sulfates, react with the lime in the set cement to form magnesium hydroxide, sodium hydroxide and calcium sulfate. The calcium sulfate reacts with the tricalcium aluminate components of cement to form sulfoaluminate, which causes expansion and ultimately disintegrates to the set cement. To increase the resistance of a cement to sulfate attack, the amount of tricalcium aluminate and free lime in the cement should be decreased. Alternatively, the amount of pozzolanic material can be increased in the cement to obtain a similar resistance. The designations of ordinary sulfate resistance, moderate sulfate resistance and high sulfate resistance in the cement classes above indicate decreasing amounts of tricalcium aluminate.

Table 9.1 gives the basic properties of the various classes of the dry API cements [7].

9.5 PROPERTIES OF CEMENT SLURRY AND SET CEMENT

In well engineering and applications, cement must be dealt with in both its slurry form and in its set form. At the surface the cement must be mixed and then pumped with surface pumping equipment through tubulars to a designated location in the well. After the cement has set, its structure must support the various static and dynamic loads placed on the well tubulars.

9.5.1 Specific Weight

Specific weight is one of the most important properties of a cement slurry. A neat cement slurry is a combination of only cement and water. The specific weight of a neat cement slurry is defined by the amount of water used with the dry cement. The specific weight range for a particular class of cement is, therefore, limited by the minimum and maximum water-to-cement ratios permissible by API standards.

The minimum amount of water for any class of cement is defined as that amount of water that can be used in the slurry with the dry cement that will still produce a neat slurry with a consistency that is below 30 Bearden units of consistency.[2] Note that the minimum water content defined in this manner is much greater than the stoichiometric minimum for cement hydration and setting.

The maximum amount of water for any class of cement is usually defined as the ratio that results in the cement particles remaining in suspension until the initial set of the slurry has taken place. If more than the maximum amount of water is used, then the cement particles will settle in such a manner that there will be pockets of free water within the set cement column.

Table 9.2 gives the API recommended optimum water-to-cement ratios and the resulting neat slurry specific weight (lb/gal) and specific volume, or yield (ft^3/sack) for the various classes of API cements [2].

Table 9.3 gives the maximum and minimum water-to-cement ratios and the resulting neat slurry specific weight and specific volume, or yield for three classes of API cement.

TABLE 9.2 Properties of Neat Cement Slurries for Various Classes of API Cements

API Class	API Rec'd Water (gal/sk)	API Rec'd Water (gal/sk)	API Rec'd Water (gal/sk)	Percent of Mixing Water
A (Portland)	5.20	15.6	1.18	46
B (Portland)	5.20	15.6	1.18	46
C (High Early)	6.32	14.8	1.32	56
D (Retarded)	4.29	16.46	1.05	38
E (Retarded)	4.29	16.46	1.05	38
F (Retarded)	4.29	16.46	1.05	38
G (Basic)	4.97	15.8	1.15	44
H (Basic)	4.29	16.46	1.05	38

TABLE 9.3 Maximum and Minimum Water-to-Cement Ratios for API Classes A, C and E Neat Cement Slurries

	Maximum Water			Minimum Water		
API Class	Water (gal/sk)	Weight (lb/gal)	Volume (ft^3/sk)	Water (gal/sk)	Weight (pb/gal)	Volume (ft^3/sk)
A	5.5	15.39	1.22	3.90	16.89	1.00
C	7.9	13.92	1.53	6.32	14.80	1.32
E	4.4	16.36	1.07	3.15	17.84	0.90

2. This was formally referred to as poise.

The specific weight of a cement slurry $\overline{\gamma}$ (lb/gal) is

$$\overline{\gamma} = \frac{\text{Cement (lb)} + \text{Water (lb)} + \text{Additive (lb)}}{\text{Cement (gal)} + \text{Water (gal)} + \text{Additive (gal)}} \tag{9.1}$$

The volumes above are absolute volumes. For example, a 94-lb sack of cement contains 1 ft^3 of bulk cement powder, yet the actual or absolute space occupied by the cement particles is only 0.48 ft^3.

For dealing with other powdered materials that are additives to cement slurries, the absolute volume must be used. The absolute volume (gal) is

$$\text{Absolute value (gal)} = \frac{\text{Additive material (lb)}}{(8.34\ \text{lb/gal})(\text{S.G. of material})} \tag{9.2}$$

The volume of slurry that results from 1 sack of cement additives is defined as the yield. The yield (ft^3/sack) is

$$\text{Yield} = \frac{\text{Cement (gal)} + \text{Water (gal)} + \text{Additives (gal)}}{7.48\ \text{gal/ft}^3} \tag{9.3}$$

where the cement, water and additives are those associated with 1 sack of cement, or 94 lb of cement.

Example 9.1

Calculate the specific weight and yield for a neat slurry of Class A cement using the maximum permissible water-to-cement ratio.

Table 9.3 gives the maximum water-to-cement ratio for Class A cement as 5.5 gal/sack. Thus using the absolute volume for Class A cement from Table 9.1 as 3.59 gal/sack, Equation 9.1 is

$$\begin{aligned}\overline{\gamma} &= \frac{94 + 5.5(8.34)}{3.59 + 5.5} \\ &= 15.39\ \text{lb/gal}\end{aligned}$$

The yield determined from Equation 9.3 is

$$\begin{aligned}\text{Yield} &= \frac{3.59 + 5.5}{7.48} \\ &= 1.22\ \text{ft}^3/\text{sack}\end{aligned}$$

It is often necessary to decrease the specific weight of a cement slurry to avoid fracturing weak formations during cementing operations. There are basically two methods for accomplishing lower specific weights. These are

1. Adding clay or chemical silicate type extenders together with their required extra water.
2. Adding large quantities of pozzolan, ceramic microspheres or nitrogen. These materials lighten the slurry because they have lower specific gravities than the cement.

When using the first method above great care must be taken not to use too much water. There is a maximum permissible water-to-cement ratio for each cement class. This amount of water can be used with the appropriate extra water required for the added clay or chemical silicate material. Using too much water will result in a very poor cement operation.

It also may be necessary to increase the specific weight of a cement slurry, particularly when cementing through high-pressure formations. There are basically two methods for accomplishing higher specific weights. These are

1. Using the minimum permissible water-to-cement ratio for the particular cement class and adding dispersants to increase the fluidity of the slurry.
2. Adding high-specific-gravity materials to the slurry together with optimal or slightly reduced (but not necessarily the minimum) water-to-cement ratio for the particular cement class.

The first method above is usually restricted to setting plugs in wells since it results in high strength cement that is rather difficult to pump. The second method is used for primary cementing, but these slurries are difficult to design since the settling velocity of the high-specific-gravity additive must be taken into consideration.

9.5.2 Thickening Time

It is important that the thickening time for a given cement slurry be known prior to using the slurry in a cementing operation. When water is added to dry cement and its additives, a chemical reaction begins that results in an increase in slurry viscosity. This viscosity increases over time, which will vary in accordance with the class of cement used, the additives placed in the dry cement prior to mixing with water and the temperature and confining pressure in the location where the cement slurry is placed. When the viscosity becomes too large, the slurry is no longer pumpable. Thus, if the slurry has not been placed in its proper location within the well prior to the cement slurry becoming unpumpable, the well and the surface equipment would be seriously damaged.

Thickening time T_t (hr) is defined as the time required for the cement slurry to reach the limit of 100 Bearden units of consistency.[3] This thickening time must be considerably longer than the time necessary to carry out the actual cementing operation. This can be accomplished by choosing the class of cement that has a sufficiently long thickening time, or placing the appropriate additives in the slurry that will retard the slurry chemical reaction and lengthen the thickening time.

3. 70 Bearden units of consistency is considered to be maximum pumpable viscosity.

It is necessary that an accurate estimate of the total time be for the actual operation. A safety factor should be added to this estimate. Usually this safety factor is from 30 min (for shallow operations) to as long as 2 hr (for deep complex operations).

The cementing operation time T_o (hr) is the time required for the cement slurry to be placed in the well:

$$T_o = T_m + T_d + T_p + T_s \tag{9.4}$$

where T_m = time required to mix the dry cement (and additives) with water in hr

T_d = the time required to displace the cement slurry (that was pumped to the well as mixing took place) by mud or water from inside the casing in hr

T_p = plug release time in hr

T_s = the safety factor of 0.5 to 2 hr

The mixing time T_m is

$$T_m = \frac{\text{Volume of dry cement}}{\text{Mixing rate}} = \frac{V_c\ \text{(sacks)}}{\text{Mixing rate (sack/min)(60)}} \tag{9.5}$$

where V_c is the dry volume of cement (sacks).

The displacement time T_d is

$$T_d = \frac{\text{Volume of fluid required to displace top plug}}{\text{Displacement rate}} = \frac{\text{Volume (ft}^3\text{)}}{\text{Displacement rate (ft}^3\text{/ min)(60)}} \tag{9.6}$$

The cement slurry chosen must have a thickening time that is greater than the estimated time obtained for the actual cementing operation using Equation 9.4. Thus, $T_t > T_o$.

Figure 9.4 gives a relationship between well depth and cementing operation time and the specifications for the various cement classes. This figure can be used to approximate cementing operation time.

The cement slurry thickening time can be increased or decreased by adding special chemicals to the dry cement prior to mixing with water. Retarders are added to increase the thickening time and thus increase the time when the cement slurry sets. Some common retarders are organic compounds such as lignosulphonate, cellulose derivatives and sugar derivatives. Accelerators are added to decrease the thickening time and thus decrease the time when the cement slurry sets. Accelerators are often used when it is required to have the cement obtain and early compressive strength, usually of the order of 500 psi (35 bar). Early setting cement

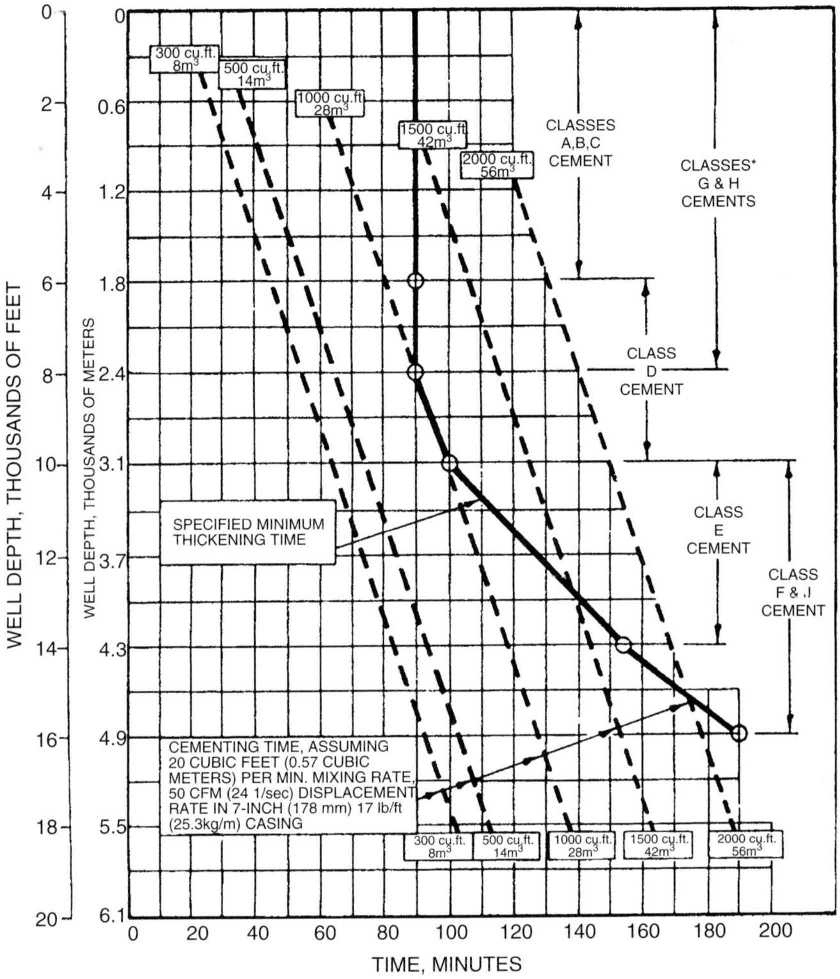

FIGURE 9.4 Well depth and cementing time relationships [5]. *Specified maximum thickening time—120 minutes.

slurries are used to cement surface casing strings or directional drilling plugs where waiting-on-cementing (WOC) must be kept to a minimum. The most common accelerators are calcium chloride to a lesser extent sodium chloride.

In general the thickening time for all neat cement slurries of all classes of cement will decrease significantly with increasing well environment temperature and/or confining pressure. Thus it is very important that the well extreme temperatures and confining pressures be defined for a particular cementing operation before the cement slurry is designed. Once the well temperature and pressure conditions are known, either from offset wells or from actual well logs during the drilling operation, and the estimated cementing operation time is known, the cement slurry with the required thickening time can be designed. After the initial cement slurry design has been made it is usually necessary to carry out laboratory tests to verify that the actual cement batch mix used will give the required thickening time (and other characteristics needed). Such tests are usually carried out by the cementing service company laboratory using the operator's specifications.

Example 9.2

Calculate the minimum thickening time required for a primary cementing operation to cement a long intermediate casing string. The intermediate casing string is a $9\frac{5}{8}$ in., 53.5-lb/ft casing set in a $12\frac{1}{2}$-in. hole. The string is 12,000 ft in length from the top of the float collar to the surface. The cementing operation will require 1,200 sacks of Class H cement. The single cementing truck has a mixing capacity of 25 sacks per minute. The rig duplex mud pump has an 18-in. stroke (2.5 in. rod) and $6\frac{1}{2}$-in. liners and will be operated at 50 strokes per minute with a 90% volumetric efficiency. The plug release time is estimated to be about 15 min.

The mixing time is obtained from Equation 9.5. This is

$$T_m = \frac{1{,}200}{25(60)}$$
$$= 0.80 \text{ hr}$$

The inside diameter of $9\frac{5}{8}$-in., 53.50-lb/ft casing is 8.535 in. The internal capacity per unit length of casing is

$$\text{Casing internal capacity} = \frac{\pi}{4}\frac{(8.535)^2}{12}$$
$$= 0.3973 \text{ ft}^3/\text{ft}$$

Thus the total internal capacity of the casing is

$$\text{Total volume} = 12{,}000(0.3973)$$
$$= 4{,}768 \text{ ft}^3$$

The volume capacity of the mud pump per stroke q (ft^3/stroke) is

$$q = \left\{2\left[\frac{\pi}{4}(6.5)^2\right](18.0) + 2\left[\frac{\pi}{4}(6.5)^2 - \frac{\pi}{4}(2.5^2)\right](18.0)\right\}\frac{(0.90)}{(12)^3}$$
$$= 1.1523 \text{ ft}^3/\text{stroke}$$

The displacement time is obtained from Equation 9.6. This is

$$T_d = \frac{4{,}768}{50(1.1523)(60)}$$
$$= 1.38 \text{ hr}$$

The plug release time is

$$T_P = \frac{15}{60}$$
$$= 0.25 \text{ hr}$$

Using a safety factor of 1 hr, the cementing operation time is obtained from Equation 9.4. This is

$$T_o = 0.80 + 1.38 + 0.25 + 1.00$$
$$= 3.43 \text{ hr}$$

Thus the minimum thickening time for the cement slurry to be used this cementing operation is

$$T_t = 3.43 \text{ hr}$$

or 3 hr and 26 min.

9.5.3 Strength of Set Cement

A properly designed cement slurry will set after it has been placed in its appropriate location within the well. Cement strength is the strength the set cement has obtained. This usually refers to compressive strength, but can also refer to tensile strength. Cement having a compressive strength of 500 psi (35 bar) is considered adequate for most well applications.

The compressive strength of set cement is dependent upon the water-to-cement ratio used in the slurry, curing time, the temperature during curing, the confining pressure during curing and the additives in the cement. As part of the cement design procedures, samples of the cement slurry to be used in the cementing operation are cured and compression strength tested and often shear bond strength tested. These tests are usually carried out by the cementing company laboratory at the request of the operator.

In the compression test, four or five sample cubes of the slurry are allowed to cure for a specified period of time. The cement cubes are placed in a compression testing machine and the compressive strength of each sample cube obtained experimentally. The average value of the samples is obtained and reported as the compressive strength of the set cement.

In the shear bond strength test, the cement slurry is allowed to cure in the annulus of two concentric steel cylinders. After curing the force to break the bond between the set cement and one of the cylinders (usually the inner one) is obtained experimentally. The shear force, F_S(lb), which can be supported on the inner cylinder, or by the casing, is [4]

$$F_S = 0.969\,\sigma_c d l \tag{9.7}$$

where σ_c = compressive strength of cement in psi
d = outside of the casing in in.
l = length of the cement column in in.

Table 9.4 shows the influence of curing time, temperature and confining pressure on the compressive strength of Class H cement [3]. Also shown is the effect of the calcium chloride accelerator. In general, for the ranges of temperature and other values considered, the compressive strength increases with increasing curing time, temperature, confining pressure and the amount of calcium chloride accelerator. In general, the other classes of cement follow the same trend in compressive strength versus curing time, temperature and confining pressure.

At curing temperatures of about 200°F (93°C) or greater, the compressive strength of nearly all classes of cement cured under pressure will decrease. This decrease in compressive strength is often denoted as strength retrogression. An increase in cement permeability accompanies this decrease in strength. This strength retrogression usually continues for up to 15 days and, thereafter, the strength level will remain constant. This strength retrogression problem can be solved by adding from 30 to 50% (by weight) of silica flour or silica sand to the cement slurry. This silica additive prevents

TABLE 9.4 Influence of Time and Temperature on the Compressive Strength of API Class H Cement [3]

Curing Time (hours)	Calcium Chloride (percent)	Compressive Strength (psi) at Curing Temperature and Pressure of			
		95°F 800 psi	110°F 1,600 psi	140°F 3,000 psi	170°F 3,000 psi
6	0	100	350	1,270	1,950
8		500	1,200	2,500	4,000
12		1,090	1,980	3,125	4,700
24		3,000	4,050	5,500	6,700
6	1	900	1,460	2,320	2,500
8	1	1,600	1,950	2,900	4,100
12	1	2,200	2,970	3,440	4,450
24	1	4,100	5,100	6,500	7,000
6	2	1,100	1,700	2,650	2,990
8	2	1,850	2,600	3,600	4,370
12	2	2,420	3,380	3,900	5,530
24	2	4,700	5,600	6,850	7,400

TABLE 9.5 Effect of Silica Flour on the Compressive Strength of Class G Cement Cured at Pressure

Curing Time (days)	Silica Flour (percent)	Compressive Strength (psi), 1 Day at 130°F, All Remaining Days at 700°F
1	0	3525
2		988
4		1012
8		1000
1	40	2670
2		3612
4		3188
8		4588

the strength retrogression and the corresponding increase in permeability of the cement.

Table 9.5 shows the effect of silica flour on the compressive strength of Class G cement cured at 700°F (371°C).

Example 9.3

For Example 9.2, determine the total weight that can be supported by the set cement bonded to the $9\frac{5}{8}$-in. casing. Assume the cement slurry yield is 1.05 ft^3/sack (see Table 9.2) and there are 120 ft of casing below the float collar. A compressive strength of 500 psi is to be used.

The total volume of the cement slurry to be pumped to the well is

$$\text{Volume of slurry} = 1{,}200(1.05)$$
$$= 1{,}260 \text{ ft}^3$$

The volume of slurry that is pumped to the annulus between the open borehole and the outside of the casing is

$$\text{Volume of annulus} = 1{,}260 - (0.3973)(120)$$
$$= 1{,}212 \text{ ft}^3$$

The height in the annulus to where the casing is cemented to the borehole well and casing is

$$h = \frac{1{,}212}{\pi/4[(12.5/12)^2 - (9.625/12)^2]} = 3{,}493 \text{ ft}$$

The total weight that can be supported is calculated from Equation 9.7. This is

$$F_s = 0.969(500)(9.625)(3{,}493)(12)$$
$$= 195.5 \times 10^6 \text{ lb}$$

9.6 CEMENT ADDITIVES

There are many chemical additives that can be used to alter the basic properties of the neat cement slurry and its resulting set cement. These additives are to alter the cement so that it is more appropriate to the surface cementing equipment and the subsurface environment.

Additives can be subdivided into six functional groups that are [8, 9]:

1. Specific weight control
2. Thickening and setting time control
3. Loss of circulation control
4. Filtration control
5. Viscosity control
6. Special problems control

9.6.1 Specific Weight Control

As in drilling mud design, the cement slurry must have a specific weight that is high enough to prevent high pore pressure formation fluids from flowing into the well. But also the specific weight must not be so high as to cause fracturing of the weaker exposed formations. In general, the specific weight of the neat cement slurry of any of the various API classes of cement are so high that most exposed formations will fracture. Thus, it is necessary to lower the specific weight of nearly all cement slurries. The slurry specific weight can be reduced by using a higher water-to-cement ratio. But this can only be accomplished within thc limits for maximum and minimum water-to-cement ratios set by API standards (see Table 9.3). The reduction in specific weight is normally accomplished by adding low-specific-gravity solids to the slurry. Table 9.6 gives the specific gravity properties of a number of the low-specific-gravity solids that are used to reduce the specific weight of cement slurries.

The most common low-specific-gravity solids used to reduce cement slurry specific weight are bentonite, diatomaceous earth, solid hydrocarbons, expanded perlite and pozzolan. It may not be possible to reduce the cement slurry specific weight enough with the above low-specific-weight materials when very weak formations are exposed. In such cases nitrogen is used to aerate the mud column above the cement slurry to assist in further decreasing the hydrostatic pressure.

Nearly all materials that are added to a cement slurry require the addition of additional water to the slurry. Table 9.7 gives the additional requirements for the various cement additives [8].

Bentonite. Bentonite without an organic polymer can be used as an additive to cement slurries. The addition of bentonite requires the use

TABLE 9.6 Physical Properties of Cementing Materials

Material	Bulk Weight (lbm/cu ft)	Specific Gravity	Weight 3.6[a] Absolute Gal	Absolute Volume	
				gal/lbm	cu ft/lbm
API cements	94	3.14	94	0.0382	0.0051
Ciment Fondu	90	3.23	97	0.0371	0.0050
Lumnite cement	90	3.20	96	0.0375	0.0050
Trinity Lite-Wate	75	2.80	75.0[e]	0.0429	0.0057
Activated charcoal	14	1.57	47.1	0.0765	0.0102
Barite	135	4.23	126.9	0.0284	0.0038
Bentonite (gel)	60	2.65	79.5	0.0453	0.0060
Calcium chloride, flake[b]	56.4	1.96	58.8	0.0612	0.0082
Calcium chloride, powder[b]	50.5	1.96	58.8	0.0612	0.0082
Cal-Seal, gypsum cement	75	2.70	81.0	0.0444	0.0059
CFR-1[b]	40.3	1.63	48.9	0.0736	0.0098
CFR-2[b]	43.0	1.30	39.0	0.0688	0.0092
DETA (liquid)	59.5	0.95	28.5	0.1258	0.0168
Diacel A[b]	60.3	2.62	78.6	0.0458	0.0061
Diacel D	16.7	2.10	63.0	0.0572	0.0076
Diacel LWL[b]	29.0	1.36	40.8	0.0882	0.0118
Diesel Oil No. 1 (liquid)	51.1	0.82	24.7	0.1457	0.0195
Diesel Oil No. 2 (liquid)	53.0	0.85	25.5	0.1411	0.0188
Gilsonite	50	1.07	32	0.1122	0.0150
HALDAD®-9[b]	37.2	1.22	36.6	0.0984	0.0131
HALDAD®-14[b]	39.5	1.31	39.3	0.0916	0.0122
Hematite	193	5.02	150.5	0.0239	0.0032
HR-4[b]	35	1.56	46.8	0.0760	0.0103
HR-7[b]	30	1.30	39	0.0923	0.0123
HR-12[b]	23.2	1.22	36.6	0.0984	0.0131
HR-L (liquid)[b]	76.6	1.23	36.9	0.0976	0.0130
Hydrated lime	31	2.20	66	0.0545	0.0073
Hydromite	68	2.15	64.5	0.0538	0.0072
LA-2 Latex (liquid)	68.5	1.10	33	0.1087	0.0145
LAP-1 Latex[b]	50	1.25	37.5	0.0960	0.0128
LR-11 Resin (liquid)	79.1	1.27	38.1	0.0945	0.0126
NF-1 (liquid)[b]	61.1	0.98	29.4	0.1225	0.0164
NF-P[b]	40	1.30	39.0	0.0923	0.0123
Perlite regular	8[c]	2.20	66.0	0.0546	0.0073
Perlite Six	38[d]	–	–	0.0499	0.0067
Pozmix® A	74	2.46	74	0.0487	0.0065
Pozmix® D	47	2.50	73.6	0.0489	0.0065
Salt (dry NaCl)	71	2.17	65.1	0.0553	0.0074
Salt (in solution at 77°F with fresh water)					
6%, 0.5 lbm/gal	–	–	–	0.0384	0.0051
12%, 1.0 lbm/gal	–	–	–	0.0399	0.0053
18%, 1.5 lbm/gal	–	–	–	0.0412	0.0055
24%, 2.0 lbm/gal	–	–	–	0.0424	0.0057
Saturated, 3.1 lbm/gal	–	–	–	0.0445	0.0059

(Continued)

TABLE 9.6 (*Continued*)

Material	Bulk Weight (lbm/cu ft)	Specific Gravity	Weight 3.6[a] Absolute Gal	Absolute Volume	
				gal/lbm	cu ft/lbm
Salt (in solution at 140°F with fresh water) saturated, 3.1 lbm/gal	–	–	–	0.0458	0.0061
Sand (Ottawa)	100	2.63	78.9	0.0456	0.0061
Silica flour (SSA-1)	70	2.63	78.9	0.0456	0.0061
Coarse silica (SSA-2)	100	2.63	78.9	0.0456	0.0061
Tuf Additive No. 1	–	1.23	36.9	0.0976	0.0130
Tuf-Plug	48	1.28	38.4	0.0938	0.0125
Water	62.4	1.00	30.0	0.1200	0.0160

[a]Equivalent to one 94-lbm sack of cement in volume.
[b]When less than 5% is used, these chemicals may be omitted from calculations without significant error.
[c]For 8 lbm of Perlite regular use a volume of 1.43 gal at zero pressure.
[d]For 38 lbm of Perlite Six use a volume of 2.89 gal at zero pressure.
[e]75 lbm = 3.22 absolute gal.
Source: Courtesy Halliburton Services.

of additional water, thereby, further reducing the specific weight of the slurry (Table 9.7). Bentonite is usually dry blended with the dry cement prior to mixing with water. High percentages of bentonite in cement will significantly reduce compressive strength and thickening time. Also, high percentages of bentonite increase permeability and lower the resistance of cement to sulfate attack. At temperatures above 200°F (93°C), the bentonite additive promotes, retrogression of strength in cements with time. Bentonite has been used in 25% by weight of cement. Such high concentrations are not recommended. In general, the bentonite additive makes a poor well cement.

Diatomaceous Earth. Diatomaceous earth has a lower specific gravity than bentonite. Like bentonite this additive also requires additional water to be added to the slurry. This additive will affect the slurry properties similar to the addition of bentonite. However, it will not increase the viscosity as bentonite will do. Diatomaceous earth concentrations as high as 40% by weight of cement have been used. This additive is more expensive than bentonite.

Solid Hydrocarbons. Gilsonite (an asphaltite) and coal are used as very-low-specific-gravity solids additives. These additives do not require a great deal of water to be added to the slurry when they are used.

Expanded Perlite. Expanded perlite requires a great deal of water to be added to the slurry when it is used to reduce the specific weight of a slurry. Often perlite as an additive is used in a blend of additives such as perlite

TABLE 9.7 Water Requirement of Cementing Materials

Material	Water Requirements
API Class A and B cements	5.2 gal (0.70 cu ft)/94-lbm sack
API Class C cement (H_1 Early)	6.3 gal (0.84 cu ft)/94-lbm sack
API Class D and E cements (retarded)	4.3 gal (0.58 cu ft)/94-lbm sack
API Class G cement	5.0 gal (0.67 cu ft)/94-lbm sack
API Class H cement	4.3 to 5.2 gal /94-lbm sack
Chem Comp cement	6.3 gal (0.84 cu ft)/94-lbm sack
Ciment Fondu	4.5 gal (0.60 cu ft)/94-lbm sack
Lumnite cement	4.5 gal (0.60 cu ft)/94-lbm sack
HLC	7.7 to 10.9 gal/87-lbm sack
Trinity Lite-Wate cement	7.7 gal (1.03 cu ft)/75-lbm sack (maximum)
Activated charcoal	none at 1 lbm/sack of cement
Barite	2.4 gal (0.32 cu ft)/100-lbm sack
Bentonite (gel)	1.3 gal (0.174 cu ft)/2% in cement
Calcium chloride	none
Gypsum hemihydrate	4.8 gal (0.64 cu ft)/100-lbm sack
CFR-1	none
CFR-2	none
Diacel A	none
Diacel D	3.3 to 7.2 gal /10% in cement (see Lt. Wt. Cement)
Diacel LWL	none (up to 0.7%)
	0.8 to 1.0 gal /1% in cement (except gel or Diacel D slurries)
Gilsonite	2.0 gal (0.267 cu ft)/50 lbm/cu ft
HALAD-9	none (up to 0.5%)
	0.4 to 0.5 gal/sack of cement at over 0.5%
HALAD-14	none
Hematite	0.36 gal (0.048 cu ft)/100-lbm sack
HR-4	none
HR-7	none
HR-12	none
HR-20	none
Hydrated lime	0.153 gal (0.020 cu ft)/lbm
Hydromite	3.0 gal (0.40 cu ft)/100-lbm sack
LA-2 Latex	0 to 0.8 gal/sack of cement
LAP-1 powdered latex	1.7 gal (0.227 cu ft)/1%-in cement
NF-P	none
Perlite regular	4.0 gal (0.535 cu ft)/8 lbm/cu ft
Perlite Six	6.0 gal (0.80 cu ft)/38 lbm/cu ft
Pozmix A	3.6 gal (0.48 cu ft)/74 lbm/cu ft
Salt (NaCl)	none
Sand. Ottawa	none
Silica flour (SSA-1)	1.5 gal (0.20 cu ft)/35% in cement (32.9 lbm)
Coarse silica (SSA-2)	none
Tuf Additive No. 1	none
Tuf Plug	none

Source: Courtesy of Halliburton Services.

with volcanic glass fines, or with pozzolanic materials, or with bentonite. Without bentonite the perlite tends to separate and float in the upper part of the slurry.

Pozzolan. Diatomaceous earth is a type of pozzolan. Pozzolan refers to a finely ground pumice or fly ash that is marketed as a cement additive under that name. The specific gravity of pozzolans is slightly less than the specific gravity of cement. The water requirements for this additive are about the same as for cements. Only a slight reduction in specific weight of a slurry can be realized by using these additives. The cost of pozzolans is very low.

Where very high formations pore pressure are present the specific weight of the cement slurry can be increased by using the minimum water-to-cement ratio and/or by adding high-specific-gravity materials to the slurry. The most common high-specific-gravity solids used to increase cement slurry specific weight are hematite, ilmenite, barite and sand. Table 9.6 gives the specific gravity properties of a number of these high-specific-gravity additives.

Hematite. This additive can be used to increase the specific weight of a cement slurry to as high as 19 lb/gal. This is an iron oxide ore with a specific gravity of about 5.02. Hematite requires the addition of some water when it is used as an additive. Hematite has minimal effect on thickening time and compressive strength of the cement.

Ilmenite. This additive has a specific gravity of about 4.67. It is a mineral composed of iron, titanium and oxygen. It requires no additional water to be added to the slurry; thus, it can yield slurry specific weights as high as the hematite additive. Ilmenite also has mineral effect on thickening time and compressive strength of the cement.

Barite. This mineral additive requires much more water to be added to the cement slurry than does hematite. This large amount of added water required will decrease the compressive strength of the cement. This additive can be used to increase the specific weight of a cement slurry to as high as 19 lb/gal.

Sand. Ottawa sand has a low specific gravity of about 2.63. But since no additional water is required when using this additive, it is possible to use sand to increase the cement slurry specific weight. The sand has little effect on the pumpability of the cement slurry. When set the cement will form a very hard surface. Sand used as an additive can be used to increase the specific weight of a cement slurry to as high as 18 lb/gal.

Example 9.4

A specific weight of 12.8 lb/gal is required for a Class A cement slurry. It is decided that the cement be mixed with bentonite to reduce the specific weight of the slurry. Determine the weight of bentonite that should be dry blended with each sack of cement. Determine the yield of the cement slurry. Determine the volume (gal) of water needed for each sack of cement.

The weight of bentonite to be blended is found by using Equation 9.1. Taking the appropriate data from Tables 9.1, 9.2, 9.6 and 9.7 and letting x be the unknown weight of bentonite per sack of cement, Equation 9.1 is

$$12.8 = \frac{94 + x + 8.34(5.20 + 0.692x)}{\left(\frac{94}{3.14(8.34)} + \frac{x}{2.65(8.34)} + 5.20 + 0.692x\right)}$$

Solving the above for x gives

$$x = 9.35\,\text{lb}$$

Thus, 9.35 lb of bentonite will need to be added for each sack of Class A cement used.

The yield can be determined from Equation 9.3. This is

$$\text{Yield} = \frac{\frac{94}{3.14(8.34)} + \frac{9.35}{2.65(8.34)} + 5.20 + 0.692(9.35)}{7.48}$$
$$= 2.10\,\text{ft}^3/\text{sack}$$

The volume of water needed is

$$\text{Volume of water} = 5.20 + 0.692(9.35)$$
$$= 11.69\,\text{gal/sack}$$

Example 9.5

A specific weight of 18.2 lb/gal is required for a Class H cement slurry. It is decided that the cement be mixed with hematite to increase the specific weight of the slurry. Determine the weight of hematite that should be dry blended with each sack of cement. Determine the yield of the cement slurry. Determine the volume (gal) of water needed for each sack of cement.

The weight of hematite to be blended is found by using Equation 9.1. Taking the appropriate data from Tables 9.1, 9.2, 9.6 and 9.7 and letting x be the unknown weight of hematite per sack of cement, Equation 9.1 is

$$18.2 = \frac{94 + x + 8.34(4.29 + 0.0036x)}{\frac{94}{3.14(8.34)} + \frac{x}{5.02(8.34)} + (4.29 + 0.0036x)}$$

Solving the above for x gives

$$x = 25.8\,\text{lb}$$

Thus, 25.8 lb of hematite will need to be added for each sack of Class H cement used.

The yield can be determined from Equation 9.3. This is

$$\text{Yield} = \frac{\dfrac{94}{3.14(8.34)} + \dfrac{25.8}{5.07(8.34)} + 4.29 + 0.0036(25.8)}{7.48}$$

$$= 1.15\,\text{ft}^3/\text{sack}$$

The volume of water needed is

$$\text{Volume of water} = 4.29 + 0.0036(25.8)$$
$$= 4.38\,\text{gal/sack}$$

Example 9.6

A specific weight of 17.1 lb/gal is required for a Class H cement slurry. It is decided that the cement be mixed with sand in order to increase the specific weight of the slurry. Determine the weight of sand that should be added with each sack of cement. Determine the yield of the cement slurry. Determine the volume (gal) of water needed for each sack of cement.

The weight of sand to be added is found by using Equation 9.1. Taking the appropriate data from Tables 9.1, 9.2, 9.6, and 9.7 and letting x be the unknown weight of sand per sack of cement, Equation 9.1 is

$$17.1 = \frac{94 + x + 8.34(4.29)}{\dfrac{94}{3.14(8.34)} + \dfrac{x}{2.63(8.34)} + 4.29}$$

Solving the above for x gives

$$x = 31.7\,\text{lb}$$

Thus, 31.7 lb of sand will need to be added for each sack of Class H cement used.

The yield can be determined from Equation 9.3. This is

$$\text{Yield} = \frac{\dfrac{94}{3.14(8.34)} + \dfrac{31.7}{2.63(8.34)} + 4.29}{7.48} = 1.25\,\text{ft}^3/\text{sack}$$

Since no additional water is needed when using sand as the additive, the volume of water needed is

$$\text{Volume of water} = 4.29\,\text{gal/sack}$$

9.6.2 Thickening Setting Time Control

It is often necessary to either accelerate, or retard the thickening and setting time of a cement slurry.

For example, when cementing a casing string run to shallow depth or when setting a directional drilling kick-off plug, it is necessary to accelerate the cement hydration so that the waiting period will be minimized. The most commonly used cement hydration accelerators are calcium chloride, sodium chloride, a hemihydrate form of gypsum and sodium silicate.

Calcium Chloride. This accelerator may be used in concentrations up to 4% (by weight of mixing water) in well having bottomhole temperatures less than 125°F. This additive is usually available in an anhydrous grade (96% calcium chloride). Under pressure curing conditions calcium chloride tends to improve compressive strength and significantly reduce thickening and setting time.

Sodium Chloride. This additive will act as an accelerator when used in cements containing no bentonite and when in concentrations of about 5% (by weight of mixing water). In concentrations above 5% the effectiveness of sodium chloride as an accelerator is reduced. This additive should be used in wells with bottom-hole temperatures less than 160°F. Saturated sodium chloride solutions act as a retarder. Such saturated sodium chloride solutions are used with cement slurries that are to be used to cement through formations that are sensitive to freshwater. However, potassium chloride is far more effective in inhibiting shale hydration. In general, up to a 5% concentration, sodium chloride will improve compressive strength while reducing thickening and setting time.

Gypsum. Special grades of gypsum hemihydrate cement are blended with Portland cement to produce a cement with reduced thickening and setting time for low-temperature applications. Such cement blends should be used in wells with bottomhole temperatures less than 140°F (regular-temperature grade) or 180°F (high-temperature grade). There is a significant additional water requirement for the addition of gypsum (see Table 9.7). When very rapid thickening and setting times are required for low-temperature conditions (i.e., primary cementing of a shallow casing string or a shallow kick-off plug), a special blend is sometimes used. This is 90 lb of gypsum hemihydrate, 10 lb of Class A Portland cement and 2 lb of sodium chloride mixed with 4.8 gal of water. Such a cement slurry and 2 lb of sodium chloride mixed with 4.8 gal of water. Such a cement slurry can develop a compression strength of about 1,000 psi in just 30 min (at 50°).

Sodium Silicate. When diatomaceous earth is used with the cement slurry, sodium silicate is used as the accelerator. It can be used in concentrations up to 7% by weight.

When it is necessary to cement casing or line strings set at great depths, additives are often used in the design of the cement slurry to retard the thickening and setting time. Usually such retarding additives are organic compounds. These materials are also referred to as thinners or dispersants. Calcium lignosulfonate is one of the most commonly used cement retarders. It is very effective at increasing thickening and setting time in cement slurries at very low concentrations (of the order of 1% by weight or less). It is necessary to add an organic acid to the calcium lignosulfonate when high-temperature conditions are encountered.

Calcium-sodium lignosulfonate is a better retarding additive when high concentrations of bentonite are to be used in the design of the cement slurry.

Also sodium tetraborate (borax) and carboxymethyl hydroxyethyl cellulose are used as retarding additives.

9.6.3 Filtration Control

Filtration control additives are added to cement slurries for the same reason that they are added to drilling muds. However, untreated cement slurries have much greater filtration rates than untreated drilling muds. It is this important to limit the loss of water filtrate from a slurry to a permeable formation. This is necessary for several reasons:

- Minimize hydration of formations containing water sensitive formations
- Limit the increase in slurry viscosity as cement is placed in the well
- Prevent annular bridges
- Allow for sufficient water to be available for cement hydration

Examples of filtration control additives are latex, bentonite with a dispersant and other various organic compounds and polymers.

9.6.4 Viscosity Control

The viscosity of the cement slurry will affect the pumping requirements for the slurry and frictional pressure gradient within the well. The viscosity must be kept low enough to ensure that the cement slurry can be pumped to the well during the entire cementing operation period. High viscosity will result in high-pressure gradients that could allow formation fractures.

Examples of commonly used viscosity control additives are calcium lignosulfonate, sodium chloride and some long-chain polymers. These additives also act as accelerators or retarders so care must be taken in designing the cement slurry with these materials.

9.6.5 Special Problems Control

There are some special problems in the design of cement slurries for which additives have been developed. These are

- Gel strength additives for the preparation of spacers (usually fragile gel additives).
- Silica flour is used to form a stronger and less permeable cement, especially for high-temperature applications.
- Hydrazine is used to control corrosion of the casing.
- Radioactive tracers are used to assist in assessing where the cement has been placed.
- Gas-bubble-producing compounds to slowly create gas bubbles in the cement as the slurry sets and hardens. It is felt that such a cement will have less tendency to leak formation gas.
- Paraformaldehyde and sodium chromate that counteract organic contaminants left in the well from drilling operations.
- Fibrous materials such as nylon are added to increase strength, in particular, resistance to impact loads.

9.7 PRIMARY CEMENTING

Primary cementing refers to the cementing of casing and liner strings in a well. The cementing of casing or liner string is carried out so that producible oil and gas, or saltwater will not escape from the producing formation to another formation, or pollute freshwater sands at shallower depth.

The running of long casing strings and liner strings to great depths and successfully cementing these strings required careful engineering design and planning.

9.7.1 Normal Single-Stage Casing Cementing

Under good rig operating conditions, casing can be run into the hole at the rate from 1,000 to 2,000 ft/hr (300 to 600 m/hr). It is often necessary to circulate the drilling mud in the hole prior to running the casing string. This is to assure that all the cutting and any borehole wall prices have been removed. However, the longer a borehole remains open, the more problems will occur with the well.

Prior to running the casing string into the well the mill varnish should be removed from the outer surface of the casing. The removal of the mill varnish is necessary to ensure that the cement will bond to the steel surface.

The casing string is run into the well with a guide shoe (Figure 9.5) or float shoe (Figure 9.6). Figure 9.4 illustrates a regular pattern guide shoe with a wireline re-entry bevel and no internal components to drill.

FIGURE 9.5 Regular pattern guide shoe [2]. (Courtesy of Weatherford International Limited.)

FIGURE 9.6 Swirl guide shoe [2]. (Courtesy of Weatherford International Limited.)

FIGURE 9.7 Ball float collar [2]. (Courtesy of Weatherford International Limited.)

FIGURE 9.8 Flapper float collar [2]. (Courtesy of Weatherford International Limited.)

One to four joints above the guide or float shoe is the float collar. The float collar acts as a back flow valve that keeps the heavier cement slurry from flowing back into the casing string after it has been placed into the annulus between the outside of the casing and the borehole wall. Figures 9.7 and 9.8 show two typical float collars. Figure 9.7 is a plunger style float collar and Figure 9.8 is a flapper float collar. Most modern float equipment is PDC drillable and constructed of composites and cement. Typically a float shoe and a float collar with both be run on one string to provide a redundant check valve capability.

Centralizers are placed along the length of the casing string in a density of between 2 per casing joint to 1 every 4 joints. The density of centralizer placement varies dependent on casing to hole size combination, fluid rheology, and well profile and is usually determined using centralizer placement software. Centralizers ensure that the casing is nearly centered in the borehole, thus allowing a more uniform distribution of cement slurry flow around the casing. This nearly uniform flow around the casing is necessary to remove the drilling mud in the annulus and provide an effective seal.

Centralizers are available in two basic varieties, rigid and bow spring. Figure 9.9 shows a typical rigid centralizer and 9.10 shows a typical bow spring centralizer. Bow spring centralizers are generally run when casing is being run through a restricted bore hole, into under reamed sections, or into washed out zones. Rigid centralizers perform best when the well internal diameter is constant.

The casing string is lowered into the drilling mud in the well using the rig drawworks and elevators. The displaced drilling mud flows to the mud tanks and is stored there for later use. Once the entire casing string is in place

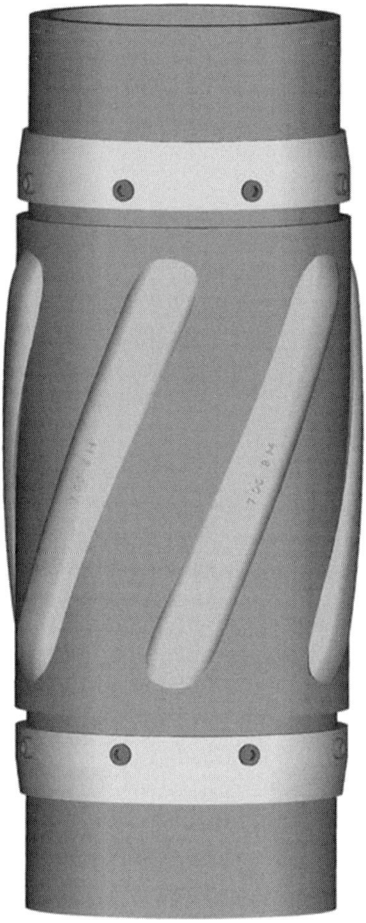

FIGURE 9.9 Scratcher [2]. (Courtesy of Weatherford International Limited.)

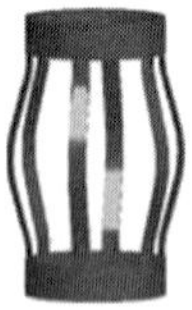

FIGURE 9.10 Centralizer [2]. (Courtesy of Weatherford International Limited.)

in the borehole, the casing string is left hanging in the elevators through the cementing operation. This allows the casing string to be reciprocated (moved up and down) and possibly rotated as the cement is placed in the annulus. This movement assists the removal of the drilling mud.

While the casing string is hanging in the elevators a cementing head is made up to the upper end of the string (see Figure 9.3). The cementing head is then connected with flow liners that come from pump truck (see Figure 9.2). The blender mixes the dry cement and additives with water. The high-pressure, low-volume triplex cement pump on the pump truck pumps the cement slurry to the cementing head. Usually a preflush or spacer is initially pumped ahead of the cement slurry. This spacer (usually about 20 bbl or 3.2 m^3) is used to assist in removing the drilling mud from the annular space between the outside of the casing and the borehole wall.

Figure 9.11 gives a series of schematics that show how the spacer and cement slurry displace drilling mud in the well. Two wiper plugs are usually used to separate the spacer and the cement slurry from the drilling mud in the well. The cementing head has two retainer valves that hold the two flexible rubber wiper plugs with two separate plug-release pins (Figure 9.11a). When the spacer and the cement slurry are to be pumped to the inside of the casing through the cementing head, the bottom plug-release pin is removed. This releases the bottom wiper plug into the initial

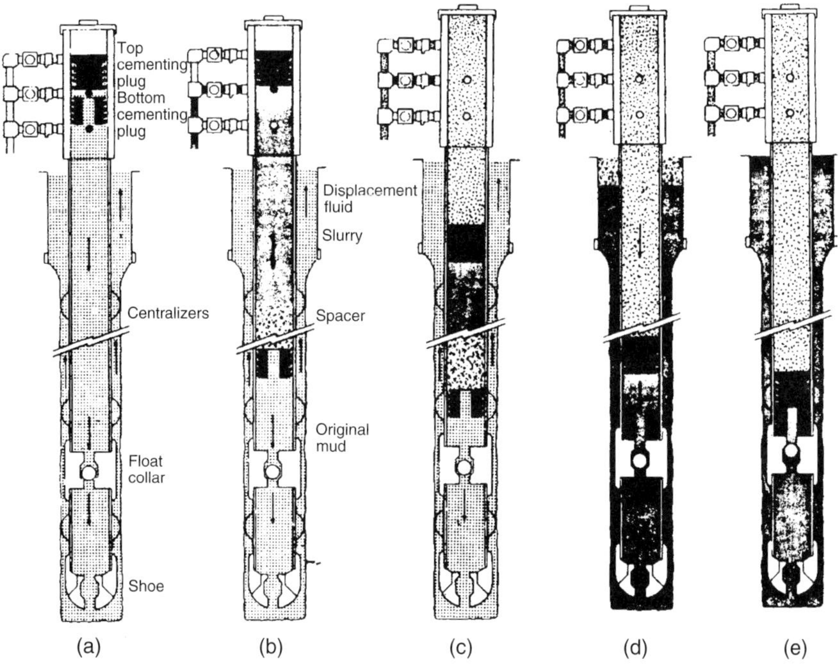

FIGURE 9.11 Single-stage cementing: (a) circulating mud; (b) pumping spacer and slurry; (c) and (d) displacing; (e) end of job. •, plug-releasing pin in; o, plug releasing pin out (Courtesy of Schlumberger).

portion of the spacer flow to the well (see Figure 9.11b). This bottom wiper plug keeps the drilling mud from contaminating the spacer and the cement slurry while they pass through the inside of the casing. When all the cement slurry has passed through the cementing head, the top plug-release pin is removed releasing the top wiper plug into the flow to the well. At this point in the cementing operation the cement pump begins to pump drilling mud through the cementing head to the well (see Figure 9.11a).

When the bottom wiper plug reached the float collar a diaphram is burst at 200 to 400 psi (14 to 28 bar). Typically this pressure event is not witnessed at surface due to the density difference between the cement and the drilling mud. This density differential is almost always large enough to rupture the diaphram. Once the diaphram is broken the spacer string and then into the annulus between the casing and the borehole wall (see Figure 9.11d). The spacer and the cement slurry displace the drilling mud below the float collar and the drilling mud in the annulus. The spacer is designed to efficiently displace nearly all the drilling mud in the annulus prior to the cement slurry entering the annulus.

When the top wiper plug reached the float collar, the pump pressure rises sharply. This wiper plug does not have a diaphram and, therefore, no further flow into the well can take place. At this point in the cementing operation the cement pump is usually shut down and the pressure released on the cementing head. The back-flow valve in the float collar stops the heavier cement slurry from flowing back into the inside of the casing string (see Figure 9.11e). The volume of spacer and cement slurry is calculated to allow the cement slurry to either completely fill the annulus, or to fill the annulus to a height sufficient to accomplish the objectives of the casing and cementing operation.

Usually a lighter drilling mud is used to follow the heavy cement slurry. In this way the casing is under compression from a higher differential pressure on the outside of the casing. Thus when the cement sets and drilling or production operations continues, the casing will always have an elastic load on the cement-casing interface. This elastic load is considered essential for maintenance of the cement-casing bond and to keep leakage between the cement and casing (i.e., the microannulus) from occurring.

Since the early 1960s there has been a great deal of discussion regarding the desirability of using a low viscosity cement slurry to improve the removal of drilling mud from the annulus [2, 3]. More recent studies have shown that low viscosity slurries do not necessarily provide effective removal of drilling mud in the annulus. In fact, there is strong evidence that spacers and slurries should have a higher gel strength and a higher specific weight than the mud that is being displaced [10]. Also, the displacement pumping rate should be low with an annular velocity of 90 ft/min or less [10]. Figure 9.12 gives the annular residual drilling mud removal efficiency as a function of differential gel strength and differential specific

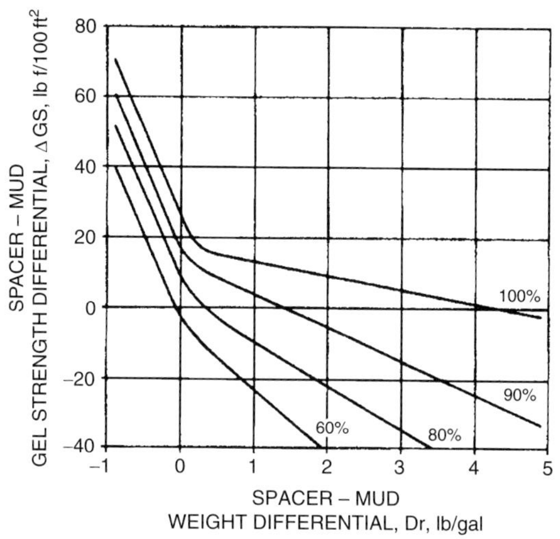

FIGURE 9.12 Spacer gel and specific weight relationships [9].

weight between the drilling mud and the spacer. The most successful spacer is an initial portion of the cement slurry that will give about 200 ft of annulus and that has a higher gel strength and a higher specific weight than the drilling mud it is to displace. The cement slurry spacer should have a gel strength that is about 10 to 15 lb/100 ft^2 greater and a specific weight that is about 2 to 4 lb/gal greater than the drilling mud to be displaced. The spacer should be a specially treated cement slurry.

It makes no sense to use a water or a drilling mud spacer. If such spacers are used, the follow-on normal cement slurry will be just as unsuccessful at removing the water or the drilling mud of the spacer as it would at removing drilling mud if no spacer were used.

The basic steps for the planning and execution of a successful cementing operation are:

1. Condition the well and the drilling mud prior to pulling the drill pipe. Typically at least one hole volume is circulated to remove cutting and debris.
2. Centralize the casing using a placement program.
3. Determine the specific weight and gel strength of drilling mud to be displaced.
4. Estimate the approximate cementing operation time.
5. Run a preflush spacer compatible with the mud and cement.
6. Select the most appropriate API class of cement that meets the depth, temperature, sulfate resistance and other well limitations. Select the cement class that has a natural thickening time that most nearly meets the cementing operation time requirement, or that will require only small amounts of retarding additives.

7. Do not add water loss control additives unless the drilling mud requires it.
8. If the major portion of the cement slurry must have a rather low specific weight, do not use excess water to lower specific weight. Try to use only the optimum recommended water-to-cement ratio. Utilize only additives that will require little or no added water to lower the specific weight of the cement slurry. Avoid using bentonite in large concentrations since it requires a great deal of added water and significantly reduces cement strength. Silicate flour should be used whenever possible to increase compressive strength and decrease permeability of set cement.
9. Design the cement slurry and its initial spacer to have the appropriate gel strength and specific weight. Have the cementing service company run laboratory tests on the cement slurry blend selected. These tests should be run at the anticipated bottomhole temperature and pressure. These tests should be carried out to verify thickening time, specific weight, gel strength (of spacer) and compressive strength.
10. Reciprocate or rotate the casing or liner while conditioning the hole and during the cementing operations. The combination of rotation and reciprocation is widely considered to provide the best results. Rotation is more effective then reciprocation if only one of the two options can be employed. Special cementing hardware may be required to rotate casing.
11. The fluid mechanics should be modeled to provide the optimum pump rates during displacement.
12. The mixed cement slurry should be closely monitored with high-quality continuous specific weight monitoring and recording equipment as the slurry is pumped to the well.
13. When the cementing operation is completed, bleed off internal pressure inside the casing and leave valve open on cementing head.

Example 9.7

A 7-in., N-80, 29-lb/ft casing string is to be run in a 13,900-ft borehole. The casing string is to be 13,890 ft in length. Thus the guide shoe will be held about 10 ft from the bottom of the hole during cementing. The float collar is to be 90 ft from the bottom of the casing string. The borehole is cased with $9\frac{5}{8}$–in., 32.30-lb/ft casing to 11,451 ft of depth. The open hole below the casing shoe of the $9\frac{5}{8}$–in. casing is $8\frac{1}{2}$ in diameter. The 7-in. casing string is to be cemented to the top of the borehole. Class E cement with 20% silica flour is to be used from the bottom of the hole to a height of 1,250 ft from the bottom of the borehole. Class G cement with at least 10% silica flour is to be used from 1,250 ft from the bottom to the top of the borehole. The drilling mud in the borehole has a specific weight of 12.2 lb/gal and an initial gel strength of 15 lb/100 ft^2. A space sufficient to give 200 ft of length in the open-hole section will be used. An excess factor of 1.2 will be applied to the Class G cement slurry volume.

1. Determine the specific weight and gel strength for the spacer.
2. Determine the number of cement sacks for each class of cement to be used.
3. Determine the volume of water to be used.
4. Determine the cementing operation time and thus the minimum thickening time. Assume a cement mixing rate of 25 sacks/min. Also assume an annular displacement rate no greater than 90 ft/min while the spacer is moving through the open-hole section and a flowrate of 300 gal/min thereafter. A safety factor of 1.0 hr is to be used.
5. Determine the pressure differential prior to bumping the plug.
6. Determine the total mud returns during the cementing operation.

The basic properties of Class E and Class G cements are given in Table 9.1 and 9.2.

1. The drilling mud has a specific weight of 12.2 lb/gal and an initial gel strength of 15 lb/100 ft^2. Thus from Figure 9.12 it will be desirable to have a cement slurry spacer with a specific weight of at least 15.2 lb/gal and an initial gel strength of about 20 lb/100 ft^2.

   The spacer will be designed with Class G cement. Taking the appropriate data from Tables 9.6 and 9.7, the weight of silica flour to be used is

$$0.10(94) = 9.4\,\text{lb/sack}$$

   Equation 9.1 becomes

$$\bar{\gamma} = \frac{94 + 9.4 + 8.34[4.97 + 0.0456(9.4)]}{\dfrac{94}{3.14(8.34)} + \dfrac{9.4}{2.63(5.34)} + 4.97 + 0.0456(9.4)}$$
$$= 16.4\,\text{lb/gal}$$

   Thus the spacer will be designed to have a specific weight of 16-4 lb/gal and sufficient fragile gel additive to have an initial gel strength of about 20 lb/100 ft^2.

2. The spacer must have a volume sufficient to give 200 ft of length in the open-hole section. The annular capacity of the open hole is

$$\frac{\pi}{4}\left[\left(\frac{8.5}{12}\right)^2 - \left(\frac{7.0}{12}\right)^2\right] = 0.1268\,\text{ft}^3/\text{ft}$$

   Equation 9.3 gives the yield of the spacer as

$$\text{Yield} = \frac{3.59 + 0.4286 + 4.97 + 0.4286}{7.48} = 1.21\,\text{ft}^3/\text{sack}$$

Thus the number of sacks to give that will be 200 ft in length in the open-hole section is

$$\text{Sacks} = \frac{0.1268(200)}{1.21} = 20.96$$

The above is rounded off to the next highest sack, or 21 sacks.

The volume of Class G cement is the sum of the spacer volume, annular volume in the cased portion of the borehole and the applicable annular volume in the open-hole section of the borehole. The annular capacity of the cased portion of the borehole is

$$\frac{\pi}{4}\left[\left(\frac{9.001}{12}\right)^2 - \left(\frac{7.0}{12}\right)^2\right] = 0.1746\,\text{ft}^3/\text{ft}$$

Thus the volume of Class G cement slurry to be used to cement the well (excluding spacer, but considering the excess factor) is

$$\begin{aligned}\text{Volume} &= [(2{,}449 - 1{,}250)(0.1268) + 11{,}451(0.1746)]1.2\\ &= 2{,}581.66\,\text{ft}^3\end{aligned}$$

The total number of Class G sacks of cement needed, including the spacer volume, is

$$\text{Total sacks} = \frac{2{,}581.66}{1.21} + 21 = 2{,}134 + 21 \approx 2{,}155$$

The volume of Class E cement is

$$\begin{aligned}\text{Volume} &= \frac{\pi}{4}\left(\frac{8.5}{12}\right)^2(10) + \frac{\pi}{4}\left(\frac{6.184}{12}\right)^2 90 + (1{,}250 - 10)(0.1268)\\ &= 3.94 + 18.77 + 157.23 = 179.94\,\text{ft}^3\end{aligned}$$

The specific weight of the Class E cement is determined from Equation 9.1. The weight of silica flour to be added is

$$0.20(94) = 18.8\,\text{lb}$$

Equation 9.1 is

$$\begin{aligned}\bar{\gamma} &= \frac{94 + 18.8 + 8.34[4.29 + 0.0456(18.8)]}{\dfrac{94}{3.14(8.34)} + \dfrac{18.8}{2.63(8.34)} + 4.29 + 0.0456(18.8)}\\ &= 16.2\,\text{lb/gal}\end{aligned}$$

Equation 9.3 is

$$\text{Yield} = \frac{3.59 + 0.8571 + 4.29 + 0.8573}{7.48} = 1.28\,\text{ft}^3/\text{sack}$$

The total number of Class E sacks of cement needed are

$$\text{Total sacks} = \frac{179.94}{1.28} \approx 141$$

3. The volume of water to be used in the cementing operation is

$$\begin{aligned}\text{Volume} &= 2{,}155(4.97 + 0.4286) + 141(4.29 + 0.8573)\\ &\approx 11{,}634 + 726\\ &\approx 12{,}360\,\text{gal}\end{aligned}$$

or about 295 bbl.

4. The total cementing operation time is somewhat complicated since it is desired to reduce the rate of flow when the spacer passes through the open hole section of the well. The mixing time is

$$T_m = \frac{2{,}155 + 141}{25(60)} = 1.53\,\text{hr}$$

During this time the volume of cement slurry pumped to the casing string is

$$\begin{aligned}\text{Volume pumped} &= 2{,}155(1.21) + 141(1.28)\\ &= 2{,}788.1\,\text{ft}^3\end{aligned}$$

The internal volume of the casing string above the float collar is

$$\begin{aligned}\text{Internal volume} &= 13{,}800(0.2086)\\ &= 2{,}878.7\,\text{ft}^3\end{aligned}$$

Therefore, the mud pumps are used to fill the remainder of the casing string at a rate of 300 gal/min. Thus, the additional time T_1 to get the bottom plug to the float collar is

$$T_1 = \frac{2{,}878.7 - 2{,}788.1}{\left(\frac{300}{7.48}\right)(60)} = 0.04\,\text{hr}$$

The time to release the plug is

$$\begin{aligned}T_p &= 15/60\\ &= 0.25\,\text{hr}\end{aligned}$$

From the time the plug is released the pumping rate is reduced to 85.4 gal/min, which gives a velocity of 90 ft/min in the openhole section of the annulus. The time for the spacer to pass through the open-hole section T_2 (including the 90 ft of internal volume of the casing below the float collar and the 10 ft of open hole below the casing string guide shoe) is

$$T_2 = \frac{3.94 + 18.77 + (2{,}449 - 10)(0.1268)}{\left(\frac{85.4}{7.48}\right)(60)} = 0.49\,\text{hr}$$

After this time period, the rate of pumping to the well can be returned to 300 gal/min rate.

Equation 9.4 becomes

$$T_0 = 1.53 + 0.04 + 0.25 + 0.49 + \frac{2{,}878.7 - 91 - 332}{\left(\frac{300}{7.48}\right)(60)} + 1.0 = 4.33\,\text{hr}$$

Therefore, the thickening time for the cement slurries to be used must be greater than 4.33 hr.

5. The pressure differential prior to the top plug reaching the float collar is

$$p = \frac{122.7}{144}(13{,}900 - 1{,}250) + \frac{121.2}{144}(1{,}250 - 90 - 10)$$
$$- \frac{91.3}{144}(13{,}900 - 90 - 10)$$
$$= 2997\,\text{psi}$$

6. The total volume of the mud returns will be

$$\text{Mud volume} = \text{Total volume well without steel} - \text{Steel volume}$$
$$= \frac{\pi}{4}\frac{(8.5)^2}{12}(13{,}900 - 11{,}451)$$
$$+ \frac{\pi}{4}\frac{(9.001)^2}{12}(11{,}451) - \frac{29.0}{490}(13{,}890)$$
$$= 5{,}203\,\text{ft}^3 \text{ or about } 927\,\text{bbl}$$

9.7.2 Large-Diameter Casing Cementing

Often when large-diameter casing strings are to be cemented, an alternate method to cementing through inside diameter of the casing is used. This alternative method requires that the inner string of drillpipe be placed inside the large-diameter casing string. The drillpipe string is centralized within the casing and a special stab-in unit is made up to the bottom of the drillpipe string. After the drillpipe string is run into the well through the casing, the special unit on the bottom of the drillpipe string is stabbed into the stab-in cementing collar located a joint or two above the casing string guide shoe, or the unit is stabbed into a stab-in cementing shoe (a combination of the stab-in cementing collar and the guide shoe). Figure 9.13 shows

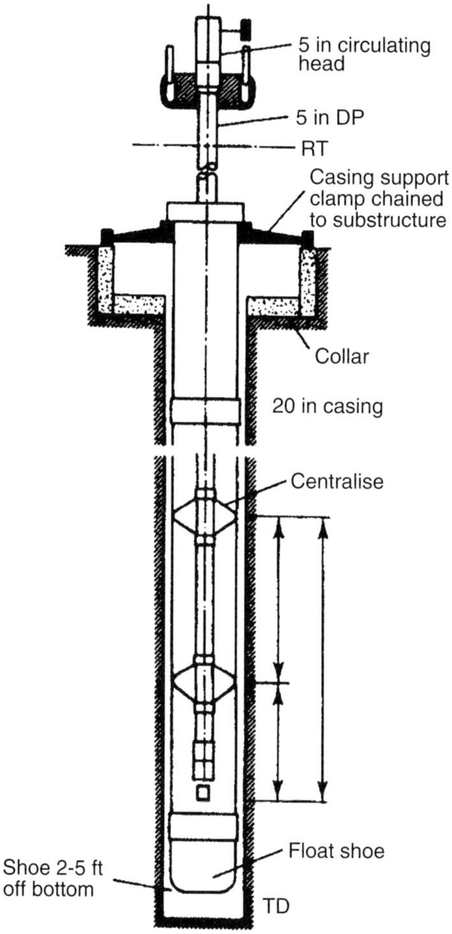

FIGURE 9.13 Cementing of large-diameter casing with and inner drillpipe string (courtesy of Dowell Schlumberger).

the schematic of a large-diameter casing with the inner drillpipe string used for cementing [4]. Figure 9.14 shows the stab-in cementing shoe and a stab-in cementing collar. Also shown is a flexible latch-in plug used to follow the cement slurry. Once the stab-in unit of the drillpipe string is seated in the stab-in cementing shoe or cementing collar, a special circulating head is made up to the top of the drillstring. Circulation is established through the circulating head to the drillpipe and up the annular space between the casing and the borehole wall. The stab-in unit may be locked into the cementing collar. The collar will act as a back-flow valve [3]. After the cement slurry has been run, the drillpipe string can be unlocked and withdrawn from the casing.

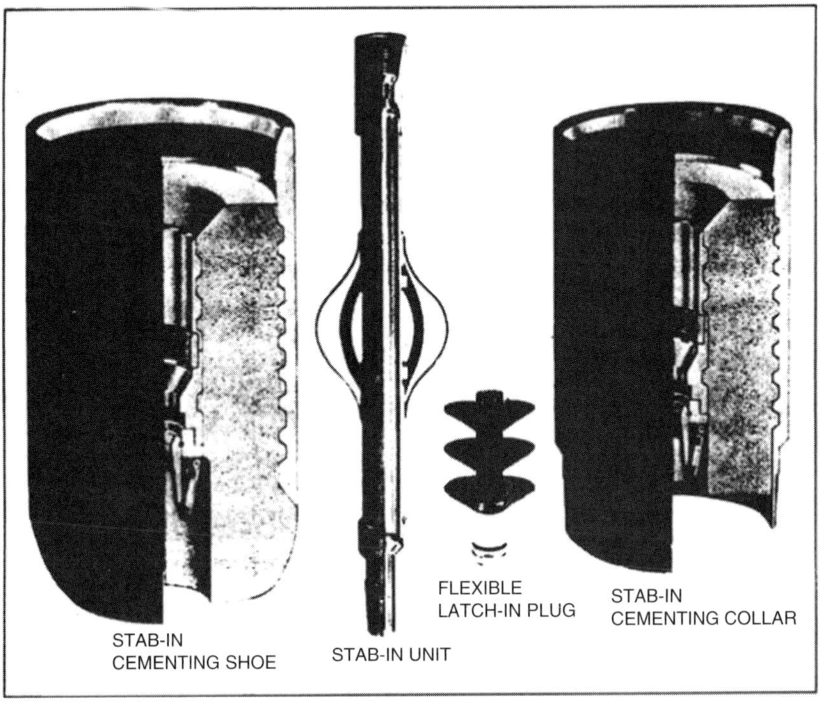

FIGURE 9.14 Stab-in unit, stab-in cementing shoe, stab-in cementing collar, and flexible latch-in plug [2].

The advantages of cementing large diameter casing using an inner drillpipe string are:

- Avoids excessive mud contaminations of the cement slurry with drilling mud prior to reaching the annulus
- Allows cement slurry to be added if wash out zones are excessive (avoids top-up job)

When cementing large-diameter casing loss of circulation can occur. The solution to such problems is to recement down the annulus. This technique of cementing is denoted as a top-up job. This type of cementing can be accomplished by running small-diameter tubing called "spaghetti" into the annulus space from the surface. Under these conditions usually low-specific-weight cement is used so that the formations will not fracture.

The hook load after a cementing operation (but prior to the cement setting) is important to know, particularly when large-diameter casings are to be cemented. Figure 9.15 shows the schematic of a casing string with the cement slurry height shown. Prior to the cement slurry setting the hook

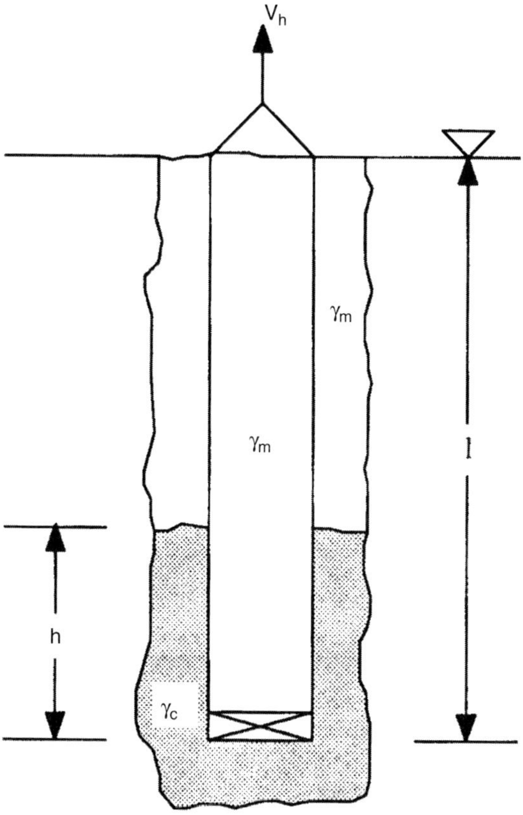

FIGURE 9.15 Schematic of casing after a cementing operation.

load V_h(lb) will be

$$V_h = w_c \left\{1 - \left[\left(\frac{\gamma_c}{\gamma_s}\right)\frac{h}{1} + \left(\frac{\gamma_m}{\gamma_s}\right)\left(\frac{1-h}{1}\right)\right]\right\} 1 - A_i[(\gamma_c - \gamma_m)h] \tag{9.8}$$

where w_c = unit weight of the casing in lb/ft
γ_c = specific weight of the cement slurry in lb/ft^3
γ_m = specific weight of the drilling mud in lb/ft^3
γ_s = specific weight of the steel, which is 490 lb/ft^3
l = length of the casing in ft
h = height the cement slurry is placed in the annular in ft
A_i = internal cross-sectional diameter of the casing in ft^2

Equation 9.8 is valid for a back-flow valve at the bottom of the casing string such as a float collar. It is also valid for a back-flow valve located at the top of the casing string.

Example 9.8

A large-diameter casing is to be cemented to the top in a 2,021 ft borehole. The casing string is 2,000 ft long and has ball float shoe at the bottom of the cement slurry. The borehole is 26 in. in diameter. The casing is to be 20 in., J-55, 94 lb/ft. The drilling mud in the borehole prior to running the casing has a specific weight of 10.0 lb/gal. The cement slurry is to have a specific weight of 17.0 lb/gal. The conventional cementing operation will be utilized, i.e., cementing through the inside of the casing. Determine the hook load after the cementing operation has been completed.

The hook load may be determined from Equation 9.8. This is

$$V_u = 94.00\left\{1+\left[\left(\frac{127.3}{490.00}\right)\frac{2{,}000}{2{,}000}+\left(\frac{75.0}{490.0}\right)\frac{0}{2{,}000}\right]\right\}2{,}000$$

$$-\frac{\pi}{4}\left(\frac{19.124}{12}\right)^2[(127.3-75.0)2{,}000]$$

$$= -69{,}557 \text{ lb}$$

This answer means that the casing and its contained drilling mud will float when the cementing operation is completed and the back-flow valve is actuated. Thus the casing should be secured down at the well head prior to initiating the cementing operation.

It should be noted that if this cementing operation were to be carried out using an inner drillpipe string to place the cement in the annulus, the above force of buoyancy would be reduced by the buoyed weight of the drillpipe. However, unless very heavy drillpipe were used, the casing and drillpipe would still float on the cement slurry.

As the cement slurry sets the hook load will decrease from its value just after the cementing operation. If the cement slurry sets up to the height (freeze point) indicated in Figure 9.14, and if the casing has been landed as cemented, then the hook load V_h will be approximately

$$V_h = W_c\left(1-\frac{\gamma_m}{\gamma_s}\right)(1-h) \tag{9.9}$$

Example 9.9

A $5\frac{1}{2}$-in., J-55, 17.00-lb/ft casing string 16,100 ft in length is to be run into a 16,115-ft deep well. The float collar is 30 ft from the bottom of the casing string. The drilling mud in the well has a specific weight of 12.0 lb/gal. The casing string is to be cemented to a height of 4,355 ft above the bottom of the borehole with a cement slurry having a specific weight of 16.4 lb/gal.

1. Determine the approximate hook load prior to the cementing operation.
2. Determine the approximate hook load just after the cement slurry has been run.

3. Determine the approximate hook load after the cement has set.

   1. The hook load prior to the cementing operation should read approximately

$$V_h = 17.00\left(1 - \frac{89.9}{490.0}\right)16{,}100 = 223{,}540\,\text{lb}$$

   2. The hook load just after the cementing operation can be approximated by Equation 9.8. This is

$$\begin{aligned} V_h = 17.00&\left\{1 - \left[\left(\frac{122.7}{490.0}\right)\frac{4{,}340}{16{,}100} + \left(\frac{89.9}{490.0}\right)\right.\right.\\ &\times\left.\left.\left(\frac{16{,}100 - 4{,}340}{16{,}100}\right)\right]\right\}16{,}100\\ &- \frac{\pi}{4}\left(\frac{4.892}{12}\right)^2[(122.7 - 89.8)4{,}340]\\ &= 199{,}943\text{ lb}\end{aligned}$$

   3. The hook load after the cement has set can be approximated by Equation 9.9. This is

$$V_h - 17.00\left(1 - \frac{89.8}{490.0}\right)(16{,}100 - 4{,}340) = 163{,}282\text{ lb}$$

9.7.3 Multistage Casing Cementing

Multistage casing cementing is used to cement long casing strings. The reasons that multistage cementing techniques are necessary are as follows:

- Reduce the pumping pressure of the cement pumping equipment
- Reduce the hydrostatic pressure on weak formations to prevent fracture
- Selected formations can be cemented
- Entire length of a long casing string may be cemented
- Casing shoe of the previous casing string may be effectively cemented to the new casing string
- Reduces cement contamination

There are methods for carrying out multistage cementing. These are

1. Regular two-stage cementing
2. Continuous two-stage cementing
3. Regular three-stage cementing

Regular two-stage cementing requires the use of stage cementing collar and plugs in addition to the conventional casing cementing equipment. The stage cementing collar is placed in the casing string at near the mid point, or at a position in the casing string where the upper cementing of the

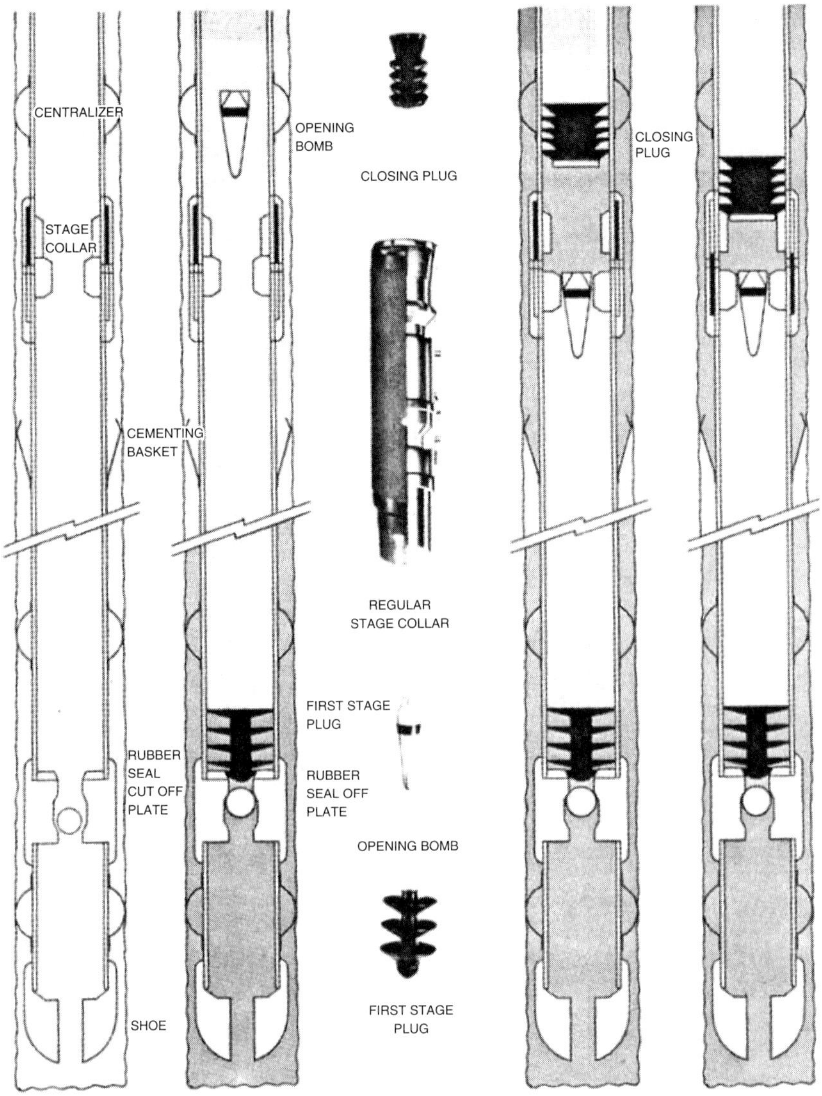

FIGURE 9.16 Regular two-stage cementing [2].

casing is to take place. Figure 9.16 shows a schematic of a regular two-stage casing cementing operation. The stage cementing collar is a special collar with ports to the annulus that can be opened and closed (sealed off) by pressure operated sleeves (see Figure 9.16).

The first stage of cementing (the lower section of the annulus) is carried out similar to a conventional single-stage casing cementing operation. The exception is that a wiper plug is generally not run in the casing prior to the spacer and the cement slurry.[4] During the pumping of the spacer and cement slurry for the lower section of the annulus, the ports on the stage cementing collar are closed. After the appropriate volume of spacer and cement slurry has been pumped to the well (lower section) the first stage plug is released. This plug is pumped to its position on the float collar at the bottom of the casing string with drilling mud or completion fluid as the displacement fluid. This first plug is designed to pass through the stage cementing collar without actuating it. When the first plug is landed on the float collar, there is a pressure rise at the pump. This plug seals the float collar such that further flow throughout the collar cannot take place.

After the first-stage cementing operation has been carried out the opening bomb can be dropped and allowed to fall by gravity to the lower seal of the stage cementing collar. This can be done immediately after the first-stage cementing has taken place, or at some later time when the cement slurry in the lower section of the well has had time to set. In the later case great care must be taken that the first-stage cement slurry has not risen in the annulus to a height above the stage cementing collar. Once the opening bomb is seated, pump pressure is applied that allows hydraulic force to be applied to the lower sleeve of the stage cementing collar. This force shears the lower sleeve retaining pins and exposes the ports to the annulus.

Once the ports are opened the well is circulated until the appropriate drilling mud or other fluid is in the well. The second-stage cementing slurry is mixed and pumped to the well (again without a wiper plug separating the spacer from the drilling mud). This cement slurry passes through the stage cementing float ports to the upper section of the annulus. The closing plug is released at the end of the second-stage cement slurry and is displaced to the stage cementing collar with drilling mud or other fluid. The closing plug seats on the upper sleeve of the stage cementing collar. A pressure of about 1,500 psi causes the retaining pins in the upper sleeve to shear thus forcing the sleeve downward to close the ports in the stage cementing collar.

Continuous two-stage cementing is an operation that requires that the cement slurry be mixed and displaced to the lower and upper sections of the annulus in sequence without stopping to wait for an opening bomb to actuate the stage cementing collar. In this operation the first-stage cement slurry is pumped to the well and a wiper plug released behind it (Figure 9.17). Displacing the wiper plug is a volume of drilling mud or completion fluid that will displace the cement slurry out of the casing and fill the inside of the casing string from the float collar at the bottom of the casing string to a height of the stage cementing collar. A bypass insert allows fluid to pass through the wiper plug and float collar after the plug is landed. The opening plug

4. A wiper plug may be accommodated with special equipment [2].

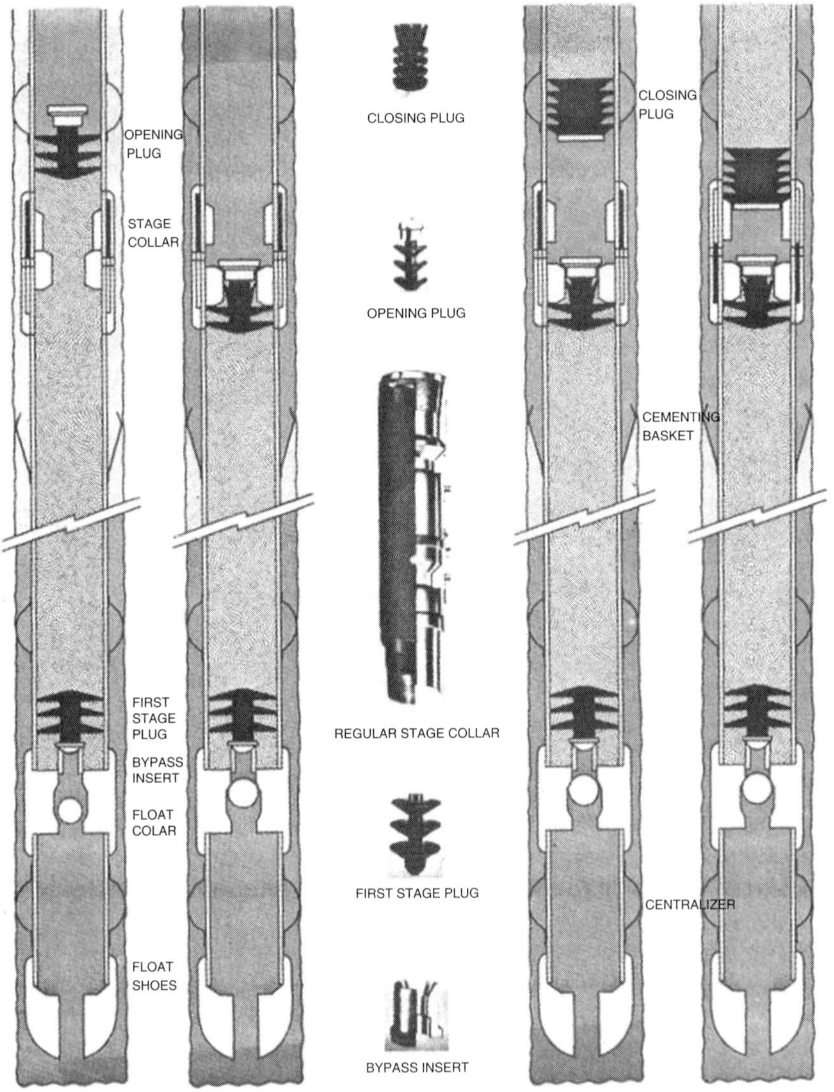

FIGURE 9.17 Continuous two-stage cementing [2].

is pumped immediately behind the volume of drilling mud. Immediately behind the opening plug is the second-stage spacer and cement slurry. The opening plug sits on the lower sleeve of the stage cementing collar, opening the ports to the annulus. At the end of the second-stage cement slurry the closing plug is run. This plug sits on the upper sleeve of the stage cementing collar and with hydraulic pressure, closing the ports in the stage cementing collar.

Three-stage cementing is carried out using the same procedure as the regular two-stage cementing operation discussed above. In this case, however, two-stage cementing collars are placed at appropriate locations in the casing string above the float collar. Each stage of cementing is carried out in sequence, the lower annulus section cemented first, the middle annulus section next and the top annulus section last. Each stage of cement can be allowed to set, but great care must be taken in not allowing the lower stage of cement to rise above the stage cementing collar of the next stage above.

9.7.4 Liner Cementing

A liner is a short string of casing that does not reach the surface. The liner is hung from the bottom of the previous casing string using a liner hanger that grips the bottom of the previous casing with a set of slips. Figure 9.18 shows typical liner types. The liner is run into the borehole on the drillpipe and the cementing operation for the liner is carried out through the same drillpipe. The placing of liners and their cementing

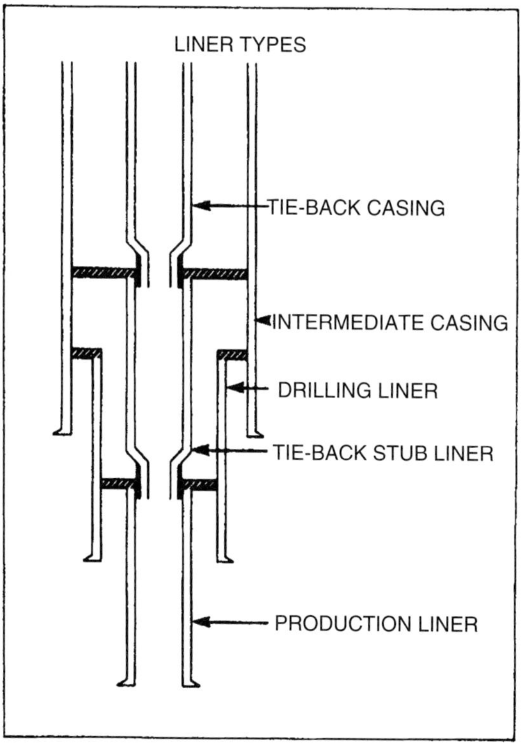

FIGURE 9.18 Liner types [2].

operations are some of the most difficult operation in well drilling and completions. Great care must be taken in designing and planning these operations to ensure a seal between the liner and the previous casing.

Figure 9.19 shows a typical linear assembly. The liner assembly is made up with the following components:

Float shoe. The float shoe may be placed at the bottom of the liner. This component is a combination of a guide shoe and a float collar.

Landing collar. This is a short sub placed in the string to provide a seat for the casing string.

Liner. This is a string of casing used to case off the open hole without bringing the end of the string to the surface. Usually the liner overlaps the previous casing string (shoe) by about 200 to 500 ft. (60 to 150 m.).

Liner hanger. This special tool is installed on the top of the liner string. The top of the liner hanger makes up to the drillpipe on which the entire liner assembly is lowered into the well. Liner hangers can be either mechanically or hydraulically actuated.

The liner hanger is the key element in the running of liners and the follow-on cementing operations. The liner hanger allows the liner to be hung from the bottom section of the previous casing. After the cementing operation the upper part of the liner hanger is retrievable, thus allowing the residual cement above the liner hanger to be cleaned out of the annulus between the drillpipe and the previous casing (by reverse circulation) and the liner left in the well.

The casing joints of the liner are placed in the well in the normal manner. The liner hanger is made up to the top of the liner. The top of the liner hanger is made up to drillpipe and the whole assembly (i.e. liner, liner hanger and drillpipe) is lowered into the well. After the liner is at the desirable location in the well, the drillpipe is connected to the rig pumps and mud circulation carried out. This allows conditioning of the drilling mud in the well prior to the cementing operation and ensures that circulation is possible before the liner is hung and cemented.

The liner hanger is set (either mechanically or hydraulically) and the drillpipe with the upper part of the liner hanger (the setting tool) released. The drillpipe and the setting tool are raised to make sure that the setting tool and the drillpipe can be released from the lower part of the liner hanger and the liner. The drillpipe and the setting tool are lowered to make a tight seal with the lower portion of the liner hanger.

After these tests of the equipment have been made, a liner cementing head is made up to the drillpipe at the surface. Figure 9.20 shows the liner cementing head and the pump-down plug. The pump-down plug is placed in the liner cementing head, but not released. The cement pump

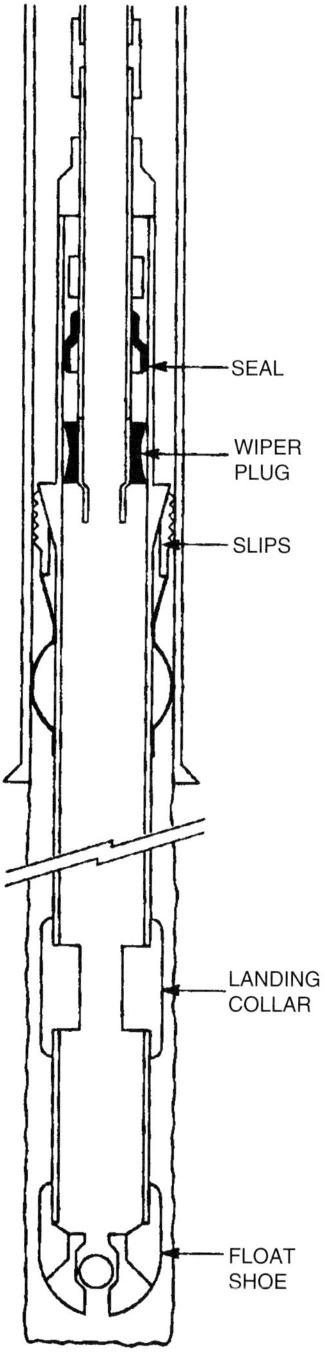

FIGURE 9.19 Liner assembly [2].

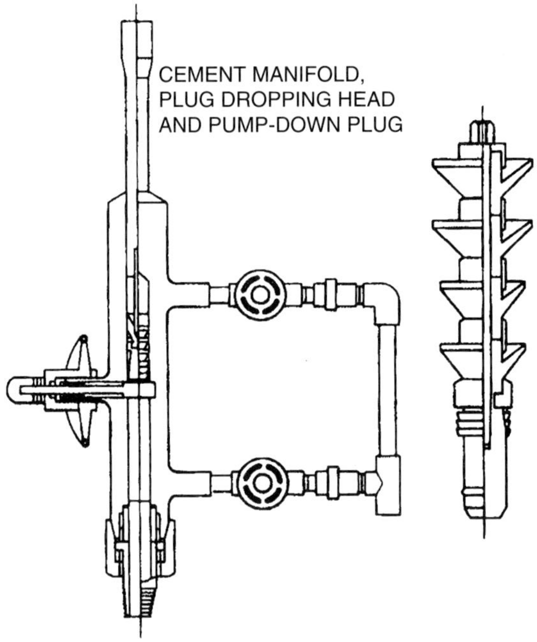

FIGURE 9.20 Liner cementing head [2].

is connected to the liner cementing head and the spacer and cement slurry pumped to the head (no wiper plug is run ahead of the spacer). The pump-down plug is released at the end of the cement slurry and separates the slurry from the follow-on drilling mud. Figure 9.21 shows the sequence of the liner cementing operation. The drilling mud displaces the pump-down plug to the liner hanger. As the pump-down plug passes through the liner hanger it latches into the liner wiper plug (see Figure 9.19). With increased surface pressure (1200 psi or 83 bar) by the pump the liner wiper by the pump the liner wiper plug with the pump-down plug coupled to it are released from the liner hanger and begin to move again downward. These two coupled plugs eventually seat on the landing collar or on the float collar. Another pressure rise indicates the cement is in place behind the liner.

Once the spacer and the cement slurry have been successfully pumped to their location in the well, the drillpipe and setting tool are released from the lower part of the liner hanger. Also, the liner cementing head is removed from the drillpipe. The drillpipe and setting tool are raised slightly and the excess cement slurry reverse circulated from around the liner hanger area. These steps should be taken immediately after the completion of the cementing operation, otherwise the excess cement slurry between the drillpipe and the previous casing could set and cause later drilling and

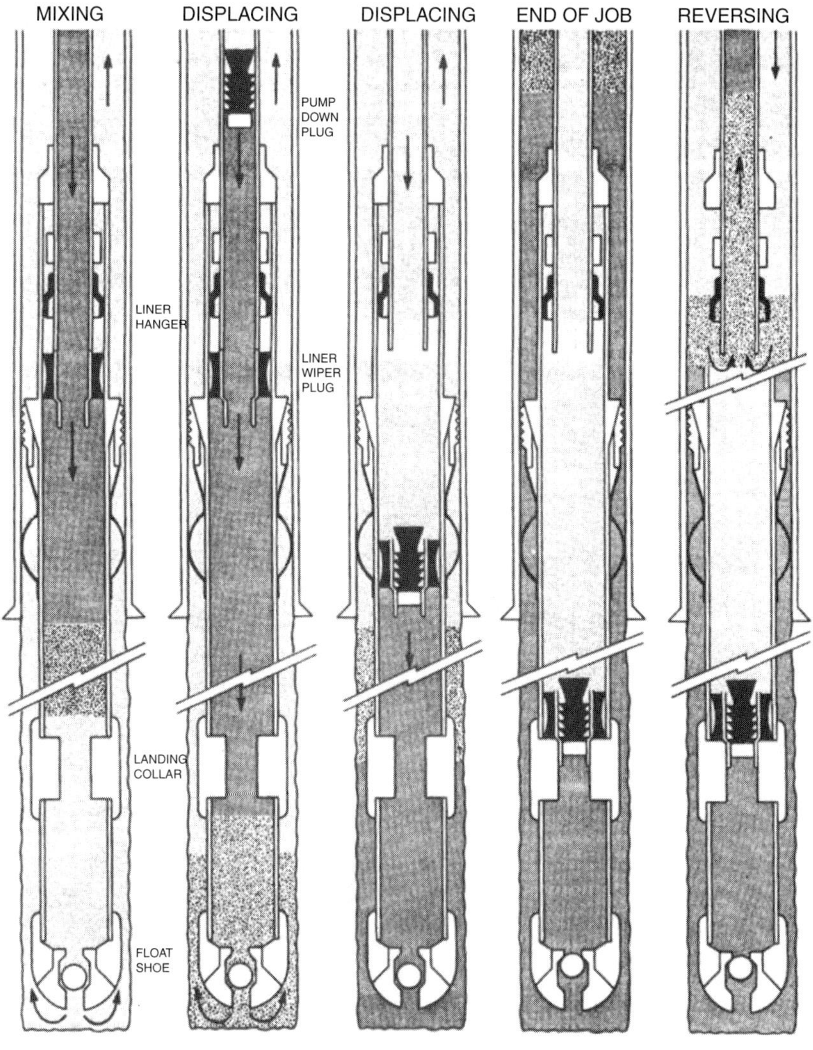

FIGURE 9.21 Liner cementing [2].

completions problems if the excess cement is too great. If reverse circulation is not planned, then the excess cement slurry that comes through the liner hanger to the annulus between the drillpipe and the previous casing must be kept to a minimum so that it can be easily drilled out after it sets.

The determination of the excess cement slurry should be carefully calculated: too little and the cement seal at the liner hanger is contaminated with drilling mud; too much and there are problems removing it.

9.8 SECONDARY CEMENTING

Secondary or remedial cementing refers to cementing operations that are intended to use cement as a means of maintaining or improving the well's operation. There are two general secondary cementing operations, squeeze cementing and plug cementing.

9.8.1 Squeeze Cementing

Squeeze cementing operations utilize the mechanical power of the cement pumps to force a cement slurry into the annular space behind a casing and/or into a formation for the following purposes:

- Repair a faulty primary cementing operation
- Stop loss of circulation in the open hole during drilling operations
- Reduce water-oil, water-gas and gas-oil ratios by selectively sealing certain fluid producing formations
- Seal abandoned or depleted formations
- Repair casing leaks such as joint leaks, split casing, parted casing or corroded casing
- Isolate a production zone by sealing off adjacent nonproductive zones

There are numerous squeeze cementing placement methods that utilize many different special downhole tools. In general, the squeeze cementing operation forces the cement slurry into fractures under high pressure, or into casing perforations using low pressure. As a cement slurry is forced into rock fractures, or through casing perforations into the rock formations, the slurry loses part of its mix water. This leaves a filter cake of cement particles at the interface of the fluid and the permeable rock. As the filter cake builds up during the squeeze cementing operation, more channels into the formation are sealed and the pressure increases. If the cement slurry is poorly designed for an intended squeeze cementing operation, the filter cake builds up too fast and the cement pump capability is reached before the cement slurry has penetrated sufficiently to accomplish its purpose. Thus the squeeze cementing operation slurry should be designed to match the characteristics of the rock formation to be squeeze cemented and the equipment to be used.

There are five important considerations regarding the cement slurry design for a squeeze cementing operation:

1. *Fluid loss control*. The slurry should be designed to match the formation to be squeezed. Low permeability formations should utilize slurries with 100–200 ml/30 min water losses. High Permeability formations should utilize slurries with 50–100 ml/30 min water losses.
2. *Slurry volume*. The cement slurry volume should be estimated prior to the squeeze operation. In general, high-pressure squeeze operations of

high-permeability formations that have relatively low fracture strengths will require large volumes of slurry. Low-pressure squeeze operation through perforations will require low volumes.

3. *Thickening time*. High-pressure squeeze operations that pump large volumes in a rather short time period usually require accelerator additives. Low-pressure slow-pumping-rate squeeze operations usually require retarder additives.
4. *Dispersion*. Thick slurries will not flow well in narrow channels. Squeeze cement slurries should be designed to be thin and have low yield points. Dispersive agents should be added to these slurries.
5. *Compressive strength*. High compressive strength is not a necessary characteristic of squeeze cement slurries.

There are basically two squeeze cementing techniques used: the high-pressure squeeze operation and the low-pressure squeeze operation.

High-pressure squeeze cementing operations are utilized where the hydraulic pressure is used to make new channels in the rock formations (by fracturing the rock) and force the cement slurry into these channels.

Low-pressure squeeze cementing operations are utilized where the existing permeability structure is sufficient to allow the cement slurry to efficiently move in formation without making new fracture surfaces with the hydraulic pressure.

Hesitation method of applying pressure is applicable to both high and low-pressure squeeze cementing operations. This method of applying pressure (and thus volume) appears to be more effective than continuous pressure application. The hesitation method is the intermittent application of pressure, separated by a period of pressure leakoff caused by the loss of filtrate into the formation. The leakoff periods are short at the beginning of an operation but get longer as the operation progresses.

Figure 9.22 shows a typical squeeze cementing downhole schematic where a retrievable packer is used. The cement slurry is spotted adjacent to the perforations in the casing. The packer raise and set against the casing wall. Pressure is then applied to the drilling mud and cement slurry below the packer forcing the cement slurry into the perforation and into the rock formation or voids in the annulus behind the casing.

Figure 9.23 shows how a retrievable packer with a tail pipe can be used to precisely spot the cement slurry at the location of the perforations.

Figure 9.24 shows how a drillable cement retainer can be used to ensure that the cement slurry will be applied directly to the perforations.

Most squeeze cementing operations take place in cased sections of a well. However, open-hole packers can be used to carry out squeeze cement operations of thief zones during drilling operations.

Another technique for carrying out a squeeze cementing operation is the Bradenhead technique. This technique can be used to squeeze a cement

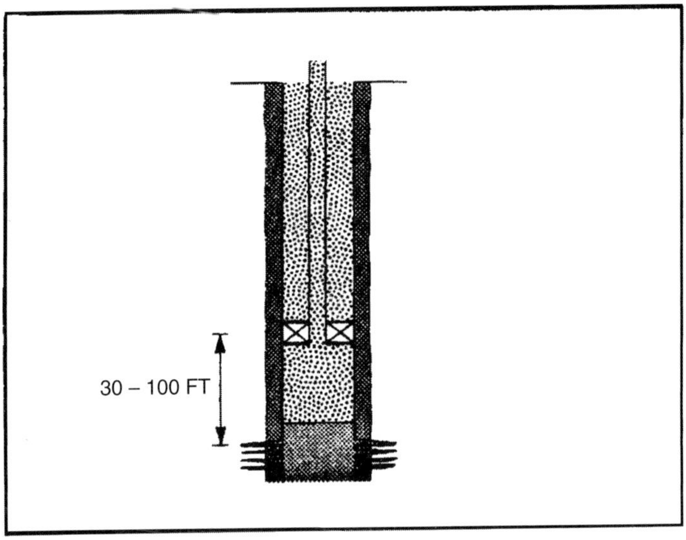

FIGURE 9.22 Retrievable packer near perforations [2].

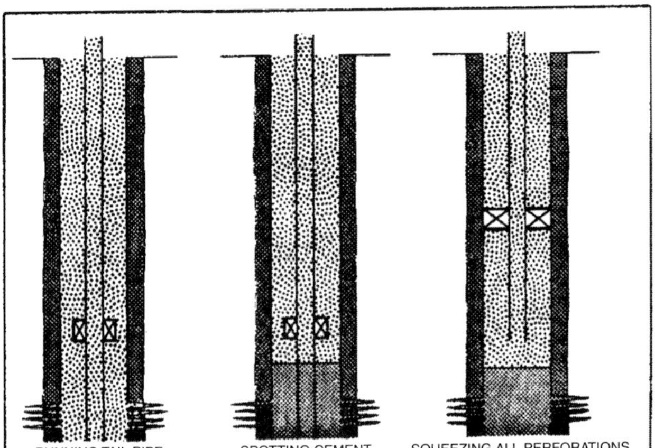

FIGURE 9.23 Retrievable packer with tail pipe [2].

slurry in a cased hole or in an open hole. Figure 9.25 shows a schematic of the technique. Instead of using a downhole packer, the cement slurry is spotted by drillpipe (or tubing) adjacent to the perforations or a thief zone. After the drillpipe has been raised, the pipe rams are closed over the drillpipe and pump pressure applied to the drilling mud which in turn forces the cement slurry into the perforations or fractures.

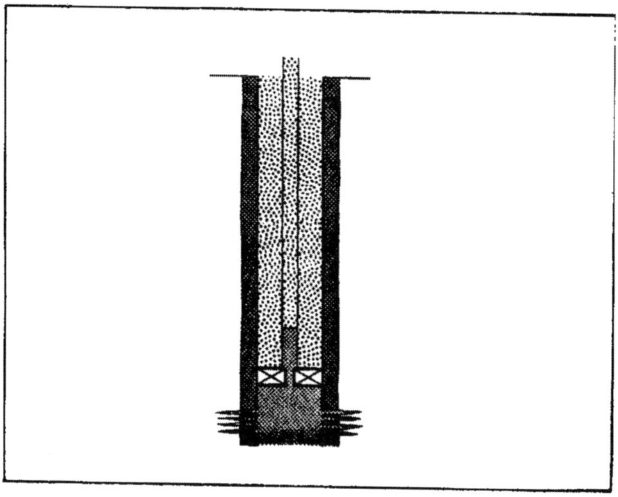

FIGURE 9.24 Drillable cement retainer [2].

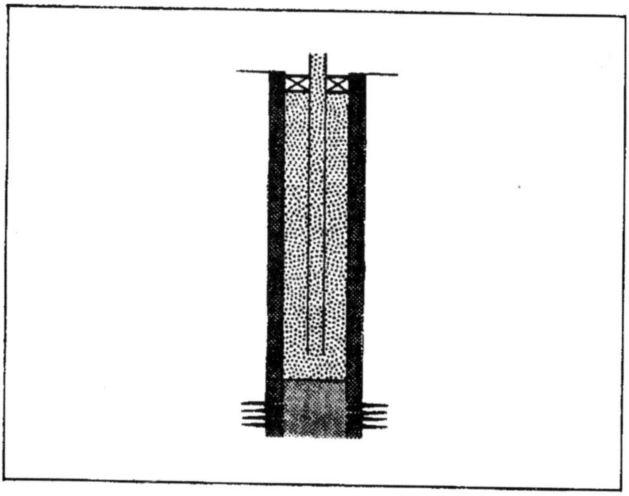

FIGURE 9.25 Bradenhead squeeze [2].

The basic squeeze cementing operation procedures are as follows [2]:

1. Lower zones are isolated by a retrievable or drillable bridge plug.
2. Perforations are washed using a perforation wash tool or they are reopened by surging. If there is no danger of damaging the lower perforations, this operation can be carried out before running the bridge plug, which may be run in one trip with the packer.
3. The perforation wash tool is retrieved and the packer run in the hole with a work string, set at the desired depth and tested. An annular

pressure test of 1,000 psi (69 bar) is usually sufficient. The packer is run with or without a tail pipe, depending on the operation to be performed. If cement is to be spotted in front of the perforations, a tail pipe that covers the length of the zone plus 10 to 15 ft (3 to 5 meters) must be run with the packer.

4. An injectivity test is performed using clean, solids-free water or brine. If a low fluid loss completion fluid is in the hole, it must be displaced from the perforations before starting the injecting. This test will give an idea of the permeability of the formation to the cement filtrate.
5. A spearhead or breakdown fluid followed by the cement slurry is circulated downhole with the packer by-pass open. This is done to avoid the squeezing of damaging fluids ahead of the slurry. A small amount of back pressure must be applied on the annulus to prevent the slurry fall caused by U tubing. If no tail pipe has been run, the packer by-pass must be closed 2 or 3 bbl (.25 to .33 m^3) before the slurry reaches the packer. If the cement is to be spotted in front of the perforations, with the packer unset, circulation is stopped as soon as the cement covers the desired zone, the tail pipe pulled out of the cement slurry and the packer set at the desired depth at which the packer is set must be carefully decided.

   If tail pipe is run, the minimum distance between perforations and packer is limited to the length of the tail pipe. The packer must not be set too close to the perforations as pressure communication through the annulus above the packer might cause the casing to collapse. A safe setting depth must be decided on after seeing the logged quality of the cement bond. Casing conditions and possible cement contamination limit the maximum spacing between packer and treated zone.
6. Squeeze pressure is applied at the surface. If high-pressure squeezing is practiced, the formation is broken down and the cement slurry pumped into the fractures before the hesitation technique is applied. If low-pressure squeezing is desired, hesitation is started as soon as the packer is set.
7. Hesitation continues until no pressure leak-off is observed. A further test of about 500 psi (35 bar) over the final injection pressure will indicate the end of the injection process. Usually, well-cementing perforations will tolerate pressures above the formation fracture pressure, but the risk of fracturing is increased.
8. Pressure is released and returns are checked. If no returns are noticed, the packer by-pass is opened and excess cement reversed out. Washing off cement in front of perforations can be performed by releasing the packer and slowly lowering the work string during the reversing.
9. Tools are pulled out and the cement is left to cure for the recommended time, usually 4 to 6 hr.

Plug Cementing

The major reasons for plug cementing are:

Abandonment. State regulations have rules on plugging and abandoning wells. Cement plugs are normally used for that purpose (see Figure 9.26).

Kick-off plug. Usually an Ottawa sand-cement plug is used to plug off a section of the borehole. This plug uses a hard surface to assist the kick-off procedure (see Figure 9.27).

Lost circulation. A cement plug can be placed adjacent to a zone of lost circulation in the hope that the cement slurry will penetrate and seal fractures (see Figure 9.28).

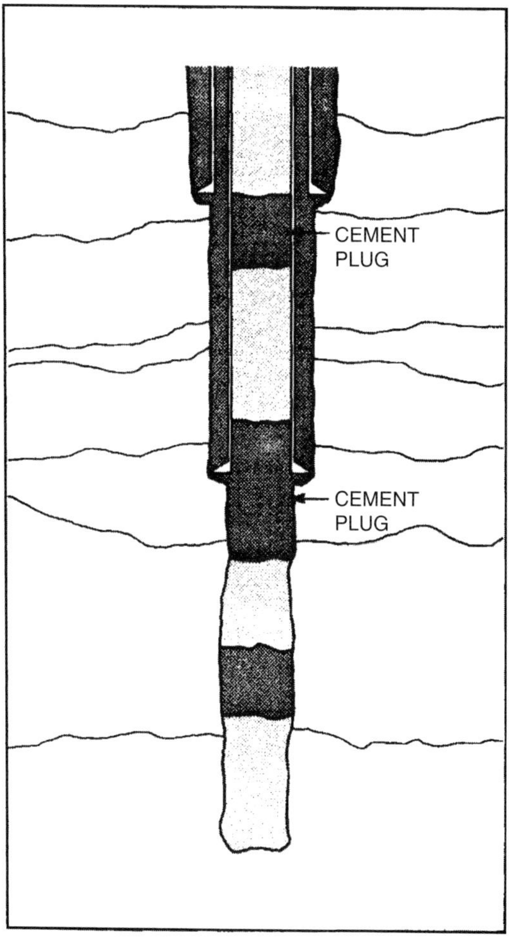

FIGURE 9.26 Abandonment plugs [2].

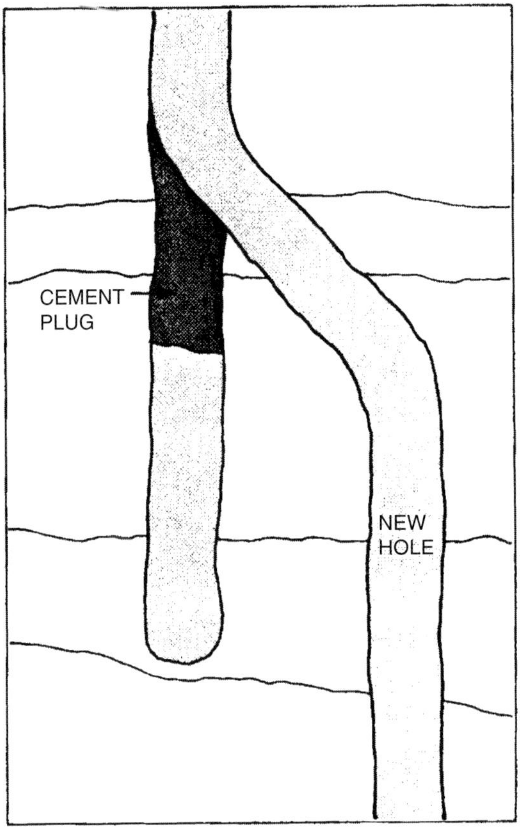

FIGURE 9.27 Kickoff plug [2].

Openhole completions. Often in openhole completions it is necessary to shut off water flows, or to provide an anchor for testing tools, or other maintenance operations (Figure 9.29).

There are three methods for placing cement plugs.

1. *Balance plug method* is the most commonly used. The cement slurry is placed at the desired depth through the drillpipe or tubing run to that depth. A spacer is placed below and above the slurry plug to avoid contamination of the cement slurry with surrounding drilling mud and to assist in balancing the mud. There are a number of tools available in the market to assist in plug balancing. Most involve drill pipe darts that provide a means of separating fluids, wiping the pipe and provide an indication of the end of displacement.
2. *Dump bailing method* utilizes a bailing device that contains a measure volume of cement slurry. The bailer is run to the appropriate depth on a

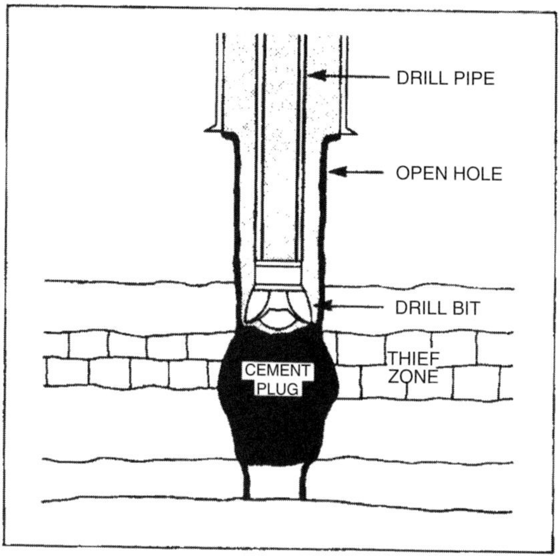

FIGURE 9.28 Lost circulation plug [2].

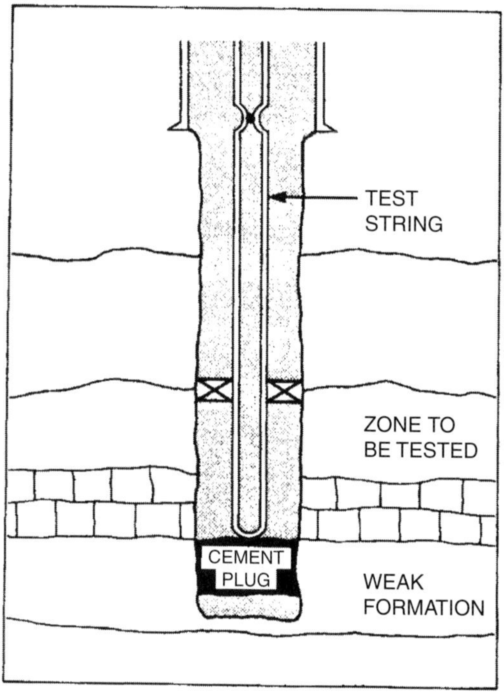

FIGURE 9.29 Openhole completions plug [2].

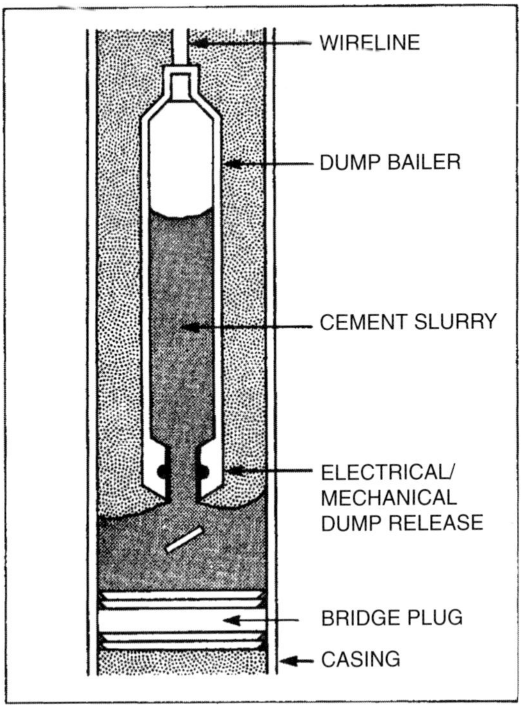

FIGURE 9.30 Dump bailer [2].

wireline and releases its load upon bumping the bottom or a permanent bridge plug set at the desired depth (see Figure 9.30).

Cement plugs fail for the following reasons:

- Lack of hardness (kick-off plug)
- Contaminated cement slurry
- Placed at wrong depth
- Plug migrates from intended depth location due to lack of balancing control

When placing a plug in well the fluid column in the tubing or drillpipe must balance the fluid column in the annulus. If the fluid in the drillpipe is heaviest, then the fluid from the drillpipe (the slurry) will continue down and out the end of the drillpipe and then up into the annulus after pumping has stopped. If the fluid in the annulus is heaviest, then the fluid from the drillpipe will continue downward in the annulus after pumping. The two columns are balanced by using a spacer (or wash) above and below the cement slurry. Since the cement slurry usually has a higher specific weight than the annular fluid, the spacers are normally water.

Example 9.10

Balance a 100-sack cement plug of Class C neat cement slurry when 15 bbl of water are to be run ahead of the slurry. The plug is to be run through 2-in., 4.716-lb/ft, EUE tubing. The depth of placement is 4,000 ft.

1. Determine the top of the plug.
2. Determine the volume of water to run behind cement slurry to balance the 15 bbl of water run ahead of the slurry.
3. Determine the volume of mud to pump to balance plug.

The annular volume per ft is 0.1997 $\mathrm{ft}^3/\mathrm{ft}$. The tubing volume per ft is 0.0217 $\mathrm{ft}^3/\mathrm{ft}$). The yield of a Class C neat cement slurry is 1.32 $\mathrm{ft}^3/\mathrm{sack}$ (see Table 9.2). Thus the volume of the cement slurry is

$$\text{Volume} = 1.32(100) = 132\,\mathrm{ft}^3$$

1. The height H (ft) of the plug is

$$H = \frac{132}{0.1997 - 0.0217} = 742\,\mathrm{ft}$$

Depth to top of plug D (ft) is

$$D = 4{,}000 - 742 = 3{,}258\,\mathrm{ft}.$$

2. The amount of water to place behind the plug to balance the 15 bbl ahead of the cement slurry is

$$\text{Volume} = 15\left(\frac{0.0217}{0.1997}\right) = 1.6\,\mathrm{bbl}$$

3. The height h (ft) of trailing water in the tubing is

$$h = \frac{1.6(42)}{(0.0217)(7.48)} = 414\,\mathrm{ft}$$

The volume of mud to pump to balance plug is

$$\text{Volume} = [4{,}000 - (742 + 414)]\frac{0.0217}{5.6146} = 11\,\mathrm{bbl}$$

References

[1] McCray, A. W., and Cole, F. W., *Oil Well Drilling Technology*, University of Oklahoma Press, Norman, 1959.
[2] *Cementing Technology*, Dowel Schlumberger Publication. 1984.
[3] Smith, D. K., *Cementing*, SPE Publications, Second Printing, 1976.
[4] Rabia, H., *Oilwell Drilling Engineering, Principles and Practices*, Graham and Trotman Ltd, 1985.

[5] API Specification 10: "API Specification for Materials and Testing for well Cementing," January 1982.
[6] ASTM C 150: Standard Specifications for Porland Cement, *Book of ASTM standards*, Part 13, ASTM Publication, 1987.
[7] API Standard 10A: "API Specifications for Oil-Well Cements and Cement Additives," April 1969.
[8] Bourgoyne, A. T., *Applied Drilling* Engineering, SPE Textbook Series, Vol. 2, 1986.
[9] Parker, P. N., Ladd, B. J., Ross, W. M., and Wahl, W. W., "An evaluation of primary cementing technique using low displacement rates," SPE Fall Meeting, 1965, SPE No. 1234.
[10] McLean, R. H., Maury, C. W., and Whitaker, W. W., "Displacement mechanics in primary cementing," *Journal of Petroleum Technology*, February 1967.

CHAPTER

10

Tubing and Tubing String Design

OUTLINE

Tubing string is installed in the hole to protect the casing from erosion and corrosion and to provide control for production. The proper selection of tubing size, steel grade, unit weight and type of connections is of critical importance in a well completion program. Tubing must be designed against failure from tensile/compressive forces, internal and external pressures and buckling.

Tubing is classified according to the outside diameter, the steel grade, unit weight (well thickness), length and type of joints. The API tubing list is given in Tables 10.1 and 10.2.

10.1 API PHYSICAL PROPERTY SPECIFICATIONS

The tubing steel grade data as given by API Specification 5A, 35th Edition (March 1981), Specification 5AX, 12th Edition (March 1982) and Specification 5AC, 13th Edition (March 1982), are listed in Table 10.3.

10.1.1 Dimensions, Weights and Lengths

The following are excerpts from API Specification 5A, 7th Edition, 2001 "API Specification for Casing, Tubing, and Drill Pipe."

TABLE 10.1 ISO/API Tubing List Sizes, Masses, Wall Thickness, Grade and Applicable End Finish

Labels					Nominal linear masses[a,b]				Type of end-finish[c]						
	2			Outside diameter	Non-upset	Ext. upset	Integ.	Wall thick-							
1	NU T&C	EU T&C	IJ	D in.	T&C lb/ft	T&C lb/ft	joint lb/ft	ness t in.	H40	J55	L80	N80 Type 1,Q	C90[d]	T95[d]	P110
1	2	3	4	5	6	7	8	9	10	11	12	13	14	15	16
1.050	1.14	1.20		1.050	1.14	1.20		0.113	PNU	PNU	PNU	PNU	PNU	PNU	
1.050	1.48	1.54		1.050	1.48	1.54		0.154	PU	PU	PU	PU	PU	PU	PU
1.315	1.70	1.80	1.72	1.315	1.70	1.80	1.72	0.133	PNUI	PNUI	PNUI	PNUI	PNUI	PNUI	
1.315	2.19	2.24		1.315	2.19	2.24		0.179	PU	PU	PU	PU	PU	PU	PU
1.660	2.09		2.10	1.660			2.10	0.125	PI	PI					
1.660	2.30	2.40	2.33	1.660	2.30	2.40	2.33	0.140	PNUI	PNUI	PNUI	PNUI	PNUI	PNUI	
1.660	3.03	3.07		1.660	3.03	3.07		0.191	PU	PU	PU	PU	PU	PU	PU
1.900	2.40		2.40	1.900			2.40	0.125	PI	PI					
1.900	2.75	2.90	2.76	1.900	2.75	2.90	2.76	0.145	PNUI	PNUI	PNUI	PNUI	PNUI	PNUI	
1.900	3.65	3.73		1.900	3.65	3.73		0.200	PU	PU	PU	PU	PU	PU	PU
1.900	4.42			1.900	4.42			0.250			P		P	P	
1.900	5.15			1.900	5.15			0.300			P		P	P	
2.063	3.24		3.25	2.063			3.25	0.156	PI	PI	PI	PI	PI	PI	
2.063	4.50			2.063	4.50			0.225	P	P	P	P	P	P	P
$2\frac{3}{8}$	4.00			2.375	4.00			0.167	PN	PN	PN	PN	PN	PN	
$2\frac{3}{8}$	4.60	4.70		2.375	4.60	4.70		0.190	PNU	PNU	PNU	PNU	PNU	PNU	PNU
$2\frac{3}{8}$	5.80	5.95		2.375	5.80	5.95		0.254			PNU	PNU	PNU	PNU	PNU
$2\frac{3}{8}$	6.60			2.375	6.60			0.295			P		P	P	
$2\frac{3}{8}$	7.35	7.45		2.375	7.35	7.45		0.336			PU		PU	PU	
$2\frac{7}{8}$	6.40	6.50		2.875	6.40	6.50		0.217	PNU	PNU	PNU	PNU	PNU	PNU	PNU
$2\frac{7}{8}$	7.80	7.90		2.875	7.80	7.90		0.276			PNU	PNU	PNU	PNU	PNU
$2\frac{7}{8}$	8.60	8.70		2.875	8.60	8.70		0.308			PNU	PNU	PNU	PNU	PNU

$2\frac{7}{8}$	9.35	9.45	2.875	9.35	9.45	0.340			PU		PU	PU	
$2\frac{7}{8}$	10.50		2.875	10.50		0.392			P		P	P	
$2\frac{7}{8}$	11.50		2.875	11.50		0.440			P		P	P	
$3\frac{1}{2}$	7.70		3.500	7.70		0.216	PN	PN	PN	PN	PN	PN	
$3\frac{1}{2}$	9.20	9.30	3.500	9.20	9.30	0.254	PNU	PNU	PNU	PNU	PNU	PNU	PNU
$3\frac{1}{2}$	10.20		3.500	10.20		0.289	PN	PN	PN	PN	PN	PN	
$3\frac{1}{2}$	12.70	12.95	3.500	12.70	12.95	0.375			PNU	PNU	PNU	PNU	PNU
$3\frac{1}{2}$	14.30		3.500	14.30		0.430			P		P	P	
$3\frac{1}{2}$	15.50		3.500	15.50		0.476			P		P	P	
$3\frac{1}{2}$	17.00		3.500	17.00		0.530			P		P	P	
4	9.50		4.000	9.50		0.226	PN	PN	PN	PN	PN	PN	
4	10.70	11.00	4.000		11.00	0.262	PU	PU	PU	PU	PU	PU	
4	13.20		4.000	13.20		0.330			P		P	P	
4	16.10		4.000	16.10		0.415			P		P	P	
4	18.90		4.000	18.90		0.500			P		P	P	
4	22.20		4.000	22.20		0.610			P		P	P	
$4\frac{1}{2}$	12.60	12.75	4.500	12.60	12.75	0.271	PNU	PNU	PNU	PNU	PNU	PNU	
$4\frac{1}{2}$	15.20		4.500	15.20		0.337			P		P	P	
$4\frac{1}{2}$	17.00		4.500	17.00		0.380			P		P	P	
$4\frac{1}{2}$	18.90		4.500	18.90		0.430			P		P	P	
$4\frac{1}{2}$	21.50		4.500	21.50		0.500			P		P	P	
$4\frac{1}{2}$	23.70		4.500	23.70		0.560			P		P	P	
$4\frac{1}{2}$	26.10		4.500	26.10		0.630			P		P	P	

P = Plain end, N = Non-upset threaded and coupled, U = External upset threaded and coupled, I = Integral joint.

[a] Nominal linear masses, threads and coupling (cols. 2, 3, 4) are shown for information only.

[b] The densities of martensitic chromium steels (L80 types 9Cr and 13Cr) are different from carbon steels. The masses shown are therefore not accurate for martensitic chromium steels. A mass correction factor of 0,989 may be used.

[c] Non-upset tubing is available with regular couplings or special bevel couplings. External-upset tubing is available with regular, special-bevel, or special-clearance couplings.

[d] Grade C90 and T95 tubing shall be furnished in sizes, masses, and wall thicknesses as listed above, or as shown on the purchase agreement.

TABLE 10.2 ISO/API Tubing List Sizes, Masses, Wall Thickness, Grade and Applicable End Finish

Labels				Outside diameter	Nominal linear masses[a,b]			Wall thickness t	Type of end-finish[c]						
	2				Non-upset	Ext. upset	Integ.								
1	NU T&C	EU T&C	IJ	D mm	T&C kg/m	T&C kg/m	joint kg/m	mm	H40	J55	L80	N80 Type 1,Q	C90[d]	T95[d]	P110
1	2	3	4	5	6	7	8	9	10	11	12	13	14	15	16
1.050	1.14	1.20	–	26.67	1.70	1.79	–	2.87	PNU	PNU	PNU	PNU	PNU	PNU	–
1.050	1.48	1.54	–	26.67	2.20	2.29	–	3.91	PU	PU	PU	PU	PU	PU	PU
1.315	1.70	1.80	1.72	33.40	2.53	2.68	2.56	3.38	PNUI	PNUI	PNUI	PNUI	PNUI	PNUI	–
1.315	2.19	2.24	–	33.40	3.26	3.33	–	4.55	PU	PU	PU	PU	PU	PU	FU
1.660	2.09	–	2.10	42.16	–	–	3.13	3.18	PI	PI	–	–	–	–	–
1.660	2.30	2.40	2.33	42.16	3.42	3.57	3.47	3.56	PNUI	PNUI	PNUI	PNUI	PNUI	PNUI	–
1.660	3.03	3.07	–	42.16	4.51	4.57	–	4.85	PU	PU	PU	PU	PU	PU	PU
1.900	2.40	–	2.40	48.26	–	–	3.57	3.18	PI	PI	–	–	–	–	–
1.900	2.75	2.90	2.76	48.26	4.09	4.32	4.11	3.68	PNUI	PNUI	PNUI	PNUI	PNUI	PNUI	–
1.900	3.65	3.73	–	48.26	5.43	5.55	–	5.08	PU	PU	PU	PU	PU	PU	PU
1.900	4.42	–	–	48.26	6.58	–	–	6.35	–	–	P	–	P	P	–
1.900	5.15	–	–	48.26	7.66	–	–	7.62	–	–	P	–	P	P	–
2.063	3.24	–	3.25	52.40	–	–	4.84	3.96	PI	PI	PI	PI	PI	PI	–
2.063	4.50	–	–	52.40	–	–	–	5.72	P	P	P	P	P	F	P
$2\frac{3}{8}$	4.00	–	–	60.32	5.95	–	–	4.24	PN	PN	PN	PN	PN	FN	–
$2\frac{3}{8}$	4.60	4.70	–	60.32	6.85	6.99	–	4.83	PNU	PNU	PNU	PNU	PNU	FNU	PNU
$2\frac{3}{8}$	5.80	5.95	–	60.32	8.63	8.85	–	6.45	–	–	PNU	PNU	PNU	PNU	PNU
$2\frac{3}{8}$	6.60	–	–	60.32	9.82	–	–	7.49	–	–	P	–	P	P	–
$2\frac{3}{8}$	7.35	7.45	–	60.32	10.94	11.09	–	8.53	–	–	PU	–	PU	PU	–
$2\frac{7}{8}$	6.40	6.50	–	73.02	9.52	9.67	–	5.51	PNU	PNU	PNU	PNU	PNU	PNU	PNU
$2\frac{7}{8}$	7.80	7.90	–	73.02	11.61	11.76	–	7.01	–	–	PNU	PNU	PNU	PNU	PNU

$2\frac{7}{8}$	8.60	8.70	–	73.02	12.80	12.95	–	7.82	–	–	PNU	PNU	PNU	PNU	PNU
$2\frac{7}{8}$	9.35	9.45	–	73.02	13.91	14.06	–	8.64	–	–	PU	–	PU	PU	–
$2\frac{7}{8}$	10.50	–	–	73.02	15.63	–	–	9.96	–	–	P	–	P	P	–
$2\frac{7}{8}$	11.50	–	–	73.02	17.11	–	–	11.18	–	–	P	–	P	P	–
$3\frac{1}{2}$	7.70	–	–	88.90	11.46	–	–	5.49	PN	PN	PN	PN	PN	PN	–
$3\frac{1}{2}$	9.20	9.30	–	88.90	13.69	13.84	–	6.45	PNU	PNU	PNU	PNU	PNU	PNU	PNU
$3\frac{1}{2}$	10.20	–	–	88.90	15.18	–	–	7.34	PN	PN	PN	PN	PN	PN	–
$3\frac{1}{2}$	12.70	12.95	–	88.90	18.90	19.27	–	9.52	–	–	PNU	PNU	PNU	PNU	PNU
$3\frac{1}{2}$	14.30	–	–	88.90	21.28	–	–	10.92	–	–	P	–	P	P	–
$3\frac{1}{2}$	15.50	–	–	88.90	23.07	–	–	12.09	–	–	P	–	P	P	–
$3\frac{1}{2}$	17.00	–	–	88.90	25.30	–	–	13.46	–	–	P	–	P	P	–
4	9.50	–	–	101.60	14.14	–	–	5.74	PN	PN	PN	PN	PN	PN	–
4	10.70	11.00	–	101.60	–	16.37	–	6.85	PU	PU	PU	PU	PU	PU	–
4	13.20	–	–	101.60	19.64	–	–	8.38	–	–	P	–	P	P	–
4	16.10	–	–	101.60	23.96	–	–	10.54	–	–	P	–	P	P	–
4	18.90	–	–	101.60	28.13	–	–	12.70	–	–	P	–	P	P	–
4	22.20	–	–	101.60	33.04	–	–	15.49	–	–	P	–	P	P	–
$4\frac{1}{2}$	12.60	12.75	–	114.30	18.75	18.97	–	6.88	PNU	PNU	PNU	PNU	PNU	PNU	–
$4\frac{1}{2}$	15.20	–	–	114.30	22.62	–	–	8.56	–	–	P	–	P	P	–
$4\frac{1}{2}$	17.00	–	–	114.30	25.30	–	–	9.65	–	–	P	–	P	P	–
$4\frac{1}{2}$	18.90	–	–	114.30	28.13	–	–	10.92	–	–	P	–	P	P	–
$4\frac{1}{2}$	21.50	–	–	114.30	32.00	–	–	12.70	–	–	P	–	P	P	–
$4\frac{1}{2}$	23.70	–	–	114.30	35.27	–	–	14.22	–	–	P	–	P	P	–
$4\frac{1}{2}$	26.10	–	–	114.30	38.84	–	–	16.00	–	–	P	–	P	P	–

P = Plain end, N = Non-upset threaded and coupled, U = External upset threaded and coupled, I = Integral joint.

[a]Nominal linear masses, threads and coupling (cols. 2, 3, 4) are shown for information only.

[b]The densities of martensitic chromium steels (L80 types 9Cr and 13Cr) are different from carbon steels. The masses shown are therefore not accurate for martensitic chromium steels. A mass correction factor of 0,989 may be used.

[c]Non-upset tubing is available with regular couplings or special bevel couplings. External-upset tubing is available with regular, special-bevel, or special-clearance couplings.

[d]Grade C90 and T95 tubing shall be furnished in sizes, masses, and wall thicknesses as listed above, or as shown on the purchase agreement.

TABLE 10.3 Tubing Steel Grade Data

Steel Grade	Yield Strength, psi Min.	Yield Strength, psi Max.	Min. Tensile Strength, psi	Elongation*
H-40	40,000	80,000	60,000	27
J-55	55,000	80,000	75,000	20
C-75	75,000	90,000	95,000	16
C-95	95,000	110,000	105,000	16
N-80/L-80	80,000	110,000	100,000	16
P-105	105,000	135,000	120,000	15

*The minimum elongation in 2 in. (50.80 mm) should be determined by the following formula:

$$e = 625{,}000 \frac{A^{0.2}}{U^{0.9}}$$

where e = minimum elongation in 2 in. (50:80 mm) in percent rounded to the nearest $\frac{1}{2}$%

A = cross-sectional area of the tensile specimen in in.^{2}, based on a specified outside diameter or nominal specimen width and specified wall thickness, rounded to the nearest 0.01 or 0.75 in.,^{2} whichever is smaller

U = specified tensile strength in psi.

1. *Dimensions and weights.* Pipe shall be furnished in the sizes, wall thicknesses and weights (as shown in Tables 10.4, 10.5 and 10.6) as specified by API Specification 5A: "Casing, Tubing and Drill Pipe."
2. *Diameter.* The outside diameter of 4 in.-tubing and smaller shall be within ±0.79 mm. Inside diameters are governed by the outside diameter and weight tolerances.
3. *Wall thickness.* Each length of pipe shall be measured for conformance to wall-thickness requirements. The wall thickness at any place shall not be less than the tabulated thickness minus the permissible undertolerance of 12.5%. Wall-thickness measurements shall be made with a mechanical caliper or with a properly calibrated nondestructive testing device of appropriate accuracy. In case of dispute, the measurement determined by use of the mechanical caliper shall govern.
4. *Weight.* Each length of tubing in sizes 1.660 in. and larger shall be weighed separately. Lengths of tubing in sizes 1.050 and 1.315 in. shall be weighed either individually or in convenient lots.

   Threaded-and-coupled pipe shall be weighed with the couplings screwed on or without couplings, provided proper allowance is made for the weight of the coupling. Threaded-and-coupled pipe, integral joint pipe and pipe shipped without couplings shall be weighed without thread protectors except for carload weighings, for which proper allowances shall be made for the weight of the thread protectors.

TABLE 10.4 Nonupset Tubing — Dimensions and Weights

1	2	3	4	5	6
Size: Outside Diameter, in. D	Nominal Weight:[1] Threads and Coupling, lb. per ft	Wall Thickness in. t	Inside Diameter in. d	Calculated Weight Plain End lb/ft w_{pe}	Calculated Weight Threads[2] and Coupling lb e_w
*1.050	1.14	0.113	0.824	1.13	0.20
1.315	1.70	0.133	1.049	1.68	0.40
1.660	2.80	0.140	1.380	2.27	0.80
1.900	2.75	0.145	1.610	2.72	0.60
$2\frac{3}{8}$	4.00	0.167	2.041	3.94	1.60
$2\frac{3}{8}$	4.60	0.190	1.995	4.43	1.60
$2\frac{3}{8}$	5.80	0.254	1.867	5.75	1.40
$2\frac{7}{8}$	6.40	0.217	2.441	6.16	3.20
$2\frac{7}{8}$	8.60	0.308	2.259	8.44	2.60
$3\frac{1}{2}$	7.70	0.216	3.068	7.58	5.40
$3\frac{1}{2}$	9.20	0.254	2.992	8.81	5.00
$3\frac{1}{2}$	10.20	0.289	2.922	9.91	4.80
$3\frac{1}{2}$	12.70	0.375	2.750	12.52	4.00
4	9.50	0.226	3.548	9.11	6.20
$4\frac{1}{2}$	12.60	0.271	3.958	12.24	6.00

[1]Nominal weights, threads and coupling (cols. 2), are shown for the purpose of identification in ordering.
[2]Weight gain due to end finishing.
*For information only.
See Figure 10.1.
From API Specification 5A, 35th ed. (March 1981).

5. Calculated weights shall be determined in accordance with the following formula:

$$W_L = (Wpe \times L) + e_w$$

where W_L = calculated weight of a piece of pipe of length L in lb (kg)
Wpe = plain-end weight in lb/ft (kg/m)
L = length of pipe, including end finish, as defined below in ft (m)
e_w = weight gain or loss due to end finishing in lb (kg) (for plain-end pipe, $e_w = 0$)

6. *Length.* Pipe shall be furnished in range lengths conforming to the following as specified on the purchase order: When pipe is furnished with threads and couplings, the length shall be measured to the outer face of the coupling, or if measured without couplings, proper allowance shall

TABLE 10.5 External-Upset tubing—Upset end dimensions and Masses for Group 1, 2 and 3 (See Figure 10.2)

Labels[a]		Size Outside diameter in	Normal linear mass, threaded and coupled[b]	Upset			
				Outside diameter[c] + 0.0625 0 in.	Length from end of pipe to start of taper[d, c] 0 −1 in.	Length from end of pipe to end of taper[e] in.	Length from end of pipe to start of pipe body[e] max. in.
1	2	D	lb/ft	D_4	L_{eu}	L_e	L_b
1	2	3	4	5	6	7	8
1.050	1.20	1.050	1.20	1.315	$2\frac{3}{8}$	–	–
1.050	1.54	1.050	1.20	1.315	$2\frac{3}{8}$	–	–
1.315	1.80	1.315	1.80	1.469	2_	–	–
1.315	2.24	1.315	1.80	1.469	2_	–	–
1.660	2.40	1.660	2.40	1.812	$2\frac{5}{8}$	–	–
1.660	3.07	1.660	2.40	1.812	$2\frac{5}{8}$	–	–
1.900	2.90	1.900	2.90	2.094	$2\frac{11}{16}$	–	–
1.900	3.73	1.900	2.90	2.094	$2\frac{11}{16}$	–	–
$2\frac{3}{8}$	4.70	2.375	4.70	2.594	4.00	6.00	10.00
$2\frac{3}{8}$	5.95	2.375	5.95	2.594	4.00	6.00	10.0C

$2\frac{3}{8}$	7.45	2.375	7.45	2.594	4.00	6.00	10.00
$2\frac{7}{8}$	6.50	2.875	6.50	3.094	$4\frac{1}{4}$	$6\frac{1}{4}$	$10\frac{1}{4}$
$2\frac{7}{8}$	7.90	2.875	7.90	3.094	$4\frac{1}{4}$	$6\frac{1}{4}$	$10\frac{1}{4}$
$2\frac{7}{8}$	8.70	2.875	8.70	3.094	$4\frac{1}{4}$	$6\frac{1}{4}$	$10\frac{1}{4}$
$2\frac{7}{8}$	9.45	2.875	9.45	3.094	$4\frac{1}{4}$	$6\frac{1}{4}$	$10\frac{1}{4}$
$3\frac{1}{2}$	9.30	3.500	9.30	3.750	$4\frac{1}{2}$	$6\frac{1}{2}$	$10\frac{1}{2}$
$3\frac{1}{2}$	12.95	3.500	12.95	3.750	$4\frac{1}{2}$	$6\frac{1}{2}$	$10\frac{1}{2}$
4	11.00	4.000	11.00	4.250	$4\frac{1}{2}$	$6\frac{1}{2}$	$10\frac{1}{2}$
$4\frac{1}{2}$	12.75	4.500	12.75	4.750	$4\frac{3}{4}$	$6\frac{3}{4}$	$10\frac{3}{4}$

[a] Labels are for information and assistance in ordering.

[b] The densities of martensitic chromium steels (L80 9Cr and 13Cr) are different from carbon steels. The masses shown are therefore not accurate for chromium steels. A mass correction factor of 0.989 may be used.

[c] The minimum outside diameter of upset D_4 is limited by the minimum length of full-crest threads. See ISO 10422 or API Spec 5B.

[d] For pup-joints only, the length tolerance on L_{eu} is: ${}^{+4}_{-1}$ in. The length on L_b may be 4 in longer than specified.

[e] For extended length upsets on external upset tubing, add 1 in to the dimensions in columns 5, 6, and 7.

TABLE 10.6 Integral Joint Tubing—Upset end dimensions and Masses for Groups 1, 2 (See Figure 10.3)

Labels				Upset dimensions, in.								
				Pin				Box				
1	2	Outside dia. in D	Nominal linear mass[a] T&C lb/ft	Outside dia.[b] +0.0625 0 D_4	Inside dia.[c] +0.015 0 d_{iu}	Length min. L_{iu}	Length of taper min. m_{iu}	Outside diameter +0.015 −0.025 W_b	Length min. L_{eu}	Length of taper m_{eu}	Diameter of recess Q	Width of face min. b
1	2	3	4	5	6	7	8	9	10	11	12	13
1.315	1.72	1.315	1.72	–	0.970	$1\frac{3}{8}$	$\frac{1}{4}$	1.550	1.750	1	1.378	$\frac{1}{32}$
1.660	2.10	1.660	2.10	–	1.301	$1\frac{1}{2}$	$\frac{1}{4}$	1.880	1.875	1	1.723	$\frac{1}{32}$
1.660	2.33	1.660	2.33	–	1.301	$1\frac{1}{2}$	$\frac{1}{4}$	1.880	1.875	1	1.723	$\frac{1}{32}$
1.900	2.40	1.900	2.40	–	1.531	$1\frac{5}{8}$	$\frac{1}{4}$	2.110	2.000	1	1.963	$\frac{1}{32}$
1.900	2.76	1.900	2.76	–	1.531	$1\frac{5}{8}$	$\frac{1}{4}$	2.110	2.000	1	1.963	$\frac{1}{32}$
2.063	3.25	2.063	3.25	2.094	1.672	$1\frac{11}{16}$	$\frac{1}{4}$	2.325	2.125	1	2.156	$\frac{1}{32}$

[a]Nominal linear masses, upset and threaded, are shown for information only.
[b]The minimum outside diameter D4 is limited by the minimum length of full-crest threads. See ISO 10422 or API Spec SB.
[c]The minimum diameter d_{iu} is limited by the drift test.

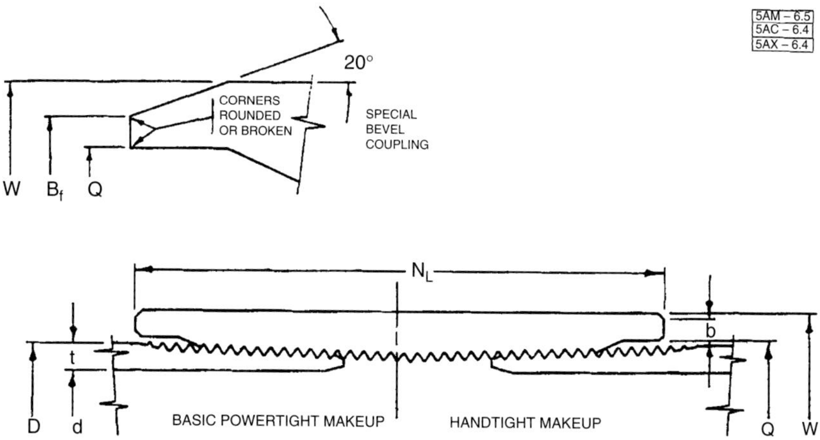

FIGURE 10.1 Non-upset tubing and coupling. (See Table 10.4 for pipe dimensions)

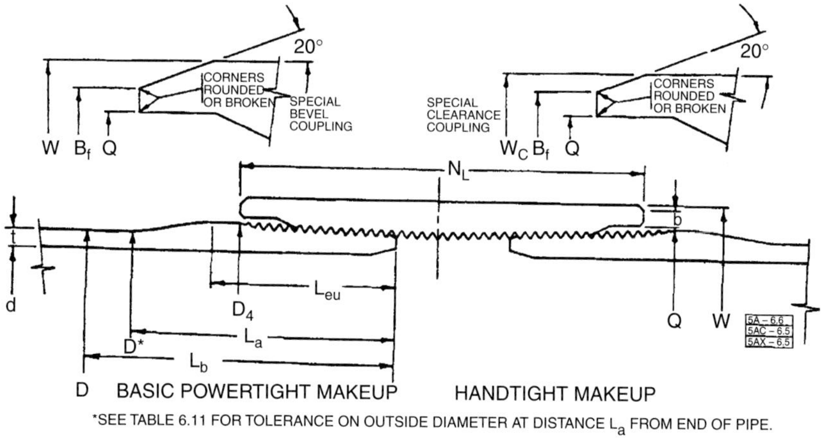

FIGURE 10.2 External-upset tubing and coupling. (See Table 10.5 for pipe dimensions)

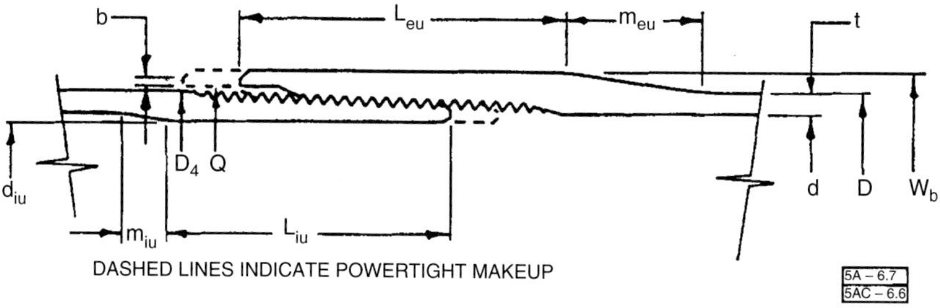

FIGURE 10.3 Integral-joint tubing. (See Table 10.6 for pipe dimensions)

be made to include the length of the coupling. For integral joint tubing, the length shall be measured to the outer face of the box end.

Pipe Range Lengths

	Range 1	Range 2
Total range length*	20–24	28–32
Range length for 100% of carload:†		
Permissible variation, max.	2	2
Permissible length, min.	20	28

*By agreement between purchaser and manufacturer, the total range length for Range 1 tubing may be 6.10–8.53 m.

†Carload tolerances shall not apply to orders less than a carload shipped from the mill. For any carload of pipe shipped from the mill to the final destination without transfer or removal from the car, the tolerance shall apply to each car. For any order consisting of more than a carload and shipped from the mill by rail, but not to the final destination in the rail cars loaded at the mill, the carload tolerances shall apply to the total order, but not to the individual carloads.

10.1.2 Performance Properties

Tubing performance properties, according to API Bulletin 5C2, 18th Edition (March 1982), are given in Table 10.7. Formulas and procedures for calculating the values in Table 10.7 are given in API Bulletin 5C3, 3rd Edition (March 1980).

10.2 RUNNING AND PULLING TUBING

The following are excerpts from "API Recommended Practice for Care and Use of Casing and Tubing," API RP 5C1, 12th Edition (March 1981).

10.3 PREPARATION AND INSPECTION BEFORE RUNNING

1. New tubing is delivered free of injurious defects as defined in API Standard 5A, 5AC and 5AX, and within the practical limits of the inspection procedures therein prescribed. Some users have found that, for a limited number of critical well applications, these procedures do not result in casing sufficiently free of defects to meet their needs for such critical applications. Various nondestructive inspection services have been employed by users to assure that the desired quality of tubing is being run. In view of this practice, it is suggested that the individual user:
   a. Familiarize her or himself with inspection practices specified in the standards and employed by the respective mills and with the definition of "injurious defect" contained in the standards.

TABLE 10.7 Dimensions and Masses for Non-upset, External Upset and Integral Joint Tubing

Labels[a]				Outside dia. D in.	Nominal linear masses[b,c]			Wall thick-ness t in.	Inside dia. d in.	Plain end W_{pe} lb/ft	Calculated mass[c] e_W mass gain or loss due to end finishing[d], lb			
	2											External upset[e]		
1	NU T&C	EU T&C	IJ		Non-upset T&C lb/ft	Ext. upset T&C lb/ft	Integral joint lb/ft				Non-upset	Regular	Special clearance	Integral joint
1	2	3	4	5	6	7	8	9	10	11	12	13	14	15
1.050	1.14	1.20	–	1.050	1.14	1.20	–	0.113	0.824	1.13	0.20	1.40	–	–
1.050	1.48	1.54	–	1.050	1.48	1.54	–	0.154	0.742	1.48	–	1.32	–	–
1.315	1.70	1.80	1.72	1.315	1.70	1.80	1.72	0.133	1.049	1.68	0.40	1.40	–	0.20
1.315	2.19	2.24	–	1.315	2.19	2.24	–	0.179	0.957	2.17	–	1.35	–	–
1.660	2.09	–	2.10	1.660	–	–	2.10	0.125	1.410	2.05	–	–	–	0.20
1.660	2.30	2.40	2.33	1.660	2.30	2.40	2.33	0.140	1.380	2.27	0.80	1.60	–	0.20
1.660	3.03	3.07	–	1.660	3.03	3.07	–	0.191	1.278	3.00	–	1.50	0.20	–
1.900	2.40	–	2.40	1.990	–	–	2.40	0.125	1.650	2.37	–	–	–	0.20
1.900	2.75	2.90	2.76	1.990	2.75	2.90	2.76	0.145	1.610	2.72	0.60	2.00	–	0.20
1.900	3.65	3.73	–	1.990	3.65	3.73	–	0.200	1.500	3.63	–	2.03	–	–
1.900	4.42	–	–	1.990	4.42	–	–	0.250	1.400	4.41	–	–	–	–
1.900	5.15	–	–	1.990	5.15	–	–	0.300	1.300	5.13	–	–	–	–
2.063	3.24	–	3.25	2.063	–	–	3.25	0.156	1.751	3.18	–	–	–	0.20
2.063	4.50	–	–	2.063	–	–	–	0.225	1.613	4.42	–	–	–	–

(Continued)

TABLE 10.7 (*Continued*)

Labels[a]				Outside dia.	Nominal linear masses[b,c]			Wall thick-ness t	Inside dia.	Plain end	Calculated mass[c]			
	2				Non-upset	Ext. upset	Integral joint				e_W mass gain or loss due to end finishing[d], lb			
												External upset[e]		
1	NU T&C	EU T&C	IJ	D in.	T&C lb/ft	T&C lb/ft	lb/ft	in.	d in.	W_{pe} lb/ft	Non-upset	Regular	Special clearance	Integral joint
$2\frac{3}{8}$	4.00	–	–	2.375	4.00	–	–	0.167	2.041	3.94	1.60	–	–	–
$2\frac{3}{8}$	4.60	4.70	–	2.375	4.60	4.70	–	0.190	1.995	4.44	1.60	4.00	2.96	–
$2\frac{3}{8}$	5.80	5.95	–	2.375	5.80	5.95	–	0.254	1.867	5.76	1.40	3.60	2.56	–
$2\frac{3}{8}$	6.60	–	–	2.375	6.60	–	–	0.295	1.785	6.56	–	–	–	–
$2\frac{3}{8}$	7.35	7.45	–	2.375	7.35	7.45	–	0.336	1.703	7.32	–	–	–	–
$2\frac{7}{8}$	6.40	6.50	–	2.875	6.40	6.50	–	0.217	2.441	6.17	3.20	5.60	3.76	–
$2\frac{7}{8}$	7.80	7.90	–	2.875	7.80	7.90	–	0.276	2.323	7.67	2.80	5.80	3.92	–
$2\frac{7}{8}$	8.60	8.70	–	2.875	8.60	8.70	–	0.308	2.259	8.45	2.60	5.00	3.16	–
$2\frac{7}{8}$	9.35	9.45	–	2.875	9.35	9.45	–	0.340	2.195	9.21	–	–	–	–
$2\frac{7}{8}$	1.050	–	–	2.875	10.50	–	–	0.392	2.091	10.40	–	–	–	–
$2\frac{7}{8}$	11.50	–	–	2.875	11.50	–	–	0.440	1.995	11.45	–	–	–	–
$3\frac{1}{2}$	7.70	–	–	3.500	7.70	–	–	0.216	3.068	7.58	5.40	–	–	–
$3\frac{1}{2}$	9.20	9.30	–	3.500	9.20	9.30	–	0.254	2.992	8.81	5.00	9.20	5.40	–

$3\frac{1}{2}$	10.20	–	–	3.500	10.20	–	–	0.289	2.922	9.92	4.80	–	–	–
$3\frac{1}{2}$	12.70	12.95	–	3.500	12.70	12.95	–	0.375	2.750	12.53	4.00	8.20	4.40	–
$3\frac{1}{2}$	14.30	–	–	3.500	14.30	–	–	0.430	2.640	14.11	–	–	–	–
$3\frac{1}{2}$	15.50	–	–	3.500	15.50	–	–	0.476	2.548	15.39	–	–	–	–
$3\frac{1}{2}$	17.00	–	–	3.500	17.00	–	–	0.530	2.440	16.83	–	–	–	–
4	9.50	–	–	4.000	9.50	–	–	0.226	3.548	9.12	6.20	–	–	–
4	10.70	11.00	–	4.000	–	11.00	–	0.262	3.476	10.47	–	10.60	–	–
4	13.20	–	–	4.000	13.20	–	–	0.330	3.340	12.95	–	–	–	–
4	16.10	–	–	4.000	16.10	–	–	0.415	3.170	15.90	–	–	–	–
4	18.90	–	–	4.000	18.90	–	–	0.500	3.000	18.71	–	–	–	–
4	22.20	–	–	4.000	22.20	–	–	0.610	2.780	22.11	–	–	–	–
$4\frac{1}{2}$	12.60	12.75	–	4.500	12.60	12.75	–	0.271	3.958	12.25	6.00	13.20	–	–
$4\frac{1}{2}$	15.20	–	–	4.500	15.20	–	–	0.337	3.826	15.00	–	–	–	–
$4\frac{1}{2}$	17.00	–	–	4.500	17.00	–	–	0.380	3.740	16.77	–	–	–	–
$4\frac{1}{2}$	18.90	–	–	4.500	18.90	–	–	0.430	3.640	18.71	–	–	–	–
$4\frac{1}{2}$	21.50	–	–	4.500	21.50	–	–	0.500	3.500	21.38	–	–	–	–
$4\frac{1}{2}$	23.70	–	–	4.500	23.70	–	–	0.560	3.380	23.59	–	–	–	–
$4\frac{1}{2}$	26.10	–	–	4.500	26.10	–	–	0.630	3.240	26.06	–	–	–	–

[a]Labels are for information and assistance in ordering.
[b]Nominal linear masses, threaded and coupled (cols. 6, 7 and 8) are shown for information only.
[c]The densities of martensitic chromium (L80 Types 9Cr and 13Cr) are different from carbon steels. The masses shown are therefore not accurate for martensitic chromium steels. A mass correction factor of 0.989 may be used.
[d]Mass gain or loss due to end finishing.
[e]The length of the upset may alter the mass gain or loss due to end finishing.

   b. Thoroughly evaluate any nondestructive inspection to be used by him or her on API tubular goods to ensure that the inspection does in fact correctly locate and differentiate injurious defects from other variables that can be and frequently are sources of misleading "defect" signals with such inspection methods.

   Caution: Due to the permissible tolerance on the outside diameter immediately behind the tubing upset, the user is cautioned that difficulties may occur when wraparound seal-type hangers are installed on tubing manufactured on the high side of the tolerance; therefore, it is recommended that the user select the joint of tubing to be installed at the top of the string.

2. All tubing, whether new, used or reconditioned, should always be handled with thread protectors in place. Tubing should be handled at all times on racks or on wooden or metal surfaces free of rocks, sand or dirt other than normal drilling mud. When lengths of tubing are inadvertently dragged in the dirt, the threads should be recleaned.
   a. Before running in the hole for the first time, tubing should be drifted with an API drift mandrel to ensure passage of pumps, swabs and packers.
   b. Elevators should be in good repair and should have links of equal length.
   c. Slip-type elevators are recommended when running special clearance couplings, especially those beveled on the lower end.
   d. Elevators should be examined to note if latch fitting is complete.
   e. Spider slips that will not crush the tubing should be used. Slips should be examined before using to see that they are working together.

      Note: Slip and tong marks are injurious. Every possible effort should be made to keep such damage at a minimum by using proper up-to-date equipment.
   f. Tubing tongs that will not crush the tubing should be used on the body of the tubing and should fit properly to avoid unnecessary cutting of the pipe wall. Tong dies should fit properly and conform to the curvature of the tubing. The use of pipe wrenches is not recommended.
   g. The following precautions should be taken in the preparation of tubing threads:
      1. Immediately before running, remove protectors from both field end and coupling end and clean threads thoroughly, repeating as additional rows become uncovered.
      2. Carefully inspect the threads. Those found damaged, even slightly, should be laid aside unless satisfactory means are available for correcting thread damage.

3. The length of each piece of tubing should be measured prior to running. A steel tape calibrated in decimal feet to the nearest 0.01 ft should be used. The measurement should be made from the outermost face of the coupling or box to the position on the externally threaded end where the coupling or the box stops when the joint is made up powertight. The total of the individual length so measured will represent the unloaded length of the tubing string.
4. Place clean protectors on the field end of the pipe so that the thread will not be damaged while rolling the pipe onto the rack and pulling it into the derrick.
5. Check each coupling for makeup. If the stand off is abnormally great, check the coupling for tightness. Loose couplings should be removed, the threads thoroughly cleaned and fresh compound should be applied over the entire thread surfaces. Then the coupling should be replaced and tightened before pulling the tubing into the derrick.
6. Before stabbing, liberally apply thread compound to the entire internally and externally threaded areas. It is recommended that high-pressure, modified thread compound as specified in API Bulletin 5A2: "Bulletin on Thread Compounds" be used except in special cases where severe conditions are encountered. In these special cases it is recommended that high-pressure silicone thread compound as specified in Bulletin 5A2 be used.

h. For high-pressure or condensate wells, additional precautions to insure tight joints should be taken as follows:

1. Couplings should be removed and both the mill-end pipe thread and coupling thread thoroughly cleaned and inspected. To facilitate this operation, tubing may be ordered with couplings handling tight, which is approximately one turn beyond hand tight, or may be ordered with the couplings shipped separately.
2. Thread compound should be applied to both the external and internal threads, and the coupling should be reapplied handling tight. Field-end threads and the mating coupling threads should have thread compounds applied just before stabbing.

10.3.1 Stabbing, Making Up and Lowering

1. Do not remove thread protector from field end of tubing until ready to stab.

2. If necessary, apply thread compound over entire surface of threads just before stabbing.
3. In stabbing, lower tubing carefully to avoid injuring threads. Stab vertically, preferably with the assistance of someone on a stabbing board. If the tubing tilts to one side after stabbing, lift up, clean and correct any damaged thread with a three-cornered file, then carefully remove any filings and reapply compound over thread surface. Intermediate supports may be placed in the derrick to limit bowing of the tubing.
4. After stabbing, start screwing the pipe together by hand or apply regular or power tubing tongs slowly. To prevent galling when making up connections in the field, the connections should be made up to a speed not to exceed 25 rpm. Power tubing tongs are recommended for high-pressure or condensate wells to ensure uniform makeup and tight joints. Joints should be made up tight, approximately two turns beyond the hand-tight position, with care being taken not to gall the threads.

10.3.2 Field Makeup

1. Joint life of tubing under repeated field makeup is inversely proportional to the field makeup torque applied. Therefore, in wells where leak resistance is not a great factor, minimum field makeup torque values should be used to prolong joint life. Table 10.8 contains recommended optimum makeup torque values for nonupset, external upset, and integral joint tubing, based on 1% of the calculated joint pullout strength determined from the joint pull-out strength formula for eight-round thread casing in Bulletin 5C3. Minimum torque values listed are 75% of optimum values, and maximum torque values listed are 125% of optimum values. All values are rounded to the nearest 10 ft-lb. The torque values listed in Table 10.8 apply only to tubing with zinc-plated couplings. When making up connections with tin-plated couplings, 80% of the listed value can be used as a guide.
2. Spider slips and elevators should be cleaned frequently, and slips should be kept sharp.
3. Finding bottom should be accomplished with extreme caution. Do not set tubing down heavily.

10.3.3 Pulling Tubing

1. A caliper survey prior to pulling a worn string of tubing will provide a quick means of segregating badly worn lengths for removal.
2. Break-out tongs should be positioned close to the coupling. Hammering the coupling to break the joint is an injurious practice.
3. Great care should be exercised to disengage all of the thread before lifting the tubing out of the coupling. Do not jump the tubing out of the coupling.

TABLE 10.8 Recommended Tubing Makeup Torque

1	2	3	4	5	6	7	8	9	10	11	12	13	14
	Nominal Weight lb. per ft.				Torque, ft-lb								
Size: Outside Diameter	Threads and Coupling				Non-Upset			Upset			Integral Joint		
in.	Non-Upset	Upset	Integral Joint	Grade	Opt.	Min.	Max.	Opt.	Min.	Max	Opt.	Min.	Max.
1.050	1.14	1.20	–	H-40	140	110	180	460	350	580	–	–	–
	1.14	1.20	–	J-55	180	140	230	600	450	750	–	–	–
	1.14	1.20	–	C-75	230	170	290	780	590	980	–	–	–
	1.14	1.20	–	L-80	240	180	300	810	610	1010	–	–	–
	1.14	1.20	–	N-80	250	190	310	830	620	1040	–	–	–
1.315	1.70	1.80	1.72	H-40	210	160	260	440	330	550	310	230	390
	1.70	1.80	1.72	J-55	270	200	340	570	430	710	400	300	500
	1.70	1.80	1.72	C-75	360	270	450	740	560	930	520	390	650
	1.70	1.80	1.72	L-80	370	280	460	760	570	950	540	400	680
	1.70	1.80	1.72	N-80	380	290	480	790	590	990	550	410	690
1.660	–	–	2.10	H-40	–	–	–	–	–	–	380	280	480
	2.30	2.40	2.33	H-40	270	200	340	530	400	660	380	280	480
	–	–	2.10	J-55	–	–	–	–	–	–	500	380	630
	2.30	2.40	2.33	J-55	350	260	440	690	520	860	500	380	630
	2.30	2.40	2.33	C-75	460	350	580	910	680	1140	650	490	810
	2.30	2.40	2.33	L-80	470	350	590	940	710	1180	670	500	850
	2.30	2.40	2.33	N-80	490	370	610	960	720	1200	690	520	860
1.900	–	–	2.40	H-40	–	–	–	–	–	–	450	340	560
	2.75	2.90	2.76	H-40	320	240	400	670	500	840	450	340	560
	–	–	2.40	J-55	–	–	–	–	–	–	580	440	730

(Continued)

TABLE 10.8 (*Continued*)

1	2	3	4	5	6	7	8	9	10	11	12	13	14
Size: Outside Diameter in.	Nominal Weight lb. per ft.			Grade	Torque, ft-lb								
	Threads and Coupling		Integral Joint		Non-Upset			Upset			Integral Joint		
	Non-Upset	Upset			Opt.	Min.	Max.	Opt.	Min.	Max	Opt.	Min.	Max.
	2.75	2.90	2.76	J-55	410	310	510	880	660	1100	580	440	730
	2.75	2.90	2.76	C-75	540	410	680	1150	860	1440	760	570	950
	2.75	2.90	2.76	L-80	560	420	700	1190	890	1490	790	590	990
	2.75	2.90	2.76	N-80	570	430	710	1220	920	1530	810	610	1010
2.063	–	–	3.25	H-40	–	–	–	–	–	–	570	430	710
	–	–	3.25	J-55	–	–	–	–	–	–	740	560	920
	–	–	3.25	C-75	–	–	–	–	–	–	970	730	1210
	–	–	3.25	L-80	–	–	–	–	–	–	1010	760	1260
	–	–	3.25	N-80	–	–	–	–	–	–	1030	770	1290
$2\frac{3}{8}$	4.00	–	–	H-40	470	350	590	–	–	–	–	–	–
	4.60	4.70	–	H-40	560	420	700	990	740	1240	–	–	–
	4.00	–	–	J-55	610	460	760	–	–	–	–	–	–
	4.60	4.70	–	J-55	730	550	910	1290	970	1610	–	–	–
	4.00	–	–	C-75	800	600	1000	–	–	–	–	–	–
	4.60	4.70	–	C-75	960	720	1200	1700	1280	2130	–	–	–
	5.80	5.95	–	C-75	1380	1040	1730	2120	1590	2650	–	–	–
	4.00	–	–	L-80	830	620	1040	–	–	–	–	–	–
	4.60	4.70	–	L-80	990	740	1240	1760	1320	2200	–	–	–
	5.80	5.95	–	L-80	1420	1070	1780	2190	1640	2740	–	–	–
	4.00	–	–	N-80	850	640	1060	–	–	–	–	–	–
	4.60	4.70	–	N-80	1020	770	1280	1800	1350	2250	–	–	–
	5.80	5.95	–	N-80	1460	1100	1830	2240	1680	2800	–	–	–

	4.60	4.70	–	P-105	1280	960	1600	2270	1700	2840	–	–	–
	5.80	5.95	–	P-105	1840	1380	2300	2830	2120	3540	–	–	–
$2\frac{7}{8}$	6.40	6.50	–	H-40	800	600	1000	1250	940	1560	–	–	–
	6.40	6.50	–	J-55	1050	790	1310	1650	1240	2060	–	–	–
	6.40	6.50	–	C-75	1380	1040	1730	2170	1630	2710	–	–	–
	8.60	8.70	–	C-75	2090	1570	2610	2850	2140	3560	–	–	–
	6.40	6.50	–	L-80	1430	1070	1790	2250	1690	2810	–	–	–
	8.60	8.70	–	L-80	2160	1620	2700	2950	2210	3690	–	–	–
	6.40	6.50	–	N-80	1470	1100	1840	2300	1730	2880	–	–	–
	8.60	8.70	–	N-80	2210	1660	2760	3020	2270	3780	–	–	–
	6.40	6.50	–	P-105	1850	1390	2310	2910	2180	3640	–	–	–
	8.60	8.70	–	P-105	2790	2090	3490	3810	2860	4760	–	–	–
$3\frac{1}{2}$	7.70	–	–	H-40	920	690	1150	–	–	–	–	–	–
	9.20	9.30	–	H-40	1120	840	1400	1730	1300	2160	–	–	–
	10.20	–	–	H-40	1310	980	1640	–	–	–	–	–	–
	7.70	–	–	J-55	1210	910	1510	–	–	–	–	–	–
	9.20	9.30	–	J-55	1480	1110	1850	2280	1710	2850	–	–	–
	10.20	–	–	J-55	1720	1290	2150	–	–	–	–	–	–
	7.70	–	–	C-75	1600	1200	2000	–	–	–	–	–	–
	9.20	9.30	–	C-75	1950	1460	2440	3010	2260	3760	–	–	–
	10.20	–	–	C-75	2270	1700	2840	–	–	–	–	–	–
	12.70	12.95	–	C-75	3030	2270	3790	4040	3030	5050	–	–	–
	7.70	–	–	L-80	1660	1250	2080	–	–	–	–	–	–
	9.20	9.30	–	L-80	2030	1520	2540	3130	2350	3910	–	–	–
	10.20	–	–	L-80	2360	1770	2950	–	–	–	–	–	–
	12.70	12.95	–	L-80	3140	2360	3930	4200	3150	5250	–	–	–
	7.70	–	–	N-80	1700	1280	2130	–	–	–	–	–	–
	9.20	9.30	–	N-80	2070	1550	2590	3200	2400	4000	–	–	–

(Continued)

TABLE 10.8 (*Continued*)

1	2	3	4	5	6	7	8	9	10	11	12	13	14
	Nominal Weight lb. per ft.				Torque, ft-lb								
Size: Outside Diameter in.	Threads and Coupling		Integral Joint	Grade	Non-Upset			Upset			Integral Joint		
	Non-Upset	Upset			Opt.	Min.	Max.	Opt.	Min.	Max	Opt.	Min.	Max.
	10.20	–	–	N-80	2410	1810	3010	–	–	–	–	–	–
	12.70	12.95	–	N-80	3210	2410	4010	4290	3220	5360	–	–	–
	9.20	9.30	–	P-105	2620	1970	3280	4050	3040	5060	–	–	–
	12.70	12.95	–	P-105	4060	3050	5080	5430	4070	6790	–	–	–
4	9.50	–	–	H-40	940	710	1180	–	–	–	–	–	–
	–	11.00	–	H-40	–	–	–	1940	1460	2430	–	–	–
	9.50	–	–	J-55	1240	930	1550	–	–	–	–	–	–
	–	11.00	–	J-55	–	–	–	2560	1920	3200	–	–	–
	9.50	–	–	C-75	1640	1230	2050	–	–	–	–	–	–
	–	11.00	–	C-75	–	–	–	3390	2540	4240	–	–	–
	9.50	–	–	L-80	1710	1280	2140	–	–	–	–	–	–
	–	11.00	–	L-80	–	–	–	3530	2650	4410	–	–	–
	9.50	–	–	N-80	1740	1310	2180	–	–	–	–	–	–
	–	11.00	–	N-80	–	–	–	3600	2700	4500	–	–	–
$4\frac{1}{2}$	12.60	12.75	–	H-40	1320	990	1650	2160	1620	2700	–	–	–
	12.60	12.75	–	J-55	1740	1310	2180	2860	2150	3580	–	–	–
	12.60	12.75	–	C-75	2300	1730	2880	3780	2840	4730	–	–	–
	12.60	12.75	–	L-80	2400	1800	3000	3940	2960	4930	–	–	–
	12.60	12.75	–	N-80	2440	1830	3050	4020	3020	5030	–	–	–

Source: From API Recommended Practice, 5C1, 16th ed. (May 1998).

4. Tubing stacked in the derrick should be set on a firm wooden platform without the bottom thread protector since the design of most protectors is not such as to support the joint or stand without damage to the field thread.
5. Protect threads from dirt or injury when the tubing is out of the hole.
6. Tubing set back in the derrick should be properly supported to prevent undue bending. Tubing that is $2\frac{3}{8}$ in. OD and larger preferably should be pulled in stands approximately 60 ft long or in doubles of range 2. Stands of tubing 1.900-in. OD or smaller and stands longer than 60 ft should have intermediate support.

10.3.4 Causes of Tubing Trouble

The more common causes of tubing troubles are as follows:

1. Improper selection for strength and life required.
2. Insufficient inspection of finished product at the mill and in the yard.
3. Careless loading, unloading and cartage.
4. Damaged threads resulting from protectors loosening and falling off.
5. Lack of care in storage to give proper protection.
6. Excessive hammering on couplings.
7. Use of worn-out and wrong types of handling equipment.
8. Nonobservance of proper rules in running and pulling tubing.
9. Coupling wear and rod cutting.
10. Excessive sucker rod breakage.
11. Fatigue that often causes failure at the last engaged thread. There is no positive remedy, but using external upset tubing in place of nonupset tubing greatly delays the start of this trouble.
12. Replacement of worn couplings with non-API couplings.
13. Dropping a string, even a short distance.
14. Leaky joints, under external or internal pressure.
15. Corrosion. Both the inside and outside of tubing can be damaged by corrosion. The damage is generally in the form of pitting, box wear, stress-corrosion cracking, and sulfide stress cracking, but localized attack like corrosion–erosion, ringworm and caliper tracks–can also occur. Since corrosion can result from many causes and influences and can take different forms, no simple and universal remedy can be given for control. Each problem must be treated individually, and the solution must be attempted in light of known factors and operating conditions.
    - **a.** Where internal or external tubing corrosion is known to exist and corrosive fluids are being produced, the following measures can be employed:
        1. In flowing wells the annulus can be packed off and the corrosive fluid confined to the inside of the tubing. The inside

of the tubing can be protected with special liners, coatings or inhibitors. Under severe conditions, special alloy steel or glass-reinforced plastics may be used. Alloys do not eliminate corrosion. When H_2S is present in the well fluids, tubing of high-yield strength may be subject to sulfide corrosion cracking. The concentration of H_2S necessary to cause cracking in different strength materials is not yet well defined. Literature on sulfide corrosion or persons competent in this field should be consulted.

2. In pumping and gas-lifting wells, inhibitors introduced via the casing-tubing annulus afford appreciable protection. In this type of completion, especially in pumping wells, better operating practices can also aid in extending the life of tubing; viz., through the use of rod protectors, rotation of tubing and longer and slower pumping strokes.

b. To determine the value and effectiveness of the above practices and measures, cost and equipment failure records can be compared before and after application of control measures.

c. In general, all new areas should be considered as being potentially corrosive, and investigations should be initiated early in the life of a field, and repeated periodically, to detect and localize corrosion before it has done destructive damage. Where conditions favorable to corrosion exist, a qualified corrosion engineer should be consulted.

10.3.5 Selection of Wall Thickness and Steel Grade of Tubing

Tubing design relies on the selection of the most economical steel grades and wall thicknesses (unit weight of tubing) that will withstand, without failure, the forces to which tubing will be exposed throughout the expected tubing life.

Tubing must be designed on:

- Collapse
- Burst
- Tension
- Possibility of permanent corkscrewing

Tubing string design is very much the same as for casing. For shallow and moderately deep holes, uniform strings are preferable; however, in deep wells, a tapered tubing can be desirable. A design factor (safety factor) for tension should be about 1.6. The collapse design factor must not be less than 1.0, assuming an annulus filled up with fluid and tubing empty inside. The design factor for burst should not be less than 1.1.

10.3.6 Tubing Elongation/Contraction Due to the Effect of Changes in Pressure and Temperature

During the service life of a well, tubing can experience various combinations of pressures and temperatures that result in tubing length changes. The four basic effects to consider are as follows:

1. Piston effect
2. Helical buckling effect
3. Ballooning and reverse ballooning effect
4. Temperature change effect

The tubing movement due to piston effect is

$$\Delta L_1 = \frac{[(A_p - A_i)\Delta P_i - (A_p - A_o)\Delta P_o]}{EA_s} L \tag{10.1}$$

Tubing movement due to helical buckling is

$$\Delta L_2 = -\frac{r^2 F_f^2}{8EI(W_s + W_i - W_o)} \tag{10.2}$$

where

$$I = \frac{\pi}{64}(OD^4 - ID^4) \qquad F_f = (\Delta P_i - \Delta P_o)A_p$$

Note: If $F_f < 0$, $\Delta L_2 = 0$.

The tubing movement due to ballooning effect is

$$\Delta L_3 = -\frac{v}{E}\frac{\Delta\rho_i - R^2\Delta\rho_o - \frac{1+2v}{2v}\delta}{R^2 - 1}L^2 - \frac{2v}{E}\frac{\Delta p_i - R^2\Delta p_o}{R^2 - 1}L \tag{10.3}$$

The tubing movement due to change in temperature is

$$\Delta L_4 = \beta L \Delta T \tag{10.4}$$

Two approaches can be used to handle tubing movement:

1. Provide seals of enough length.
2. Slack off enough weight of tubing to prevent movement.

For practical purposes a combination of the approaches mentioned above can be applicable.

$$\Delta L_5 = \frac{LF_s}{EA_s} + \frac{r^2 F_s^2}{8EI(W_s + W_i - W_o)} \tag{10.5}$$

Notations used in the above equations are

A_i = area corresponding to tubing ID in in.^{2}
A_o = area corresponding to tubing OD in in.^{2}
A_p = area corresponding to packer bore in in.^{2}
A_s = cross-sectional area of the tubing wall in in.^{2}

F_f = fictitious force in presence of no restraint in the packer in lb_f
I = moment of inertia of tubing cross-section with respect to its diameter in $in.^4$
L = length of tubing in in.
P_i, P_o = pressure inside and outside the tubing at the packer level respectively in psi
$\Delta P_i, \Delta P_o$ = change in pressure inside and outside of tubing at the packer level in psi
$\Delta p_i, \Delta p_o$ = change in pressure inside and outside of tubing at the surface in psi
R = ratio OD/ID of the tubing
r = tubing and casing radial clearance in in.
$\Delta \rho_i$ = change in density of liquid in the tubing in $lbm/in.^3$
$\Delta \rho_o$ = change in density of liquid in annulus in $lbm/in.^3$
β = coefficient of thermal expansion of the tubing material (for steel, $\beta = 6.9 \times 10^{-6}/°F$)
W_s = average (i.e., including coupling) weight of tubing per unit length in lb/in.
W_i = weight of liquid in the tubing per unit length in lb/in.
W_o = weight of outside liquid displaced per unit length
δ = drop of pressure in the tubing due to flow per unit length in psi/in.
ΔT = change in average tubing temperature in.°F
ν = Poisson's ratio of the tubing material (for steel, $\nu = 0.3$)
E = Young's modulus (for steel, $E = 30 \times 10^6$ psi)

Nomenclature used is in the paper by A. Lubinski et al. API Bulletin 5C2, 20th ed. (May 1987).

Example 10.1

Calculate the expected movement of tubing under conditions as specified below. Initially both tubing and annulus are filled with a crude of 30°API. Thereafter, the crude in the tubing is replaced by a 15-lb/gal cement slurry to perform a squeeze cementing operation. While the squeeze cementing job is performed, pressures $p_i = 5{,}000$ psi and $p_o = 1{,}000$ psi are applied at the surface on the tubing and annulus respectively.

Tubing: $2\frac{7}{8}$ in.; 6.5 lb/ft
Casing: 7 in.; 32 lb/ft (r = 1.61 in.)
$A_o = 6.49\,in.^2$, $A_i = 4.68\,in.^2$, $A_s = 1.81\,in.^2$
Ratio of OD/ID of tubing, R = 1.178
$A_p = 8.30\,in.^2$
Length of tubing = 10,000 ft (120,000 in.)
Average change in temperature: −20 degF
Pressure drop due to flow is disregarded ($\delta = 0$)

Solution

Pressure changes are the following. At the surface the pressures are $\Delta p_i = 5{,}000$ psi and $\Delta p_o = 1{,}000$ psi. At the packer level, $\Delta P_i = 9{,}000$ psi and $\Delta P_o = 1{,}000$ psi. Liquid density changes are $\Delta \rho_i = 0.0332$ psi/in. and $\Delta \rho_o = 0.0$ psi/in. The unit weight of tubing is

$$W = W_s + W_i - W_o = \frac{6.5}{12} + \frac{1.5 \times 4.68}{231} - \frac{0.876 \times 8.34 \times 6.49}{231} = 0.64\,\text{lb/in.}$$

The moment of inertia is $I = 1.61$ in.^{4} The tubing movement due to piston effect is

$$\begin{aligned}\Delta L_1 &= \frac{120{,}000}{30 \times 10^6}[(8.3-4.68)9{,}000 - (8.3-6.49)1{,}000] \\ &= -68.0\,\text{in.}\end{aligned}$$

The tubing movement due to buckling effect is

$$\begin{aligned}\Delta L_2 &= \frac{(1.61)^2 \times (8.3)^2(9{,}000-1{,}000)^2}{8 \times 30 \times 10^6 \times 1.61 \times 0.64} \\ &= -46.23\,\text{in.}\end{aligned}$$

The tubing movement due to ballooning effect is

$$\begin{aligned}L_3 &= \frac{0.3}{30 \times 10^6}\frac{0.0332-0}{1.178^2-1}(120{,}000)^2 - \frac{2 \times 0.3}{30 \times 10^6} \times \frac{5{,}000-(1.178^2)1{,}000}{1.178^2-1}120{,}000 \\ &= -34.69\,\text{in.}\end{aligned}$$

The temperature effect is

$$\Delta L_4 = -6.9 \times 10^{-6} \times 120{,}000 \times 20 = -16.56\,\text{in.}$$

The total expected tubing movement is

$$\Delta L_6 = \Delta L_1 + \Delta L_2 + \Delta L_3 + \Delta L_4 = -165.48\,\text{in.}$$

10.3.7 Packer-To-Tubing Force

Certain types of packers permit no tubing motion in either direction. Depending upon operational conditions, a tubing can be landed either in compression (slack off) or tension (pull up). Landing tubing in compression is desirable if the expected tubing movement would produce tubing shortening while landing in tension to compensate the expected tubing elongation.

Restraint of the tubing in the packer results in a packer-to-tubing force. To find the expected packer-to-tubing force, the following sequence of calculations is applicable:

$$F_f = A_p(P_i - P_o) \tag{10.6}$$

$$F_a = (A_p - A_i)P_i - (A_p - A_o)P_o \tag{10.7}$$

$$\Delta L_p = -\Delta L_6 \tag{10.8}$$

$$\Delta L_f = -\frac{LF_f}{EA_s} - \frac{r^2 F_f^2}{8EI(W_s + W_i - W_o)} \tag{10.9}$$

$$\Delta\widehat{L}_f = \Delta L_f + \Delta L_p \tag{10.10}$$

If $\Delta\widehat{L}_f$ is positive, then

$$\widehat{F}_f = -\frac{\Delta\widehat{L}_f EA_s}{L} \tag{10.11}$$

If $\Delta\widehat{L}_f$ is negative, then

$$\widehat{F}_f = \frac{4I(W_s + W_i - W_o)}{A_s r^2} \times \left[-L \pm \left(L^2 - \frac{A_s^2 r^2 E \Delta\widehat{L}_f}{2I(W_s + W_i - W_o)} \right)^{0.5} \right] \tag{10.12}$$

and finally

$$F_p = \widehat{F}_f - F_f \tag{10.13}$$

Upon determining the packer to tubing force F_p, the actual force $\widehat{F}_a$ immediately above the packer is given by

$$\widehat{F}_a = F_a + F_p \tag{10.14}$$

The symbols used are

F_a = actually existing pressure force at the lower end of tubing subjected to no restraint in the packer in lb_f
F_f = fictitious force in presence of no restraint in the packer in lb_f
$\widehat{F}_a$ = actually existing force at the lower end of tubing in lb_f
$\widehat{F}_f$ = fictitious force in presence of packer restraint in lb_f
ΔL_6 = overall tubing length change in in.
ΔL_p = length change necessary to bring the end of the tubing to the packer

Other symbols are as previously used.

Example 10.2

The operating conditions are the same as those described in Example 10.1. Assume that the packer does not permit any tubing movement at the packer setting depth (10,000 ft). Since tubing shortening is expected, a 20,000-lb force is slacked off before the squeeze job. Find the tubing-to-packer force.

Solution

Length change due to slack-off force is

$$\Delta L_s = \frac{120{,}000 \times 20{,}000}{30 \times 10^6 \times 1.81} + \frac{(1.61)^2 \times (20{,}000)^2}{8 \times 30 \times 10^6 \times 1.61 \times 0.64}$$
$$= 48.39 \text{ in.}$$

The overall tubing length change is

$$\Delta L_6 = -165.48 + 48.39 = -117.09\,\text{in.}$$

The fictitious and actual forces are

$$F_f = 8.3(12{,}800 - 4{,}800) = 66{,}400\,\text{lb}_f$$

$$F_a = 12{,}800(8.3 - 4.68) - 4{,}800(8.3 - 6.49) = 37{,}648\,\text{lb}_f$$

Note: A positive sign indicates a compressive-type force, a negative sign indicates a tensional force.

$$\Delta L_p = -\Delta L_6 = 115.5\,\text{in.}$$

$$\Delta\widehat{L}_f = -\frac{120{,}000 \times 66{,}400}{30 \times 10^6 \times 1.81} - \frac{(1.61^2) \times (66{,}400^2)}{8 \times 30 \times 10^6 \times 1.61 \times 0.64} = -192.9\,\text{in.}$$

$$\Delta\widehat{L}_f = -192.9 + 115.5 = -77.4\,\text{in.}$$

Since $\Delta\widehat{L}_f$ is negative, the force $\widehat{F}_f$ is calculated from Equation 10.12:

$$\widehat{F}_f = \frac{4 \times 1.61 \times 0.66}{1.81 \times (1.61^2)}\left[-120{,}000 + \left((120{,}000)^2 - \frac{(1.81)^2(1.61^2) \times 30 \times 10^6(-77.4)}{2 \times 1.61 \times 0.64}\right)^{0.5}\right] = 30{,}050.45\,\text{lb}_f$$

Since the force $\widehat{F}_f$ is positive (compressive), a helical buckling of tubing is expected above the packer.

The tubing-to-packer force is

$$F_p = \widehat{F}_f - F_f = 30{,}050.45 - 66{,}400 = 36{,}349.55\,\text{lb}_f$$

10.3.8 Permanent Corkscrewing

To ensure that permanent corkscrewing will not occur, the following inequalities must be satisfied:

$$\left[3\left(\frac{P_i - P_o}{R^2 - 1}\right)^2 + \left(\frac{P_i - R^2 P_o}{R^2 - 1} + \sigma_a \pm \sigma_b\right)^2\right]^{0.5} \leq Ym \quad (10.15)$$

$$\left[3\left(\frac{R^2(P_i - P_o)}{R^2 - 1}\right)^2 + \left(\frac{P_i - R^2 P_o}{R^2 - 1} + \sigma_a \pm \frac{\sigma_b}{R}\right)^2\right]^{0.5} \leq Ym \quad (10.16)$$

where

$$\sigma_a - \frac{\widehat{F}_a}{A_s} \quad \text{and} \quad \sigma_b = \frac{D \times r}{4I}\widehat{F}_f$$

All symbols in Equations 10.15 and 10.16 are as previously stated.

10.4 PACKERS

A packer is simply a mechanical means of forming a seal between two strings of pipe or between the pipe and the open hole. This generated seal isolates the tubing from the casing or open hole, preventing vertical flow of fluids or gases.

To qualify as a packer, a tool needs only a sealing means and a method of maintaining an opening through the middle. Typically, most packers available today incorporate the following components in their design.

Slips to anchor the packer in the casing or open hole

Cone to support the slips and create a wedge to enhance anchoring capability

Elastomer elements which expand to seal the area between the packer and the casing or open hole

Mandrel which maintains a consistent bore and typically has an ID compatible with the tubular used. Outer body components are supported by the mandrel.

Setting mechanism which initiates the setting sequence when desired during installation.

Typically, packers are identified by their classification (permanent or retrievable), application (production or service), and setting method (mechanical, hydraulic or wireline).

Permanent packers do not have releasing mechanisms and are not retrievable from surface. When a permanent packer is set, the tubing can be removed from the well without releasing the packer. Permanent packers can only be removed by destroying their integrity through milling or drilling.

Retrievable packers are in integral part of the tubing string. The packer is set, released and pulled with the tubing string. The tubing string cannot be removed from the well bore without pulling the packer unless a on/off tool or similar mechanism is used with the packer.

Production packers are designed to meet general or specific well conditions and provide a long operational life in those conditions. They are manufactured from a variety of materials to meet operational life expectancy in different environments. Production packers are used for a number of reasons, which can be summarized into four categories, which are: protecting the casing string, maximum safety and control, energy conservation, and improve productivity.

10.4.1 Protecting the Casing

Since the casing string is cemented in the hole, it is considered permanent and non-replaceable. A string of replaceable tubing is normally run inside the casing to control and provide the conduit necessary to bring the producible fluids to the surface.

Damaged tubing is relatively easy to remedy as it can be pulled and replaced. Damage to the casing, on the other hand, creates a serious problem, which jeopardize the productivity of the well. Casing is difficult and expensive to repair.

Where a packer is not used, the casing is exposed to produced fluids, which are often corrosive. In order to maintain a controlled inhibited atmosphere, a packer is necessary to seal the bottom of the casing annulus, which prevents the release of produced fluids into the casing annulus.

The casing may also be exposed to pressure differentials high enough to cause casing damage. The tubing string normally has higher-pressure ratings than the casing. Using a packer to seal the casing annulus above the productive zone will confine the high differentials to the tubing string.

10.4.2 Safety

A packer acts as a downhole blowout preventer or valve to provide maximum control of the formation at the formation.

During completion operations, the packer can be used as a plug to completely shut off the production zone while work is being done up the hole, or during nippling up operations. High-pressure zones can be tested or produced without exposing the casing or wellhead equipment to unnecessary high pressures.

During producing operations, the packer confines high pressures to the tubing string where they are more easily controlled. Safety shut-in devices and other accessory equipment are often installed in the tubing for maximum control.

10.4.3 Energy Conservation

One of the primary reasons for using a smaller tubing string to produce the formation fluids is to maintain sufficient velocity to take full advantage of available energy. With high velocities, oil, water and gas are intermingled so that the gas energy is used, preventing the heavier water from falling back and killing the well.

When a packer is not used, gas tends to break out into the annulus, accumulating at the top of the casing. This results in a loss of energy. A packer funnels all of the production into the tubing string, so all of the available gas energy is utilized to lift the fluid.

If two different zones are perforated, interflow between the zones results in a loss of energy for the more prolific zone and could artificial lift for both zones. Formation damage and/or loss of hydrocarbon reserves can also result from this uncontrolled formation flow.

10.4.4 Improve Productivity

A packer allows the casing to be used as an energy storage area and/or conduit in artificial lift operations; such as, gas lift and hydraulic pumping.

Sealing between two zones allows both to be produced simultaneously with maximum control by using two strings of tubing.

In some cases, it is desirable to perforate a second zone for production at a later date. To keep this alternate zone isolated from the producing zone, a packer is set between the two. With the proper equipment, the zones can be switched without pulling the tubing string.

Service packers are used to perform specialized functions. When the function is completed, the packer is removed from the well. Service packers are commonly designed with higher differential pressure ratings than permanent packers, and are more likely to have their operational capabilities tested to the fullest extent. Service packers are commonly used to pressure test casing and repair if a leak exists; temporarily suspend producing intervals; and protect the casing string from high pressures during hydraulic fracturing.

In a mechanical set application, tubing movement is used to create the force necessary to set the packer and energize the element.

In hydraulic set application, a force is generated internally within the packer by applying fluid pressure to set the slips and energize the element.

For wireline set applications, an electric wireline pressure setting tool is used. An electric current ignites a powder charge to create a gas pressure within the setting tool. The gas pressure then acts on a hydraulic piston to create a force sufficient to set the packer slips, energize the element and shear the setting tool from the packer.

Choosing and successfully installing the best packer for a job depends on the proper evaluation and analysis of the forces that may act on the tubing and the packer during the lifetime of the installation. The effect of the forces can only be determined by an accurate force analysis. The installation can proceed with a high degree of confidence if no detrimental effects are shown during analysis. However, if analysis indicates that damage to the tubing or packer could occur, reassessment and revision of the installation procedures will reduce the risk of costly recovery.

The anticipated change in well conditions that may occur after the packer has been set must be analyzed to evaluate the forces affecting the packer and tubing string.

Temperature and pressure variations occur within the well throughout its lifecycle. Changes in these conditions can lead to a change in length of the tubing string or a change in force through out the tubing string depending on the type of packer chosen and the packer to tubing configuration. The tubing will either lengthen or shorten when temperature, tubing pressure or annulus pressure is changed in the well. If the tubing is anchored, the change in force will occur at the packer as well as over the entire length of the tubing string. If the tubing is free to move, the tubing string will experience a net change in length.

The following effects will lead to a change in length or force:

Piston effect
Buckling effect
Ballooning effect
Temperature effect

The piston, buckling and ballooning effects result from pressure changes within the well. The temperature effect is the result of a change in temperature and is not related to pressure. Each effect must be calculated separately even though some of the effects may be related. The magnitude and direction of each effect must then be combined to determine the total change in force or change in tubing length.

10.4.5 Piston Effect

The piston effect is the result of pressure changes, which occur inside the tubing string and annulus at the packer.

In retrievable packer installations, pressure changes that occur inside the tubing at the packer act on the difference in areas between the packer valve area (A_p) and the tubing ID area (A_i). Pressure changes that occur in the annulus act on the difference in areas between the packer valve area (A_p) and the tubing OD area (A_o).

In permanent packer installations, pressure changes that occur inside the tubing act on the difference in areas between the seal bore (packer valve, A_p) and the tubing ID area (A_i). Pressure changes that occur in the annulus act on the difference in areas between the seal bore (packer valve, A_p) and the tubing OD area (A_o).

The result of the piston effect is either an up or down force at the packer. Upward forces are designated as negative (−) forces, and downward forces are designated as positive (+) forces.

In "anchored" installations (the tubing is not allowed to move), the piston effect leads to a change in force either upward (tension) or downward (compression) at the packer. Upward forces are designated as negative (−), and downward forces are designated as positive (+).

In "free" installations (the tubing-to-packer configuration allows movement of the tubing), the piston effect leads to a change in length either upward (shortening) or downward (lengthening) of the tubing string.

In retrievable packer installations where the tubing is attached to the packer via a coupling on the packer mandrel, the piston effect will always result in a force change on the packer. The resulting force change will either be a change in tubing compression, or a change in tubing tension (either positive or negative) unless an expansion joint is installed in the tubing string above the packer.

If the tubing pressure is greater than the annular pressure, the affected area is calculated on the knife seal OD. If the annular pressure is greater than the tubing pressure, the affected area is calculated on the knife seal ID.

When the packer and tubing string are installed in the well, the hydrostatic pressure in the well acts on the end area of the tubing string. This creates an upward pressure-related effect, resulting in a reduction in the weight of the tubing string as well as a shortening of the overall length of the tubing string. This effect is known as force due to buoyancy or the buoyancy effect. Although the tubing is shortened as a result of the buoyancy effect, the packer-to-tubing relationship is not affected until the packer is set. Once the packer is set and the packer valve is closed, a change in pressure in either the tubing or the annulus will result in a change in the tubing force at the packer, and will affect the tubing-to-packer relationship.

When calculating the piston effect, always consider the tubing-to-packer relationship as "balanced" when the packer is set. Any changes in well condition that occur after the packer is set will affect the packer-to-tubing relationship.

When calculating the force due to the piston effect, it is the change in pressure rather than the absolute pressure that is important. To determine this, both initial and final well conditions need to be known. The initial conditions are those that existed when the packer was set, or when the seal assembly was stung into the packer sealbore. Final conditions are those conditions that are to be expected during production, stimulation, or well work-overs.

Use the following formula to calculate piston force:

$$\Delta F_1 = (A_p - A_o)\Delta P_o - (A_p - A_i)\Delta P_i \tag{10.17}$$

where ΔF_1 = Force change due to Piston Effect, lbs.
A_p = Packer valve area, in^2
A_o = Area of the tubing O.D., in^2
A_i = Area of the tubing I.D., in^2
ΔP_o = Change in the total annular pressure at the packer, psi
ΔP_i = Change in the total tubing pressure at the packer, psi

To calculate the length change due to the Piston Effect (ΔL_1), use this formula:

$$\Delta L_1 = \frac{LF_1}{EA_s} \tag{10.18}$$

where ΔL_1 = Length change due to Piston Effect, in.
L = Length of the tubing string, in.

F_1 = Force change due to Piston Effect, lbs.
E = Modulus of elasticity, 30×10^6 psi
A_s = Cross sectional area of the tubing, in^2

10.4.6 Buckling Effect

The buckling effect is the most difficult of the four basic effects to understand. Tubing buckling will occur due to two different force distributions. One is the compressive force mechanically applied to the end of the tubing, and the other is the internal pressure that results in a force distribution on the ID area of the tubing.

A mechanically applied compressive force acting on the end area of the tubing will cause tubing to buckle. When compressive force is applied to a tubing string that is, essentially, relatively flexible, the tubing will bow outward or will buckle. For example, tubing that is standing in a derrick will buckle or bow outward due to its own weight.

The second force that will cause a tubing string to buckle is pressure. Pressure buckling will occur when the pressure inside a tubing string is greater than the pressure outside the tubing.

The pressure within the tubing string creates a force distribution that acts over the ID area of the tubing, while pressure outside the tubing is creating a force distribution that acts over the OD area of the tubing. Because the pressure inside is higher than the pressure outside, these force distributions produce burst stresses within the tubing. The pressure within the tubing acts perpendicular to the inside tubing wall just as the annulus pressure acts perpendicular to the outside tubing wall. Because of the variance in tubing wall thickness, which can occur during the manufacturing process, the resulting burst stresses cannot be distributed evenly within the tubing. The uneven stress distribution causes the tubing to buckle.

If tubing was manufactured with precisely controlled inside and outside dimensions, the net resulting force due to a pressure differential would be zero, and the tubing would not buckle due to pressure. Dimensional tolerances for oilfield tubulars are outlined in the current edition of API Specification 5CT.

Buckled tubing is defined as tubing that is bowed from its original straight condition. The tubing will buckle or bow outward until it contacts the casing wall at which point it will begin to coil. This coiling of the tubing is referred to as "corkscrewing" the tubing. Corkscrewed tubing is a form of buckled tubing. As long as the stresses that occur because of the buckling do not exceed the yield strength of the tubing, the tubing will return to its original shape when the force is removed. Should the stresses created due to the buckled condition of the tubing exceed the yield strength of the tubing, the tubing will not return to its original shape: It will become permanently corkscrewed.

Buckling will only shorten a tubing string. Pressure buckling will only exert a negligible change in force on a packer and will only occur if the final pressure within the tubing string is greater than the final pressure outside the tubing string.

It should be noted that pressure buckling can occur in a tubing string even if the tubing is in tension throughout its entire length. Although in tension and considered as being in a "straight" condition within the wellbore, the tubing string will buckle due to the uneven force distribution.

Tubing buckling is most severe at the bottom of the tubing string where the pressure is the greatest. It lessens further up the hole until, normally, a point is reached where tubing buckling does not occur. This point is known as the neutral point. Notice that the tubing below the neutral point is buckled, while the tubing above remains straight.

The factors that have the most influence on the amount of tubing buckling that will occur are the radial clearance (r) between the tubing OD and the casing ID., the magnitude of the pressure differential from the tubing ID to the tubing OD, and the size of the packer valve. Because these factors have a direct effect on buckling, as any one of these factors increases, the length change due to buckling will increase.

Use this formula to calculate the length change due to buckling:

$$\Delta L_2 = \frac{r^2 A_p^2 (\Delta P_i - \Delta P_o)^2}{-8EI(W_s + W_i - W_o)} \tag{10.19}$$

where ΔL_2 = Length change in inches due to buckling
r = radial clearance between the tubing and the casing, inches
A_p = Packer valve area, in.^{2}
ΔP_i = Change in total tubing pressure at the packer, psi
ΔP_o = Change in total annulus pressure at the packer, psi
E = Modulus of elasticity, 30×10^6 psi for carbon steels
I = Moment of inertia of the tubing, in.^{4}
$I = [(\text{Tubing OD})^4 - (\text{Tubing ID})^4] \times \pi \div 64$
W_s = Weight of the tubing, lbs./in.
W_i = Weight of the final fluid displaced in the tubing per unit length, psi
W_i = (Tubing ID)2 $\times$ 0.0034 $\times$ Final tubing fluid wt., lbs./gal.
W_o = Weight of the final fluid displaced in the annulus per unit length, psi
W_o = (Tubing OD)2 $\times$ 0.0034 $\times$ Final annulus fluid wt., lbs./gal.

Use this formula to determine length from packer to neutral point:

$$n = \frac{Ap(P_{ifinal} - P_{ofinal})}{(W_s + W_i - W_o)} \tag{10.20}$$

where n = The distance from the Packer to the Neutral Point, in.
A_p = Packer valve area, in.^{2}
P_{ifinal} = Total tubing pressure at the packer that will exist for the given conditions, psi.
P_{ofinal} = Total annulus pressure at the packer that will exist for the given conditions, psi.
W_s = Weight of the tubing, lbs./in.
W_i = Weight of the final fluid displaced in the tubing per unit length, psi
W_i = (Tubing ID)2 $\times$ 0.0034 $\times$ Final tubing fluid wt., lbs./gal.
W_o = Weight of the final fluid displaced in the annulus per unit length, psi
W_o = (Tubing OD)2 $\times$ 0.0034 $\times$ Final annulus fluid wt., lbs./gal.

Use this formula for Corrected Length Change due to Buckling:

$$\Delta L_2' = \Delta L_2 \times \frac{L}{n} \times \frac{[2-(L)]}{n} \tag{10.21}$$

where $\Delta L_2'$ = Length change due to buckling when neutral point is above wellhead, in.
ΔL_2 = Length change due to buckling, in.
L = Length of the tubing string, in.
n = The distance from the packer to the neutral point

10.4.7 Ballooning Effect

The third effect, which must be considered in packer installations, is the ballooning effect. Tubing ballooning occurs when pressure is applied to the inside of a tubing string: The pressure differential creates forces within the tubing string, which are trying to burst the tubing. The burst forces created within the tubing string due to the differential pressure causes the tubing to swell.

As the tubing swells, the tubing string will shorten if it is free to move. If the tubing is not free to move, the swelling will create a tension force (−force) on the packer. This shortening of the tubing due to internal differential pressure is referred to as ballooning.

If the pressure differential exists on the outside of the tubing string, the force created is attempting to collapse the tubing. As the tubing tries to collapse, it lengthens if it is free to move. If the tubing is not free to move, it will create a compressive force (+ force) on the packer. This lengthening of the tubing string due to collapse forces is referred to as reverse ballooning.

The ballooning effect is directly proportional to the area over which the pressure is acting, and as a result, the effect of reverse ballooning is slightly greater than that of ballooning.

Unlike the piston and buckling effects, the ballooning effect occurs over the entire length of the tubing string. Ballooning effect calculations are based on changes in the average pressures inside and outside the tubing.

The average pressure is based on the half the sum of the surface pressure plus the pressure at the packer. Increasing the bottom hole pressure, by changing the fluid gradient, would have half the effect of making the same change by applying added surface pressure.

Well conditions can affect the average pressure inside and outside the tubing: The effects due to ballooning and reverse ballooning are calculated together, then the net effect is expressed as either a negative force (ballooning), or a positive force (reverse ballooning), or change in length of the tubing string, depending on the tubing string's ability to move at the packer.

Pressure increases in the tubing tend to swell, or balloon the tubing, and will result in a shortening of the tubing string. Similarly, pressure increases in the casing cause the tubing diameter to contract, resulting in reverse ballooning, and an elongation of the string. As treating pressures increase, and depths become greater, the resultant changes in length can be considerable.

In considering the results of ballooning and reverse ballooning, two point to consider are

1. The pressure that is responsible for ballooning or reverse ballooning is the pressure change in the tubing and/or annulus, and from the conditions that prevailed when the well was completed. The fact that a differential pressure exists between tubing and casing does not necessarily mean that ballooning exists.
2. Pressure calculations are based on changes in the surface tubing or casing pressure (applied) and changes in fluid density (hydrostatic).

Use this formula to calculate the effects due to ballooning:

$$\Delta F_3 = 0.6[(\Delta P_{oa} A_o) - (\Delta P_{ia} A_i)] \tag{10.22}$$

where ΔF_3 = Ballooning effect force change, lbs.
ΔP_{oa} = Change in average annulus pressure, psi.
A_o = Area of the tubing OD, in.^{2}
ΔP_{ia} = Change in average tubing pressure, psi.
A_i = Area of the tubing ID, in.^{2}

To calculate the length change (DL_3) due to the ballooning effect use this formula:

$$\Delta L_3 = \frac{2L\gamma}{E} \times \frac{(R^2 \Delta P_{oa}) - \Delta P_{ia}}{R^2 - 1} \tag{10.23}$$

where ΔL_3 = Change in length (in.)
L = Length of the tubing string (in.)
γ = Poisson's ratio: 0.3 for steel
E = Modulus of elasticity: 30×10^6 psi for steel
ΔP_{ia} = Change in average tubing pressure (psi)
ΔP_{oa} = Change in average annulus pressure (psi)
R^2 = Ratio of the tubing OD to the tubing ID (where $R = OD \div ID$)

10.4.8 Temperature Effect

The last basic effect is the temperature effect. It is the only one of the four basic effects that is not pressure related: The length and force changes that occur are functions of a change in the average tubing temperature.

The basic principles of expansion and contraction apply when the average temperature of the tubing string is increased or decreased. When the average temperature of the tubing is decreased by injecting cool fluids, the tubing string will shorten in length if it is free to move. If the tubing string is not free to move, then a tension force will be created on the packer.

When the average temperature of the tubing string is increased either by injecting warm fluids, or by producing warm oil and/or gas, the tubing will lengthen, or elongate, if it is free to move. If the tubing is not free to move, a compressive force on the packer will result, caused by the increase in the average tubing temperature. In many packer installations, the temperature effect will be the largest of the four basic effects.

Because the temperature change in a tubing string occurs over the entire length of the tubing string, the change in average tubing temperature must be calculated to determine the resulting change in force or length.

To calculate the average temperature of the tubing the bottom hole temperature and the surface temperature must be known. The following formula is used to calculate the average tubing temperature:

$$\text{Avg. tbg. temp. } (^\circ\text{F}) = [(\text{Surface temp. } (^\circ\text{F}) + \text{Bottom hole temp. } (^\circ\text{F}) \div 2 \qquad (10.24)$$

If reliable temperature data is not available, it is strongly recommended that any and all assumptions made be conservative in nature to prevent possible equipment failure.

Although some assumptions are made, the magnitude and direction of the temperature effect can still be determined. It should be noted, however, that the temperature effect is not felt immediately at the packer. In installations where pressure changes occur within the well, the effects of the pressure change are felt immediately at the packer. When temperature effects occur, it may take from several minutes to several hours for the effects to be felt at the packer. However, it is typically assumed, when performing force analysis calculations, that the resulting effect due to a change in the tubing temperature is an immediate effect.

Treating temperature as an immediate effect allows it to be added to the pressure effects, so that all the effects can be considered, to determine the net effect on the tubing string. In some instances, this assumption can result in indications of potential equipment, or tubular, failure. Each installation should be considered separately to minimize the risks of tubing or equipment failure.

Use this formula to calculate the change in Force due to temperature effect:

$$\Delta F_4 = (207)(As)(\Delta T) \tag{10.25}$$

where ΔF_4 = Force change due to temperature effect, lbs.
As = Tubing cross sectional area, in.^{2}
ΔT = Change in average temperature, °F.
ΔT = Final average tubing temperature — Initial average tubing temperature

Use this formula to calculate change in tubing length due to temperature effect:

$$\Delta L_4 = (L)(\beta)(\Delta T) \tag{10.26}$$

where ΔL_4 = Lengthchangeduetotemperatureeffect, in.
L = Lengthoftubingstring, in.
β = Coefficient of thermal expansion for steel, 0.0000069 in/in/°F
ΔT = Changeinaveragetemperature, °F.

10.4.9 Total Effect

The four basic effects plus any mechanically applied forces are used to determine the total effect of the tubing on the packer. The method used to determine the total effect at the packer depends on the type of packer being considered for the installation and the tubing-to-packer relationship.

In packer installations that permit movement of the tubing, the degree and direction of the tubing length change expected (total effect) must be determined after considering each of the four basic effects independently, then collectively.

In packer installations that do not permit movement of the tubing, the degree and direction of the tubing force change expected (total effect) must be determined after considering the four basic effects independently, then collectively.

The total effect is then determined by adding together each basic effect and any mechanically applied effect and can be expressed mathematically.

Use this formula to calculate total length changes with slack off weight applied to packer:

$$\Delta L_{total1} = \Delta L_1 + \Delta L_2 + \Delta L_3 + \Delta L_4 + \Delta L_s \quad (10.27)$$

where ΔL_{total1} = Total length change, in.
ΔL_1 = Length change of the tubing due to piston effect, in.
ΔL_2 = Length change of the tubing due to buckling effect, in.
ΔL_3 = Length change of the tubing due to ballooning effect, in.
ΔL_s = Length change of the tubing due to slack off force, in.

Use this formula to calculate total length changes with tension applied to packer:

$$\Delta L_{total1} = \Delta L_1 + \Delta L_2 + \Delta L_3 + \Delta L_4 + \Delta L_t \quad (10.28)$$

where ΔL_{total1} = Total length change, in.
ΔL_1 = Length change of the tubing due to piston effect, in.
ΔL_2 = Length change of the tubing due to buckling effect, in.
ΔL_3 = Length change of the tubing due to ballooning effect, in.
ΔL_t = Length change of the tubing due to tension force, in.

When determining length changes of the tubing in an installation where the tubing is stung through the packer, there will always be a change (ΔL_{total}) in the total length of the tubing string if a change in pressure or temperature occurs, and this change must be calculated.

In a packer installation where the tubing is landed on the packer and the change in total length (ΔL_{total}) calculated is a downward movement (a positive value), the total length change will not occur due to the tubing being restrained from downward movement because of its landed configuration. The change in tubing length will result in a change in force (compressive) on the packer.

If the tubing is in a landed condition and the total change in length (ΔL_{total}) calculated is an upward movement (a negative value), the tubing is not restrained and is free to move upward.

If the tubing is latched to the packer, the total length change (ΔL_{total}) will be zero because the tubing is restrained from upward or downward movement and compressive or tension forces will result.

Use this formula to calculate the total force effect with slack off weight applied to the packer:

$$\Delta F_p = \Delta F_1 + \Delta F_3 + \Delta F_4 + \Delta F_s \quad (10.29)$$

where ΔF_p = Total force change effect, tubing-to-packer, lb
ΔF_1 = Force change due to piston effect, lb
ΔF_3 = Force change due to ballooning effect, lb
ΔF_4 = Force change due to temperature effect, lb
ΔF_s = Force change due to slack off force, lb

Use this formula to calculate the total force effect with tension applied to the packer:

$$\Delta F_p = \Delta F_1 + \Delta F_3 + \Delta F_4 + \Delta F_t \quad (10.30)$$

where ΔF_p = Total force change effect, tubing-to-packer, lb
ΔF_1 = Force change due to piston effect, lb
ΔF_3 = Force change due to ballooning effect, lb
ΔF_4 = Force change due to temperature effect, lb
ΔF_t = Force change due to tension force, lb

When determining force changes in the tubing string in an installation where the tubing is stung through the packer, the tubing-to-packer force (F_p) will always equal zero because the tubing can never exert a force on the packer.

In a packer installation where the tubing is landed onto the packer and the calculated change in force is a downward force (a positive value), the change in force will result in a compressive force (F_p) on the packer.

If the tubing is landed and the change in force calculated is an upward force (a negative value), the tubing-to-packer force (F_p) will be zero, given that tension cannot be pulled through a landed tubing-to-packer configuration.

If the tubing is latched to the packer, the tubing-to-packer force (F_p) will always result in either compressive or tension forces on the packer.

A sufficient amount of accurate data is important in the complete analysis. The validity of the results depends largely on the accuracy of the data used, and this should weigh heavily on the final selection of the procedure, equipment to be used and the safety factor required.

TABLE 10.9 Grade Yield and Ultimate Strengths, Ultimate Elongations, and Hardness (courtesy of Precision Tube Technology, Inc.)

Grade	Min. Yield	Min. Ultimate	Min. Elongation	Hardness Range
HS 70	70 ksi (483 MPa)	80 ksi (552 MPa)	25%	85-94 RH B
HS 80	80 ksi (552 MPa)	88 ksi (607 MPa)	28%	90-98 RH B
HS 90	90 ksi (621 MPa)	97 ksi (669 MPa)	25%	94 RH B-22 RH C
HS 110	110 ksi (758 MP)	115 ksi (793 MPa)	22%	22-28 RH C

TABLE 10.10 HS70 Grade USC Units (courtesy of Precision Tube Technology, Inc.)

DIMENSIONS (Inches)						Nominal	TUBE LOAD BODY (Lbs.)		INTERNAL PRESSURE (psi)		TUBING AREA (sq. in.)		TORSIONAL YIELD (ft.-lbs.)		INTERNAL CAPACITY per 1000 ft.		EXTERNAL DISPLACEMENT per 1000 ft.	
O.D. Specified	O.D. (mm)	Wall Specified	Wall (mm)	Wall Minimum	I.D. Calculated	Weight Lbs./ft.	Yield Minimum	Tensile Minimum	Hydro Test 90%	Internal Yield Min	w/min. Wall	Internal Min	Yield	Ultimate	Gallons	Barrels	Gallons	Barrels
1.000	*25.4*	**0.080**	*2.03*	0.076	0.840	0.788	16,200	18,500	9,400	10,500	0.221	0.565	319	344	28.79	0.69	40.80	0.97
1.000	*25.4*	**0.087**	*2.21*	0.083	0.826	0.850	17,500	20,000	10,300	11,400	0.239	0.546	341	370	27.84	0.66	40.80	0.97
1.000	*25.4*	**0.095**	*2.41*	0.090	0.810	0.920	18,900	21,600	11,100	12,300	0.257	0.528	362	395	26.77	0.64	40.80	0.97
1.000	*25.4*	**0.102**	*2.59*	0.097	0.796	0.981	20,100	23,000	11,900	13,300	0.275	0.510	382	420	25.85	0.62	40.80	0.97
1.000	*25.4*	**0.109**	*2.77*	0.104	0.782	1.040	21,400	24,400	12,700	14,200	0.293	0.493	401	443	24.95	0.59	40.80	0.97
1.250	*31.8*	**0.080**	*2.03*	0.076	1.090	1.002	20,600	23,500	7,600	8,400	0.280	0.947	522	555	48.47	1.15	63.75	1.52
1.250	*31.8*	**0.087**	*2.21*	0.083	1.076	1.083	22,300	25,400	8,300	9,200	0.304	0.923	561	599	47.24	1.12	63.75	1.52
1.250	*31.8*	**0.095**	*2.41*	0.090	1.060	1.175	24,100	27,600	9,000	9,900	0.328	0.899	598	642	45.84	1.09	63.75	1.52
1.250	*31.8*	**0.102**	*2.59*	0.097	1.046	1.254	25,800	29,400	9,600	10,700	0.351	0.876	633	683	44.64	1.06	63.75	1.52
1.250	*31.8*	**0.109**	*2.77*	0.104	1.032	1.332	27,400	31,300	10,300	11,400	0.374	0.853	668	724	43.45	1.03	63.75	1.52
1.250	*31.8*	**0.116**	*2.95*	0.111	1.018	1.408	28,900	33,100	11,000	12,200	0.397	0.830	700	764	42.28	1.01	63.75	1.52
1.250	*31.8*	**0.125**	*3.18*	0.118	1.000	1.506	30,900	35,300	11,600	12,900	0.420	0.808	732	802	40.80	0.97	63.75	1.52
1.250	*31.8*	**0.134**	*3.40*	0.128	0.982	1.601	32,900	37,600	12,600	14,000	0.451	0.776	775	856	39.34	0.94	63.75	1.52
1.250	*31.8*	**0.145**	*3.68*	0.138	0.960	1.715	35,200	40,300	13,500	15,000	0.482	0.745	815	907	37.60	0.90	63.75	1.52
1.250	*31.8*	**0.156**	*3.96*	0.148	0.938	1.827	37,500	42,900	14,400	16,000	0.512	0.715	853	956	35.90	0.85	63.75	1.52
1.250	*31.8*	**0.175**	*4.45*	0.167	0.900	2.014	41,400	47,300	16,100	17,900	0.568	0.659	919	1,044	33.05	0.79	63.75	1.52
1.500	*38.1*	**0.095**	*2.41*	0.090	1.310	1.429	29,400	33,500	7,500	8,300	0.399	1.368	893	947	70.02	1.67	91.80	2.19
1.500	*38.1*	**0.102**	*2.59*	0.097	1.296	1.527	31,400	35,800	8,100	9,000	0.428	1.340	949	1,011	68.53	1.63	91.80	2.19
1.500	*38.1*	**0.109**	*2.77*	0.104	1.282	1.623	33,300	38,100	8,600	9,600	0.456	1.311	1,003	1,074	67.06	1.60	91.80	2.19
1.500	*38.1*	**0.116**	*2.95*	0.111	1.268	1.719	35,300	40,300	9,200	10,200	0.484	1.283	1,055	1,135	65.60	1.56	91.80	2.19
1.500	*38.1*	**0.125**	*3.18*	0.118	1.250	1.840	37,800	43,200	9,800	10,800	0.512	1.255	1,106	1,194	63.75	1.52	91.80	2.19
1.500	*38.1*	**0.134**	*3.40*	0.128	1.232	1.960	40,300	46,000	10,600	11,700	0.552	1.215	1,175	1,278	61.93	1.47	91.80	2.19
1.500	*38.1*	**0.145**	*3.68*	0.138	1.210	2.104	43,200	49,400	11,300	12,600	0.590	1.177	1,242	1,358	59.74	1.42	91.80	2.19
1.500	*38.1*	**0.156**	*3.96*	0.148	1.188	2.245	46,100	52,700	12,100	13,500	0.629	1.139	1,305	1,436	57.58	1.37	91.80	2.19
1.500	*38.1*	**0.175**	*4.45*	0.167	1.150	2.483	51,000	58,300	13,600	15,100	0.699	1.068	1,416	1,577	53.96	1.28	91.80	2.19
1.500	*38.1*	**0.190**	*4.83*	0.180	1.120	2.665	54,700	62,600	14,600	16,200	0.746	1.021	1,486	1,668	51.18	1.22	91.80	2.19

(Continued)

TABLE 10.10 (*Continued*)

DIMENSIONS (Inches)						Nominal	TUBE LOAD BODY (Lbs.)		INTERNAL PRESSURE (psi)		TUBING AREA (sq. in.)		TORSIONAL YIELD (ft.-lbs.)		INTERNAL CAPACITY per 1000 ft.		EXTERNAL DISPLACEMENT per 1000 ft.	
O.D. Specified	O.D. (mm)	Wall Specified	Wall (mm)	Wall Minimum	I.D. Calculated	Weight Lbs./ft.	Yield Minimum	Tensile Minimum	Hydro Test 90%	Internal Yield Min	w/min. Wall	Internal Min	Yield	Ultimate	Gallons	Barrels	Gallons	Barrels
1.750	*44.5*	**0.109**	*2.77*	0.104	1.532	1.915	39,300	45,000	7,400	8,200	0.538	1.867	1,407	1,492	95.76	2.28	124.95	2.97
1.750	*44.5*	**0.116**	*2.95*	0.111	1.518	2.029	41,700	47,600	7,900	8,800	0.572	1.834	1,483	1,579	94.02	2.24	124.95	2.97
1.750	*44.5*	**0.125**	*3.18*	0.118	1.500	2.175	44,700	51,100	8,400	9,300	0.605	1.800	1,558	1,665	91.80	2.19	124.95	2.97
1.750	*44.5*	**0.134**	*3.40*	0.128	1.482	2.318	47,600	54,400	9,100	10,100	0.652	1.753	1,660	1,784	89.61	2.13	124.95	2.97
1.750	*44.5*	**0.145**	*3.68*	0.138	1.460	2.492	51,200	58,500	9,800	10,900	0.699	1.706	1,759	1,901	86.97	2.07	124.95	2.97
1.750	*44.5*	**0.156**	*3.96*	0.148	1.438	2.662	54,700	62,500	10,500	11,600	0.745	1.660	1,854	2,014	84.37	2.01	124.95	2.97
1.750	*44.5*	**0.175**	*4.45*	0.167	1.400	2.951	60,600	69,300	11,700	13,100	0.831	1.575	2,024	2,221	79.97	1.90	124.95	2.97
1.750	*44.5*	**0.190**	*4.83*	0.180	1.370	3.173	65,200	74,500	12,600	14,000	0.888	1.517	2,132	2,356	76.58	1.82	124.95	2.97
1.750	*44.5*	**0.204**	*5.18*	0.195	1.342	3.377	69,400	79,300	13,600	15,100	0.953	1.453	2,250	2,506	73.48	1.75	124.95	2.97
2.000	*50.8*	**0.109**	*2.77*	0.104	1.782	2.207	45,300	51,800	6,500	7,200	0.619	2.522	1,879	1,979	129.56	3.08	163.20	3.89
2.000	*50.8*	**0.116**	*2.95*	0.111	1.768	2.340	48,100	54,900	6,900	7,700	0.659	2.483	1,985	2,097	127.53	3.04	163.20	3.89
2.000	*50.8*	**0.125**	*3.18*	0.118	1.750	2.509	51,500	58,900	7,400	8,200	0.698	2.444	2,088	2,213	124.95	2.97	163.20	3.89
2.000	*50.8*	**0.134**	*3.40*	0.128	1.732	2.677	55,000	62,800	8,000	8,900	0.753	2.389	2,230	2,375	122.39	2.91	163.20	3.89
2.000	*50.8*	**0.145**	*3.68*	0.138	1.710	2.880	59,200	67,600	8,600	9,500	0.807	2.334	2,368	2,534	119.30	2.84	163.20	3.89
2.000	*50.8*	**0.156**	*3.96*	0.148	1.688	3.080	63,300	72,300	9,200	10,200	0.861	2.280	2,501	2,690	116.25	2.77	163.20	3.89
2.000	*50.8*	**0.175**	*4.45*	0.167	1.650	3.419	70,200	80,300	10,300	11,500	0.962	2.180	2,741	2,975	111.08	2.64	163.20	3.89
2.000	*50.8*	**0.190**	*4.83*	0.180	1.620	3.682	75,600	86,400	11,100	12,300	1.029	2.112	2,897	3,163	107.08	2.55	163.20	3.89
2.000	*50.8*	**0.204**	*5.18*	0.195	1.592	3.923	80,600	92,100	12,000	13,300	1.106	2.036	3,067	3,372	103.41	2.46	163.20	3.89
2.375	*60.3*	**0.125**	*3.18*	0.118	2.125	3.011	61,900	70,700	6,200	6,900	0.837	3.593	3,028	3,181	184.24	4.39	230.14	5.48
2.375	*60.3*	**0.134**	*3.40*	0.128	2.107	3.215	66,000	75,500	6,700	7,500	0.904	3.527	3,243	3,421	181.13	4.31	230.14	5.48
2.375	*60.3*	**0.145**	*3.68*	0.138	2.085	3.462	71,100	81,300	7,300	8,100	0.970	3.460	3,452	3,656	177.37	4.22	230.14	5.48
2.375	*60.3*	**0.156**	*3.96*	0.148	2.063	3.706	76,100	87,000	7,800	8,600	1.035	3.395	3,655	3,886	173.64	4.13	230.14	5.48
2.375	*60.3*	**0.175**	*4.45*	0.167	2.025	4.122	84,700	96,800	8,700	9,700	1.158	3.272	4,025	4,313	167.31	3.98	230.14	5.48
2.375	*60.3*	**0.190**	*4.83*	0.180	1.995	4.445	91,300	104,300	9,400	10,500	1.241	3.189	4,266	4,595	162.38	3.87	230.14	5.48
2.375	*60.3*	**0.204**	*5.18*	0.195	1.967	4.742	97,400	111,300	10,200	11,300	1.335	3.095	4,533	4,913	157.86	3.76	230.14	5.48
2.375	*60.3*	**0.224**	*5.69*	0.214	1.927	5.159	106,000	121,100	11,100	12,400	1.453	2.977	4,855	5,301	151.50	3.61	230.14	5.48

Test pressure value equals 90% of internal yield pressure rating. Maximum working pressure is a function of tube condition and is determined by user.

* Available as continuously milled tubing (CM™) or conventional butt-welded tubing sections (W™). All data is for new tubing at minimum strength. Other sizes and wall thicknesses available on request. See individual size sheets for additional wall thicknesses.

TABLE 10.11 HS70 Grade USC Units (courtesy of Precision Tube Technology, Inc.)

DIMENSIONS (mm)						Nominal	TUBE LOAD BODY (Newtons)		INTERNAL PRESSURE (kPa)		TUBING AREA (sq. cm)		TORSIONAL YIELD (N-m)		INTERNAL CAPACITY per meter	EXTERNAL DISPLACEMENT per meter
OD Specified	OD (inches)	Wall Specified	Wall (inches)	Wall Minimum	ID Calculated	Weight Kg/m	Yield Minimum	Tensile Minimum	Hydro Test 90%	Internal Yield Min	w/min. Wall	Internal Min	Yield	Ultimate	Liters	Liters
25.4	*1.000*	**2.03**	*0.080*	1.93	21.3	1.17	72,100	82,300	64,800	72,400	1.42	3.64	430	470	0.36	0.51
25.4	*1.000*	**2.21**	*0.087*	2.11	21.0	1.26	77,800	89,000	71,000	78,600	1.54	3.52	460	500	0.35	0.51
25.4	*1.000*	**2.41**	*0.095*	2.29	20.6	1.37	84,100	96,100	76,500	84,800	1.66	3.41	490	540	0.33	0.51
25.4	*1.000*	**2.59**	*0.102*	2.46	20.2	1.46	89,400	102,300	82,100	91,700	1.78	3.29	520	570	0.32	0.51
25.4	*1.000*	**2.77**	*0.109*	2.64	19.9	1.55	95,200	108,500	87,600	97,900	1.89	3.18	540	600	0.31	0.51
31.8	*1.250*	**2.03**	*0.080*	1.93	27.7	1.49	91,600	104,500	52,400	57,900	1.81	6.11	710	750	0.60	0.79
31.8	*1.250*	**2.21**	*0.087*	2.11	27.3	1.61	99,200	113,000	57,200	63,400	1.96	5.95	760	810	0.59	0.79
31.8	*1.250*	**2.41**	*0.095*	2.29	26.9	1.75	107,200	122,800	62,100	68,300	2.12	5.80	810	870	0.57	0.79
31.8	*1.250*	**2.59**	*0.102*	2.46	26.6	1.86	114,800	130,800	66,200	73,800	2.27	5.65	860	930	0.55	0.79
31.8	*1.250*	**2.77**	*0.109*	2.64	26.2	1.98	121,900	139,200	71,000	78,600	2.42	5.50	910	980	0.54	0.79
31.8	*1.250*	**2.95**	*0.116*	2.82	25.9	2.09	128,500	147,200	75,800	84,100	2.56	5.35	950	1,040	0.53	0.79
31.8	*1.250*	**3.18**	*0.125*	3.00	25.4	2.24	137,400	157,000	80,000	88,900	2.71	5.21	990	1,090	0.51	0.79
31.8	*1.250*	**3.40**	*0.134*	3.25	24.9	2.38	146,300	167,200	86,900	96,500	2.91	5.01	1,050	1,160	0.49	0.79
31.8	*1.250*	**3.68**	*0.145*	3.51	24.4	2.55	156,600	179,300	93,100	103,400	3.11	4.81	1,110	1,230	0.47	0.79
31.8	*1.250*	**3.96**	*0.156*	3.76	23.8	2.72	166,800	190,800	99,300	110,300	3.31	4.61	1,160	1,300	0.45	0.79
31.8	*1.250*	**4.45**	*0.175*	4.24	22.9	2.99	184,100	210,400	111,000	123,400	3.67	4.25	1,250	1,420	0.41	0.79
38.1	*1.500*	**2.41**	*0.095*	2.29	33.3	2.12	130,800	149,000	51,700	57,200	2.57	8.83	1,210	1,280	0.87	1.14
38.1	*1.500*	**2.59**	*0.102*	2.46	32.9	2.27	139,700	159,200	55,800	62,100	2.76	8.64	1,290	1,370	0.85	1.14
38.1	*1.500*	**2.77**	*0.109*	2.64	32.6	2.41	148,100	169,500	59,300	66,200	2.94	8.46	1,360	1,460	0.83	1.14
38.1	*1.500*	**2.95**	*0.116*	2.82	32.2	2.55	157,000	179,300	63,400	70,300	3.12	8.28	1,430	1,540	0.81	1.14
38.1	*1.500*	**3.18**	*0.125*	3.00	31.8	2.73	168,100	192,200	67,600	74,500	3.31	8.10	1,500	1,620	0.79	1.14
38.1	*1.500*	**3.40**	*0.134*	3.25	31.3	2.91	179,300	204,600	73,100	80,700	3.56	7.84	1,590	1,730	0.77	1.14
38.1	*1.500*	**3.68**	*0.145*	3.51	30.7	3.13	192,200	219,700	77,900	86,900	3.81	7.59	1,680	1,840	0.74	1.14
38.1	*1.500*	**3.96**	*0.156*	3.76	30.2	3.34	205,100	234,400	83,400	93,100	4.06	7.35	1,770	1,950	0.72	1.14
38.1	*1.500*	**4.45**	*0.175*	4.24	29.2	3.69	226,800	259,300	93,800	104,100	4.51	6.89	1,920	2,140	0.67	1.14
38.1	*1.500*	**4.83**	*0.190*	4.57	28.4	3.96	243,300	278,400	100,700	111,700	4.82	6.59	2,020	2,260	0.64	1.14

(Continued)

TABLE 10.11 *(Continued)*

DIMENSIONS (mm)						Nominal	TUBE LOAD BODY (Newtons)		INTERNAL PRESSURE (kPa)		TUBING AREA (sq. cm)		TORSIONAL YIELD (N-m)		INTERNAL CAPACITY per meter	EXTERNAL DISPLACEMENT per meter
OD Specified	OD (inches)	Wall Specified	Wall (inches)	Wall Minimum	ID Calculated	Weight Kg/m	Yield Minimum	Tensile Minimum	Hydro Test 90%	Internal Yield Min	w/min. Wall	Internal Min	Yield	Ultimate	Liters	Liters
44.5	*1.750*	**2.77**	*0.109*	2.64	38.9	2.85	174,800	200,200	51,000	56,500	3.47	12.05	1,910	2,020	1.19	1.55
44.5	*1.750*	**2.95**	*0.116*	2.82	38.6	3.02	185,500	211,700	54,500	60,700	3.69	11.83	2,010	2,140	1.17	1.55
44.5	*1.750*	**3.18**	*0.125*	3.00	38.1	3.23	198,800	227,300	57,900	64,100	3.90	11.61	2,110	2,260	1.14	1.55
44.5	*1.750*	**3.40**	*0.134*	3.25	37.6	3.45	211,700	242,000	62,700	69,600	4.21	11.31	2,250	2,420	1.11	1.55
44.5	*1.750*	**3.68**	*0.145*	3.51	37.1	3.70	227,700	260,200	67,600	75,200	4.51	11.01	2,390	2,580	1.08	1.55
44.5	*1.750*	**3.96**	*0.156*	3.76	36.5	3.96	243,300	278,000	72,400	80,000	4.81	10.71	2,510	2,730	1.05	1.55
44.5	*1.750*	**4.45**	*0.175*	4.24	35.6	4.39	269,500	308,200	80,700	90,300	5.36	10.16	2,740	3,010	0.99	1.55
44.5	*1.750*	**4.83**	*0.190*	4.57	34.8	4.72	290,000	331,400	86,900	96,500	5.73	9.79	2,890	3,190	0.95	1.55
44.5	*1.750*	**5.18**	*0.204*	4.95	34.1	5.02	308,700	352,700	93,800	104,100	6.15	9.37	3,050	3,400	0.91	1.55
50.8	*2.000*	**2.77**	*0.109*	2.64	45.3	3.28	201,500	230,400	44,800	49,600	4.00	16.27	2,550	2,680	1.61	2.03
50.8	*2.000*	**2.95**	*0.116*	2.82	44.9	3.48	213,900	244,200	47,600	53,100	4.25	16.02	2,690	2,840	1.58	2.03
50.8	*2.000*	**3.18**	*0.125*	3.00	44.5	3.73	229,100	262,000	51,000	56,500	4.50	15.77	2,830	3,000	1.55	2.03
50.8	*2.000*	**3.40**	*0.134*	3.25	44.0	3.98	244,600	279,300	55,200	61,400	4.86	15.41	3,020	3,220	1.52	2.03
50.8	*2.000*	**3.68**	*0.145*	3.51	43.4	4.28	263,300	300,700	59,300	65,500	5.21	15.06	3,210	3,440	1.48	2.03
50.8	*2.000*	**3.96**	*0.156*	3.76	42.9	4.58	281,600	321,600	63,400	70,300	5.56	14.71	3,390	3,650	1.44	2.03
50.8	*2.000*	**4.45**	*0.175*	4.24	41.9	5.08	312,200	357,200	71,000	79,300	6.20	14.06	3,720	4,030	1.38	2.03
50.8	*2.000*	**4.83**	*0.190*	4.57	41.1	5.47	336,300	384,300	76,500	84,800	6.64	13.63	3,930	4,290	1.33	2.03
50.8	*2.000*	**5.18**	*0.204*	4.95	40.4	5.83	358,500	409,700	82,700	91,700	7.13	13.13	4,160	4,570	1.28	2.03
60.3	*2.375*	**3.18**	*0.125*	3.00	54.0	4.47	275,300	314,500	42,700	47,600	5.40	23.18	4,110	4,310	2.29	2.86
60.3	*2.375*	**3.40**	*0.134*	3.25	53.5	4.78	293,600	335,800	46,200	51,700	5.83	22.75	4,400	4,640	2.25	2.86
60.3	*2.375*	**3.68**	*0.145*	3.51	53.0	5.14	316,300	361,600	50,300	55,800	6.26	22.32	4,680	4,960	2.20	2.86
60.3	*2.375*	**3.96**	*0.156*	3.76	52.4	5.51	338,500	387,000	53,800	59,300	6.68	21.90	4,960	5,270	2.16	2.86
60.3	*2.375*	**4.45**	*0.175*	4.24	51.4	6.13	376,700	430,600	60,000	66,900	7.47	21.11	5,460	5,850	2.08	2.86
60.3	*2.375*	**4.83**	*0.190*	4.57	50.7	6.61	406,100	463,900	64,800	72,400	8.01	20.57	5,780	6,230	2.02	2.86
60.3	*2.375*	**5.18**	*0.204*	4.95	50.0	7.05	433,200	495,100	70,300	77,900	8.62	19.97	6,150	6,660	1.96	2.86
60.3	*2.375*	**5.68**	*0.224*	5.44	48.9	7.67	471,500	538,700	76,500	85,500	9.37	19.21	6,580	7,190	1.88	2.86

Test pressure value equals 90% of internal yield pressure rating. Maximum working pressure is a function of tube condition and is determined by user.

*Available as continuously milled tubing (CM™) or conventional butt-welded tubing sections (W™). All data is for new tubing at minimum strength. Other sizes and wall thicknesses available on request. See individual size sheets for additional wall thicknesses.

TABLE 10.12 HS80 Grade USC Units (courtesy of Precision Tube Technology, Inc.)

DIMENSIONS (Inches)						Nominal	TUBE LOAD BODY (Lbs.)		INTERNAL PRESSURE (psi)		TUBING AREA (sq. in.)		TORSIONAL YIELD (ft.-lb.)		INTERNAL CAPACITY per 1,000 ft.		EXTERNAL DISPLACEMENT per 1,000 ft.	
OD Specified	OD (mm)	Wall Specified	Wall (mm)	Wall Minimum	ID Calculated	Weight lb./ft.	Yield Minimum	Tensile Minimum	Hydro Test 90%	Internal Yield Min	w/min. Wall	Internal Min	Yield	Ultimate	Gallons	Barrels	Gallons	Barrels
1.000	*25.4*	**0.080**	*2.03*	0.076	0.840	0.788	18,500	20,300	10,800	12,000	0.221	0.565	365	393	28.79	0.69	40.80	0.97
1.000	*25.4*	**0.087**	*2.21*	0.083	0.826	0.850	20,000	22,000	11,700	13,000	0.239	0.546	390	423	27.84	0.66	40.80	0.97
1.000	*25.4*	**0.095**	*2.41*	0.090	0.810	0.920	21,600	23,800	12,700	14,100	0.257	0.528	414	452	26.77	0.64	40.80	0.97
1.000	*25.4*	**0.102**	*2.59*	0.097	0.796	0.981	23,000	25,300	13,600	15,200	0.275	0.510	437	480	25.85	0.62	40.80	0.97
1.000	*25.4*	**0.109**	*2.77*	0.104	0.782	1.040	24,400	26,800	14,600	16,200	0.293	0.493	458	507	24.95	0.59	40.80	0.97
1.250	*31.8*	**0.080**	*2.03*	0.076	1.090	1.002	23,500	25,900	8,700	9,600	0.280	0.947	597	634	48.47	1.15	63.75	1.52
1.250	*31.8*	**0.087**	*2.21*	0.083	1.076	1.083	25,400	28,000	9,500	10,500	0.304	0.923	641	684	47.24	1.12	63.75	1.52
1.250	*31.8*	**0.095**	*2.41*	0.090	1.060	1.175	27,600	30,300	10,200	11,400	0.328	0.899	683	733	45.84	1.09	63.75	1.52
1.250	*31.8*	**0.102**	*2.59*	0.097	1.046	1.254	29,400	32,400	11,000	12,200	0.351	0.876	724	781	44.64	1.06	63.75	1.52
1.250	*31.8*	**0.109**	*2.77*	0.104	1.032	1.332	31,300	34,400	11,800	13,100	0.374	0.853	763	828	43.45	1.03	63.75	1.52
1.250	*31.8*	**0.116**	*2.95*	0.111	1.018	1.408	33,100	36,400	12,500	13,900	0.397	0.830	800	873	42.28	1.01	63.75	1.52
1.250	*31.8*	**0.125**	*3.18*	0.118	1.000	1.506	35,300	38,900	13,300	14,800	0.420	0.808	836	917	40.80	0.97	63.75	1.52
1.250	*31.8*	**0.134**	*3.40*	0.128	0.982	1.601	37,600	41,300	14,400	16,000	0.451	0.776	885	978	39.34	0.94	63.75	1.52
1.250	*31.8*	**0.145**	*3.68*	0.138	0.960	1.715	40,300	44,300	15,400	17,100	0.482	0.745	931	1,036	37.60	0.90	63.75	1.52
1.250	*31.8*	**0.156**	*3.96*	0.148	0.938	1.827	42,900	47,200	16,500	18,300	0.512	0.715	975	1,093	35.90	0.85	63.75	1.52
1.250	*31.8*	**0.175**	*4.45*	0.167	0.900	2.014	47,300	52,000	18,400	20,400	0.568	0.659	1,050	1,193	33.05	0.79	63.75	1.52
1.500	*38.1*	**0.095**	*2.41*	0.090	1.310	1.429	33,500	36,900	8,600	9,500	0.399	1.368	1,020	1,083	70.02	1.67	91.80	2.19
1.500	*38.1*	**0.102**	*2.59*	0.097	1.296	1.527	35,800	39,400	9,200	10,200	0.428	1.340	1,084	1,156	68.53	1.63	91.80	2.19
1.500	*38.1*	**0.109**	*2.77*	0.104	1.282	1.623	38,100	41,900	9,900	11,000	0.456	1.311	1,146	1,227	67.06	1.60	91.80	2.19
1.500	*38.1*	**0.116**	*2.95*	0.111	1.268	1.719	40,300	44,400	10,500	11,700	0.484	1.283	1,206	1,297	65.60	1.56	91.80	2.19
1.500	*38.1*	**0.125**	*3.18*	0.118	1.250	1.840	43,200	47,500	11,200	12,400	0.512	1.255	1,264	1,365	63.75	1.52	91.80	2.19
1.500	*38.1*	**0.134**	*3.40*	0.128	1.232	1.960	46,000	50,600	12,100	13,400	0.552	1.215	1,343	1,460	61.93	1.47	91.80	2.19
1.500	*38.1*	**0.145**	*3.68*	0.138	1.210	2.104	49,400	54,300	13,000	14,400	0.590	1.177	1,419	1,552	59.74	1.42	91.80	2.19
1.500	*38.1*	**0.156**	*3.96*	0.148	1.188	2.245	52,700	58,000	13,900	15,400	0.629	1.139	1,491	1,641	57.58	1.37	91.80	2.19
1.500	*38.1*	**0.175**	*4.45*	0.167	1.150	2.483	58,300	64,100	15,500	17,300	0.699	1.068	1,618	1,802	53.96	1.28	91.80	2.19
1.500	*38.1*	**0.190**	*4.83*	0.180	1.120	2.665	62,600	68,800	16,700	18,500	0.746	1.021	1,699	1,907	51.18	1.22	91.80	2.19

(*Continued*)

TABLE 10.12 (*Continued*)

DIMENSIONS (Inches)						Nominal	TUBE LOAD BODY (Lbs.)		INTERNAL PRESSURE (psi)		TUBING AREA (sq. in.)		TORSIONAL YIELD (ft.-lb.)		INTERNAL CAPACITY per 1,000 ft.		EXTERNAL DISPLACEMENT per 1,000 ft.	
OD Specified	OD (mm)	Wall Specified	Wall (mm)	Wall Minimum	ID Calculated	Weight lb./ft.	Yield Minimum	Tensile Minimum	Hydro Test 90%	Internal Yield Min	w/min. Wall	Internal Min	Yield	Ultimate	Gallons	Barrels	Gallons	Barrels
1.750	*44.5*	**0.109**	*2.77*	0.104	1.532	1.915	45,000	49,500	8,500	9,400	0.538	1.867	1,608	1,705	95.76	2.28	124.95	2.97
1.750	*44.5*	**0.116**	*2.95*	0.111	1.518	2.029	47,600	52,400	9,000	10,000	0.572	1.834	1,695	1,804	94.02	2.24	124.95	2.97
1.750	*44.5*	**0.125**	*3.18*	0.118	1.500	2.175	51,100	56,200	9,600	10,700	0.605	1.800	1,780	1,902	91.80	2.19	124.95	2.97
1.750	*44.5*	**0.134**	*3.40*	0.128	1.482	2.318	54,400	59,900	10,400	11,500	0.652	1.753	1,898	2,039	89.61	2.13	124.95	2.97
1.750	*44.5*	**0.145**	*3.68*	0.138	1.460	2.492	58,500	64,300	11,200	12,400	0.699	1.706	2,011	2,172	86.97	2.07	124.95	2.97
1.750	*44.5*	**0.156**	*3.96*	0.148	1.438	2.662	62,500	68,700	12,000	13,300	0.745	1.660	2,119	2,302	84.37	2.01	124.95	2.97
1.750	*44.5*	**0.175**	*4.45*	0.167	1.400	2.951	69,300	76,200	13,400	14,900	0.831	1.575	2,313	2,538	79.97	1.90	124.95	2.97
1.750	*44.5*	**0.190**	*4.83*	0.180	1.370	3.173	74,500	81,900	14,400	16,000	0.888	1.517	2,437	2,693	76.58	1.82	124.95	2.97
1.750	*44.5*	**0.204**	*5.18*	0.195	1.342	3.377	79,300	87,200	15,500	17,300	0.953	1.453	2,571	2,864	73.48	1.75	124.95	2.97
2.000	*50.8*	**0.109**	*2.77*	0.104	1.782	2.207	51,800	57,000	7,400	8,300	0.619	2.522	2,148	2,261	129.56	3.08	163.20	3.39
2.000	*50.8*	**0.116**	*2.95*	0.111	1.768	2.340	54,900	60,400	7,900	8,800	0.659	2.483	2,268	2,396	127.53	3.04	163.20	3.89
2.000	*50.8*	**0.125**	*3.18*	0.118	1.750	2.509	58,900	64,800	8,400	9,400	0.698	2.444	2,386	2,529	124.95	2.97	163.20	3.89
2.000	*50.8*	**0.134**	*3.40*	0.128	1.732	2.677	62,800	69,100	9,100	10,100	0.753	2.389	2,549	2,715	122.39	2.91	163.20	3.89
2.000	*50.8*	**0.145**	*3.68*	0.138	1.710	2.880	67,600	74,400	9,800	10,900	0.807	2.334	2,706	2,896	119.30	2.84	163.20	3.89
2.000	*50.8*	**0.156**	*3.96*	0.148	1.688	3.080	72,300	79,500	10,500	11,700	0.861	2.280	2,858	3,074	116.25	2.77	163.20	3.89
2.000	*50.8*	**0.175**	*4.45*	0.167	1.650	3.419	80,300	88,300	11,800	13,100	0.962	2.180	3,133	3,400	111.08	2.64	163.20	3.89
2.000	*50.8*	**0.190**	*4.83*	0.180	1.620	3.682	86,400	95,100	12,700	14,100	1.029	2.112	3,310	3,614	107.08	2.55	163.20	3.89
2.000	*50.8*	**0.204**	*5.18*	0.195	1.592	3.923	92,100	101,300	13,700	15,200	1.106	2.036	3,505	3,854	103.41	2.46	163.20	3.89
2.375	*60.3*	**0.125**	*3.18*	0.118	2.125	3.011	70,700	77,800	7,100	7,900	0.837	3.593	3,461	3,635	184.24	4.39	230.14	5.48
2.375	*60.3*	**0.134**	*3.40*	0.128	2.107	3.215	75,500	83,000	7,700	8,600	0.904	3.527	3,707	3,909	181.13	4.31	230.14	5.48
2.375	*60.3*	**0.145**	*3.68*	0.138	2.085	3.462	81,300	89,400	8,300	9,200	0.970	3.460	3,945	4,178	177.37	4.22	230.14	5.48
2.375	*60.3*	**0.156**	*3.96*	0.148	2.063	3.706	87,000	95,700	8,900	9,900	1.035	3.395	4,177	4,442	173.64	4.13	230.14	5.48
2.375	*60.3*	**0.175**	*4.45*	0.167	2.025	4.122	96,800	106,400	10,000	11,100	1.158	3.272	4,600	4,929	167.31	3.98	230.14	5.48
2.375	*60.3*	**0.190**	*4.83*	0.180	1.995	4.445	104,300	114,800	10,800	12,000	1.241	3.189	4,876	5,252	162.38	3.87	230.14	5.48
2.375	*60.3*	**0.204**	*5.18*	0.195	1.967	4.742	111,300	122,400	11,600	12,900	1.335	3.095	5,181	5,614	157.86	3.76	230.14	5.48
2.375	*60.3*	**0.224**	*5.69*	0.214	1.927	5.159	121,100	133,200	12,700	14,100	1.453	2.977	5,548	6,058	151.50	3.61	230.14	5.48

Test pressure value equals 90% of internal yield pressure rating. Maximum working pressure is a function of tube condition and is determined by user.
*Available as continuously milled tubing (CM™) or conventional butt-welded tubing sections (W™). All data is for new tubing at minimum strength. Other sizes and wall thicknesses available on request. See individual size sheets for additional wall thicknesses.

TABLE 10.13 HS80 Grade USC Units (courtesy of Precision Tube Technology, Inc.)

DIMENSIONS (mm)						Nominal	TUBE LOAD BODY (Newtons)		INTERNAL PRESSURE (kPa)		TUBING AREA (sq. cm)		TORSIONAL YIELD (N-m)		INTERNAL CAPACITY per meter	EXTERNAL DISPLACEMENT per meter
OD Specified	OD (inches)	Wall Specified	Wall (inches)	Wall Minimum	ID Calculated	Weight Kg/m	Yield Minimum	Tensile Minimum	Hydro Test 90%	Internal Yield Min	w/min. Wall	Internal Min	Yield	Ultimate	Liters	Liters
25.4	*1.000*	**2.03**	*0.080*	1.93	21.3	1.17	82,300	90,300	74,500	82,700	1.42	3.64	490	530	0.36	0.51
25.4	*1.000*	**2.21**	*0.087*	2.11	21.0	1.26	89,000	97,900	80,700	89,600	1.54	3.52	530	570	0.35	0.51
25.4	*1.000*	**2.41**	*0.095*	2.29	20.6	1.37	96,100	105,900	87,600	97,200	1.66	3.41	560	610	0.33	0.51
25.4	*1.000*	**2.59**	*0.102*	2.46	20.2	1.46	102,300	112,500	93,800	104,800	1.78	3.29	590	650	0.32	0.51
25.4	*1.000*	**2.77**	*0.109*	2.64	19.9	1.55	108,500	119,200	100,700	111,700	1.89	3.18	620	690	0.31	0.51
31.8	*1.250*	**2.03**	*0.080*	1.93	27.7	1.49	104,500	115,200	60,000	66,200	1.81	6.11	810	860	0.60	0.79
31.8	*1.250*	**2.21**	*0.087*	2.11	27.3	1.61	113,000	124,500	65,500	72,400	1.96	5.95	870	930	0.59	0.79
31.8	*1.250*	**2.41**	*0.095*	2.29	26.9	1.75	122,800	134,800	70,300	78,600	2.12	5.80	930	990	0.57	0.79
31.8	*1.250*	**2.59**	*0.102*	2.46	26.6	1.86	130,800	144,100	75,800	84,100	2.27	5.65	980	1,060	0.55	0.79
31.8	*1.250*	**2.77**	*0.109*	2.64	26.2	1.98	139,200	153,000	81,400	90,300	2.42	5.50	1,030	1,120	0.54	0.79
31.8	*1.250*	**2.95**	*0.116*	2.82	25.9	2.09	147,200	161,900	86,200	95,800	2.56	5.35	1,090	1,180	0.53	0.79
31.8	*1.250*	**3.18**	*0.125*	3.00	25.4	2.24	157,000	173,000	91,700	102,000	2.71	5.21	1,130	1,240	0.51	0.79
31.8	*1.250*	**3.40**	*0.134*	3.25	24.9	2.38	167,200	183,700	99,300	110,300	2.91	5.01	1,200	1,330	0.49	0.79
31.8	*1.250*	**3.68**	*0.145*	3.51	24.4	2.55	179,300	197,000	106,200	117,900	3.11	4.81	1,260	1,410	0.47	0.79
31.8	*1.250*	**3.96**	*0.156*	3.76	23.8	2.72	190,800	209,900	113,800	126,200	3.31	4.61	1,320	1,480	0.45	0.79
31.8	*1.250*	**4.45**	*0.175*	4.24	22.9	2.99	210,400	231,300	126,900	140,700	3.67	4.25	1,420	1,620	0.41	0.79
38.1	*1.500*	**2.41**	*0.095*	2.29	33.3	2.12	149,000	164,100	59,300	65,500	2.57	8.83	1,380	1,470	0.87	1.14
38.1	*1.500*	**2.59**	*0.102*	2.46	32.9	2.27	159,200	175,300	63,400	70,300	2.76	8.64	1,470	1,570	0.85	1.14
38.1	*1.500*	**2.77**	*0.109*	2.64	32.6	2.41	169,500	186,400	68,300	75,800	2.94	8.46	1,550	1,660	0.83	1.14
38.1	*1.500*	**2.95**	*0.116*	2.82	32.2	2.55	179,300	197,500	72,400	80,700	3.12	8.28	1,640	1,760	0.81	1.14
38.1	*1.500*	**3.18**	*0.125*	3.00	31.8	2.73	192,200	211,300	77,200	85,500	3.31	8.10	1,710	1,850	0.79	1.14
38.1	*1.500*	**3.40**	*0.134*	3.25	31.3	2.91	204,600	225,100	83,400	92,400	3.56	7.84	1,820	1,980	0.77	1.14
38.1	*1.500*	**3.68**	*0.145*	3.51	30.7	3.13	219,700	241,500	89,600	99,300	3.81	7.59	1,920	2,100	0.74	1.14
38.1	*1.500*	**3.96**	*0.156*	3.76	30.2	3.34	234,400	258,000	95,800	106,200	4.06	7.35	2,020	2,230	0.72	1.14
38.1	*1.500*	**4.45**	*0.175*	4.24	29.2	3.69	259,300	285,100	106,900	119,300	4.51	6.89	2,190	2,440	0.67	1.14
38.1	*1.500*	**4.83**	*0.190*	4.57	28.4	3.96	278,400	306,000	115,100	127,600	4.82	6.59	2,300	2,590	0.64	1.14

(Continued)

TABLE 10.13 *(Continued)*

DIMENSIONS (mm)						Nominal	TUBE LOAD BODY (Newtons)		INTERNAL PRESSURE (kPa)		TUBING AREA (sq. cm)		TORSIONAL YIELD (N-m)		INTERNAL CAPACITY per meter	EXTERNAL DISPLACEMENT per meter
OD Specified	OD (inches)	Wall Specified	Wall (inches)	Wall Minimum	ID Calculated	Weight Kg/m	Yield Minimum	Tensile Minimum	Hydro Test 90%	Internal Yield Min	w/min. Wall	Internal Min	Yield	Ultimate	Liters	Liters
44.5	*1.750*	**2.77**	*0.109*	2.64	38.9	2.85	200,200	220,200	58,600	64,800	3.47	12.05	2,180	2,310	1.19	1.55
44.5	*1.750*	**2.95**	*0.116*	2.82	38.6	3.02	211,700	233,100	62,100	69,000	3.69	11.83	2,300	2,450	1.17	1.55
44.5	*1.750*	**3.18**	*0.125*	3.00	38.1	3.23	227,300	250,000	66,200	73,800	3.90	11.61	2,410	2,580	1.14	1.55
44.5	*1.750*	**3.40**	*0.134*	3.25	37.6	3.45	242,000	266,400	71,700	79,300	4.21	11.31	2,570	2,760	1.11	1.55
44.5	*1.750*	**3.68**	*0.145*	3.51	37.1	3.70	260,200	286,000	77,200	85,500	4.51	11.01	2,730	2,950	1.08	1.55
44.5	*1.750*	**3.96**	*0.156*	3.76	36.5	3.96	278,000	305,600	82,700	91,700	4.81	10.71	2,870	3,120	1.05	1.55
44.5	*1.750*	**4.45**	*0.175*	4.24	35.6	4.39	308,200	338,900	92,400	102,700	5.36	10.16	3,140	3,440	0.99	1.55
44.5	*1.750*	**4.83**	*0.190*	4.57	34.8	4.72	331,400	364,300	99,300	110,300	5.73	9.79	3,300	3,650	0.95	1.55
44.5	*1.750*	**5.18**	*0.204*	4.95	34.1	5.02	352,700	387,900	106,900	119,300	6.15	9.37	3,490	3,880	0.91	1.55
50.8	*2.000*	**2.77**	*0.109*	2.64	45.3	3.28	230,400	253,500	51,000	57,200	4.00	16.27	2,910	3,070	1.61	2.03
50.8	*2.000*	**2.95**	*0.116*	2.82	44.9	3.48	244,200	268,700	54,500	60,700	4.25	16.02	3,080	3,250	1.58	2.03
50.8	*2.000*	**3.18**	*0.125*	3.00	44.5	3.73	262,000	288,200	57,900	64,800	4.50	15.77	3,240	3,430	1.55	2.03
50.8	*2.000*	**3.40**	*0.134*	3.25	44.0	3.98	279,300	307,400	62,700	69,600	4.86	15.41	3,460	3,680	1.52	2.03
50.8	*2.000*	**3.68**	*0.145*	3.51	43.4	4.28	300,700	330,900	67,600	75,200	5.21	15.06	3,670	3,930	1.48	2.03
50.8	*2.000*	**3.96**	*0.156*	3.76	42.9	4.58	321,600	353,600	72,400	80,700	5.56	14.71	3,880	4,170	1.44	2.03
50.8	*2.000*	**4.45**	*0.175*	4.24	41.9	5.08	357,200	392,800	81,400	90,300	6.20	14.06	4,250	4,610	1.38	2.03
50.8	*2.000*	**4.83**	*0.190*	4.57	41.1	5.47	384,300	423,000	87,600	97,200	6.64	13.63	4,490	4,900	1.33	2.03
50.8	*2.000*	**5.18**	*0.204*	4.95	40.4	5.83	409,700	450,600	94,500	104,800	7.13	13.13	4,750	5,230	1.28	2.03
60.3	*2.375*	**3.18**	*0.125*	3.00	54.0	4.47	314,500	346,100	49,000	54,500	5.40	23.18	4,690	4,930	2.29	2.86
60.3	*2.375*	**3.40**	*0.134*	3.25	53.5	4.78	335,800	369,200	53,100	59,300	5.83	22.75	5,030	5,300	2.25	2.86
60.3	*2.375*	**3.68**	*0.145*	3.51	53.0	5.14	361,600	397,700	57,200	63,400	6.26	22.32	5,350	5,670	2.20	2.86
60.3	*2.375*	**3.96**	*0.156*	3.76	52.4	5.51	387,000	425,700	61,400	68,300	6.68	21.90	5,660	6,020	2.16	2.86
60.3	*2.375*	**4.45**	*0.175*	4.24	51.4	6.13	430,600	473,300	69,000	76,500	7.47	21.11	6,240	6,680	2.08	2.86
60.3	*2.375*	**4.83**	*0.190*	4.57	50.7	6.61	463,900	510,600	74,500	82,700	8.01	20.57	6,610	7,120	2.02	2.86
60.3	*2.375*	**5.18**	*0.204*	4.95	50.0	7.05	495,100	544,400	80,000	88,900	8.62	19.97	7,030	7,610	1.96	2.86
60.3	*2.375*	**5.69**	*0.224*	5.44	48.9	7.67	538,700	592,500	87,600	97,200	9.37	19.21	7,520	8,210	1.88	2.85

Test pressure value equals 90% of internal yield pressure rating. Maximum working pressure is a function of tube condition and is determined by user.
*Available as continuously milled tubing (CM™) or conventional butt-welded tubing sections (W™). All data is for new tubing at minimum strength. Other sizes and wall thicknesses available on request. See individual size sheets for additional wall thicknesses.

TABLE 10.14 HS90 Grade USC Units (courtesy of Precision Tube Technology, Inc.)

DIMENSIONS (Inches)						Nominal	TUBE LOAD BODY (Lbs.)		INTERNAL PRESSURE (psi)		TUBING AREA (sq. in.)		TORSIONAL YIELD (ft.-lb.)		INTERNAL CAPACITY per 1,000 ft.		EXTERNAL DISPLACEMENT per 1,000 ft.	
OD Specified	OD (mm)	Wall Specified	Wall (mm)	Wall Minimum	ID Calculated	Weight lb./ft.	Yield Minimum	Tensile Minimum	Hydro Test 90%	Internal Yield Min.	w/min. Wall	Internal Min.	Yield	Ultimate	Gallons	Barrels	Gallons	Barrels
1.000	*25.4*	**0.087**	*2.21*	0.083	0.826	0.850	22,500	24,200	13,200	14,700	0.239	0.546	439	476	27.84	0.66	40.80	0.97
1.000	*25.4*	**0.095**	*2.41*	0.090	0.810	0.920	24,300	26,200	14,300	15,900	0.257	0.528	466	508	26.77	0.64	40.80	0.97
1.000	*25.4*	**0.102**	*2.59*	0.097	0.796	0.981	25,900	27,900	15,300	17,000	0.275	0.510	491	540	25.85	0.62	40.80	0.97
1.000	*25.4*	**0.109**	*2.77*	0.104	0.782	1.040	27,500	29,600	16,400	18,200	0.293	0.493	515	570	24.95	0.59	40.80	0.97
1.250	*31.8*	**0.087**	*2.21*	0.083	1.076	1.083	28,600	30,800	10,600	11,800	0.304	0.923	721	770	47.24	1.12	63.75	1.52
1.250	*31.8*	**0.095**	*2.41*	0.090	1.060	1.175	31,000	33,400	11,500	12,800	0.328	0.899	769	825	45.84	1.09	63.75	1.52
1.250	*31.8*	**0.102**	*2.59*	0.097	1.046	1.254	33,100	35,700	12,400	13,800	0.351	0.876	814	879	44.64	1.06	63.75	1.52
1.250	*31.8*	**0.109**	*2.77*	0.104	1.032	1.332	35,200	37,900	13,200	14,700	0.374	0.853	858	931	43.45	1.03	63.75	1.52
1.250	*31.8*	**0.116**	*2.95*	0.111	1.018	1.408	37,200	40,100	14,100	15,700	0.397	0.830	900	982	42.28	1.01	63.75	1.52
1.250	*31.8*	**0.125**	*3.18*	0.118	1.000	1.506	39,800	42,900	14,900	16,600	0.420	0.808	941	1,032	40.80	0.97	63.75	1.52
1.250	*31.8*	**0.134**	*3.40*	0.128	0.982	1.601	42,300	45,600	16,200	17,900	0.451	0.776	996	1,100	39.34	0.94	63.75	1.52
1.250	*31.8*	**0.145**	*3.68*	0.138	0.960	1.715	45,300	48,800	17,300	19,300	0.482	0.745	1,048	1,166	37.60	0.90	63.75	1.52
1.250	*31.8*	**0.156**	*3.96*	0.148	0.938	1.827	48,300	52,000	18,500	20,600	0.512	0.715	1,097	1,229	35.90	0.85	63.75	1.52
1.250	*31.8*	**0.175**	*4.45*	0.167	0.900	2.014	53,200	57,300	20,700	23,000	0.568	0.659	1,181	1,342	33.05	0.79	63.75	1.52
1.500	*38.1*	**0.095**	*2.41*	0.090	1.310	1.429	37,700	40,700	9,600	10,700	0.399	1.368	1,148	1,218	70.02	1.67	91.80	2.19
1.500	*38.1*	**0.102**	*2.59*	0.097	1.296	1.527	40,300	43,500	10,400	11,500	0.428	1.340	1,220	1,300	68.53	1.63	91.80	2.19
1.500	*38.1*	**0.109**	*2.77*	0.104	1.282	1.623	42,900	46,200	11,100	12,300	0.456	1.311	1,289	1,380	67.06	1.60	91.80	2.19
1.500	*38.1*	**0.116**	*2.95*	0.111	1.268	1.719	45,400	48,900	11,800	13,100	0.484	1.283	1,357	1,459	65.60	1.56	91.80	2.19
1.500	*38.1*	**0.125**	*3.18*	0.118	1.250	1.840	48,600	52,400	12,500	13,900	0.512	1.255	1,422	1,536	63.75	1.52	91.80	2.19
1.500	*38.1*	**0.134**	*3.40*	0.128	1.232	1.960	51,800	55,800	13,600	15,100	0.552	1.215	1,511	1,643	61.93	1.47	91.80	2.19
1.500	*38.1*	**0.145**	*3.68*	0.138	1.210	2.104	55,600	59,900	14,600	16,200	0.590	1.177	1,596	1,746	59.74	1.42	91.80	2.19
1.500	*38.1*	**0.156**	*3.96*	0.148	1.188	2.245	59,300	63,900	15,600	17,300	0.629	1.139	1,677	1,846	57.58	1.37	91.80	2.19
1.500	*38.1*	**0.175**	*4.45*	0.167	1.150	2.483	65,600	70,700	17,500	19,400	0.699	1.068	1,821	2,028	53.96	1.28	91.80	2.19
1.500	*38.1*	**0.190**	*4.83*	0.180	1.120	2.665	70,400	75,800	18,700	20,800	0.746	1.021	1,911	2,145	51.18	1.22	91.80	2.19
1.750	*44.5*	**0.109**	*2.77*	0.104	1.532	1.915	50,600	54,500	9,500	10,600	0.538	1.867	1,809	1,918	95.76	2.28	124.95	2.97
1.750	*44.5*	**0.116**	*2.95*	0.111	1.518	2.029	53,600	57,800	10,200	11,300	0.572	1.834	1,907	2,030	94.02	2.24	124.95	2.97
1.750	*44.5*	**0.125**	*3.18*	0.118	1.500	2.175	57,400	61,900	10,800	12,000	0.605	1.800	2,003	2,140	91.80	2.19	124.95	2.97

(Continued)

TABLE 10.14 (*Continued*)

DIMENSIONS (Inches)						Nominal	TUBE LOAD BODY (Lbs.)		INTERNAL PRESSURE (psi)		TUBING AREA (sq. in.)		TORSIONAL YIELD (ft.-lb.)		INTERNAL CAPACITY per 1,000 ft.		EXTERNAL DISPLACEMENT per 1,000 ft.	
OD Specified	OD (mm)	Wall Specified	Wall (mm)	Wall Minimum	ID Calculated	Weight lb./ft.	Yield Minimum	Tensile Minimum	Hydro Test 90%	Internal Yield Min.	w/min. Wall	Internal Min.	Yield	Ultimate	Gallons	Barrels	Gallons	Barrels
1.750	*44.5*	**0.134**	*3.40*	0.128	1.482	2.318	61,200	66,000	11,700	13,000	0.652	1.753	2,135	2,294	89.61	2.13	124.95	2.97
1.750	*44.5*	**0.145**	*3.68*	0.138	1.460	2.492	65,800	70,900	12,600	14,000	0.699	1.706	2,262	2,444	86.97	2.07	124.95	2.97
1.750	*44.5*	**0.156**	*3.96*	0.148	1.438	2.662	70,300	75,800	13,500	14,900	0.745	1.660	2,384	2,589	84.37	2.01	124.95	2.97
1.750	*44.5*	**0.175**	*4.45*	0.167	1.400	2.951	77,900	84,000	15,100	16,800	0.831	1.575	2,602	2,855	79.97	1.90	124.95	2.97
1.750	*44.5*	**0.190**	*4.83*	0.180	1.370	3.173	83,800	90,300	16,200	18,000	0.888	1.517	2,741	3,029	76.58	1.82	124.95	2.97
1.750	*44.5*	**0.204**	*5.18*	0.195	1.342	3.377	89,200	96,100	17,500	19,400	0.953	1.453	2,893	3,222	73.48	1.75	124.95	2.97
2.000	*50.8*	**0.109**	*2.77*	0.104	1.782	2.207	58,300	62,800	8,400	9,300	0.619	2.522	2,416	2,544	129.56	3.08	163.20	3.89
2.000	*50.8*	**0.116**	*2.95*	0.111	1.768	2.340	61,800	66,600	8,900	9,900	0.659	2.483	2,552	2,696	127.53	3.04	163.20	3.89
2.000	*50.8*	**0.125**	*3.18*	0.118	1.750	2.509	66,300	71,400	9,500	10,500	0.698	2.444	2,684	2,845	124.95	2.97	163.20	3.89
2.000	*50.8*	**0.134**	*3.40*	0.128	1.732	2.677	70,700	76,200	10,300	11,400	0.753	2.389	2,867	3,054	122.39	2.91	163.20	3.89
2.000	*50.8*	**0.145**	*3.68*	0.138	1.710	2.880	76,100	82,000	11,000	12,300	0.807	2.334	3,045	3,258	119.30	2.84	163.20	3.89
2.000	*50.8*	**0.156**	*3.96*	0.148	1.688	3.080	81,300	87,700	11,800	13,100	0.861	2.280	3,216	3,458	116.25	2.77	163.20	3.89
2.000	*50.8*	**0.175**	*4.45*	0.167	1.650	3.419	90,300	97,300	13,300	14,800	0.962	2.180	3,525	3,825	111.08	2.64	163.20	3.89
2.000	*50.8*	**0.190**	*4.83*	0.180	1.620	3.682	97,200	104,800	14,300	15,900	1.029	2.112	3,724	4,066	107.08	2.55	163.20	3.89
2.000	*50.8*	**0.204**	*5.18*	0.195	1.592	3.923	103,600	111,600	15,400	17,100	1.106	2.036	3,943	4,335	103.41	2.46	163.20	3.89
2.375	*60.3*	**0.125**	*3.18*	0.118	2.125	3.011	79,500	85,700	8,000	8,900	0.837	3.593	3,894	4,090	184.24	4.39	230.14	5.48
2.375	*60.3*	**0.134**	*3.40*	0.128	2.107	3.215	84,900	91,500	8,700	9,600	0.904	3.527	4,170	4,398	181.13	4.31	230.14	5.48
2.375	*60.3*	**0.145**	*3.68*	0.138	2.085	3.462	91,400	98,500	9,300	10,400	0.970	3.460	4,438	4,700	177.37	4.22	230.14	5.48
2.375	*60.3*	**0.156**	*3.96*	0.148	2.063	3.706	97,900	105,500	10,000	11,100	1.035	3.395	4,699	4,997	173.64	4.13	230.14	5.48
2.375	*60.3*	**0.175**	*4.45*	0.167	2.025	4.122	108,900	117,300	11,200	12,500	1.158	3.272	5,175	5,545	167.31	3.98	230.14	5.48
2.375	*60.3*	**0.190**	*4.83*	0.180	1.995	4.445	117,400	126,500	12,100	13,400	1.241	3.189	5,485	5,908	162.38	3.87	230.14	5.48
2.375	*60.3*	**0.204**	*5.18*	0.195	1.967	4.742	125,200	135,000	13,100	14,500	1.335	3.095	5,829	6,316	157.86	3.76	230.14	5.48

Test pressure value equals 90% of internal yield pressure rating. Maximum working pressure is a function of tube condition and is determined by user.

* Available as continuously milled tubing (CM™) or conventional butt-welded tubing sections (W™). All data is for new tubing at minimum strength. Other sizes and wall thicknesses available on request. See individual size sheets for additional wall thicknesses.

TABLE 10.15 HS90 Grade Metric Units (courtesy of Precision Tube Technology, Inc.)

DIMENSIONS (mm)						Nominal	TUBE LOAD BODY (Newtons)		INTERNAL PRESSURE (kPa)		TUBING AREA (sq. cm)		TORSIONAL YIELD (N-m)		INTERNAL CAPACITY per meter	EXTERNAL DISPLACEMENT per meter
OD Specified	OD (inches)	Wall Specified	Wall (inches)	Wall Minimum	ID Calculated	Weight Kg/m	Yield Minimum	Tensile Minimum	Hydro Test 90%	Internal Yield Min.	w/min. Wall	Internal Min.	Yield	Ultimate	Liters	Liters
25.4	*1.000*	**2.21**	*0.087*	2.11	21.0	1.26	100,100	107,600	91,000	101,400	1.54	3.52	590	650	0.35	0.51
25.4	*1.000*	**2.41**	*0.095*	2.29	20.6	1.37	108,100	116,500	98,600	109,600	1.66	3.41	630	690	0.33	0.51
25.4	*1.000*	**2.59**	*0.102*	2.46	20.2	1.46	115,200	124,100	105,500	117,200	1.78	3.29	670	730	0.32	0.51
25.4	*1.000*	**2.77**	*0.109*	2.64	19.9	1.55	122,300	131,700	113,100	125,500	1.89	3.18	700	770	0.31	0.51
31.8	*1.250*	**2.21**	*0.087*	2.11	27.3	1.61	127,200	137,000	73,100	81,400	1.96	5.95	980	1,040	0.59	0.79
31.8	*1.250*	**2.41**	*0.095*	2.29	26.9	1.75	137,900	148,600	79,300	88,300	2.12	5.80	1,040	1,120	0.57	0.79
31.8	*1.250*	**2.59**	*0.102*	2.46	26.6	1.86	147,200	158,800	85,500	95,200	2.27	5.65	1,100	1,190	0.55	0.79
31.8	*1.250*	**2.77**	*0.109*	2.64	26.2	1.98	156,600	168,600	91,000	101,400	2.42	5.50	1,160	1,260	0.54	0.79
31.8	*1.250*	**2.95**	*0.116*	2.82	25.9	2.09	165,500	178,400	97,200	108,300	2.56	5.35	1,220	1,330	0.53	0.79
31.8	*1.250*	**3.18**	*0.125*	3.00	25.4	2.24	177,000	190,800	102,700	114,500	2.71	5.21	1,280	1,400	0.51	0.79
31.8	*1.250*	**3.40**	*0.134*	3.25	24.9	2.38	188,200	202,800	111,700	123,400	2.91	5.01	1,350	1,490	0.49	0.79
31.8	*1.250*	**3.68**	*0.145*	3.51	24.4	2.55	201,500	217,100	119,300	133,100	3.11	4.81	1,420	1,580	0.47	0.79
31.8	*1.250*	**3.96**	*0.156*	3.76	23.8	2.72	214,800	231,300	127,600	142,000	3.31	4.61	1,490	1,670	0.45	0.79
31.8	*1.250*	**4.45**	*0.175*	4.24	22.9	2.99	236,600	254,900	142,700	158,600	3.67	4.25	1,600	1,820	0.41	0.79
38.1	*1.500*	**2.41**	*0.095*	2.29	33.3	2.12	167,700	181,000	66,200	73,800	2.57	8.83	1,560	1,650	0.87	1.14
38.1	*1.500*	**2.59**	*0.102*	2.46	32.9	2.27	179,300	193,500	71,700	79,300	2.76	8.64	1,650	1,760	0.85	1.14
38.1	*1.500*	**2.77**	*0.109*	2.64	32.6	2.41	190,800	205,500	76,500	84,800	2.94	8.46	1,750	1,870	0.83	1.14
38.1	*1.500*	**2.95**	*0.116*	2.82	32.2	2.55	201,900	217,500	81,400	90,300	3.12	8.28	1,840	1,980	0.81	1.14
38.1	*1.500*	**3.18**	*0.125*	3.00	31.8	2.73	216,200	233,100	86,200	95,800	3.31	8.10	1,930	2,080	0.79	1.14
38.1	*1.500*	**3.40**	*0.134*	3.25	31.3	2.91	230,400	248,200	93,800	104,100	3.56	7.84	2,050	2,230	0.77	1.14
38.1	*1.500*	**3.68**	*0.145*	3.51	30.7	3.13	247,300	266,400	100,700	111,700	3.81	7.59	2,160	2,370	0.74	1.14
38.1	*1.500*	**3.96**	*0.156*	3.76	30.2	3.34	263,800	284,200	107,600	119,300	4.06	7.35	2,270	2,500	0.72	1.14
38.1	*1.500*	**4.45**	*0.175*	4.24	29.2	3.69	291,800	314,500	120,700	133,800	4.51	6.89	2,470	2,750	0.67	1.14
38.1	*1.500*	**4.83**	*0.190*	4.57	28.4	3.96	313,100	337,200	128,900	143,400	4.82	6.59	2,590	2,910	0.64	1.14

(Continued)

TABLE 10.15 (*Continued*)

DIMENSIONS (mm)						Nominal	TUBE LOAD BODY (Newtons)		INTERNAL PRESSURE (kPa)		TUBING AREA (sq. cm)		TORSIONAL YIELD (N-m)		INTERNAL CAPACITY per meter	EXTERNAL DISPLACEMENT per meter
OD Specified	OD (inches)	Wall Specified	Wall (inches)	Wall Minimum	ID Calculated	Weight Kg/m	Yield Minimum	Tensile Minimum	Hydro Test 90%	Internal Yield Min.	w/min. Wall	Internal Min.	Yield	Ultimate	Liters	Liters
44.5	*1.750*	**2.77**	*0.109*	2.64	38.9	2.85	225,100	242,400	65,500	73,100	3.47	12.05	2,450	2,600	1.19	1.55
44.5	*1.750*	**2.95**	*0.116*	2.82	38.6	3.02	238,400	257,100	70,300	77,900	3.69	11.83	2,590	2,750	1.17	1.55
44.5	*1.750*	**3.18**	*0.125*	3.00	38.1	3.23	255,300	275,300	74,500	82,700	3.90	11.61	2,720	2,900	1.14	1.55
44.5	*1.750*	**3.40**	*0.134*	3.25	37.6	3.45	272,200	293,600	80,700	89,600	4.21	11.31	2,890	3,110	1.11	1.55
44.5	*1.750*	**3.68**	*0.145*	3.51	37.1	3.70	292,700	315,400	86,900	96,500	4.51	11.01	3,070	3,310	1.08	1.55
44.5	*1.750*	**3.96**	*0.156*	3.76	36.5	3.96	312,700	337,200	93,100	102,700	4.81	10.71	3,230	3,510	1.05	1.55
44.5	*1.750*	**4.45**	*0.175*	4.24	35.6	4.39	346,500	373,600	104,100	115,800	5.36	10.16	3,530	3,870	0.99	1.55
44.5	*1.750*	**4.83**	*0.190*	4.57	34.8	4.72	372,700	401,700	111,700	124,100	5.73	9.79	3,720	4,110	0.95	1.55
44.5	*1.750*	**5.18**	*0.204*	4.95	34.1	5.02	396,800	427,500	120,700	133,800	6.15	9.37	3,920	4,370	0.91	1.55
50.8	*2.000*	**2.77**	*0.109*	2.64	45.3	3.28	259,300	279,300	57,900	64,100	4.00	16.27	3,280	3,450	1.61	2.03
50.8	*2.000*	**2.95**	*0.116*	2.82	44.9	3.48	274,900	296,200	61,400	68,300	4.25	16.02	3,460	3,660	1.58	2.03
50.8	*2.000*	**3.18**	*0.125*	3.00	44.5	3.73	294,900	317,600	65,500	72,400	4.50	15.77	3,640	3,860	1.55	2.03
50.8	*2.000*	**3.40**	*0.134*	3.25	44.0	3.98	314,500	338,900	71,000	78,600	4.86	15.41	3,890	4,140	1.52	2.03
50.8	*2.000*	**3.68**	*0.145*	3.51	43.4	4.28	338,500	364,700	75,800	84,800	5.21	15.06	4,130	4,420	1.48	2.03
50.8	*2.000*	**3.96**	*0.156*	3.76	42.9	4.58	361,600	390,100	81,400	90,300	5.56	14.71	4,360	4,690	1.44	2.03
50.8	*2.000*	**4.45**	*0.175*	4.24	41.9	5.08	401,700	432,800	91,700	102,000	6.20	14.06	4,780	5,190	1.38	2.03
50.8	*2.000*	**4.83**	*0.190*	4.57	41.1	5.47	432,300	466,200	98,600	109,600	6.64	13.63	5,050	5,510	1.33	2.03
50.8	*2.000*	**5.18**	*0.204*	4.95	40.4	5.83	460,800	496,400	106,200	117,900	7.13	13.13	5,350	5,880	1.28	2.03
60.3	*2.375*	**3.18**	*0.125*	3.00	54.0	4.47	353,600	381,200	55,200	61,400	5.40	23.18	5,280	5,550	2.29	2.86
60.3	*2.375*	**3.40**	*0.134*	3.25	53.5	4.78	377,600	407,000	60,000	66,200	5.83	22.75	5,650	5,960	2.25	2.86
60.3	*2.375*	**3.68**	*0.145*	3.51	53.0	5.14	406,500	438,100	64,100	71,700	6.26	22.32	6,020	6,370	2.20	2.86
60.3	*2.375*	**3.96**	*0.156*	3.76	52.4	5.51	435,500	469,300	69,000	76,500	6.68	21.90	6,370	6,780	2.16	2.86
60.3	*2.375*	**4.45**	*0.175*	4.24	51.4	6.13	484,400	521,800	77,200	86,200	7.47	21.11	7,020	7,520	2.08	2.86
60.3	*2.375*	**4.83**	*0.190*	4.57	50.7	6.61	522,200	562,700	83,400	92,400	8.01	20.57	7,440	8,010	2.02	2.86
60.3	*2.375*	**5.18**	*0.204*	4.95	50.0	7.05	556,900	600,500	90,300	100,000	8.62	19.97	7,900	8,560	1.96	2.86

Test pressure value equals 90% of internal yield pressure rating. Maximum working pressure is a function of tube condition and is determined by user.

*Available as continuously milled tubing (CM™) or conventional butt-welded tubing sections (W™). All data is for new tubing at minimum strength. Other sizes and wall thicknesses available on request. See individual size sheets for additional wall thicknesses.

TABLE 10.16 HS110 Grade USC Units (courtesy of Precision Tube Technology, Inc.)

DIMENSIONS (Inches)						Nominal	TUBE LOAD BODY (Lbs.)		INTERNAL PRESSURE (psi)		TUBING AREA (sq. in.)		TORSIONAL YIELD (ft.-lb.)		INTERNAL CAPACITY per 1,000 ft.		EXTERNAL DISPLACEMENT per 1,000 ft.	
OD Specified	OD (mm)	Wall Specified	Wall (mm)	Wall Minimum	ID Calculated	Weight lb./ft.	Yield Minimum	Tensile Minimum	Hydro Test 90%	Internal Yield Min	w/min. Wall	Internal Min	Yield	Ultimate	Gallons	Barrels	Gallons	Barrels
1.000	*25.4*	**0.109**	*2.77*	0.104	0.782	1.040	33,000	35,100	19,700	21,900	0.293	0.493	618	684	24.95	0.59	40.80	0.97
1.250	*31.8*	**0.109**	*2.77*	0.104	1.032	1.332	42,200	44,900	15,900	17,700	0.374	0.853	1,030	1,117	43.45	1.03	63.75	1.52
1.250	*31.8*	**0.116**	*2.95*	0.111	1.018	1.408	44,600	47,500	16,900	18,800	0.397	0.830	1,081	1,178	42.28	1.01	63.75	1.52
1.250	*31.8*	**0.125**	*3.18*	0.118	1.000	1.506	47,700	50,800	17,900	19,900	0.420	0.808	1,129	1,238	40.80	0.97	63.75	1.52
1.250	*31.8*	**0.134**	*3.40*	0.128	0.982	1.601	50,700	54,000	19,400	21,500	0.451	0.776	1,195	1,320	39.34	0.94	63.75	1.52
1.250	*31.8*	**0.145**	*3.68*	0.138	0.960	1.715	54,400	57,900	20,800	23,100	0.482	0.745	1,257	1,399	37.60	0.90	63.75	1.52
1.250	*31.8*	**0.156**	*3.96*	0.148	0.938	1.827	57,900	61,700	22,200	24,700	0.512	0.715	1,316	1,475	35.90	0.85	63.75	1.52
1.250	*31.8*	**0.175**	*4.45*	0.167	0.900	2.014	63,800	68,000	24,800	27,600	0.568	0.659	1,417	1,610	33.05	0.79	63.75	1.52
1.500	*38.1*	**0.109**	*2.77*	0.104	1.282	1.623	51,400	54,800	13,300	14,800	0.456	1.311	1,547	1,656	67.06	1.60	91.80	2.19
1.500	*38.1*	**0.116**	*2.95*	0.111	1.268	1.719	54,500	58,000	14,200	15,800	0.484	1.283	1,628	1,751	65.60	1.56	91.80	2.19
1.500	*38.1*	**0.125**	*3.18*	0.118	1.250	1.840	58,300	62,100	15,100	16,700	0.512	1.255	1,706	1,843	63.75	1.52	91.80	2.19
1.500	*38.1*	**0.134**	*3.40*	0.128	1.232	1.960	62,100	66,100	16,300	18,100	0.552	1.215	1,813	1,971	61.93	1.47	91.80	2.19
1.500	*38.1*	**0.145**	*3.68*	0.138	1.210	2.104	66,700	71,000	17,500	19,400	0.590	1.177	1,916	2,095	59.74	1.42	91.80	2.19
1.500	*38.1*	**0.156**	*3.96*	0.148	1.188	2.245	71,100	75,700	18,700	20,800	0.629	1.139	2,013	2,216	57.58	1.37	91.80	2.19
1.500	*38.1*	**0.175**	*4.45*	0.167	1.150	2.483	78,700	83,800	21,000	23,300	0.699	1.068	2,185	2,433	53.96	1.28	91.80	2.19
1.500	*38.1*	**0.190**	*4.83*	0.180	1.120	2.665	84,400	89,900	22,500	25,000	0.746	1.021	2,293	2,574	51.18	1.22	91.80	2.19
1.750	*44.5*	**0.109**	*2.77*	0.104	1.532	1.915	60,700	64,600	11,400	12,700	0.538	1.867	2,170	2,301	95.76	2.28	124.95	2.97
1.750	*44.5*	**0.116**	*2.95*	0.111	1.518	2.029	64,300	68,500	12,200	13,600	0.572	1.834	2,288	2,436	94.02	2.24	124.95	2.97
1.750	*44.5*	**0.125**	*3.18*	0.118	1.500	2.175	68,900	73,400	13,000	14,400	0.605	1.800	2,403	2,568	91.80	2.19	124.95	2.97
1.750	*44.5*	**0.134**	*3.40*	0.128	1.482	2.318	73,500	78,200	14,000	15,600	0.652	1.753	2,562	2,753	89.61	2.13	124.95	2.97
1.750	*44.5*	**0.145**	*3.68*	0.138	1.460	2.492	79,000	84,100	15,100	16,800	0.699	1.706	2,714	2,932	86.97	2.07	124.95	2.97
1.750	*44.5*	**0.156**	*3.96*	0.148	1.438	2.662	84,400	89,800	16,100	17,900	0.745	1.660	2,860	3,107	84.37	2.01	124.95	2.97
1.750	*44.5*	**0.175**	*4.45*	0.167	1.400	2.951	93,500	99,600	18,100	20,100	0.831	1.575	3,122	3,426	79.97	1.90	124.95	2.97
1.750	*44.5*	**0.190**	*4.83*	0.180	1.370	3.173	100,600	107,100	19,500	21,600	0.888	1.517	3,290	3,635	76.58	1.82	124.95	2.97

(Continued)

TABLE 10.16 (Continued)

DIMENSIONS (Inches)						Nominal	TUBE LOAD BODY (Lbs.)		INTERNAL PRESSURE (psi)		TUBING AREA (sq. in.)		TORSIONAL YIELD (ft.-lb.)		INTERNAL CAPACITY per 1,000 ft.		EXTERNAL DISPLACEMENT per 1,000 ft.	
OD Specified	OD (mm)	Wall Specified	Wall (mm)	Wall Minimum	ID Calculated	Weight lb./ft.	Yield Minimum	Tensile Minimum	Hydro Test 90%	Internal Yield Min	w/min. Wall	Internal Min	Yield	Ultimate	Gallons	Barrels	Gallons	Barrels
2.000	*50.8*	**0.109**	*2.77*	0.104	1.782	2.207	69,900	74,500	10,000	11,200	0.619	2.522	2,900	3,053	129.56	3.08	163.20	3.89
2.000	*50.8*	**0.116**	*2.95*	0.111	1.768	2.340	74,200	79,000	10,700	11,900	0.659	2.483	3,062	3,235	127.53	3.04	163.20	3.89
2.000	*50.8*	**0.125**	*3.18*	0.118	1.750	2.509	79,500	84,700	11,400	12,600	0.698	2.444	3,221	3,414	124.95	2.97	163.20	3.89
2.000	*50.8*	**0.134**	*3.40*	0.128	1.732	2.677	84,800	90,300	12,300	13,700	0.753	2.389	3,441	3,665	122.39	2.91	163.20	3.89
2.000	*50.8*	**0.145**	*3.68*	0.138	1.710	2.880	91,300	97,200	13,300	14,700	0.807	2.334	3,653	3,910	119.30	2.84	163.20	3.89
2.000	*50.8*	**0.156**	*3.96*	0.148	1.688	3.080	97,600	103,900	14,200	15,800	0.861	2.280	3,859	4,150	116.25	2.77	163.20	3.89
2.000	*50.8*	**0.175**	*4.45*	0.167	1.650	3.419	108,400	115,400	15,900	17,700	0.962	2.180	4,230	4,590	111.08	2.64	163.20	3.89
2.000	*50.8*	**0.190**	*4.83*	0.180	1.620	3.682	116,700	124,200	17,100	19,000	1.029	2.112	4,469	4,879	107.08	2.55	163.20	3.89
2.375	*60.3*	**0.125**	*3.18*	0.118	2.125	3.011	95,400	101,600	9,600	10,700	0.837	3.593	4,672	4,908	184.24	4.39	230.14	5.48
2.375	*60.3*	**0.134**	*3.40*	0.128	2.107	3.215	101,900	108,500	10,400	11,600	0.904	3.527	5,004	5,277	181.13	4.31	230.14	5.48
2.375	*60.3*	**0.145**	*3.68*	0.138	2.085	3.462	109,700	116,800	11,200	12,400	0.970	3.460	5,326	5,640	177.37	4.22	230.14	5.48
2.375	*60.3*	**0.156**	*3.96*	0.148	2.063	3.706	117,500	125,100	12,000	13,300	1.035	3.395	5,639	5,996	173.64	4.13	230.14	5.48
2.375	*60.3*	**0.175**	*4.45*	0.167	2.025	4.122	130,600	139,100	13,500	15,000	1.158	3.272	6,210	6,654	167.31	3.98	230.14	5.48
2.375	*60.3*	**0.190**	*4.83*	0.180	1.995	4.445	140,900	150,000	14,500	16,100	1.241	3.189	6,582	7,090	162.38	3.87	230.14	5.48
2.625	*66.7*	**0.134**	*3.40*	0.128	2.357	3.574	113,300	120,600	9,400	10,500	1.004	4.408	6,209	6,516	226.66	5.40	281.14	5.69
2.625	*66.7*	**0.145**	*3.68*	0.138	2.335	3.850	122,000	129,900	10,100	11,300	1.078	4.334	6,617	6,970	222.45	5.30	281.14	6.69
2.625	*66.7*	**0.156**	*3.96*	0.148	2.313	4.124	130,700	139,200	10,900	12,100	1.152	4.260	7,014	7,416	218.28	5.20	281.14	6.69
2.625	*66.7*	**0.175**	*4.45*	0.167	2.275	4.590	145,500	154,900	12,200	13,600	1.290	4.122	7,742	8,243	211.17	5.03	281.14	6.69
2.625	*66.7*	**0.190**	*4.83*	0.180	2.245	4.953	157,000	167,100	13,200	14,600	1.383	4.029	8,220	8,793	205.63	4.90	281.14	6.69
2.875	*73.0*	**0.156**	*3.96*	0.148	2.563	4.541	143,900	153,200	9,900	11,000	1.268	5.224	8,541	8,987	268.01	6.38	337.24	8.03
2.875	*73.0*	**0.175**	*4.45*	0.167	2.525	5.059	160,300	170,700	11,200	12,400	1.421	5.071	9,445	10,002	260.12	6.19	337.24	8.03
2.875	*73.0*	**0.190**	*4.83*	0.180	2.495	5.462	173,100	184,300	12,100	13,400	1.524	4.968	10,041	10,680	253.98	6.05	337.24	8.03
3.500	*88.9*	**0.175**	*4.45*	0.167	3.150	6.230	197,400	210,200	9,200	10,200	1.749	7.872	14,447	15,146	404.84	9.64	499.80	11.90
3.500	*88.9*	**0.190**	*4.83*	0.180	3.120	6.733	213,400	227,200	9,900	11,000	1.877	7.744	15,397	16,200	397.16	9.46	499.80	11.90

Test pressure value equals 90% of internal yield pressure rating. Maximum working pressure is a function of tube condition and is determined by user.

*Available as continuously milled tubing (CM™) or conventional butt-welded tubing sections (W™). All data is for new tubing at minimum strength. Other sizes and wall thicknesses available on request. See individual size sheets for additional wall thicknesses.

TABLE 10.17 HS110 Grade Metric Units (courtesy of Precision Tube Technology, Inc.)

DIMENSIONS (mm)						Nominal	TUBE LOAD BODY (Newtons)		INTERNAL PRESSURE (kPa)		TUBING AREA (sq. cm)		TORSIONAL YIELD (N-m)		INTERNAL CAPACITY per meter	EXTERNAL DISPLACEMENT per meter
OD Specified	OD (inches)	Wall Specified	Wall (inches)	Wall Minimum	ID Calculated	Weight Kg/m	Yield Minimum	Tensile Minimum	Hydro Test 90%	Internal Yield Min.	w/min. Wall	Internal Min.	Yield	Ultimate	Liters	Liters
25.4	*1.000*	**2.77**	*0.109*	2.64	19.9	1.55	146,800	156,100	135,800	151,000	1.89	3.18	840	930	0.31	0.51
31.8	*1.250*	**2.77**	*0.109*	2.64	26.2	1.98	187,700	199,700	109,600	122,000	2.42	5.50	1,400	1,510	0.54	0.79
31.8	*1.250*	**2.95**	*0.116*	2.82	25.9	2.09	198,400	211,300	116,500	129,600	2.56	5.35	1,470	1,600	0.53	0.79
31.8	*1.250*	**3.18**	*0.125*	3.00	25.4	2.24	212,200	226,000	123,400	137,200	2.71	5.21	1,530	1,680	0.51	0.79
31.8	*1.250*	**3.40**	*0.134*	3.25	24.9	2.38	225,500	240,200	133,800	148,200	2.91	5.01	1,620	1,790	0.49	0.79
31.8	*1.250*	**3.68**	*0.145*	3.51	24.4	2.55	242,000	257,500	143,400	159,300	3.11	4.81	1,700	1,900	0.47	0.79
31.8	*1.250*	**3.96**	*0.156*	3.76	23.8	2.72	257,500	274,400	153,100	170,300	3.31	4.61	1,780	2,000	0.45	0.79
31.8	*1.250*	**4.45**	*0.175*	4.24	22.9	2.99	283,800	302,500	171,000	190,300	3.67	4.25	1,920	2,180	0.41	0.79
38.1	*1.500*	**2.77**	*0.109*	2.64	32.6	2.41	228,600	243,800	91,700	102,000	2.94	8.46	2,100	2,250	0.83	1.14
38.1	*1.500*	**2.95**	*0.116*	2.82	32.2	2.55	242,400	258,000	97,900	108,900	3.12	8.28	2,210	2,370	0.81	1.14
38.1	*1.500*	**3.18**	*0.125*	3.00	31.8	2.73	259,300	276,200	104,100	115,100	3.31	8.10	2,310	2,500	0.79	1.14
38.1	*1.500*	**3.40**	*0.134*	3.25	31.3	2.91	276,200	294,000	112,400	124,800	3.56	7.84	2,460	2,670	0.77	1.14
38.1	*1.500*	**3.68**	*0.145*	3.51	30.7	3.13	296,700	315,800	120,700	133,800	3.81	7.59	2,600	2,840	0.74	1.14
38.1	*1.500*	**3.96**	*0.156*	3.76	30.2	3.34	316,300	336,700	128,900	143,400	4.06	7.35	2,730	3,000	0.72	1.14
38.1	*1.500*	**4.45**	*0.175*	4.24	29.2	3.69	350,100	372,700	144,800	160,700	4.51	6.89	2,960	3,300	0.67	1.14
38.1	*1.500*	**4.83**	*0.190*	4.57	28.4	3.96	375,400	399,900	155,100	172,400	4.82	6.59	3,110	3,490	0.64	1.14
44.5	*1.750*	**2.77**	*0.109*	2.64	38.9	2.85	270,000	287,300	78,600	87,600	3.47	12.05	2,940	3,120	1.19	1.55
44.5	*1.750*	**2.95**	*0.116*	2.82	38.6	3.02	286,000	304,700	84,100	93,800	3.69	11.83	3,100	3,300	1.17	1.55
44.5	*1.750*	**3.18**	*0.125*	3.00	38.1	3.23	306,500	326,500	89,600	99,300	3.90	11.61	3,260	3,480	1.14	1.55
44.5	*1.750*	**3.40**	*0.134*	3.25	37.6	3.45	326,900	347,800	96,500	107,600	4.21	11.31	3,470	3,730	1.11	1.55
44.5	*1.750*	**3.68**	*0.145*	3.51	37.1	3.70	351,400	374,100	104,100	115,800	4.51	11.01	3,680	3,980	1.08	1.55
44.5	*1.750*	**3.96**	*0.156*	3.76	36.5	3.96	375,400	399,400	111,000	123,400	4.81	10.71	3,880	4,210	1.05	1.55
44.5	*1.750*	**4.45**	*0.175*	4.24	35.6	4.39	415,900	443,000	124,800	138,600	5.36	10.16	4,230	4,650	0.99	1.55
44.5	*1.750*	**4.83**	*0.190*	4.57	34.8	4.72	447,500	476,400	134,500	148,900	5.73	9.79	4,460	4,930	0.95	1.55
44.5	*1.750*	**2.77**	*0.109*	2.64	38.9	2.85	270,000	287,300	78,600	87,600	3.47	12.05	2,940	3,120	1.19	1.55
44.5	*1.750*	**2.95**	*0.116*	2.82	38.6	3.02	286,000	304,700	84,100	93,800	3.69	11.83	3,100	3,300	1.17	1.55
44.5	*1.750*	**3.18**	*0.125*	3.00	38.1	3.23	306,500	326,500	89,600	99,300	3.90	11.61	3,260	3,480	1.14	1.55
44.5	*1.750*	**3.40**	*0.134*	3.25	37.6	3.45	326,900	347,800	96,500	107,600	4.21	11.31	3,470	3,730	1.11	1.55
44.5	*1.750*	**3.68**	*0.145*	3.51	37.1	3.70	351,400	374,100	104,100	115,800	4.51	11.01	3,680	3,980	1.08	1.55

(Continued)

TABLE 10.17 (*Continued*)

DIMENSIONS (mm)						Nominal	TUBE LOAD BODY (Newtons)		INTERNAL PRESSURE (kPa)		TUBING AREA (sq. cm)		TORSIONAL YIELD (N-m)		INTERNAL CAPACITY per meter	EXTERNAL DISPLACEMENT per meter
OD Specified	OD (inches)	Wall Specified	Wall (inches)	Wall Minimum	ID Calculated	Weight Kg/m	Yield Minimum	Tensile Minimum	Hydro Test 90%	Internal Yield Min.	w/min. Wall	Internal Min.	Yield	Ultimate	Liters	Liters
44.5	*1.750*	**3.96**	*0.156*	3.76	36.5	3.96	375,400	399,400	111,000	123,400	4.81	10.71	3,880	4,210	1.05	1.55
44.5	*1.750*	**4.45**	*0.175*	4.24	35.6	4.39	415,900	443,000	124,800	138,600	5.36	10.16	4,230	4,650	0.99	1.55
44.5	*1.750*	**4.83**	*0.190*	4.57	34.8	4.72	447,500	476,400	134,500	148,900	5.73	9.79	4,460	4,930	0.95	1.55
50.8	*2.000*	**2.77**	*0.109*	2.64	45.3	3.28	310,900	331,400	69,000	77,200	4.00	16.27	3,930	4,140	1.61	2.03
50.8	*2.000*	**2.95**	*0.116*	2.82	44.9	3.48	330,000	351,400	73,800	82,100	4.25	16.02	4,150	4,390	1.58	2.03
50.8	*2.000*	**3.18**	*0.125*	3.00	44.5	3.73	353,600	376,700	78,600	86,900	4.50	15.77	4,370	4,630	1.55	2.03
50.8	*2.000*	**3.40**	*0.134*	3.25	44.0	3.98	377,200	401,700	84,800	94,500	4.86	15.41	4,670	4,970	1.52	2.03
50.8	*2.000*	**3.68**	*0.145*	3.51	43.4	4.28	406,100	432,300	91,700	101,400	5.21	15.06	4,950	5,300	1.48	2.03
50.8	*2.000*	**3.96**	*0.156*	3.76	42.9	4.58	434,100	462,100	97,900	108,900	5.56	14.71	5,230	5,630	1.44	2.03
50.8	*2.000*	**4.45**	*0.175*	4.24	41.9	5.08	482,200	513,300	109,600	122,000	6.20	14.06	5,740	6,220	1.38	2.03
50.8	*2.000*	**4.83**	*0.190*	4.57	41.1	5.47	519,100	552,400	117,900	131,000	6.64	13.63	6,060	6,620	1.33	2.03
60.3	*2.375*	**3.18**	*0.125*	3.00	54.0	4.47	424,300	451,900	66,200	73,800	5.40	23.18	6,340	6,650	2.29	2.86
60.3	*2.375*	**3.40**	*0.134*	3.25	53.5	4.78	453,300	482,600	71,700	80,000	5.83	22.75	6,790	7,160	2.25	2.86
60.3	*2.375*	**3.68**	*0.145*	3.51	53.0	5.14	487,900	519,500	77,200	35,500	6.26	22.32	7,220	7,650	2.20	2.36
60.3	*2.375*	**3.96**	*0.156*	3.76	52.4	5.51	522,600	556,400	82,700	91,700	6.68	21.90	7,650	8,130	2.16	2.86
60.3	*2.375*	**4.45**	*0.175*	4.24	51.4	6.13	580,900	618,700	93,100	103,400	7.47	21.11	8,420	9,020	2.08	2.86
60.3	*2.375*	**4.83**	*0.190*	4.57	50.7	6.61	626,700	667,200	100,000	111,000	8.01	20.57	8,930	9,610	2.02	2.86
66.7	*2.625*	**3.40**	*0.134*	3.25	59.9	5.31	504,000	536,400	64,800	72,400	6.48	28.44	8,420	8,840	2.81	3.49
66.7	*2.625*	**3.68**	*0.145*	3.51	59.3	5.72	542,700	577,800	69,600	77,900	6.96	27.96	8,970	9,450	2.76	3.49
66.7	*2.625*	**3.96**	*0.156*	3.76	58.8	6.13	581,400	619,200	75,200	83,400	7.43	27.49	9,510	10,060	2.71	3.49
66.7	*2.625*	**4.45**	*0.175*	4.24	57.8	6.82	647,200	689,000	84,100	93,800	8.32	26.60	10,500	11,180	2.62	3.49
66.7	*2.625*	**4.83**	*0.190*	4.57	57.0	7.36	698,300	743,300	91,000	100,700	8.92	26.00	11,150	11,920	2.55	3.49
73.0	*2.875*	**3.96**	*0.156*	3.76	65.1	6.75	640,100	681,400	68,300	75,800	8.18	33.70	11,580	12,190	3.33	4.19
73.0	*2.875*	**4.45**	*0.175*	4.24	64.1	7.52	713,000	759,300	77,200	85,500	9.17	32.72	12,810	13,560	3.23	4.19
73.0	*2.875*	**4.83**	*0.190*	4.57	63.4	8.12	769,900	819,800	83,400	92,400	9.83	32.05	13,620	14,480	3.15	4.19
89.9	*3.500*	**4.45**	*0.175*	4.24	80.0	9.26	878,000	935,000	63,400	70,300	11.28	50.79	19,590	20,540	5.03	6.12
89.9	*3.500*	**4.45**	*0.175*	4.57	79.2	10.01	949,200	1,010,600	68,300	75,800	12.11	49.96	20,880	21,970	4.93	6.21

Test pressure value equals 90% of internal yield pressure rating. Maximum working pressure is a function of tube condition and is determined by user.

* Available as continuously milled tubing (CM™) or conventional butt-welded tubing sections (W™). All data is for new tubing at minimum strength. Other sizes and wall thicknesses available on request. See individual size sheets for additional wall thicknesses.

The formulas used do not intentionally include any safety factor and the safety factor must be added. Magnitude of the safety factor should be based on the severity of the application and the validity of the data used.

When it is necessary to assume some of the data, the company representative or serviceman most familiar with the area should be able to supply the best "guesstimate" and some idea to its accuracy. Considering the guesstimates used, the same personnel should be able to establish the safety margin required and often the amount of detail required in analyzing. Obviously it is not practical to analyze data, which is highly questionable.

10.4.10 Coiled Tubing

Coiled tubing is a continuous length of steel tubing with no joints. In general, the coiled tubing is fabricated to the length needed for a particular well application and rolled onto a reel for transport to a well site for placement in the well. The tubing can be butt-welded or continuously-milled as the tubing is placed on the roll. Coiled tubing is used in a variety of oil and gas well production, completions, and drilling operations. Like conventional tubing, coiled tubing in fabricated in a four material grades. These are HS 70, HS 80, HS 90, and HS 110.

Table 10.9 gives the minimum steel strength, elongation and hardness properties for each of the above grades of coiled tubing materials. The same calculation techniques that were discussed above in this section are applicable to the design of coiled tubing strings for placement in oil and gas production wells.

Tables 10.10 (USC units) and 10.11 (Metric units) give the geometry and performance limitations for the HS 70 grade coiled tubing.

Tables 10.12 (USC units) and 10.13 (Metric units) give the geometry and performance limitations for the HS 80 grade coiled tubing.

Tables 10.14 (USC units) and 10.15 (Metric units) give the geometry and performance limitations for the HS 90 grade coiled tubing.

Tables 10.16 (USC units) and 10.17 (Metric units) give the geometry and performance limitations for the HS 110 grade coiled tubing.

CHAPTER

11

Environmental Considerations for Drilling Operations

OUTLINE

11.1 INTRODUCTION

Planning for drilling should include environmental considerations. Environmental management at the well site involves thoughtful planning at the onset of exploration or development. In today's world, a project may be postponed or terminated because of these issues. Plans must be developed and permits applied for and received before moving any equipment onto the location. Obtaining the Construction General Permit is the first step in complying with the Storm Water Pollution Prevention Plan (SWPPP). After the plans and permits are obtained, the pre-spud meeting, in addition to discussing well depth, casing points, and rig selection, should cover topics pertinent to the environmental management of the drilling and completion operation. Regulating agencies are most concerned with these issues.

A site will have particular factors that are imminently apparent and those perhaps not so apparent except to groups exhibiting a certain interest.

It is the role of the regulating agency to protect the public and the public domain from detrimental effects caused by industrial operations. Compliance with the regulator's requirements is usually simple in nature. These requirements are documented and should be followed in a particular order. Concerns disseminated by special interest groups are often unexpected. In an attempt to forego stalling a project by these parties, public disclosure of a project should be made as early as possible, even if it is still pending. A well-informed public disclosure statement released to the immediate community may prevent any future surprises.

11.2 WELL SITE

In planning a drilling operation, the location and access are primarily keyed to environmental decisions. The access and location must be able to maintain the traffic load and mitigate any impact on local resources such as the flora, fauna, and cultural and aesthetic sites. In certain instances, the preparation of an environmental assessment followed by an environmental impact statement may be considered warranted because of proximate

- Wildlife refuges
- Historic or cultural resources
- Recreation areas
- Land containing threatened or endangered species

In the United States, drilling plans are submitted to the Bureau of Land Management (BLM) or state oil and gas commissions in the form of an Application for Permit to Drill (APD), depending on ownership of mineral resources. Other countries have similar requirements. In addition to these agencies, other appropriate surface management agencies may have to be considered, including the BLM, National Park Service, Tribal Authorities, State Environmental Program, and County Environmental Program.

In the permitting process, it is important to have the entire operation planned. It is a necessary component of most APDs. In a federal APD, the 13-point surface use plan must address the following items before approval:

1. Location of site and existing roads
2. Planned access roads
3. Location of existing wells
4. Location of existing or proposed production facilities
5. Location and type of water supply
6. Source of construction materials

7. Methods of handling waste disposal
8. Ancillary facilities
9. Well site layout
10. Plans for restoration of the surface
11. Surface ownership
12. Information such as proximity to water, inhabited dwellings, and archeological, historic or cultural sites
13. Certification of liability

Details of the actual drilling program are not considered in detailed in an APD except for the casing and cementing program and how they are designed to protect any underground sources of drinking water (USDW). The bulk of the APD permit is designed to address the impact on surface resources and the mitigating procedures the operator plans to take to lessen these impacts [1].

11.3 ENVIRONMENTAL REGULATIONS

In the United States, the Environmental Protection Agency (EPA) sets policy concerning environmental regulations. The states are then allowed primacy over jurisdiction of environmental compliance if their regulations are at least as stringent as the federal regulations. Several bodies of government may be involved in environmental decisions concerning the well site. It was at one time common in most oil and gas producing states that the oil and gas division was allowed to administer all matters covering exploration and development. However, because of increased regulations, the responsibility is usually spread among several agencies, with air quality being one that is usually separate from daily oil and gas operations. It is primarily the responsibility of state agencies to protect the integrity of the state's land and water supplies from contamination due to oil and gas activity. The state oil and gas division's environmental regulations usually closely mirror those provided by the EPA.

Congress, in an attempt to promote mineral development in the United States, has exempted most hazardous wastes produced at the well site under the Resource Conservation and Recovery Act (RCRA) Subtitle C regulations. Hazardous waste are listed by inherent characteristics of

- Toxicity
- Ignitability
- Corrosiveness
- Reactivity

Although a number of waste products at the well site are considered characteristic hazardous wastes, some wastes fall under the nonhazardous description. The regulation of these fall under RCRA Subtitle D. Initially, Subtitle D wastes were regulated to control dumping of domestic trash and city runoff. The EPA is considering promulgating regulation of certain oil and gas waste under Subtitle D [2].

Under the RCRA exemption, wastes intrinsically associated with exploration and development of oil and gas do not have to follow Subtitle C regulations for disposal. Under Subtitle C, hazardous wastes must follow strict guidelines for storage, treatment, transportation, and disposal. The cost of handling materials under the Subtitle C scenario is overwhelming. Under the exemption, the operator is allowed to dispose of well site waste in a prudent manner and is not obligated to use licensed hazardous waste transporters and licensed treatment, storage, and disposal facilities (TSDF).

Covered by the exemption are drilling fluids, cuttings, completion fluids, and rig wash. Not covered by the exemption are motor and chain oil wastes, thread cleaning solvents, painting waste, trash, and unused completion fluids.

A waste product, whether exempt or not, should always be recycled if economically possible. Oil-based drilling mud is typically purchased back by the vendor for reuse. Unused chemicals are similarly taken back for resale. Arrangements should be made with the mud company for partial drums or sacks of chemical. Muds also may be used on more than one hole. With the advent of closed-system drilling, the muds be moved off location in the event of a producing well.

If a waste is generated that is a listed or characteristic item, the operator must follow certain guidelines [3]. A listed hazardous waste (e.g., mercury, benzene) is considered hazardous if the concentrations in which they naturally occur above certain limitations (40 CFR 261.31–261.33). The listed hazardous waste may not be diluted to achieve a lesser concentration and thus become nonhazardous. A characteristic hazardous waste (40 CFR 261.21–261.24) may be diluted to a nonhazardous status.

Most nonexempt, nonacute hazardous waste generated on location is considered a small quantity. In this case, the waste may remain on location for 90 days. At that time, a Department of Transportation (DOT)–licensed motor carrier must transfer the waste to an EPA-certified TSDF for disposal. Appropriate documentation and packaging must conform to regulations. The operator continues to be liable for the waste as denoted by the cradle-to-grave concept [4].

Exempt wastes are usually disposed of on location after gaining permission from the state oil and gas division. Liquid wastes, if not evaporated or fixated on location, are usually injected into Class II injection wells. Solid wastes, if not acceptable to local landfills, are remediated onsite or buried in some instances. Table 11.1 shows exempt and nonexempt waste, Figure 11.1 shows determination of exempt and nonexempt waste, and

TABLE 11.1 Exempt and Nonexempt Oil and Gas Production Related Wastes

Exempt Waste	**Nonexempt Waste**
Produced water	Unused fracturing fluids or acids
Drilling fluids	Cooling tower cleaning wastes
Drill cuttings	Used well completion/stimulation fluids
Rigwash	Radioactive traces wastes
Geothermal production fluids	Painting wastes
Hydrogen sulfide abatement wastes	Used lubrication oils
Well completion/stimulation fluids	Vacuum truck and drum rinsate containing nonexempt wastes
Basic sediment and water (BS&W) and other tank bottom material from storage facilities that hold product and exempt waste.	Refinery wastes
Accumulated materials such as hydrocarbons, solids, sand and emulsions from production separators, fluid treating vessels, and production impoundments.	Service company wastes such as empty drums, drum rinsate, vacuum truck rinsate, sandblast media, spent solvents, spilled chemicals, and waste acids
Pit sludges and contaminated bottoms from storage or disposal of exempt waste	Waste compressor oil, filters and blowdown
Workover wastes	Used hydraulic oil
Gas plant dehydration and sweetening waste	Waste solvents
Cooling tower blowdown	Caustic or acid cleaners
Spent filters, filter media, and backwash (assuming the filter itself is not hazardous and the residue in it is from an exempt waste stream)	Waste generated in transportation pipeline related pits
Packer fluids	Laboratory wastes
Produced sand	Boiler cleaning wastes
Pipe scale, hydrocarbon solids, hydrates, and other deposits removed from piping and equipment before transportation	Boiler refractory bricks
Hydrocarbon bearing soil contaminated from exempt waste streams	Boiler scrubbing fluids, sludges and ash
Pigging waste from gathering lines	Incinerator wastes
Constituents removed from produced water before it is injected or otherwise disposed of	Industrial wastes from activities other than oil and gas E&P
Waste from subsurface gas storage and retrieval	Pesticide wastes
Liquid hydrocarbons removed from the production stream but not from oil refining	Gas plant cleaning wastes
Gases from the production stream, such as H_2S, CO_2, and volatized hydrocarbons	Drums, insulation and miscellaneous solids
Materials ejected from a producing well during blowdown	Manufacturing wastes
Waste crude from primary operations and production	Contamination from refined products
Light organics volatized from exempt waste in reserve pits, impoundments, or production equipment	

(*Continued*)

TABLE 11.1 (*Continued*)

Exempt Waste	Nonexempt Waste
Liquid and solid wastes generated by crude oil and crude tank bottom reclaimers	
Stormwater runoff contaminated by exempt materials	

Although non-E&P wastes generated from crude oil and tank bottom reclamation operations (e.g., waste equipment cleaning solvent) are nonexempt, residuals derived from exempt wastes (e.g., produced water separated from tank bottoms) are exempt.

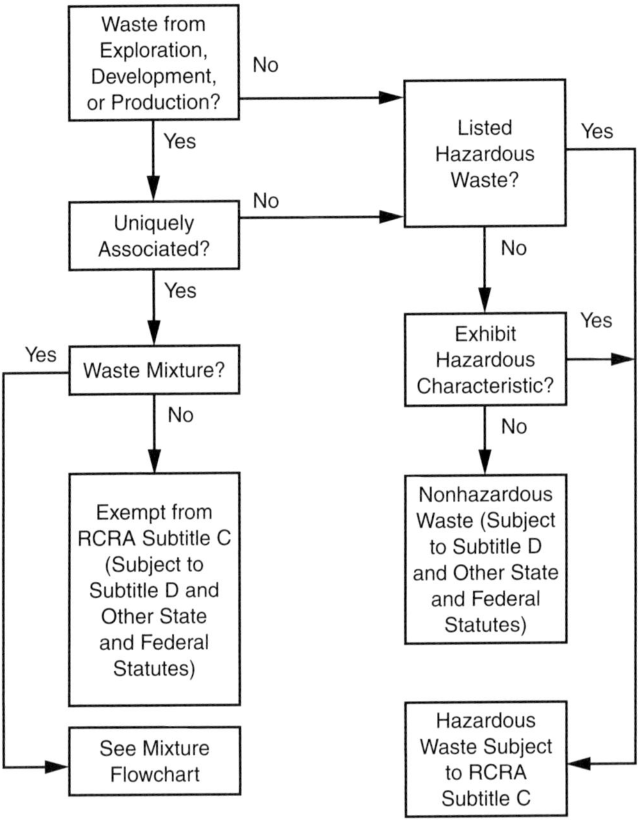

FIGURE 11.1 Exempt/nonexempt waste flowchart.

Figure 11.2 shows the possible waste mixtures and their exempt and nonexempt status [5].

Generally, waste that must be produced to complete the work involved in the exploration and production of oil and gas is considered exempt, allowing the operator the option of disposing of the waste in a prudent manner.

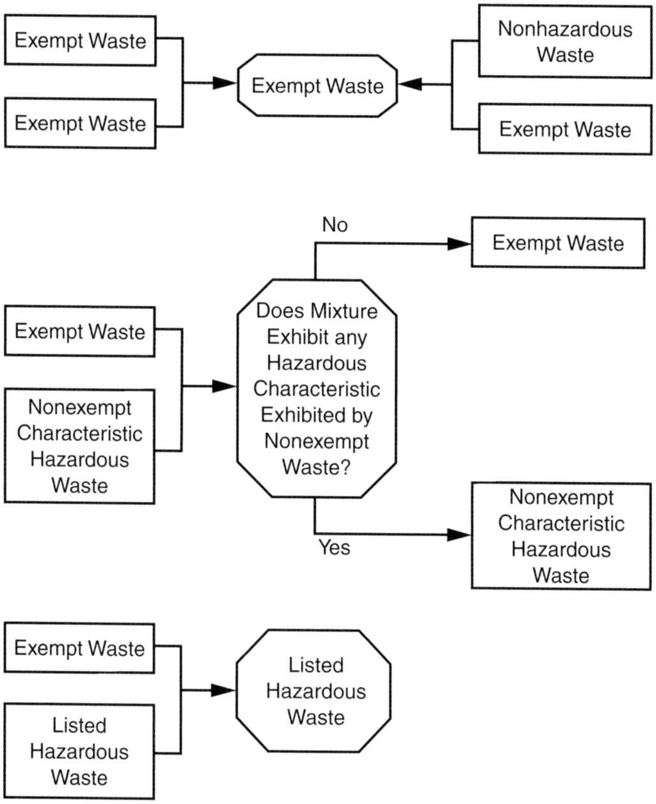

FIGURE 11.2 Possible waste mixtures and their exempt and nonexempt status.

A nonexempt waste (e.g., radioactive tracer waste) should be avoided if possible. These wastes must be disposed of according to federal and state regulations. Because these regulations are becoming more complicated with time, the operator should consult the primary regulator in questionable circumstances.

11.4 SITE ASSESSMENT AND CONSTRUCTION

11.4.1 Access and Pad

The road to the wellpad should be constructed to prevent erosion and be of dimensions suitable for traffic. A 16-ft top-running width with a 35-ft bottom width has been shown to sustain typical oil field traffic. The road should be crowned to facilitate drainage, and culverts should be placed

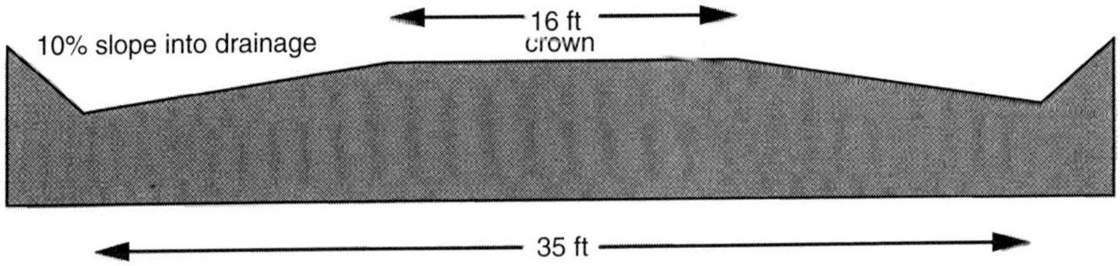

FIGURE 11.3 Typical oilfield access dimensions.

at intersections of major runoff (Figure 11.3). Before the pad and access is surveyed for final layout, preliminary routes should be studied and walked out. Trained personnel should note any significant attributes of the area such as archeological finds, plants, and animals. Because of this survey, the most favorable access is designated weighing and primarily linked to

- Drilling fluid program
- Periodic operations
- Completions

The rig selection will dictate the basic layout of the pad. Based on the necessary area needed to support its functions, ancillary equipment may be added in space-conservative measures. In addition to the placement of various stationary rig site components, other operations such as logging, trucking, and subsequent completion operations must be provided for. The most environmentally sensitive design will impact the least amount of area and will therefore be the most economic. Potential pad sites and access routes should be laid out on a topographic map before the actual survey. At this time, construction costs can be estimated and compared. Figure 11.4 shows such a layout. The cost of building a location includes the cost of reclamation such as any remediation, recontouring, and reseeding of native plants. In the event of a producing well, only that area needed to support the production operations is left in place. The reserve pit and all outlying areas are reclaimed.

11.4.2 Rig Considerations

A rig layout diagram provided from the contractor gives the dimensions necessary for planning the drill site. Figure 11.5 depicts a layout for a 10,000-ft mobile drilling rig. The main components to the rig are the drive-in unit, substructure, mud pits, pumps, light plant, catwalk, pipe racks, and fuel supply. This type of rig, with a telescoping derrick, may be laid on less than 0.37 acre. The outstanding area is taken by the positioning of the deadmen. The deadmen to which the guy wires are connected are specified by safety considerations for each rig model. A standard drilling rig with a

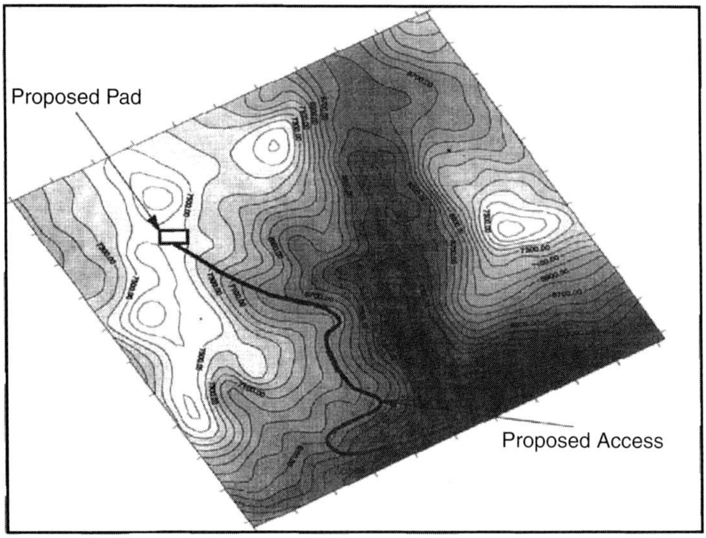

FIGURE 11.4 Topographic layout of proposed access and pad.

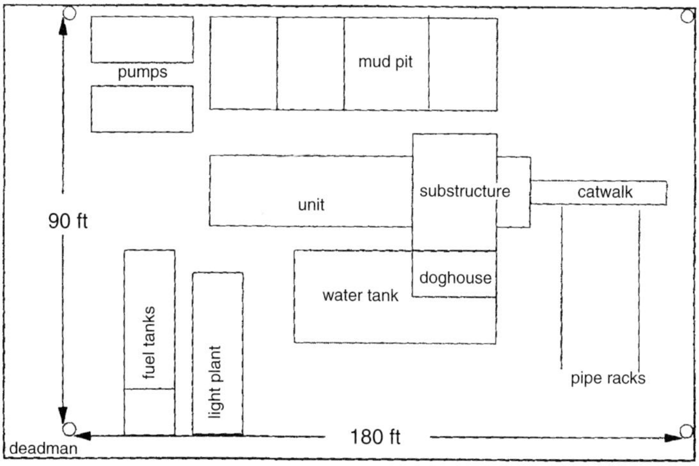

FIGURE 11.5 Basic mobile 10,000-ft drilling rig layout.

10,000-ft capacity will have a longer laydown side, as it is measured from the well center. During rig up, the derrick is assembled on the ground and then lifted onto the A legs. The laydown side of the site must allow for this operation. As depicted in Figure 11.6, the standard rig exceeds the overall aerial dimensions of the mobile rig at 0.41 acre with the laydown dimension of 130 ft, compared with 90 ft for the telescoping mast unit. A standard

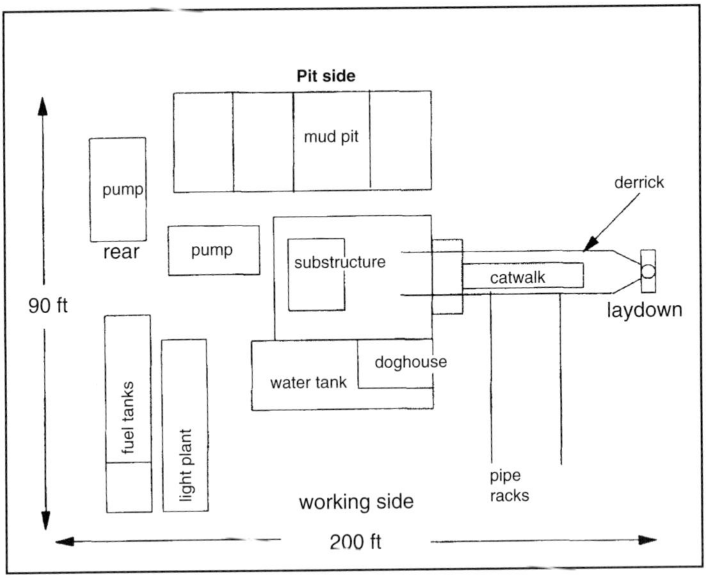

FIGURE 11.6 Basic layout for drilling rig with standard derrick.

drilling rig with a 20,000-ft capacity may run a laydown of 185 ft, with all other dimensions proportionally increased.

With the basic rig layout defined, the ancillary equipment may be determined and laid out adjacent to the structures in place. Access usually is determined where casing, mud, and other equipment may be delivered without disturbance of the infrastructure. When a crane (or forklift) is used, this may mean a simple loop from the rear to the laydown on the working side of the location. A loop is the optimum arrangement whereby multiple truck interference may be avoided. In the event a crane is not available, extra space is needed to accommodate the activity of gin trucks positioning materials. Figure 11.7 provides an overall location plan view.

11.4.3 Drilling Fluid Considerations

The drilling fluid program defines to some extent a major portion of the pad, including the reserve pit, blow pit, and equipment space. The program may include a closed system or a conventional one. The fluids can be oil-based mud, air, foam, water, or other media. Although the basic rig layout considers most mud drilling activities, it does not figure in mud storage, additional water storage, or air drilling systems. The overall layout may even be reduced in some cases. A closed mud system or air drilling eliminates the requirement for a large reserve pit.

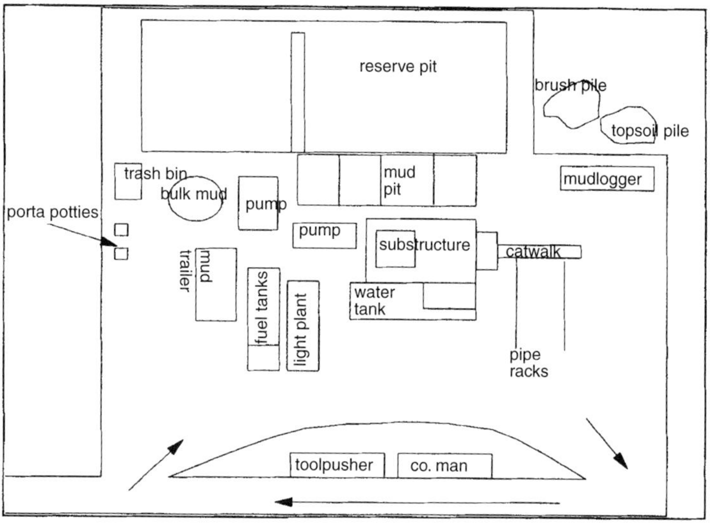

FIGURE 11.7 Overall location plan.

11.4.3.1 *Air Drilling*

In the event of air drilling, the blooie line will exit from under the substructure and away from the rig. Depending on the nature of the pay zone, the blooie line will extend different lengths from the rig. For example, a well producing 20 MMCFD in addition to the injection of 2,000 cfm air will require an extension of at least 150 ft because of heat and dust accumulation. The blow pit with berm will often extend another 40 ft beyond that to include both the blow pit and berm. The blow pit is then connected to the reserve pit. This allows the transfer of any injected or produced fluid to a storage area away from the blow pit. The blow pit and the berm should be designed with considerable contingency. During operations, the well may be producing oil or natural gas, or both. In the case of oil, the well will have to be mudded up after the well is under control. In the process of getting the well under control, the oil may be sprayed a great distance if the blow pit and berm are not properly constructed. If natural gas is encountered, drilling is usually advanced with a flare in place. The pit and the berm must then be able to protect the surrounding area from fire and to sustain the impact from a steady bombardment of drilling particles. A distance of 30 ft should be allowed from the end of the blooie line to the edge of the berm, with only 5 ft of depth needed at the pit sloping toward the reserve pit built to maintain a flowing velocity of 2 fps. The berm itself should be 20 ft high and composed primarily of 12-ft or larger-diameter stones on the face and backed with a soil having poor permeability (Figure 11.8). A lip should overhang from the top of the berm whereby the ejected materials are diverted back to the pit. The air compressors and boosters are typically

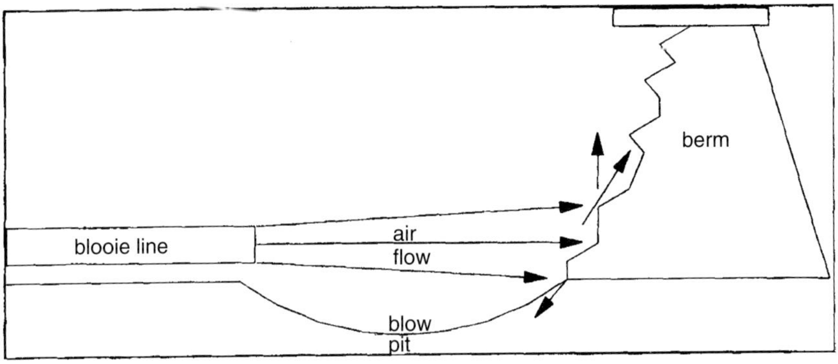

FIGURE 11.8 Side view of blow pit.

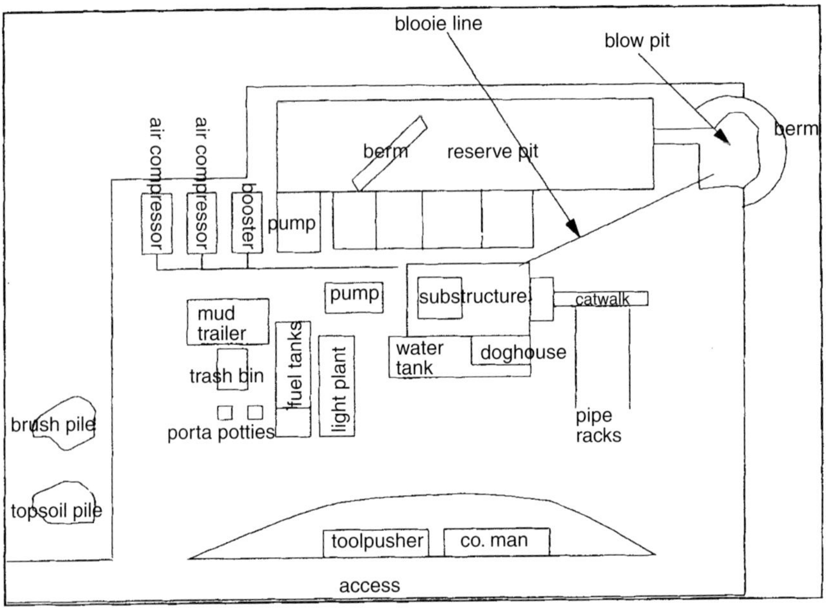

FIGURE 11.9 Air drilling pad.

located at the rear of the location, with the air piped from that point to the standpipe to alleviate additional piping. Figure 11.9 shows an air-drilling layout. Notice that the location may be compacted somewhat because of the reduced reserve pit.

11.4.3.2 *Mud Drilling*

The conventional drill site includes mud pits, mud cleaning equipment, water storage, mud storage, pumps, and mixing facilities. Often, the reserve pit associated with the conventional mud system is as large as the leveled location. A large reserve pit allows for contingency, but it also increases cost of reclamation and construction. Generally, a 3-ft freeboard is maintained for safety, and it may be relied on for contingency. A good field estimate for a reserve pit size needed for a conventional operation is

$$V_t = 2(V_s + Vi_I + V_p) + 281D_d + 16.84MV + 3WL \tag{11.1}$$

$$WL = \frac{V_r}{H} \tag{11.2}$$

where V_t = volume of reserve pit in ft^3
V_s = volume of surface hole (ft^3)
V_i = volume of intermediate hole (ft^3)
V_p = volume of production hole (ft^3)
D_d = forecasted drilling time in days
MV = mud pit volume in bbl
L = pit length: 0.5–0.75 ft of location in ft
W = pit width: 0.25–0.5 ft of pit in ft
H = pit depth $\leq$ 10 ft

Equation 11.1 assumes that for each operation, a volume of cuttings is put into the reserve pit plus a mud volume equivalent to the circulation system. The additional volumes are attributed to mud dilution and maintenance. An additional contingency is added through 3 ft of freeboard. The reserve pit should be located in an area where capacity may be increased in an emergency or additional fluid trucked away.

11.4.3.3 Closed Mud System

The closed mud system is the modern solution to an environmentally sensitive drilling operation. The circulation system on the rig is fully self-supporting, requiring only the discharge of drill cuttings. On a simple system, this may only necessitate the addition of a mud cleaner or centrifuge to the drilling rig's conventional mud system. A bid package to the drilling contractor may stipulate a closed system and the requirements, thereby letting the burden of design fall to the contractor, although the liability still rests on the operator. With a closed system, the additional area is sometimes needed within the basic layout for mud cleaning equipment and mud storage. A trench located at the pit side of the mud pits may be allowed then for cutting disposal. In the event of oil-based mud, the cuttings may be collected in a sloped container, where residual fluid is allowed to drain from the cuttings before disposal.

11.4.4 Periodic Operations

Periodic operations include cementing, running casing, and logging. Room must be allocated for the storage of casing. Running 10,000 ft of range 3, $5\frac{1}{2}$ in. casing requires a 40 × 115 ft area. The casing may be stacked, although never in excess of three layers and preferably only two layers. Cementing operations may include placement of bulk tanks in addition to pump trucks and bulk trailers. These are usually located near the water source and rig floor. Laying down drill pipe and logging both necessitate approximately 30 ft of space in front of the catwalk.

11.4.5 Completions

If a well proves productive, the ensuing completion operation may require an area in excess of the drilling area. This may mean allocation for frac tank placement, blenders, pump trucks, bulk trucks, and nitrogen trucks. In today's economic climate, the operator should weigh the probability of success, Bayes theorem (Equation 11.3), with the cost of constructing and reclaiming an additional area needed for stimulation (Equation 11.4). Plans such as these should be considered and proposed in the APD. The operator may then construct the additional space without a permitting delay.

$$P_s = PR(E/R) = \frac{PR(E)PR(R/E)}{PR(E)PR(R/E) + PR(\text{not}\,E)PR(R/\text{not}\,E)} \tag{11.3}$$

$$P_s C_b \leq C_a \tag{11.4}$$

where P_s = probability of success
$PR(E/R)$ = probability of event E given information R
C_b = construction cost of additional space before drilling in $
C_a = construction cost of additional space after drilling in $

11.4.6 Pad Construction

Once access and size of the pad have been determined, the areas may be physically flagged to further reduce the amount of dirt moved. Important features of the area should be noted at this time, including

- Depth of water table
- Natural drainage patterns
- Vegetation types and abundance
- Surface water
- Proximate structures

These features are quantified, as well as testing for contamination in soil and water to prevent unnecessary litigation over previous pollution.

In construction of the pad, brush and trees are pushed to one area. Topsoil is then removed from the site and stockpiled for respreading during subsequent seeding operations. The leveled pad is slightly crowned to move fluids collected on the pad to the perimeter, where drainage ditches divert this fluid to the reserve pit. In the event a subsurface pit is not possible, the drainage will run into a small sump. This sump is used as a holding tank for pumping of collected fluids to an elevated reserve pit.

When possible, the reserve pit is placed on the low side of the location to reduce dirt removal. In this event, the pit wall should be keyed into the earth and the summation of forces and moments on the retaining wall calculated to prevent failure. Pits most often fail because of leaking liners undercutting the retaining wall and sliding out. A minimum horizontal-to-vertical slope of 2 is recommended for earthen dams [6]. The pit bottom should be soft filled to prevent liner tears. Other key factors of the location provide for the drainage of all precipitation away from the location. This prevents the operator from unduly managing it as a waste product. Any water landing on the location must be diverted to the reserve pit.

A new scheme for location management had developed whereby wastes are diverted to separate holding facilities according to the hazard imposed by the waste. Separate pits are created to hold rig washing and precipitation wastes, solid wastes, and drilling fluids [7]. The waste is then reused, disposed on site, or hauled away for offsite treatment. The system reduces contamination of less hazardous materials with more hazardous materials, thereby reducing disposal costs.

11.5 ENVIRONMENTAL CONCERNS WHILE IN OPERATION

11.5.1 Drilling

A blowout primarily consisting of oil presents the greatest environmental hazard while drilling. During normal drilling situations, downhole drilling fluids are usually the greatest potential threat to the environment. In the case of oil-based mud, the cuttings also present a problem though absorption of the diesel base. These cuttings are presently landfarmed, landspread, reinjected, or thermally treated to drive off the hydrocarbons.

Most wells encounter shales when drilling. The oil-based muds are quite effective in reducing the swelling tendencies of the shale in addition to presenting less intrusive invasion characteristics to the reservoir. The oil-based muds may be sold back to the distributor, where they are recycled.

An alternative to oil-based muds may be found in using a synthetic-based material [8]. Saline mud is also used to reduce the shale's swelling characteristics. Chlorides may be found to be within toxic levels in the drilling fluid, making its use less desirable. Because the swelling of the clay is attributed to the cation exchange capacity (CEC), the chloride levels may be reduced by replacement with another anion. Calculation of the shale's CEC estimates the amount of cations that can be added to the drilling mud to effectively reduce swelling tendencies. Generally, a multivalent cation is more effective in reducing hydration of the clay. An equilibrium equation is used to define the CEC of a shale (Equation 11.5). Figure 11.10 shows the CEC relationships among several multivalent cations and sodium in attapulgite clay [9].

While drilling, the drilled cutting's CEC is usually tested with methylene blue (Equation 11.6). Table 11.2 shows CEC of several clays encountered in drilling situations [10]. In reducing the amount of any one type cation or anion present in the drilling mud, the environmental risks are reduced. A mixed salt solution containing several salt combinations, each of which

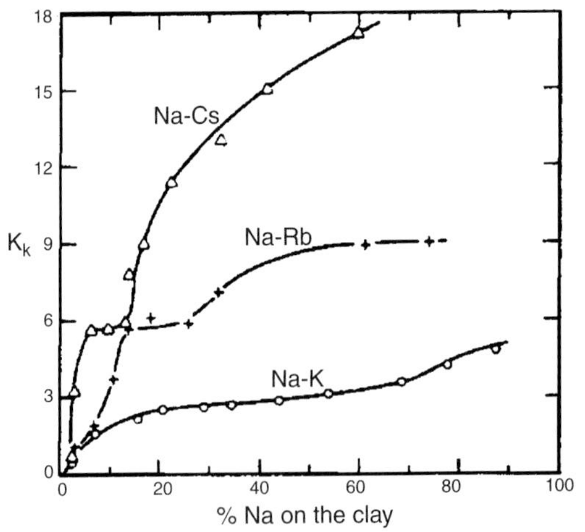

FIGURE 11.10 Selectivity number K_k vs. exchangeable ion composition on an attapulgite clay.

TABLE 11.2 Cation Exchange Capacity of Materials Encountered in Drilling

Material	mEq/100 g – Moisture Free
Wyoming bentonite	75
Soft shale	45
Kaolinite	10
Drilled cuttings	8

is below the toxic limits, may produce the desired effects. Thus,

$$K_k = \frac{(NaX)^c(M)}{(MX)(Na)^c} \tag{11.5}$$

where NaX = sodium on the shale in mEq
M = multivalent cation in mEq
c = charge of multivalent cation

$$MBC = \frac{MB}{M_{shale}} = CEC_{shale} \tag{11.6}$$

where MBC = methylene blue capacity in mEq/100 g
MB = methylene blue capacity in mEq
M_{shale} = weight of shale in 100 g

There are a variety of toxic chemicals used in the drilling fluid makeup. Chromates and asbestos were once commonly used but are now off the market. A mud inventory should be kept for all drilling additives. Included in the inventory is the Material Safety Data Sheet (MSDS) that describes each material's pertinent characteristics. The chemicals found on the MSDS should be compared with the priority pollutants, and any material should be eliminated if a match is found. The chemicals should also be checked on arrival for breakage and returned to the vendor if defective packaging is found. All mud additives should be housed in a dry area and properly card for to prevent waste. Chemicals should always be mixed in packaged proportions. Wasted chemicals, ejected to the reserve pit by untrained personnel, can present future liabilities to the operator, who is responsible for them from cradle to grave.

11.5.2 Rig Practice

All drilling fluids should be contained in the mud pits for reserve pit. The cellar should have a conduit linking it to the reserve pit such that accumulation of mud from connections does not spread over the location. In the case of toxic mud, a bell should be located beneath the rotary table to direct such fluids back into the drilling nipple or into the mud pit in the case of a rotating head. The use of a mud bucket is also recommended in situations where the pipe cannot be shaken dry.

Many operators are now requiring that absorbent or catch pans be placed beneath the drilling rig. Oil and grease leaks or spills are common to the drilling rig operation. Even if the drilling rig is new, leaks and spills are inevitable. Fuel and oil racks should be provide a spill-resistant pour device for direct placement of the lubricant or fuel into equipment. Oil changed on location must be caught in barrels and recycled by the contractor. Pipe dope should be environmentally sound and not a metal type. Thread cleaning on casing is now frequently done with machines that catch the solvent for reuse on subsequent operations.

Drilling contractors should be advised before the operations about what is trying to be accomplished environmentally on location. Contractors know that the standard practices of throwing dope buckets, and everything else into the reserve pit is no longer acceptable, but an occasional drilling crew may not take directives seriously. Because of this, drilling contractors should line item the liabilities associated with imprudent practices.

11.5.3 Completions

Inherent to the completion operations are the stimulation fluids used to carry sand or otherwise enhance producing qualities of the well. These stimulations fluids may or may not be toxic in nature. The stimulation fluid, although sometimes batch mixed at the service company's facility, may also be mixed on location. The latter system is preferable unless the service company is willing to let form the operator any liability common to the fluid in case of excess.

Many frac jobs are mini-fraced before the actual operation and the design criteria established before the actual frac job is accomplished. The method helps prevent the screening out of the well beforehand. The screened-out well causes potential problems not only in productivity but also in waste management. The unused mixed chemicals, such as KCl makeup water in frac tanks leftover due to the screenout, must be properly disposed. An alternative to this is the mixing of chemicals on the fly instead of preblending and stocking in frac tanks. Some chemicals, such as KCl in water, must be mixed well in advance on location to attain heightened concentrations. In theses in instances, a properly designed frac job, based on the mini-frac, will allow for some certainty in getting the designed job away. Frac flow back may be introduced to the reserve pit after separation from any hydrocarbons. Current frac fluids are composed of primarily of natural organics such as guar gum, but they may contain other components that may be harmful to the environment. Containment of flowback in a lined reserve pit before disposal is a prudent practice.

Acid jobs may also be designed according to prior investigation, although most service companies accept unused acid back into their facilities. Spent acid flowback may be introduced to the lined reserve pit with little consequence. Often, the residual contains salts such as $CaCl_2$ preciously introduced to the pit. Live acid is occasionally flowed back to surface. This acid may be flowed to the lined pit, given acceptance of the liner material for low pH. The buffer capacity of most drilling fluids is significant, and it is able to assimilate the excess hydrogen ions introduced to the system. The buffer capacity of the fluid may be calculated from Equations 11.7 and 11.8 [11]. From a water analysis, pH and alkalinity are determined, and the remaining parameters may then be calculated.

$$\beta = \frac{dC}{dpH} \tag{11.7}$$

$$\beta = 2.3\left[\frac{\alpha_1([ALK]-[OH^-]+[H^+])\left([H^+]+\frac{(K_1'K_2')}{[H^+]}+4K_2'\right)}{K'\left(1+\frac{2K_2'}{[H^+]}\right)}+[H^+]+[OH^-]\right] \tag{11.8}$$

$$[ALK] = [HCO_3^-] + 2[CO_3^{-2}] + [OH^-] - [H^+] \tag{11.9}$$

$$[Ct] = [H_2CO_3^*] + [HCO_3^-] + 2[CO_3^{-2}] \tag{11.10}$$

$$K_1' = \frac{\gamma[HCO_3^-](H^+)}{[H_2CO_3^*]} \tag{11.11}$$

$$K_2' = \frac{\gamma^4[HCO_3^{-2}](H^+)}{\gamma[HCO_3^*]} \tag{11.12}$$

$$\alpha_1 = \frac{1}{1+\frac{K_2'}{(H*)}+\frac{(H^+)}{K_1'}} \tag{11.13}$$

$$\text{Log}\,\gamma = -AZ^2\left[\frac{[\sqrt{I}]}{[1+\sqrt{I}]}\right] \tag{11.14}$$

where β = buffer capacity, (equivalents/unit pH change)

C = equivalents of buffer available = assumed equal to [ALK]

[ALK] = equivalents of alkalinity

K = equilibrium coefficients

γ = the monovalent activity coefficient

I = the ionic strength

Z = the charge of the species of interest

$A = 1.82 \times 106(DT)^{-1.5}$

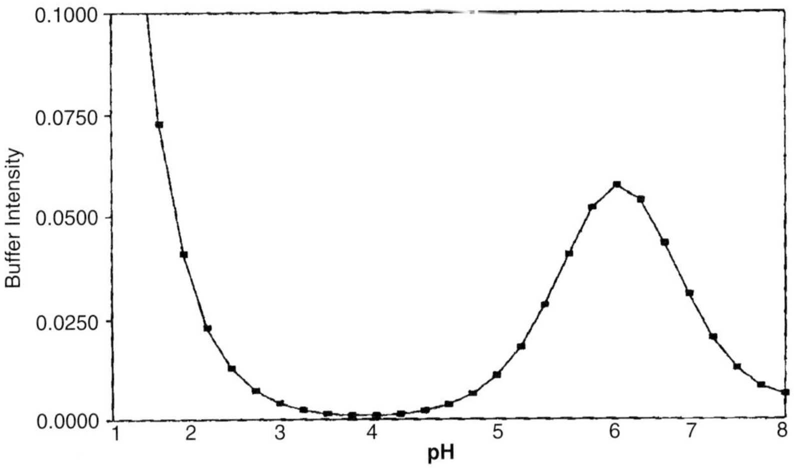

FIGURE 11.11 Nonideal buffer characteristics of a 0.10 M carbonate reserve pit fluid.

D = the dielectric constant for water, 78.3 at 25°C

T = °K

( ) = activity of the ion

[ ] = concentration of the ion

The buffer capacity of the pit fluid is equal to the change in alkalinity of the system per unit change of pH. Figure 11.11 shows the buffer intensity (capacity) of a 0.1 M carbonate pit fluid [10]. Calculating the initial buffer capacity of the pit fluid allows for prediction of the pH change on introduction of live acid and any addition of buffer, such as sodium bicarbonate, required to neutralize the excess hydrogen ions.

Care should be taken in every stimulation circumstance to allow fluids to drain to the reserve pit. In the completion operation, it is exceedingly difficult to accomplish this because of traffic, and the service company should therefore provide leak-free hoses, lines, and connections. On completion of the job, the hoses should be drained to a common area for holding subsequent to introduction to the reserve pit. Every precaution should be taken to prevent accumulation of fluids on the pad proper, thereby posing a potential risk to groundwater and runoff of location.

As with the drilling operation, the equipment on location providing the completion service may leak oil. The use of absorbents and catch pans is advised.

In the case of produced liquid hydrocarbons and other chemicals spilled during operations, subsequent remediation may be necessary. This section details some remediation techniques currently employed.

11.5.4 Reclamation of the Drill Site

In the event of a dry hole, the reserve pit water usage should be maximized to prepare the mud spacers between plugs. Water in excess of this may be pumped into the hole, including solids. All USDWs must be protected in this event. Once the hole has been properly plugged and the drilling rig removed, the mousehole and rathole should be backed filled immediately to preclude any accidents. Trash is removed from the location and adjacent area and is hauled to permit facilities.

11.5.5 Reserve Pit Closure

The reserve pit commonly holds all fluids introduced to the wellbore during drilling and completion operations. This includes the drilling and completion fluids in the event the well is stimulated for production and those cuttings produced during the drilling operation. The reserve pit, on completion of the initial rig site activities, must be reclaimed. On removal of the drilling rig, the reserve pit is fenced to prevent wildlife and livestock from watering. The fence is removed on initiation of reclamation.

The fluids from the reserve pit may be hauled away from location for disposal, reclaimed in situ, or pumped into the wellbore given a dry hole. The operator of the well site is responsible for the transportation offsite of the drilling fluids. The fluids may be considered hazardous in nature due to the toxic characteristics of most drilling and completion fluids.

11.5.6 Evaporation

Evaporation of the water held in the pit is often the first step in the reserve pit remediation because of economic considerations about trucking and disposal. The evaporation may be mechanically driven or take place naturally. Natural evaporation is very effective in semi-arid regions. The Meyer equation (Equation 11.15), as derived from Dalton's law, may be used to estimate the local natural evaporation [6].

$$E = C(e_w - e_a)\psi \tag{11.15}$$

$$\Psi = 1 + 0.1w \tag{11.16}$$

where E = evaporation rate (in 30 days)
C = empirical coefficient equal to 15 for small shallow pools and 11 for large deep reservoirs
e_w = saturation vapor pressure corresponding to the monthly mean temperature of air for small bodies and monthly mean temperature of water for reservoirs (in.Hg)

e_a = actual vapor pressure corresponding to the monthly mean temperature of air and relative humidity 30 ft above the body of water (in.Hg)

Ψ = wind factor

w = monthly mean wind velocity measured at 30 ft above body of water (miles/hr)

Some mechanically driven systems include heated vessels or spraying of the water to enhance the natural evaporation rate. In heating, the energy needed to evaporate the water is equal to that needed to bring the water to the temperature of vaporization plus the energy required for the evaporation, where for constant volume this is

$$\Delta E = CpdT + \Delta H_{vap} \tag{11.17}$$

The heat capacity of ΔH_{vap} of pure water at 14.7 pisa are commonly taken as 1 btu/lbm (°F) and 970 btu/lbm [12]. Ionic content in the pit fluid raises the energy necessary to evaporate the fluid. Figure 11.12 shows this relationship for brine water containing primarily NaCl [10].

In field evaporative units using natural gas as the fuel source, the primary driving force is the heat supplied to the water. The theoretical evaporation rate for these units may be expressed as

$$\frac{H_c Q_g}{\Delta E \rho_w} = Q_{evap} \tag{11.18}$$

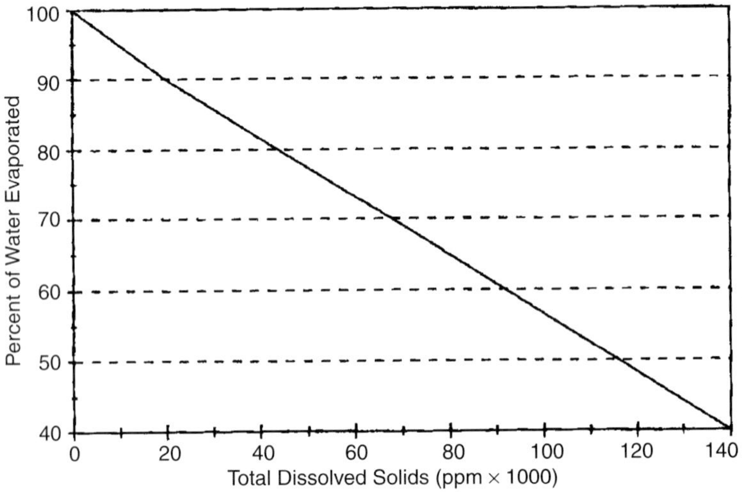

FIGURE 11.12 Maximum limit of evaporation as defined by TDS (NaCl).

where H_c = natural gas heating value (btu/mcf)
Q_c = natural gas flow rate to burner (mcfpd)
ρ_W = density of water (ppg)
Q_{evap} = evaporation rate (gpd)

Mechanical efficiency may range from 25% to 75% of the theoretical evaporation rate. Efficiencies may be raised with the application of multi-effect or vapor compression evaporators. The more complicated the systems can seldom be warranted due to the short service offered.

Spray systems rely on forming minuscule droplets of water and allowing the vaporization thereof while in suspension over the reserve pit. Allowance for wind carriage of the droplets beyond the pit must be made to prevent salting damage to the surrounding area. The shear force extended on each droplet in combination with the relative humidity provides the driving force for the operation. Neglecting the shear component, driving force is actual and saturation vapor pressure differential. A derivation of Fick's law may be used to express the molar flux of water in air.

$$Na = \frac{DA}{1000RT}\frac{dV_P}{dx} \tag{11.19}$$

where Na = moles of water diffused to the air (mol/sec)
D = diffusivity of water in air (0.256 cm^2/secat 25°C, 1 atm)
A = the area perpendicular to the flux (cm^2)
V_ρ = vapor pressure of water in air (atm)
R = gas law constant (0.0821 atm L/mol °K)
X = the thickness of the film where dV_ρ exists (cm)

Inspection of Equation 11.19 shows that increasing the area of the active water surface will allow for greater evaporation rates. In the case of the spray systems,

$$Na = 0.01256\frac{nD}{RT}\frac{dV_P}{dx} \tag{11.20}$$

Because of the differences in determining x, the thickness of the film between the two vapor pressures, an overall transfer coefficient is introduced. Based on the two-film theory, the overall transfer coefficient is used. In the case of water evaporation, the gas film is the controlling mechanism and the resulting equation is

$$Na = \frac{Kga}{RT}(V_{Psat} - V_{Pact}) \tag{11.21}$$

where Kga = the overall mass transfer coefficient (t^{-1})

Service companies offering evaporation services can supply the operator with values of Kga maybe used comparatively between all systems for economic analysis.

Using the Meyer equation (Equation 11.15), the evaporative rate from a 5000 ft^2 pit is estimated. The average temperature in an area in the winter is 40°F, and the corresponding saturated vapor pressure is 0.26 in.Hg, with the actual average vapor pressure residing at 0.19 in.Hg. Wind velocity reaches a peak at 40 mph with a time weighted mean velocity of 5 mph, such that the evaporation rate may be estimated as

$$E = 15(0.26 - 0.19)(1 + 0.1(5)) = 1.58\,\text{in./mo or}$$

$$111\,\text{bbl/mo}\ (0.00083\,\text{mol sec}^{-1})$$

Given this evaporation rate, the overall mass transfer coefficient may then be calculated from Equation 11.21:

$$Kga = \frac{NaRT}{(V_{Psat} - V_{Pact})} = 7.94\ \text{sec}^{-1}$$

11.5.7 Fixation of Reserve Pit Water and Solids

Another method of reclaiming the reserve pit involves combining water-absorbing materials to the water and mud. Usually, the pit contents are pumped through tanks where sorbent is combined with the pit fluid and solids. The mixture is dried and subsequently buried. Care must be taken with this method such that any harmful containment is immobilized to prevent contamination to the surroundings. Studies have shown that for muds, once most of the water has been evaporated or pulled from the pit, the remainder may be solidified to comply with existing regulations. This may be done with cement, fly ash, pozzolan, or any number of absorbents. Polymers have been developed to handle high pH, salt, and oil contents, for which the previous mixtures fell short. The mixture is then allowed to dry and the bulk mass is then buried. This method requires the forethought on pit construction whereby complete mixing of the slurry is accomplished. If primarily bentonite and water are used, evidence has shown minor or no migration from the pit [13].

$$W = \Sigma M_{at\ldots n} R_{wi\ldots n} \tag{11.22}$$

where W = water available in pit (bbls)
M_r = mass of absorbent (lbm)
R_W = water required by absorbent (bbl/lbm)

Even though materials such as bentonite can absorb tremendous amounts of water, they cannot solidify to an extent that the pit may be reclaimed. In moist instances, a dozer must be able to walk out to the center of the pit under a load of pushed dirt. In the event the pit materials are wet, the dozer may become mired and unable to complete the work. It is often better to pick a sorbent that will harden sufficiently for this purpose.

11.5.8 Final Closure

On elimination of the fluids, the liner to the pit is folded over the residual solids in a way to prevent fluid migration. The liner is then buried in place. The operator may choose to remove the liner contents completely to preclude any future contamination. In the case of a producing well, the location is reclaimed up to the deadmen. The adjacent areas are contoured to provide drainage away from the production facilities. In the case of a dry hole, the entire location is reclaimed to the initial condition. All of the reclaimed are should be ripped to enhance soil conductivity. The topsoil is then spread over the reclaimed area, followed by seeding. Local seed mixtures are broadcast to quicken reintroduction of native plants.

References

[1] Bureau of Land Management, "Onshore Oil and Gas Order No. 1: Approval of Operations on Onshore Federal and Indian Oil and Gas Leases," United States Department of Interior, Bureau of Land Management, Washington, D.C., 1983.
[2] Fitzpatrick, M., "Common Misconceptions about the RCRA Subtitle C Exemption from Crude Oil and Natural Gas Exploration, Development and Production," Proceedings from the First International Symposium on Oil and Gas Exploration Waste Management Practices, pp. 169–179, 1990.
[3] USEPA, "RCRA Information on Hazardous Wastes for Publicly Owned Treatment Works" Office of Water Enforcement Permits, Washington, D.C., 1985.
[4] Wentz, C., *Hazardous Waste Management*, McGraw-Hill, New York, 1989.
[5] EPA Exemption of Oil and Gas Exploration and Production Wastes from Federal Hazardous Waste Regulations, EPAA530-K-01-004 January 2002 (www.epa.gov/epaoswer/other/oil/oil-gas.pdf).
[6] Merritt, F.S., *Standard Handbook for Civil Engineers*, McGraw-Hill, New York, 1983.
[7] Pontiff, D., and Sammons, J., "Theory, Design and Operation of an Environmentally managed Pit System," First International Symposium on Oil and Gas Exploration Waste Management Practices, pp. 997–987, 1990.
[8] Carlson, T., "Finding Suitable Replacement for Petroleum Hydrocarbons in Oil Muds," SPE Paper 23062, 1992 American Association of Drilling Engineers New Advancements in Drilling Fluids Technology Conference, Houston, TX, 1992.
[9] Marshall, C., and Garcia, G., *Journal of Physical Chemistry*, 1959.
[10] Bariod, N.L., *Manual of Drilling Fluids Technology*, NL Industries Inc., Houston, 1979.
[11] Russell, C., M.S. Thesis, Desalination of Bicarbonate Brine Water: Experimental Finding Leading to an Ion Exchange Process: New Mexico Tech, Socorro, NM, 1994.
[12] Engineering Data Book, Gas Processors Suppliers Association, Tulsa, 1981.
[13] Grimme, S. J., and Erb, J. E., "Solidification of Residual Waste Pits as an Alternative Disposal Practice in Pennsylvania," Proceedings from the First International Symposium on Oil and Gas Exploration Waste Management Practices, pp. 873–881, 1990.

Index

Notes: Page numbers followed by "f" refer to figures; page numbers followed by "t" refer to tables.

J

K

J

M

N

O

T

NOTES

NOTES